Mechanism-Based Enzyme Inactivation: Chemistry and Enzymology

Volume II

Author

Richard B. Silverman

Professor of Chemistry and of
Biochemistry, Molecular Biology, and Cell Biology
Northwestern University
Evanston, Illinois

CRC Press, Inc.
Boca Raton, Florida

02957498

CHEMISTRY

Library of Congress Cataloging-in-Publication Data

Silverman, Richard B.
  Mechanism-based enzyme inactivation.

  Includes bibliographies and index.
  1. Enzyme inhibitors. I. Title. [DNLM: 1. Enzyme
Inhibitors. QU 143 S587m]
QP601.5.S55   1988   591'.1925   87-13825

ISBN 0-8493-4543-X (v. 1)
ISBN 0-8493-4544-8 (v. 2)

International Standard Book Number 0-8493-4543-X (v. 1)
International Standard Book Number 0-8493-4544-8 (v.2)

Library of Congress Card Number 87-13825
Printed in the United States

PREFACE

This book is intended to be used as a reference text by those in the field interested in applying mechanism-based inactivation approaches to studies of a particular enzyme or by those with an interest in this exciting area of enzyme inactivation in order to gain an overview of the field and to learn the fundamentals of the method. The text also can be used in a graduate-level course concerned with mechanism-based enzyme inactivation or as one of several texts in an advanced undergraduate or graduate course on enzyme inhibition in general. Since it is assumed in this text that the reader is familiar with enzyme mechanisms, it may be helpful to precede a course that utilizes this text by a course on enzyme mechanisms. At least, the students should be familiar with writing organic chemical mechanisms.

It may be useful to the reader to know how the information in this text was collected. It was observed that a computer search was not very useful, since most of the papers in the field do not have the words ''mechanism-based,'' ''enzyme-activated,'' or ''suicide'' in their titles or abstracts (and I did not have the inclination to peruse every paper that mentioned ''enzyme inhibition or inactivation''). The approach taken was a more direct one, namely, that letters were sent requesting reprints (or publication lists) from 146 different principal authors of papers published in the field. Of the 146 requests, all but six were answered (I have restrained myself from revealing the names of those six individuals). These references were used as primary sources from which other relevant references were extracted. This process continued, i.e., obtaining references from within the references, until all possibilities were exhausted. I am indebted to the 140 individuals who responded to my request; much time in collecting references was saved by their assistance. When these sources were expended, the references in available reviews were checked to be certain that no obvious omissions were made. The literature through the first half of 1987 is covered here (the work cited from mid-1986 and references therein were added to the galley proofs). These endeavors should have been sufficient to uncover most of the relevant references. If I missed your contribution to this field, please accept my apology. Rather than sending me a nasty note regarding my oversight, kindly send me copies of your relevant publications; maybe, some day an addendum will be published.

Richard B. Silverman
Evanston, IL
August, 1986

THE AUTHOR

Richard B. Silverman, Ph.D., is Professor of Chemistry and Professor of Biochemistry, Molecular Biology, and Cell Biology at Northwestern University.

Dr. Silverman received his B.S. degree in chemistry at the Pennsylvania State University in 1968, then spent two years in the U.S. Army as a medicinal chemist at the Walter Reed Army Institute of Research in Washington, D.C. He obtained his M.A. degree in 1972 and Ph.D. degree in 1974 in chemistry from Harvard University. Prior to joining the faculty of Northwestern University in 1976, Dr. Silverman was a National Institutes of Health post-doctoral research fellow in the Graduate Department of Biochemistry at Brandeis University.

Dr. Silverman's current research interests are the design and synthesis of mechanism-based enzyme inactivators and their mechanisms of enzyme inactivation, the elucidation of enzyme mechanisms, and the design of potential new pharmaceutical agents.

ACKNOWLEDGMENTS

I am grateful to Dr. Stephen J. Hoffman and especially to Dr. Mark A. Levy for their critical comments on the first chapter of the original draft (and for not discouraging the writing of the remainder of the book), to Carol Lewis for typing the manuscript, to Karen Heneghan and Dianne Deplewski for drawing most of the schemes and figures, and to the National Institutes of Health (grants GM 32634, HL 27108, NS 15703, and GM 35844) for financial support of my research during the writing of this work.

To loving Barbara for her devotion, her understanding, and her radiance, to Matt and Marggie for their love, their wit and their exuberance, and to Philly for his giant smiles, his hugs, and his belly laughs.

MECHANISM-BASED
ENZYME INACTIVATION:
CHEMISTRY
AND
ENZYMOLOGY

Volume I

Introduction
Protonation and Deprotonation Reactions
Phosphorylation Reactions
Addition Reactions
Acylation Reactions
Elimination Reactions

Volume II

Isomerization Reactions
Decarboxylation Reactions
Oxidation Reactions
Oxygenation Reactions
Polymerization Reactions

TABLE OF CONTENTS

Chapter 11
Polymerization Reactions

ABBREVIATIONS AND SHORTHAND NOTATIONS
USED IN THE BOOK

Because of the myriad of abbreviations commonly used, particularly in synthetic organic chemistry, this alphabetized compilation of abbreviations is included for easy referral while reading the book.

ACC	1-aminocyclopropane-1-carboxylate
Ad	adenine
Ado	adenosine
B:	active site base
BHT	butylated hydroxytoluene (2,6,-di-*tert*-butyl-4-methylphenol)
Bn	benzyl
Boc	*tert*-butoxycarbonyl
Bz	benzoyl
Cbz	carbobenzoxyl
CH_2H_4folate	5,10-methylenetetrahydrofolate
m-CPBA	*meta*-chloroperoxybenzoic acid
Cy	cyclohexyl
Dabco	1,4-diazabicyclo[2.2.2]octane
DBN	1,5-diazabicyclo[4.3.0]non-5-ene
DBU	1,8-diazabicyclo[5.4.0]undec-7-ene
DCC	dicyclohexylcarbodiimide
DDEP	3,5-dicarbethoxy-2,6-dimethyl-4-ethyl-1,4-dihydropyridine
DEAD	diethyl azodicarboxylate
Dibal	diisobutyl aluminum hydride
DMAP	4-dimethylaminopyridine
DMSO	dimethyl sulfoxide
DON	6-diazo-5-oxo-*L*-norleucine
DONV	5-diazo-4-oxo-*L*-norvaline
dR	deoxyribosyl
dRP	deoxyribose phosphate
DTNB	5,5'-dithiobis(2-nitrobenzoate)
DTT	dithiothreitol
EDTA	ethylenediaminetetraacetic acid
EPR	electron paramagnetic resonance
ESR	electron spin resonance
FdUMP	5-fluoro-2'-deoxyuridine monophosphate
Fl	oxidized flavin
FlH· or Fl$^{\div}$	flavin semiquinone
FlH$^-$ or FlH$_2$	reduced flavin
GABA	γ-aminobutyric acid
GTP	guanosine triphosphate
HPLC	high performance liquid chromatography
Im	imidazole
IR	infrared
LDA	lithium diisopropylamide
MAO	monoamine oxidase
MPDP$^+$	1-methyl-4-phenyl-2,3-dihydropyridinium ion
MPP$^+$	1-methyl-4-phenylpyridinium ion

MPTP	1-methyl-4-phenyl-1,2,3,6-tetrahydropyridine
Ms	mesyl (methanesulfonyl)
NAC	N-acetylcysteamine
NAD^+	nicotinamide adenine dinucleotide
NADH	reduce form of NAD^+
$NaDodSO_4$	sodium dodecyl sulfate
NBS	N-bromosuccinimide
NMR	nuclear magnetic resonance

	porphyrin ring system (usually protoporphyrin IX)
Np	naphthyl
NXS	N-halosuccinimide
ORD	optical rotatory dispersion
PAGE	polyacrylamide gel electrophoresis
Pant	pantetheine
PhthN	phthalimido
P_i	phosphate ion
PLP	pyridoxal 5′-phosphate
PMP	pyridoxamine 5′-phosphate
PPA	polyphosphoric acid
PQQ	pyrroloquinoline quinone
Pyr	the substituted pyridine nucleus of PLP or PMP
RBL	rat basophilic leukemia
TFA	trifluoroacetic acid
THF	tetrahydrofuran
THP	tetrahydropyranyl
TMEDA	N,N,N′,N′-tetramethylethylenediamine
TMS	trimethylsilyl
TPCK	L-1-chloro-3-tosylamido-4-phenyl-2-butanone
TPP	thiamin pyrophosphate
Ts	tosyl (p-toluenesulfonyl)

X:	active site nucleophile

Chapter 7

ISOMERIZATION REACTIONS*

I. ISOMERIZATION

A. S-Adenosylmethionine and Derivatives

S-Adenosylmethionine[1] and several of its analogues[2,3] (Structural Formulas 7.1a, Scheme 1 and 7.1b, Scheme 2) are time-dependent inactivators of S-adenosylmethionine decarboxylase from rat liver; inactivation results from transamination. The analogues (**7.1b**) were synthesized by Kolb et al.[3] by the route shown in Scheme 2.

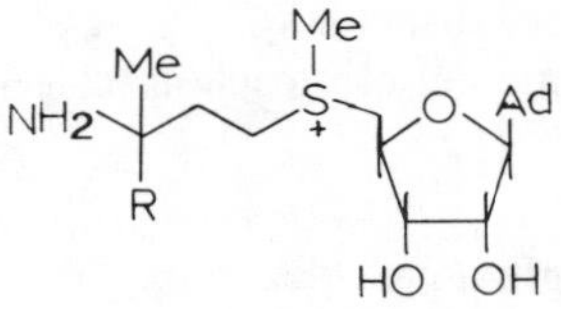

Scheme 1.   Structural Formula 7.1a.

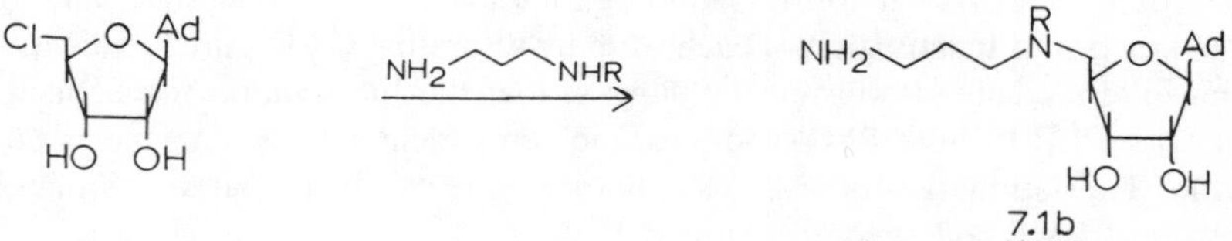

Scheme 2.   Containing Structural Formula 7.1b.

B. 3,5-Dioxocyclohexanecarboxylic Acid

On the basis of the principle of microscopic reversibility, 3,5-dioxocyclohexanecarboxylic acid (Structural Formula 7.2, Scheme 4) was designed by Alston et al.[4] as an inactivator for GABA aminotransferase in the PMP form. GABA aminotransferase from *Pseudomonas fluorescens* and from pig brain is stable to treatment with 3,5-dioxocyclohexanecarboxylate until GABA is added; then time-dependent inactivation of both enzymes occurs. The GABA is required to convert the PLP form of the cofactor to its PMP form. After complete inactivation, addition of α-ketoglutaric acid, which converts the PMP to PLP, results in a very slow reoxidation of the cofactor. The mechanism proposed is shown in Scheme 3. Reversal of the mechanism gives PMP, which rapidly is converted to PLP. The inactivation mechanism, however, is not initiated by the normal mechanism of the enzyme, so it is not a true mechanism-based enzyme inactivator. Compound **7.2** was synthesized by the method of Tobin et al.[5] (Scheme 4).

* A list of abbreviations and shorthand notations can be found prior to Chapter 7.

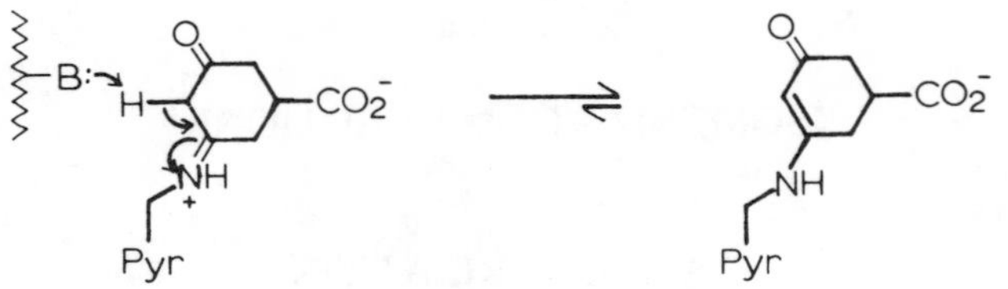

Scheme 3.

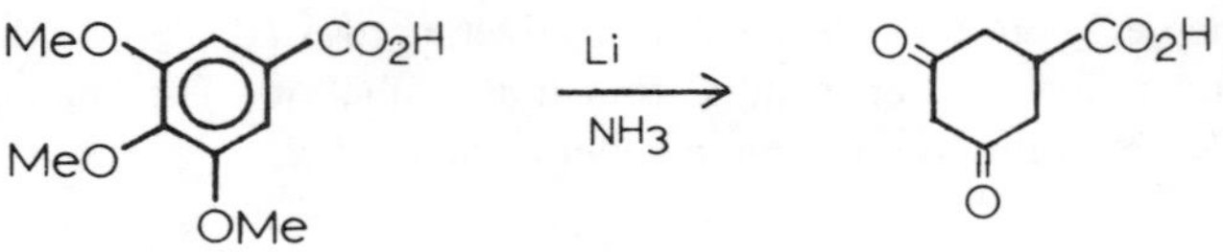

Scheme 4.   Containing Structural Formula 7.2.

C. γ-Aminobutyric Acid

All three forms of rat brain glutamate decarboxylase are inactivated by GABA in the absence of PLP in a time- and concentration-dependent process; added PLP results in reactivation of the enzyme.[5a] GABA also is a time- and concentration-dependent inactivator of the three isozymic forms of pig brain glutamate decarboxylase with the order of inactivation being γ-form > β-form > α-form.[5b] Although the inactivation rates differ among the three forms, the kinetics of inactivation of each form by saturating GABA are identical, suggesting a common intermediate. Porter et al.[5] suggest that the inactivation mechanism is transamination with PLP to give PMP and succinic semialdehyde; holoenzyme is converted to apoenzyme. Transamination is reversible; holoenzyme can be prepared from apoenzyme in the presence of PMP and succinic semialdehyde.

II. ISOMERIZATION/PROTONATION

A. β-D-Galactopyranosylmethyl (*p*-nitrophenyl)triazene and Related

Affinity labeling approaches to the inactivation of *Escherichia coli* β-galactosidase, such as synthesis of galactose analogues containing α-halo carbonyl and epoxy groups, require properly disposed active site nucleophiles; photoaffinity-labeling agents would generate highly reactive carbenes and nitrenes that are nonspecific alkylators. Carbocation intermediates, however, should have intermediate reactivity. Therefore, three approaches were considered by Sinnott and Smith[6] for the generation of β-D-galactopyranosylmethyl cation: from a diazo derivative, a nitrosoamide, and a triazene. The nitrosoamide rapidly decomposes in water and the diazo compound results in enzyme precipitation. β-D-Galactopyranosyl-methyl (*p*-nitrophenyl)triazene (Structural Formula 7.3), however, is relatively stable (t[1/2] = 9.5 hr at 25°C in water, but less stable in buffer), and can be synthesized by the route in Scheme 5 (Ar = *p*-nitrophenyl).[7] Furthermore, it is a time-dependent inactivator of *E. coli* β-galactosidase.[6,7] The mechanism proposed by Sinnott and Smith[6] for carbocation formation and inactivation is shown in Scheme 6. Since this inactivation mechanism is not related to the normal catalytic mechanism of the enzyme, this is not a true mechanism-based enzyme inactivator. In support of this pathway of inactivation of β-galactosidase by β-D-galactopyranosylmethyl(*p*-nitrophenyl)triazene, it is known that although both triazene tautomers of 3-substituted 1-aryltriazenes exist in solution, the slow step in chemical studies is proton donation and departure of the alkyl cation.[8] Proton and carbon-13 NMR spectroscopy were used by Kelly et al.[9] to observe the tautomerization of unconjugated to conjugated alkylaryltriazenes (Scheme 7).

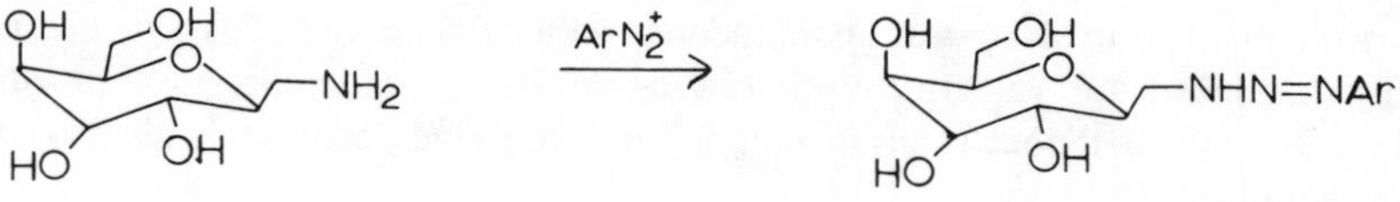

Scheme 5.   Containing Structural Formula 7.3.

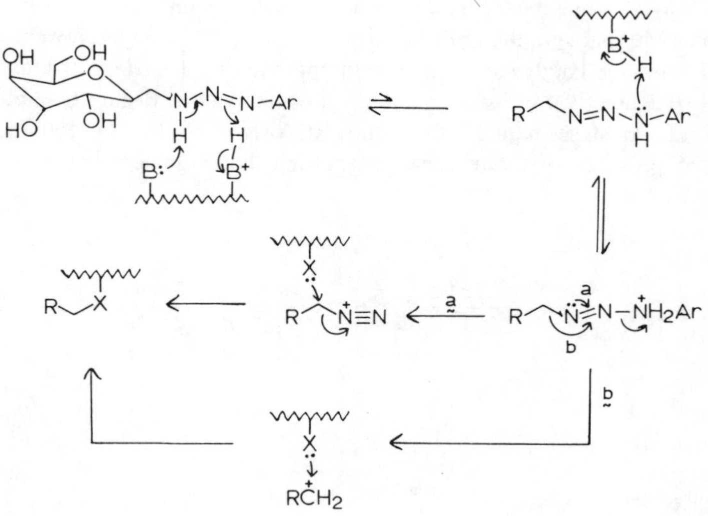

Scheme 6.

$$R-N=N-NH-Ar \rightleftharpoons R-NH-N=N-Ar$$

Scheme 7.

The position of the equilibrium appears to depend only on the substituent effect on the aryl group; there is little effect of solvent or the structure of the alkyl group. Evidence was presented by Jones et al.[10] to indicate that alkylaryltriazenes, including β-D-galactopyranosylmethyl(*p*-nitrophenyl)triazene, decompose with proton donation by a unimolecular reaction to the diazonium ion and arylamine (Scheme 8).

$$R-NH-N=NAr \rightleftharpoons RN=N-NHAr \xrightarrow{H^+} R\overset{+}{N}\equiv N + NH_2Ar$$

Scheme 8.

This implies that the mechanism of inactivation proceeds via a diazonium ion, rather than a concerted elimination to the carbocation. The mechanism of decomposition of primary alkylaryltriazenes in water to give alkanediazonium ions and anilines under a variety of conditions was shown by Jones et al.[11] to involve two processes. One involves a transition state in which the proton of the catalyzing acid is completely transferred prior to breakage of the N—N bond, the other involves spontaneous departure of aniline anion without acid assistance. The substituent effects on inactivation of Mg(II)-free *lacZ*-β-galactosidase of *E. coli* by various aryl-substituted galactosylmethylaryltriazenes indicated to Sinnott et al.[12]

that the mechanism follows that of spontaneous, rather than acid-catalyzed, decomposition to the alkyldiazonium ion (electron-withdrawing substituents increase the rate of inactivation). Therefore, further confirmation that these compounds are mechanism-based inactivators was sought.

One approach was a study of inactivation of β-galactosidase by 2-(β-D-galactopyranosyl)ethyltriazenes (Structural Formula 7.4), synthesized as shown in Scheme 9. If it is not enzyme-catalyzed, then extending the alkyl chain should either not affect the rate of inactivation or increase the rate, since nonenzymatic decomposition of the corresponding ethyl derivative is 5 to 10 times faster than the parent methyl compound. However, inactivation by 2-(β-D-galactopyranosyl)ethylaryltriazenes occurs at 3 to 13 times slower rate. Therefore, inactivation is enzyme catalyzed. The second approach to determine if these triazenes are enzyme-catalyzed inactivators was to attempt to inactivate a β-galactose-binding enzyme other than β-galactosidase, namely, the lectin, RCA ricin. No loss of lectin-active sites was observed in the presence of the triazene, suggesting that β-galactosidase catalyzes its own inactivation.

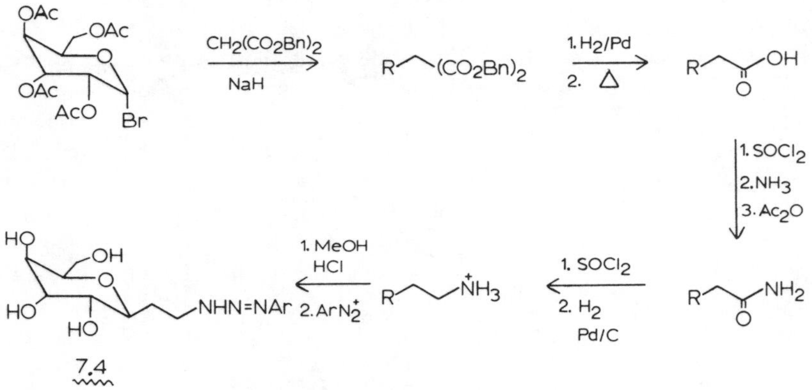

Scheme 9. Containing Structural Formula 7.4.

β-D-Galactopyranosyl [14C]methyl(*p*-nitrophenyl)triazene inactivates *lacZ* β-galactosidase of *E. coli* with incorporation of 0.91 mol of radioactivity per mole of enzyme protomer[6,7] with a partition ratio of 4.[6] Tryptic digestion of the labeled enzyme gives two labeled peptides, one in much greater amount than the other. Hydrolysis of the major peptide for 14 hr gives a labeled dipeptide, which upon further hydrolysis liberates proline. The molecular weight of the labeled dipeptide is compatible with (prolyl)galactosylmethylhomocysteine if cyclization of the proline to a dioxopiperazine (Structure Formula 7.5) has taken place. Homocysteine, however, is not one of the amino acids in the native enzyme. Demethylation of an *S*-(galactosylmethyl)methionine residue by 2-mercaptoethanol used to carboxymethylate the enzyme prior to tryptic digestion could account for the homocysteine. The mechanism proposed by Sinnott and Smith[6] is shown in Scheme 10. Cyanogen bromide cleavage of β-galactosidase inactivated with β-D-galactosyl-[14C]methyl(*p*-nitrophenyl)triazene gives two peptides, one of 100 residues (major) and one of 60 residues (minor); the major peptide was sequenced by Fowler et al.[13] and shown to be residues 442 to 540 of the enzyme. Trypsin digestion of the pooled cyanogen bromide peptides gives a 26 residue peptide corresponding to residues 478 to 503; radioactivity was detected at methionine-500. Treatment of this peptide with glutamic acid-specific *Staphylococcus aureus* protease gives an 18-residue peptide of residues 486 to 503. Thermolysin hydrolysis of this peptide gives a radioactive peptide comprised of residues 496 to 501; the radioactivity corresponds to Met-500.

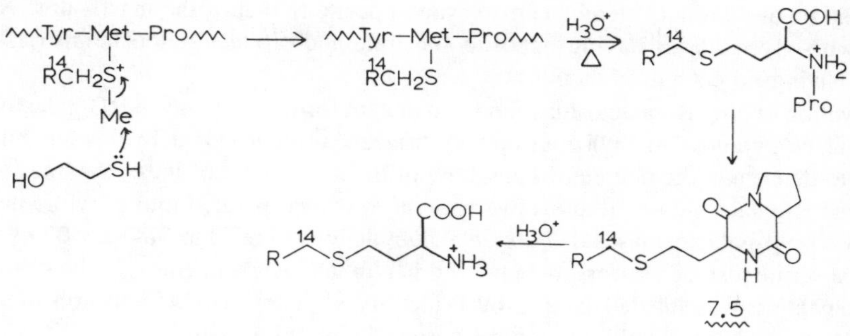

Scheme 10. Containing Structural Formula 7.5.

β-D-Galactopyranosylmethyl(p-nitrophenyl)triazene specifically inactivates liposomal β-galactosidase in cultured lumin skin fibroblasts; no change in α-galactosidase and α- and β-glucosidase activities nor in cell viability was detected by Van Diggelen et al.[14] β-D-Glucopyranosylmethyl(p-nitrophenyl)triazene inactivates β-glucosidase and β-xylosidase which is to be expected, since the former enzyme is indifferent to the position of the substituent at C-5 of the pyranose ring. There also is some inactivation of α-glucosidase at higher concentrations. Since the viability of the cells is unaltered, the restitution of β-galactosidase and β-glucosidase after inactivation was followed. Turnover times for the two enzymes are 10 and 5 days, respectively. Because these inactivators are not toxic to the cell, they may have use in the induction of lysosomal storage diseases for animal model studies. There are several genetic diseases characterized by a deficiency of β-galactosidase activity (different forms of G_{MI}-gangliosidosis). β-D-Galactopyranosylmethyl(p-nitrophenyl)triazene was used by Van Diggelen et al.[15] to determine the turnover time of β-galactosidase in fibroblasts from patients with different types of genetic β-galactosidase deficiencies. The rate of β-galactosidase synthesis in normal and mutant cells was 0.4 to 0.5 pmol/day/mg of cellular protein. Cells from patients with G_{MI}-gangliosidosis contained the normal amount of β-galactosidase/mg of protein and had the normal turnover time of 10 days; however, only 10% of β-galactosidase activity per enzyme molecule was observed. For those patients having both β-galactosidase and neuraminidase deficiency, their cells contained only 0.3 pmol of β-galactosidase per milligram of protein with a turnover time of only 1 day and normal hydrolytic activity.

Inactivations of a variety of glycosidases by the corresponding glycosylmethyl(p-nitrophenyl)triazenes were surveyed by Marshall et al.[16] β-D-Galactopyranosylmethyl(p-nitrophenyl)triazene inactivates the *lacZ-* and *ebg*° β-galactosidase of *E. coli*, human liver lysosomal β-galactosidase, and, to a lesser extent, α-galactosidase from green coffee beans. No inactivation of the *lac* repressor of *E. coli* was observed. β-D-Xylopyranosylmethyl(p-nitrophenyl)triazene inactivates β-xylosidase of *Penicillium wortmanni*. Sweet almond β-glucosidase B, which also has β-galactosidase activity, is inactivated by β-D-glucopyranosylmethyl(p-nitrophenyl)triazene and very slowly by the corresponding galacto inactivator. Human lysosomal α-glucosidase is slightly sensitive to the gluco inactivator, but the yeast enzyme is inert. α-L-Arabinofuranosidases from the fungus *Monilinia fructigena* are not inactivated by α-L-arabinofuranosylmethyl(p-nitrophenyl)triazene at their optimal pH values because of the instability of the inactivator at low pH. At pH 7.0, inactivation of one of the arabinofuranosidases is evident. The inverting enzymes, glucoamylase of *Aspergillus niger* and β-xylosidase of *Bacillus pumilus,* are inert to the corresponding triazenes. Therefore, the glycosylmethyl(p-nitrophenyl)triazenes are specific inactivators for glycosidases for which both substrate and product stereochemistries at the anomeric carbon are the same as that of the inactivator; however, they can act with much lower efficiency on glycosidases for which both substrate and product stereochemistry are opposite to that of the inactivator. No effect is seen with glycosidases having opposite substrate and product stereochemistries or with proteins having no catalytic function.

Inactivation of *ebg°* β-galactosidase and two mutant forms, *ebg*[a] and *ebg*[b] β-galactosidases, by β-D-galactopyranosylmethyl(*p*-nitrophenyl)triazene was carried out by Fowler and Smith[17] in order to determine the degree of homology of the active sites of these enzymes with that of the *lacZ* β-galactosidase. Radioactive triazene was incorporated into a tryptic peptide of *ebg°* and *ebg*[a] β-galactosidases that are 38% homologous to residues 483 to 503 of the *lacZ* enzyme; a serine residue appears to be labeled in *ebg°* and *ebg*[a] enzymes. In the *ebg*[b] enzyme a tryptic peptide was labeled having 69% identity with residues 457 to 468 of the *lacZ* enzyme; both a glutamic acid and a tyrosine residue are alkylated.

III. ISOMERIZATION/ADDITION

Many of the compounds in this section are acetylene-containing compounds that undergo isomerization to an allene that is in conjugation with a carbonyl group (Structural Formula 7.6). This activated allene then reacts with an active-site nucleophile (Scheme 11). Because of the generality of this mechanism, Scheme 11 will be referred to throughout this chapter.

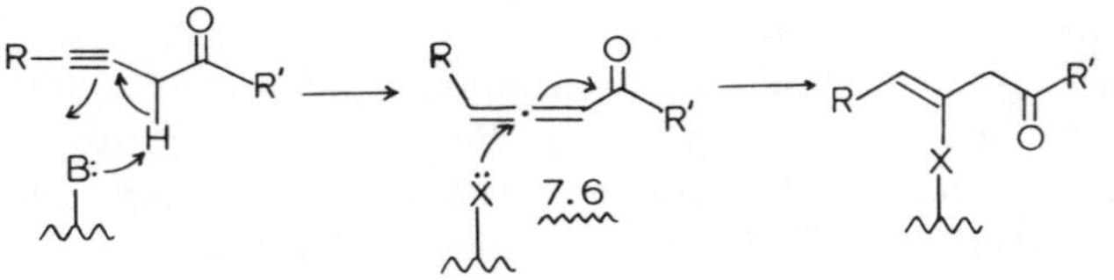

Scheme 11.   Containing Structural Formula 7.6.

A. Acetylene-Containing Inactivators
1. 3-Alkynoylthioesters
a. 3-Decynoyl-N-acetylcysteamine

Preincubation of the *E. coli* fatty acid synthetase system with 3-decynoyl-*N*-acetylcysteamine (**7.7**, R = $CH_3[CH_2]_5$; see Scheme 12) abolishes formation of long-chain unsaturated fatty acids; β-hydroxydecanoic acid is the major metabolite produced.[18] The enzyme involved was shown by Helmkamp et al.[19] to be β-hydroxydecanoyl thioester dehydrase, a component enzyme of the fatty acid synthetase system that catalyzes the reversible interconversions of the thioesters of β-hydroxydecanoic acid, *trans*-α,β-decenoic acid, and *cis*-β,γ-decenoic acid. β,γ-Acetylenic thioesters with varying chain lengths from 8 to 12 carbons are potent inhibitors of the enzyme; the inhibitory activities parallel the corresponding substrate activities for the substrates with 8 to 12 carbons. Compound **7.7** irreversibly inactivates the enzyme in a time-dependent fashion. $[1-^{14}C]$3-Decynoyl-*N*-acetylcysteamine covalently binds to the enzyme after gel filtration and the inactivator becomes attached to an active site histidine residue.[20]

$$O$$
$$\|$$
$$R–C{\equiv}CCH_2CSCH_2CH_2NHAc$$

Scheme 12.   Structural Formula 7.7.

2,3-Decadienoyl-*N*-acetylcysteamine, prepared by base isomerization of 3-decynoic acid followed by conversion to the thioester or by direct isomerization of 3-decynoyl-NAC with sodium ethoxide, was used by Endo et al.[21] to study the inactivation mechanism. The allenic thioester, an affinity-labeling agent, is a much more potent inactivator of β-hydroxydecanoyl thioester dehydrase than is 3-decynoyl-NAC. 2,2-Dideuterio-3-decynoyl-NAC inactivates the enzyme with a rate constant 2.6 times smaller than that for the undeuterated compound, whereas 2-deuterio-2,3-decadienoyl-NAC inactivates the enzyme at the same rate as the undeuterated molecule. Inactivation of the dehydrase with either the acetylenic or allenic thioester results in loss of 2 histidyl residues. These results led to the proposed inactivation mechanism shown in Scheme 13 (X = His).

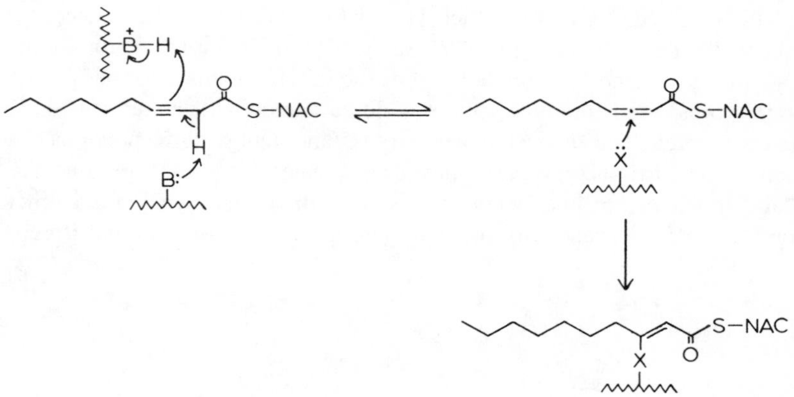

Scheme 13.

A series of acetylenic thioesters of varying chain lengths showed time-dependent inactivation of the enzyme; the C_{10} compound was the most potent.[22] 2,3-Decadienoic acid was resolved into its two enantiomeric forms; only the +-isomer is an inactivator (i.e., an affinity-labeling agent). One mole of the (+)-isomer completely inactivated one mole of enzyme. Stereochemical studies with *E. coli* β-hydroxydecanoylthioester dehydrase by Schwab and Klassen[23] showed that the pro-4*R* hydrogen of thioesters of (*E*)-2-decenoic acid is stereospecifically removed during its isomerization to thioesters of (*Z*)-3-decenoic acid. The proton is added to the *si* face of C-2 during this reaction. Therefore, the allylic rearrangement is a suprafacial process and is consistent with a single base mechanism. If this is also the pathway followed by the mechanism-based inactivator, 3-decynoyl-NAC, then it was predicted that the pro-2*S* proton would be removed from the inactivator and the acetylene would be protonated at C-4 on the same face. This would result in an allenic thioester with the *S* configuration, namely (*S*)-(+)-2,3-decadienoyl NAC. This prediction was confirmed by X-ray analysis of the known inactivator of the enzyme,[24] and indicated the isomerization stereochemistry in Scheme 14.

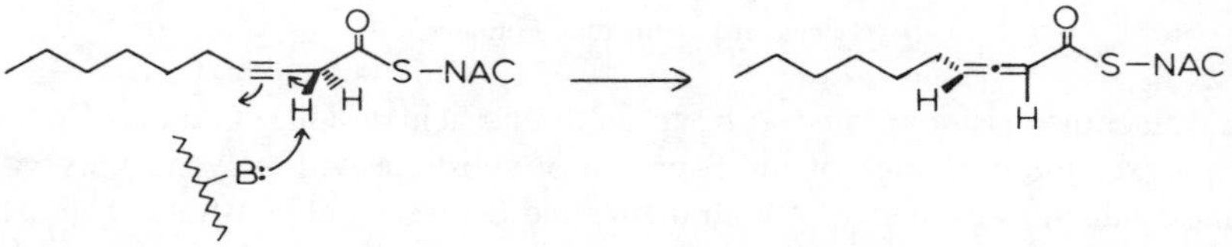

Scheme 14.

The reactions of 2,3-decadienoylthioesters with histidine derivatives were studied by Morisaki and Bloch[25] as models for the reaction of acetylenic and allenic thioesters with the dehydrase. Depending upon the histidine analogue used, a β,γ-enamine (**7.8**) or α,β-enamine (**7.9**) is obtained (see Scheme 15). Carbon-13 NMR spectroscopy was used by Schwab et al.[26] to show that upon inactivation of β-hydroxydecanoylthioester dehydrase by one equivalent of 3-decynoyl-NAC, the covalent active site adduct is a β,γ-unsaturated thioester (**7.8**) not an α,β-unsaturated thioester (**7.9**). Furthermore, when a second equivalent of inactivator was added, no free inactivator was observed, and the total integrated area of the bound species doubled. Further ^{13}C NMR spectroscopy studies of *E. coli* β-hydroxydecanoyl thioester dehydrase inactivated by 3-[2-^{13}C]decynoyl-NAC indicates that isomerization to 2,3-decadienoyl-NAC, and histidine attack occurs rapidly and gives the (*E*)-3-(N[im]-histidinyl)-3-decenoyl thioester adduct (**7.8**), which is slowly converted to the 2-decenoyl thioester congener (**7.9**).[26a] The rationale given by Schwab et al.[26a] for initial formation of the 3-decenoyl adduct is that the 1,2-double bond of the initial dienolate formed is orthogonal to the 3,4-double bond, and, therefore, there is no carbanionic character at C-4. Protonation at C-2, then is favored. The (*E*)-stereochemistry is consistent with the previous observation[23] that the active-site histidine residue removes or donates the *pro-2S* proton. These results suggest that 2 molecules of inactivator are bound to the dimeric enzyme, a conclusion that differs from the half-sites reactivity previously proposed by Morisaki and Bloch.[22]

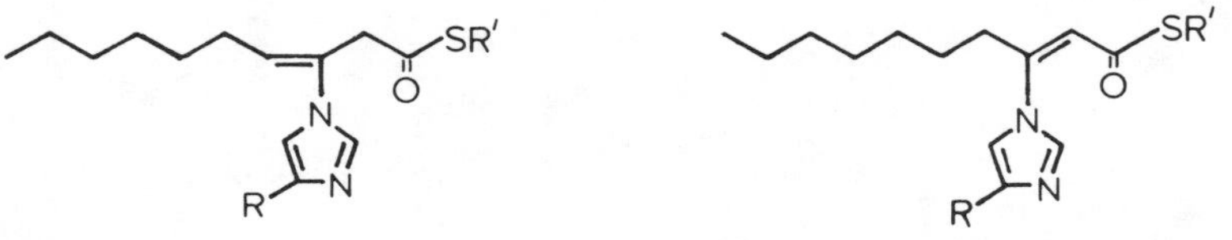

Scheme 15.   Structural Formulas 7.8 and 7.9.

Further support for an enzyme-catalyzed isomerization mechanism is the isolation of an enzyme from hog liver that catalyzes the conversion of 3-decynoyl-NAC to 2,3-decadienoyl-NAC without enzyme inactivation.[27,28]

b. 3-Butynoylpantetheine and 3-Pentynoylpantetheine

Glutaryl-CoA dehydrogenase from *Pseudomonas fluorescens* and butyryl-CoA dehydrogenase from *Megasphaera elsdenii* are irreversibly inactivated in a pseudo first-order process by 3-pentynoyl- (**7.10**, R = CH₃) and 3-butynoylpantetheine (**7.10**, R = H; see Scheme 16).[29] Inactivation does not result in a change of the flavin spectrum;

$$R-C{\equiv}CCH_2\overset{\displaystyle O}{\overset{\displaystyle \|}{C}}SCoA$$

Scheme 16.   Structural Formula 7.10

therefore, attachment is suggested to be to an amino acid residue. Glutaryl-CoA dehydrogenase catalyzes the exchange of the β-proton of substrate with solvent; butyryl-CoA dehydrogenase catalyzes exchange of both the α- and β-protons of substrate. The fact that the acetylenic compounds (**7.10**) inactivate these enzymes provides further evidence that proton removal is involved prior to the oxidation of substrates by these enzymes. The inactivation mechanism proposed by Gomes et al.[29] is similar to that reported for β-hydroxydecanoylthioester dehydrase (see Scheme 11, R = CH$_3$ or H,R' = pantetheine). The inactivators were synthesized[29] by the routes shown in Schemes 17 and 18. Further studies on the

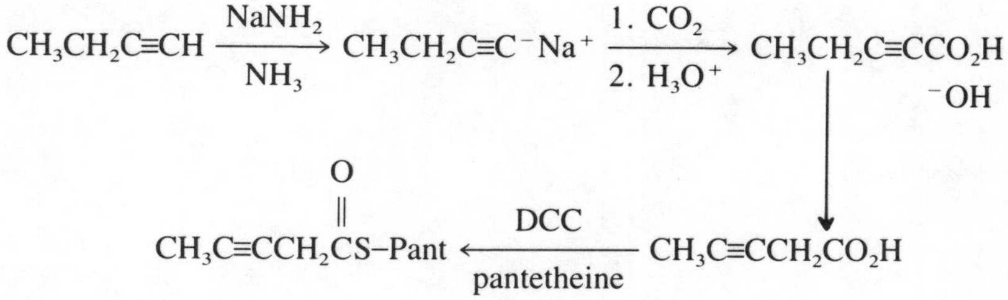

$$CH_3CH_2C\equiv CH \xrightarrow[NH_3]{NaNH_2} CH_3CH_2C\equiv C^-Na^+ \xrightarrow[2.\ H_3O^+]{1.\ CO_2} CH_3CH_2C\equiv CCO_2H$$

$$^-OH$$

$$CH_3C\equiv CCH_2\overset{\overset{\displaystyle O}{\|}}{C}S\text{--Pant} \xleftarrow[\text{pantetheine}]{DCC} CH_3C\equiv CCH_2CO_2H$$

Scheme 17.

$$HC\equiv CCH_2CH_2OH \xrightarrow[H_2SO_4]{CrO_3} HC\equiv CCH_2CO_2H \xrightarrow[\text{pantetheine}]{DCC} HC\equiv CCH_2\overset{\overset{\displaystyle O}{\|}}{C}S\text{--Pant}$$

Scheme 18.

inactivation of butyryl CoA dehydrogenase from *Megasphaera elsdenii* were carried out be Fendrich and Abeles.[30] Inactivation with [1-^{14}C]3-pentynoylpantetheine gives a stoichiometry of 0.61 ± 0.1 mol of radioactivity per mole of enzyme. It is not known why such a low stoichiometry results with complete inactivation. The green color of the enzyme is bleached upon inactivation, suggesting a disruption of a charge-transfer interaction between the flavin and a CoA persulfide at the active site. A carboxylate was suggested as the active site base that becomes alkylated because of the lability of the adduct toward hydroxylamine, borohydride, and base: treatment with hydroxylamine produces a hydroxamic acid on the protein; borohydride produces 1,3-pentanediol from the adduct; and treatment with hydroxide produces 3-oxopentanoic acid. Acid hydrolysis after reduction results in formation of 2-amino-5-hydroxyvaleric acid, the product of reduction of a glutamyl ester. Because there is no 260-nm absorbance for an α,β-unsaturated thioester after inactivation, the β,γ-unsaturated adduct (Structural Formula 7.11, Scheme 19) was preferred. (3-Chloro-3 butenoyl)pantetheine also inactivates the enzyme (see Chapter 6 in Volume I, Section II.D.3).

c. 3-Butynoyl CoA and 3-Octynoyl-CoA

3-Butynoyl-CoA and 3-octynoyl-CoA are time-dependent inactivators of pig liver mitochondrial general acyl-CoA dehydrogenase; dialysis does not regenerate enzyme activity.[31] The flavin spectrum is not bleached by inactivation, suggesting that an active site residue, rather than the flavin, is modified. Inactivation by 3-octynoyl-CoA is nearly stoichiometric; 83% inactivation occurs with one equivalent of inactivator. The rate of inactivation increases with decreasing pH, suggesting that protonation of the acetylene may be important, as would be expected for an acetylene-allene isomerization mechanism of inactivation (Scheme 11). In accord with this hypothesis, 2,3-octadienoyl-CoA was shown by Frerman et al.[31] to be an irreversible inactivator, independent of pH. These studies indicate the importance of α-deprotonation in the normal catalytic mechanism.

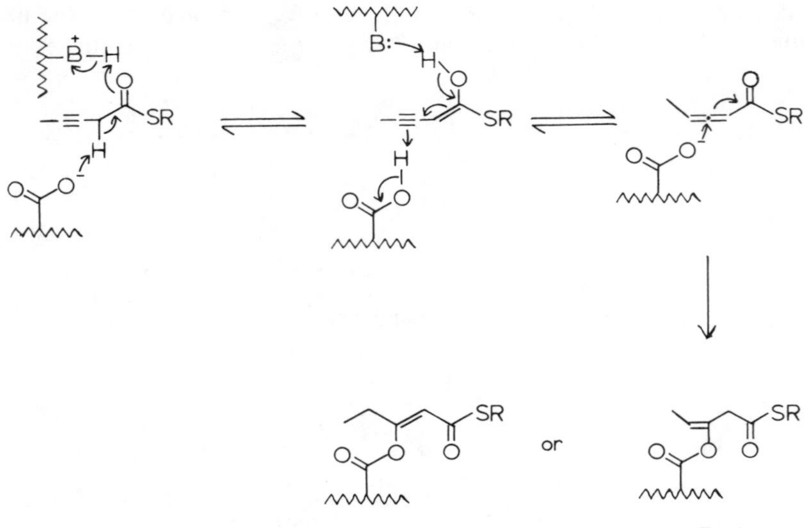

Scheme 19.   Containing Structural Formula 7.11.

2. Acetylene-Containing 5,10-Secosteroids

Δ^5-3-Ketosteroid isomerase catalyzes the conversion of Δ^5-3-ketosteroids to the corresponding Δ^4-3-ketosteroids. The enzyme from *Pseudomonas testosteroni* is inactivated by 5,10-secoester-5-yne-3,10,17-trione (**7.12**, R = O) and 5,10-seco-19-norpregn-5-yne-3,10,20-trione (**7.12**, R = β-Ac, α-H; see Scheme 20); dialysis does not restore enzyme activity.[32] 5,10-Seco[7-^{3}H]estr-5-yne-3,10,17-trione ([7-^{3}H]-**7.12**, R = O) was used to titrate the active site of the enzyme.[33] The rate of inactivation coincides with the rate of incorporation of radioactivity into the enzyme. With complete inactivation, one equivalent of tritium is incorporated after dialysis or gel filtration. Further studies by Penning et al.[34] indicated that the two acetylenic inactivators (**7.12**) are excellent substrates for the enzyme and are isomerized 6×10^5 times faster than they inactivate the enzyme. After inactivation, 1 molecule of the [7-^{3}H]-**7.12** (R = O) remains bound per subunit of the dimeric enzyme. Two radioactive metabolites were isolated after inactivation, (4R)-5,10-secoestra-4,5-diene-3,10,17-trione (**7.13**) and a cyclized 3,5-diketone (**7.14**; see Scheme 21). Evidence for active-site

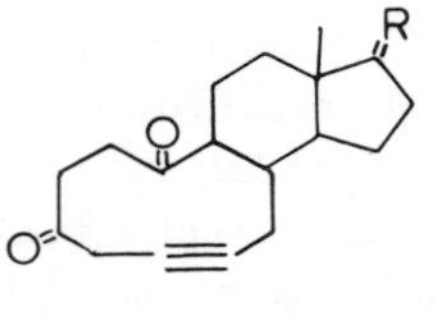

Scheme 20.   Structural Formula 7.12.

7.13   **7.14**

Scheme 21.   Structural Formulas 7.13 and 7.14.

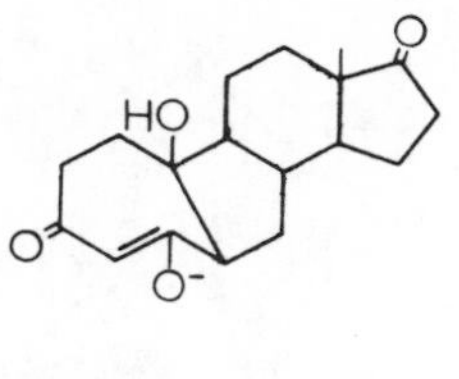

Scheme 22. Structural Formula 7.15.

inactivation is as follows: (1) 19-nortestosterone, a competitive inhibitor, protects the enzyme; (2) the inactivated isomerase is not retained on a 19-nortestosterone-agarose affinity column, suggesting an obliterated steroid binding site; and (3) similar quenching of tyrosine fluorescence was observed with saturated concentrations of 19-nortestosterone and by the acetylenic inactivators. The covalent linkage formed to the enzyme is extremely labile to acid and to base. In base, a new ultraviolet chromophore appears, shown to be the result of formation of the enolate (Structural Formula 7.15, Scheme 22).

On the basis of these results, the mechanism shown in Scheme 23 was proposed. The structure of the adduct was characterized by Penning and Talalay.[35] A tetrapeptide containing the radioactivity was isolated. On the basis of the composition of this peptide (Tyr–Ala–Asn–Ser) and the known sequence of the enzyme, the structure was deduced to be that of residues 55 to 58. Cleavage of the secosteroid-peptide bond produces a peptide with a net positive charge. Further hydrolysis gives Tyr, Ala, Asp, and Ser. These results were rationalized by attack of the asparagine amide on the reactive intermediate (Scheme 24). This is one of the few cases where an asparagine residue may be implicated as an active site base involved in a catalytic mechanism. Electron impact and fast atom bombardment mass spectrometry of the tetrapeptide isolated after cleavage of the secosteroid from it were utilized to delineate the structure as the cyclic basic oxazine, 5,6-dihydro-6-immonio-4H-1,3-oxazine (**7.16**) the product resulting from pathway b in Scheme 24.[36]

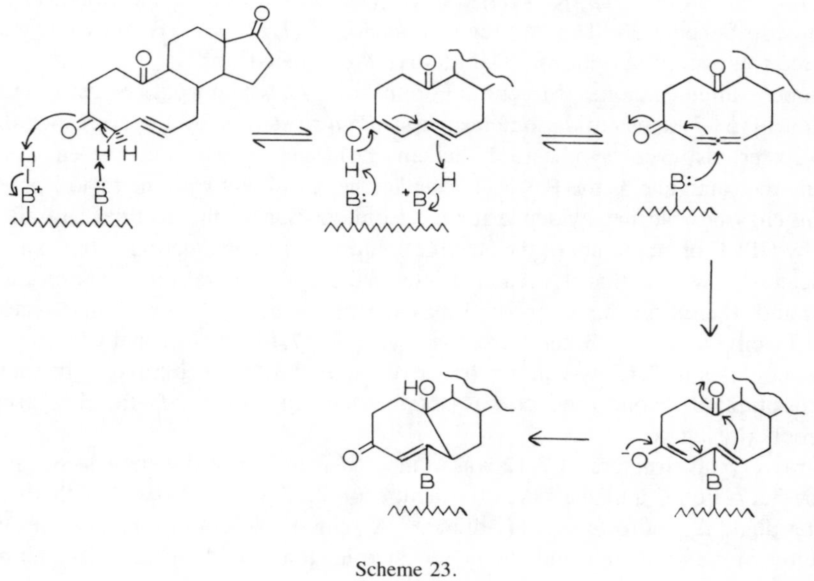

Scheme 23.

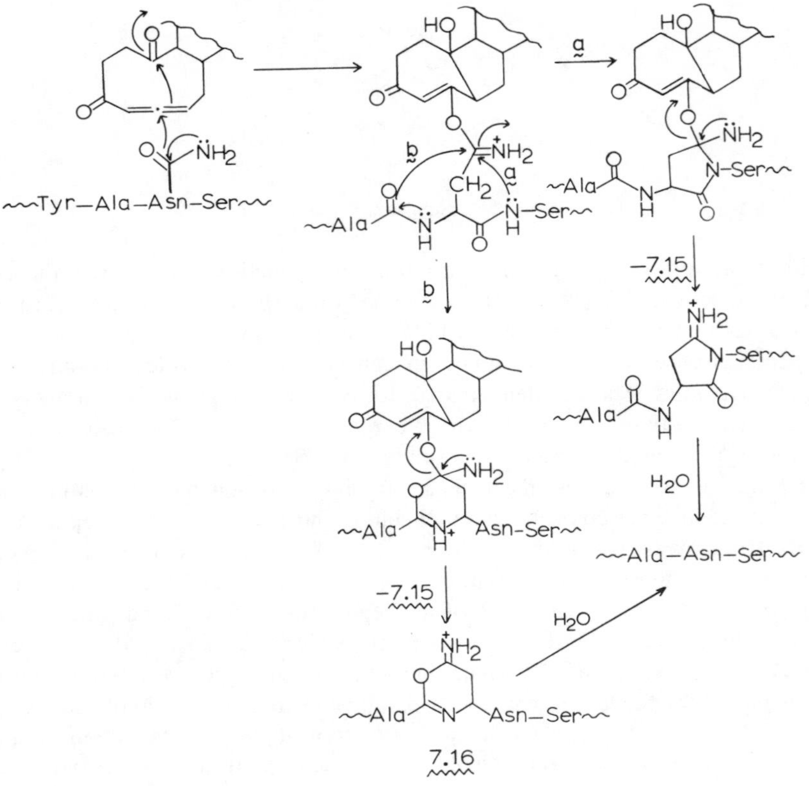

Scheme 24.   Containing Structural Formula 7.16.

The 5,10-secosteroid (**7.12**, R = O) was synthesized by Batzold and Robinson[37] by the route shown in Scheme 25. The norpregnone analogue (**7.12**, R = β-Ac, α-H) was made by the same route starting with 3β,20-diacetoxy-5-pregnen-19-ol.

The isomeric allenic ketones (Structural Formulas 7.17), which would be generated during inactivation of Δ^5-3-ketosteroid isomerase from *P. testosteroni* by the corresponding acetylenic secosteroids, were synthesized (Scheme 26).[38] They inactivate the enzyme at approximately the same rate as the β,γ-acetylenic ketone, consistent with the rapid isomerization rate of the enzyme. Further evidence for the isomerization of the acetylene to allene was obtained by HPLC of an aliquot of the enzyme solution during incubation with the acetylene. This indicates, however, that the reactive intermediate is released into solution during inactivation and, therefore, this compound may not be a mechanism-based inactivator. As a model for inactivation of Δ^5-3-ketosteroid isomerase by **7.17**, the Michael addition of S, O, and N nucleophiles to **7.17** was shown by Covey et al.[39] to proceed rapidly. In some cases the α,β-unsaturated ketone product was obtained and in other cases the β,γ-unsaturated ketone product resulted.

The X-ray crystal structure of **7.12** was compared with that of the enzyme substrate, Δ^5-androstene-3,17-dione, and the crystal structure of **7.17** was compared with that of the enzyme product, Δ^4-androstene-3,17,-dione.[40] A remarkable conformational similarity of the structures of the acetylenic and allenic secosteroids to those of the substrate and product, respectively, was observed. From these structure determinations it was suggested by Carrell et al.[40] that the enzyme substrates are positioned in the active site principally by the C and D rings. Following C-4 proton abstraction, a conformational change may occur in order to facilitate reprotonation at C-6 instead of at C-4.

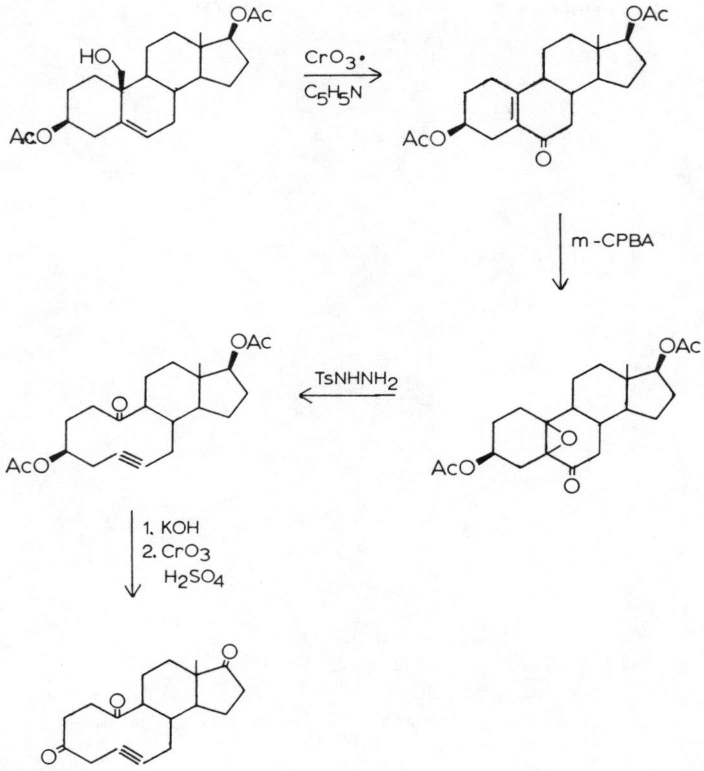

Scheme 25.

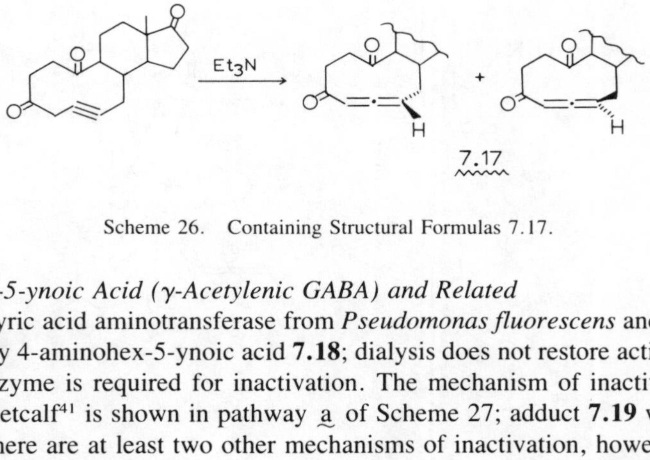

Scheme 26.   Containing Structural Formulas 7.17.

3. 4-Aminohex-5-ynoic Acid (γ-Acetylenic GABA) and Related

γ-Aminobutyric acid aminotransferase from *Pseudomonas fluorescens* and from pig brain is inactivated by 4-aminohex-5-ynoic acid **7.18**; dialysis does not restore activity.[41] The PLP form of the enzyme is required for inactivation. The mechanism of inactivation proposed by Jung and Metcalf[41] is shown in pathway a of Scheme 27; adduct **7.19** was proposed as the product. There are at least two other mechanisms of inactivation, however, that can be envisioned. One mechanism is pathway b in Scheme 27 leading to adduct **7.20**. The other route is based on the enamine mechanism[42,43] of inactivation of aspartate aminotransferase and glutamate decarboxylase by L-serine *O*-sulfate discussed in Chapter 6 of Volume 1 (Section II.G.3.a.) which would involve an allenamine intermediate (Scheme 28).

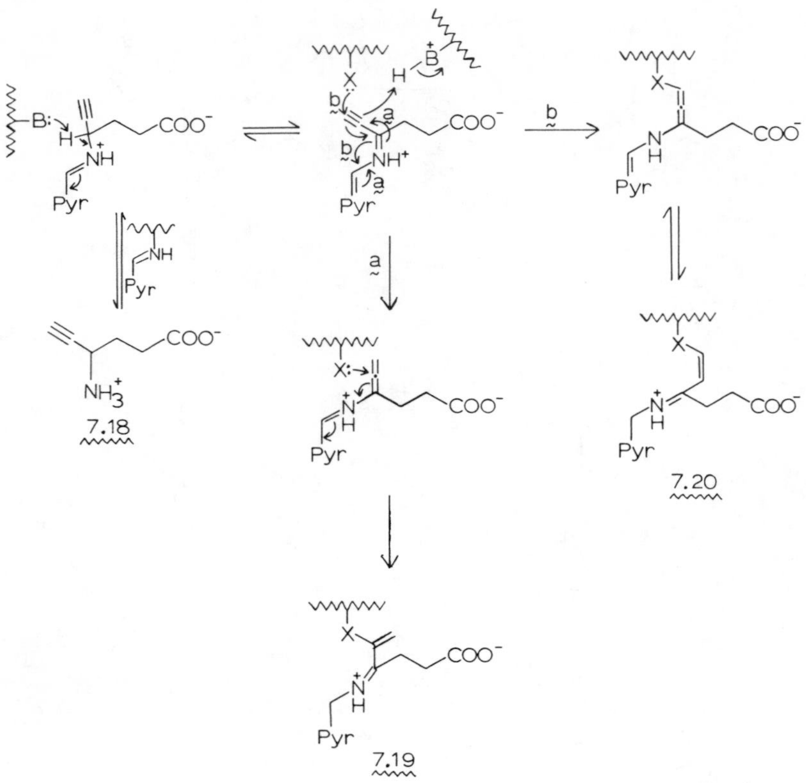

Scheme 27.   Containing Structural Formulas 7.18, 7.19, and 7.20.

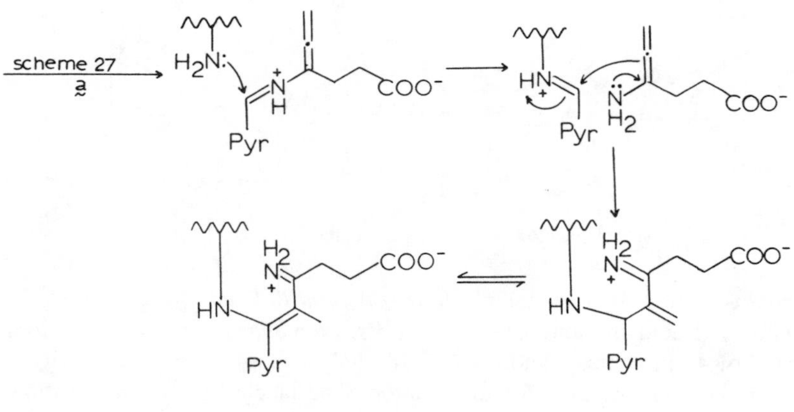

Scheme 28.

Lippert et al.[44] found that γ-deuterio-γ-acetylenic GABA exhibits a primary kinetic isotope effect of 2.47 on the apparent Michaelis constant (K_D/K_H) for GABA aminotransferase. This indicates that proton abstraction occurs before inactivation, but that it is not fully rate determining in the overall process. With 4-[2,3-^{14}C]aminohex-5-ynoic acid, time-dependent loss of enzyme activity was coincident with incorporation of radioactivity into the enzyme. After complete inactivation, 0.9 equivalent of γ-acetylenic GABA was incorporated per subunit. GABA aminotranferase abstracts the pro-4S proton from GABA and the analogous protons from (S)-(+)-γ-acetylenic GABA. During in vivo studies with (R)-(−)-4-aminohex-

5-ynoic acid, it was noted by Jung et al.[45] that there also was a decrease in glutamate decarboxylase activity with time. This compound was, then, shown to be a time-dependent irreversible inactivator of glutamate decarboxylase from *E. coli*. Only the *(R)*-(−)-isomer is active as an inactivator, but the *(S)*-(+)-isomer protects the enzyme from inactivation by the *(R)*-(−)-isomer, suggesting that it also binds to the active site. Substitution of a proton at C-4 by a deuterium results in an isotope effect of 2 on the K_I and of 1.35 on the rate constant at infinite concentration. Inactivation with (±)-[2-³H]-4-aminohex-5-ynoic acid results in the incorporation of 0.95 mol of radioactivity per PLP binding site. Since the mechanism of glutamate decarboxylase is believed to be reversible, except for the step involving the loss of CO_2, it was suggested that the principle of microscopic reversibility is valid and that the enzyme catalyzes deprotonation at C-4, a step that is the reverse of protonation following decarboxylation. The mechanism of inactivation, then, would be identical to that for inactivation of GABA aminotransferase by 4-aminohex-5-ynoic acid (see Scheme 27). The spectrum of the PLP is converted to that for PMP during inactivation. The fact that the *R*-(−) isomer is the only active one is consistent with a decarboxylation mechanism involving retention of configuration during protonation.

The stereospecificity of the inactivations of GABA aminotransferase and glutamate decarboxylase by 4-aminohex-5-ynoic acid was studied by Bouclier et al.[46] Only the *(S)*-(+)-enantiomer of 4-aminohex-5-ynoic acid irreversibly inactivates rat brain glutamate decarboxylase and pig brain and *P. fluorescens* GABA aminotransferase. Bacterial *(E. coli)* glutamate decarboxylase, however, is inactivated by the *(R)*-(−)-enantiomer.[45] GABA aminotransferase from mammalian and bacterial sources catalyze the abstraction of the pro-*S* proton at C-4. Mammalian and bacterial glutamate decarboxylase catalyzes the stereospecific decarboxylation of L-glutamate with retention of configuration (i.e., *pro-R* protonation). Therefore, it appeared to be an anomaly that the *(S)*-enantiomer of 4-aminohex-5-ynoic acid should inactivate the mammalian glutamate decarboxylase. In this case a proton is removed that would correspond to transamination. In order to rationalize this observation, it was proposed that the electron withdrawing effect of the acetylenic group renders the *(S)*-proton acidic enough for abstraction, thereby leading to inactivation. The inactivation of mammalian and bacterial GABA aminotransferase and glutamate decarboxylase by both enantiomers of 4-aminohex-5-ynoic acid were reinvestigated by Danzin et al.[47] Both rat brain and bacterial GABA aminotransferase are irreversibly inactivated by *(S)*-(+)- and *(R)*-(−)-4-aminohex-5-ynoic acid in a time-dependent fashion. In the case of the rat brain enzyme, the *(S)*-isomer shows saturation kinetics with a $K_I = 75$ μM and $k_{inact} = 0.20$ min⁻¹; the *(R)*-isomer, however, indicates no saturation, although substrate does protect. Both isomers also inactivate the *P. fluorescens* enzyme. Rat brain glutamate decarboxylase is inactivated by the *(S)*-isomer and bacterial glutamate decarboxylase is inactivated by the *(R)*-isomer. The basis for the mistaken earlier report[46] that only the *(S)*-isomer inactivates GABA aminotransferase was the observation that this isomer inactivates both the pig brain and *P. fluorescens* enzymes at the same rate as the racemate at double the concentration. To explain the inactivation by both isomers, it was proposed that GABA aminotransferase has two active site bases that could potentially remove the γ-proton; ordinarily, only the one on the *pro-S* side is operative. Because of the increased acidity of the γ-proton adjacent to the ethynyl group, however, the *pro-R* proton becomes acidic enough for removal by the second base. The *(R)*-isomer does not inactivate mammalian glutamate decarboxylase, but does inactivate GABA aminotransferase; therefore, it is a selective inhibitor.

4-Aminohex-5-ynoic acid also inactivates rat liver ornithine aminotransferase; only the *(S)*-isomer is active.[48] 4-Aminohex-5-enoic acid, 3-aminopent-5-ynoic acid, and 5-hexyne-1,4-diamine have no effect on the enzyme; it appears that the reason that 4-aminohex-5-enoic acid is not an inactivator is that it is mostly transaminated, forming the PMP form of the enzyme. A deuterium isotope effect of 3 to 4 on the rate of inactivation was observed for [4-²H]4-aminohex-5-ynoic acid.

Since γ-acetylenic GABA inactivates several enzymes, namely, GABA aminotransferase, ornithine δ-aminotransferase, and glutamate decarboxylase,[45,48] and gabaculine inactivates GABA aminotransferase and ornithine δ-aminotransferase, but not glutamate decarboxylase (see Section VII.A.1.), then treatment of tissue first with gabaculine followed by [α-³H]γ-acetylenic GABA should result in specific radioactive labeling of glutamate decarboxylase. This proved to be the case in crude mouse brain; loss of glutamate decarboxylase activity correlates with incorporation of radioactivity.[49] In the absence of gabaculine pretreatment, about twice as much radioactivity is incorporated.

4-Aminohex-5-ynoate inactivates rabbit brain GABA aminotransferase, rat liver ornithine aminotransferase, and the α-subform of pig heart cytoplasmic aspartate aminotransferase; interpretation of absorption spectral changes was used by John et al.[50] as a basis for their conclusions. The acetylenic analogue is most active against ornithine aminotransferase, but in the absence of an α-ketoacid, transamination is an important reaction in protecting the enzyme from inactivation by conversion of the cofactor to the PMP form. Ornithine aminotransferase from rat liver is inactivated by DL-4-amino[2,3-¹⁴C]hex-5-ynoic acid with the incorporation of 1 mol of radioactivity per mole of enzyme after gel filtration, ammonium sulfate precipitation, DEAE cellulose chromatography, and heating to 60°C.[51] Apoenzyme reconstituted with [³H]pyridoxal phosphate was inactivated with 4-aminohex-5-ynoic acid; after extensive dialysis under conditions where 95% of the [³H] was released from the uninactivated control, only 12% of the [³H] was released from inactivated enzyme. Therefore, both inactivator and cofactor are bound irreversibly. The 330-nm absorption of the inactivated enzyme was used by Jones et al.[51] to support structure **7.19** (Scheme 27); however, these results also suggest inactivation by the enamine mechanism[42,43] (see Scheme 28). Radioactively labeled inactivated enzyme was inseparable from native enzyme by DEAE-cellulose chromatography, by crystallization from ammonium sulfate, or during purification of the enzyme from sonicated rate liver mitrochondria. This indicates that inactivated enzyme is quite similar in size and charge to native enzyme.

5-Hexyne-1,4-diamine is metabolized to 4-aminohex-5-ynoic acid in vivo (see Section III.A.4. for details).

A series of analogues of β-alanine, γ-aminobutyrate, and δ-aminopentanoate containing a β-, γ-, and δ-ethynyl group, respectively, was shown by Bey et al.[52] to inactivate pig brain GABA aminotransferase. Unlike the corresponding fluoromethyl compounds (see Chapter 6 of Volume I, Section II.A.2.), the potency of the acetylenic inactivators increases with increasing chain length.

4-Aminohex-5-ynoic acid was synthesized by Metcalf and Casara[53] by a method that utilizes a propargylamine anion synthon (Scheme 29). A related approach by

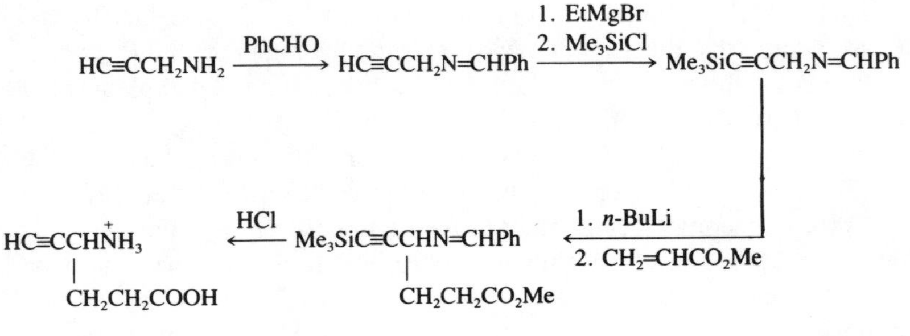

HC≡CCH₂NH₂ $\xrightarrow{\text{PhCHO}}$ HC≡CCH₂N=CHPh $\xrightarrow[\text{2. Me}_3\text{SiCl}]{\text{1. EtMgBr}}$ Me₃SiC≡CCH₂N=CHPh

$$\text{HC} \equiv \text{CCH} \overset{+}{\text{N}} \text{H}_3 \quad \xleftarrow{\text{HCl}} \quad \text{Me}_3\text{SiC} \equiv \text{CCHN} = \text{CHPh} \quad \xleftarrow[\text{2. CH}_2=\text{CHCO}_2\text{Me}]{\text{1. } n\text{-BuLi}}$$

with lower groups:

$\underset{\text{CH}_2\text{CH}_2\text{COOH}}{|}$ $\underset{\text{CH}_2\text{CH}_2\text{CO}_2\text{Me}}{|}$

Scheme 29.

Metcalf and Casara[54] utilizes Boc protection of the amino group and the formation of a dianion.

4. 5-Hexyne-1,4-diamine and Related

On the basis of the principle of microscopic reversibility, the putrescine analogue, 5-hexyne-1,4-diamine (Structural Formula 7.21, Scheme 30) was found by Metcalf et al.[55] to inactivate thioacetamide-treated rat liver ornithine decarboxylase. Dialysis does not regenerate enzyme activity and dithiothreitol in the preincubation buffer does not prevent inactivation. Only the (−)-isomer is active. No kinetic isotope effect on the rate was detected with 4-deuterio-5-hexyne-1,4-diamine, but rather a primary isotope effect of $K_H/K_D = 1.9$ was observed on the apparent Michaelis constant. Proton abstraction, therefore, must occur, but not in a rate-determining step. Compound **7.21** does not inactivate glutamate- or aromatic amino acid decarboxylase, GABA aminotransferase, or diamine oxidase. The mechanism proposed was a typical acetylene-allene isomerization mechanism (Scheme 30).

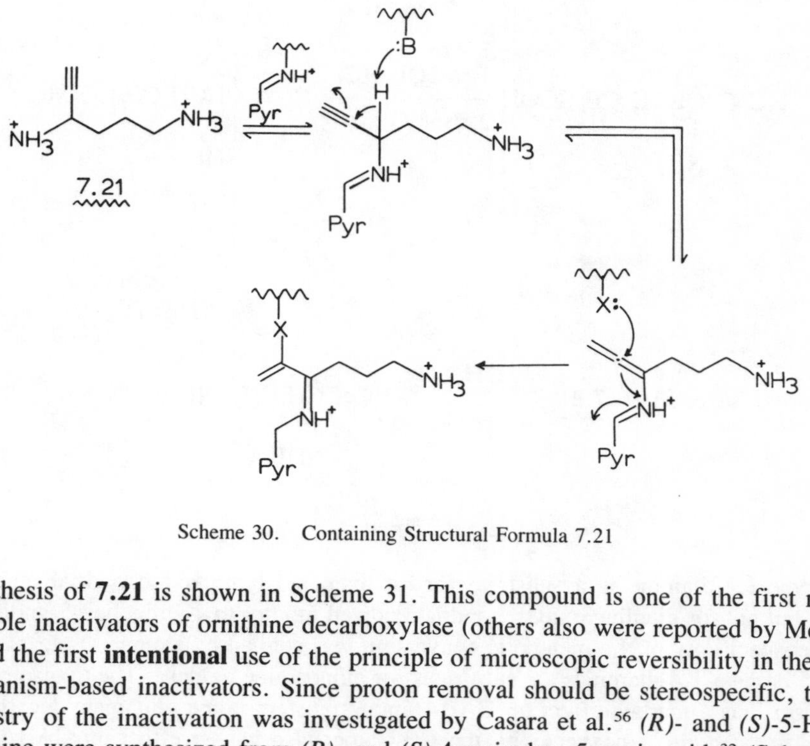

Scheme 30.   Containing Structural Formula 7.21

The synthesis of **7.21** is shown in Scheme 31. This compound is one of the first reported irreversible inactivators of ornithine decarboxylase (others also were reported by Metcalf et al.[55]) and the first **intentional** use of the principle of microscopic reversibility in the **design** of mechanism-based inactivators. Since proton removal should be stereospecific, the stereochemistry of the inactivation was investigated by Casara et al.[56] *(R)-* and *(S)-*5-Hexyne-1,4-diamine were synthesized from *(R)-* and *(S)-*4-aminohex-5-ynoic acids[32] (Scheme 32). Only the *(R)-*(−)-isomer is a time-dependent inactivator of the enzyme from liver of thioacetamide-treated rats; this is consistent with an enzyme-catalyzed stereospecific proton removal. At the time this was published, the enzyme was not purified and it was not known what the stereospecificity was with orithine or putrescine. The *(S)-*(+)-isomer is a competitive reversible inhibitor. These results suggest that decarboxylation of ornithine proceeds with retention of configuration. Compound **7.21** also is a time-dependent inactivator of *E. coli* ornithine decarboxylase.[57]

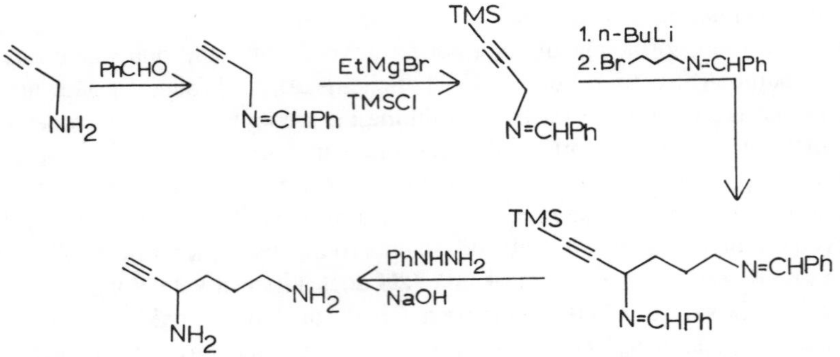

Scheme 31.

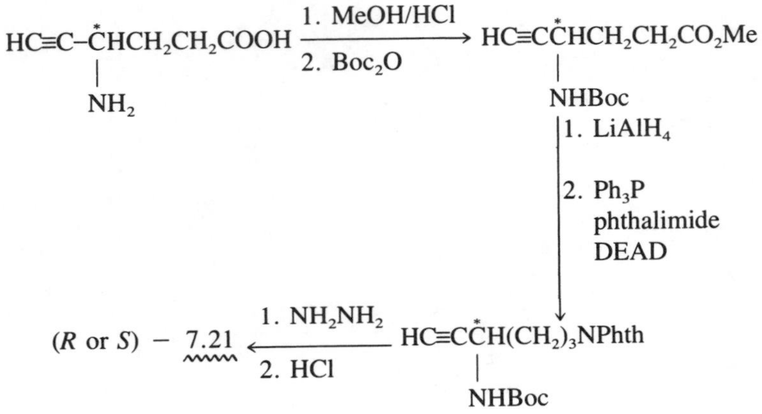

Scheme 32.

5-Hexyne-1,4-diamine is a substrate for rat liver and brain mitochondrial monoamine oxidase, but not for diamine oxidase, and is oxidized in vitro to 4-aminohex-5-ynoic acid.[58] The stereospecificity of the metabolic conversion was studied by Danzin et al.[59] Both (R)- and (S)-5-hexyne-1,4-diamines are substrates for monoamine oxidase. The (S)-isomer, after metabolism, has a marked effect on GABA aminotransferase and glutamate decarboxylase activities in rat brain. Unexpectedly, the (R)-isomer also has an effect on GABA amino-transferase activity. These results were responsible for the reinvestigation of the stereospecificity of inactivation of GABA aminotransferase by the isomers of 4-aminohex-5-ynoic acid (see Section III.A.3.).

Since monoamine oxidase does not oxidize α-substituted amines to any great extent, the approach by Danzin et al.[60] to avoid the in vivo metabolism of 5-hexyne-1,4-diamine was to α-methylate it. 2-Methyl-6-heptyne-2,5-diamine (**7.22**, R = R′ = CH₃), 6-heptyne-2,5-diamine (**7.22**, R = H, R′ = CH₃), and (2R,5R)-6-heptyne-2,5-diamine were prepared (see Scheme 33) and tested as substrates for monoamine oxidase and activators of ornithine decarboxylase. All are mechanism-based inactivators of ornithine decarboxylase from liver of thioacetamide-treated rats with K_I values of 1300, 13, and 3 μM, respectively; prolonged dialysis results in partial return of enzyme activity. (2R,5R)-6-Heptyne-2,5-diamine is not a substrate for monoamine oxidase and does not inhibit GABA aminotransferase activity in vivo in brain. Also, it is 10 times more potent than α-difluoromethylornithine (see Chapter 8, Section II.B.5.) in vitro and in vivo as an inactivator of ornithine decarboxylase; unlike α-difluoromethylornithine, however, it depletes not only putrescine and spermidine levels in vivo, but also spermine, thus causing complete cessation of tumor cell replication.[61] (2R,5R)-6-Heptyne-2,5-diamine also is a potent (K_I = 5 μM) time-dependent inactivator of ornithine decarboxylase from *Trypanasoma brucei brucei*.[61a] 6-Heptyne-2,5-diamine was synthesized by Casara et al.[62] as shown in Scheme 34. The four stereoisomers of the product were separated, after bis-*N*-trifluoroacetate derivatization, by gas-liquid chromatography with a chiral polysiloxane type stationary phase (Chirasil-Val®).

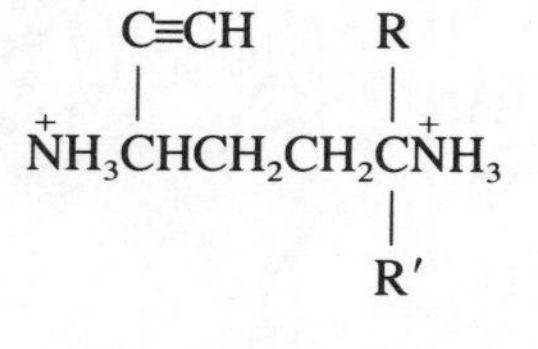

$$
\begin{array}{cc}
C\equiv CH & R \\
| & | \\
\overset{+}{N}H_3CHCH_2CH_2C\overset{+}{N}H_3 \\
& | \\
& R'
\end{array}
$$

Scheme 33. Structural Formula 7.22.

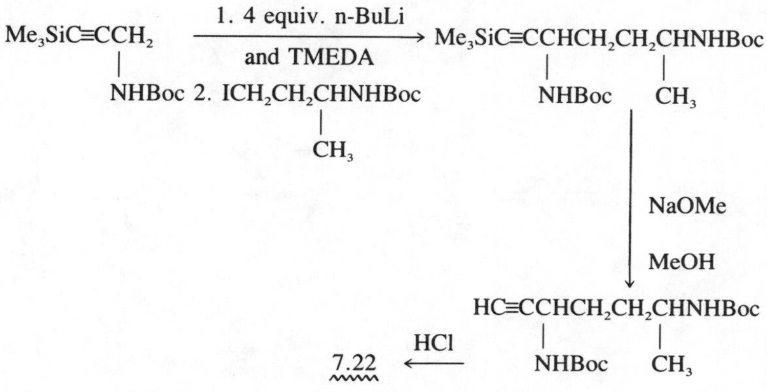

Scheme 34.

The four stereoisomers were synthesized as shown in Scheme 35. All 4 stereoisomers are time-dependent irreversible inactivators of rat liver ornithine decarboxylase; however, the (2R,5R)-isomer is, by far, the most potent.

Another analogue of 6-hexyne-1,4-diamine that was prepared by Metcalf et al.[55] is *trans*-2-hexen-5-yne-1,4-diamine (Structural Formula 7.23, Scheme 36). This compound is a time-dependent inactivator of thioacetamide-treated rat liver ornithine decarboxylase[55] and the corresponding *E. coli* enzyme.[57]

Arginine decarboxylase from *E. coli*, oats, and barley is inactivated by α-ethynylagmatine (Structural Formula 7.23a, Scheme 36A),[62a] which is more potent than α-difluoromethyl-larginine. Inactivation by **7.23a** was found by Bitonti et al.[62a] to be partially reversed upon dialysis.

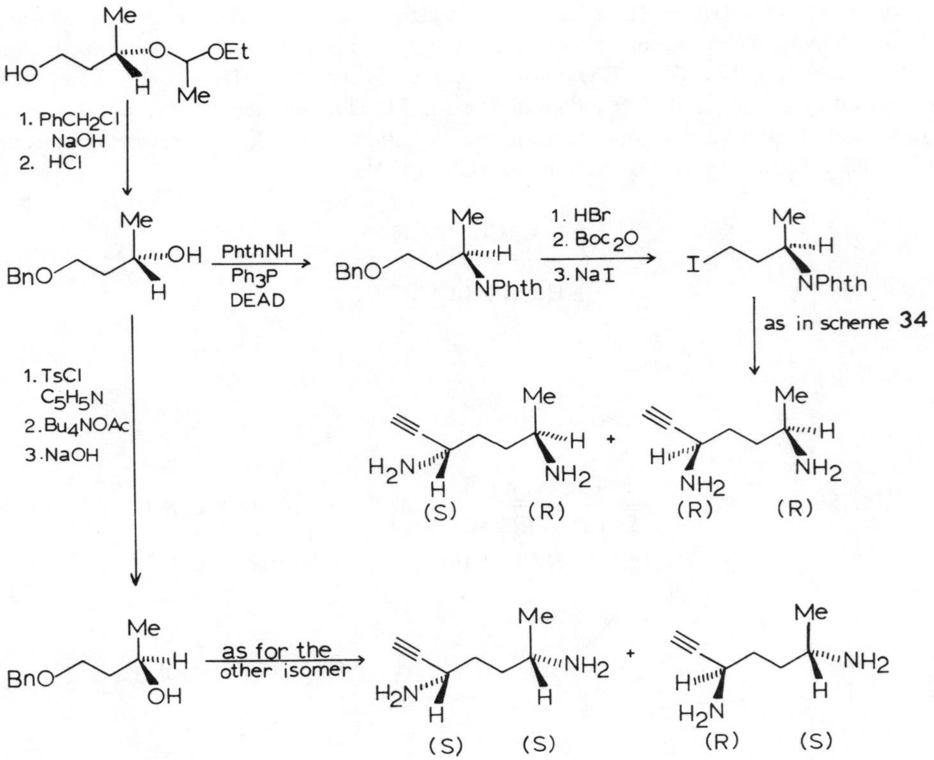

Scheme 35.

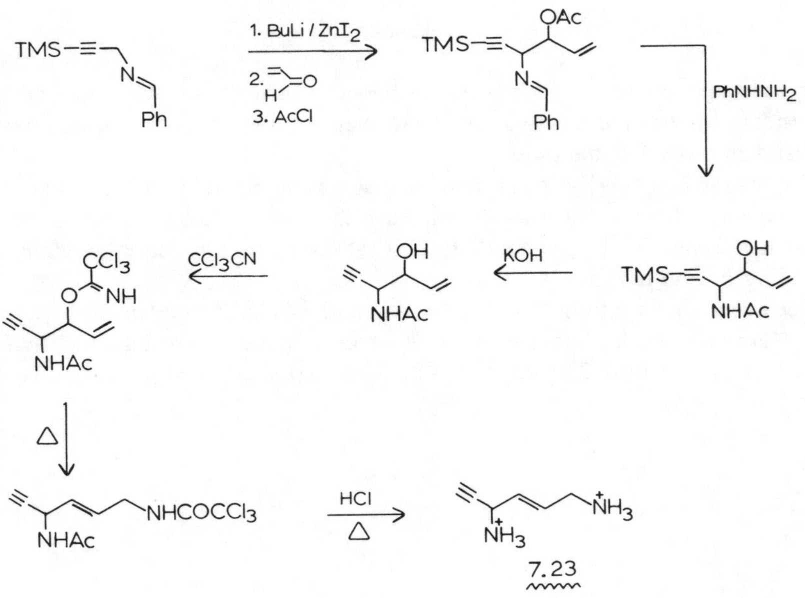

Scheme 36.   Containing Structural Formula 7.23.

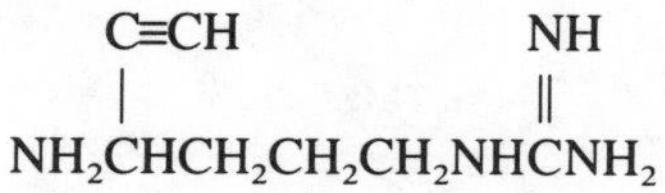

$$C\equiv CH \qquad\qquad NH$$
$$| \qquad\qquad\qquad ||$$
$$NH_2CHCH_2CH_2CH_2NHCNH_2$$

Scheme 36A.   Structural Formula 7.23a.

5. D,L-2,6-Diamino-4-hexynoate (D,L-Lysyne)

D,L-2,6-Diamino-4-hexynoate (lysyne) (**7.24**, Scheme 37) is both a substrate and time-dependent inactivator of L-lysine ε-aminotransferase from *Achromobacter liquidum*.[63] The partition ratio is 35 to 40 as measured by conversion of [^{14}C]α-ketoglutarate to [^{14}C]glutamate. [2-^{14}C]Lysyne inactivation results in the incorporation of about one equivalent of radioactivity into the enzyme, but the inactivated enzyme adduct is unstable; dialysis at pH 8.5 is sufficient to cause complete reactivation overnight. Lysine α-racemase is not inactivated by **7.24**, indicating the specificity of PLP enzymes that act at the ε-position. Neither rat liver ornithine δ-aminotransferase nor GABA aminotransferase is inactivated. The mechanisms of inactivation of L-lysine ε-aminotransferase suggested by Shannon et al.[63] are the two standard isomerization/addition mechanisms for acetylenic inactivators shown in Scheme 27 (Section III.A.3.). Compound **7.24** can be synthesized by the route shown in Scheme 38.

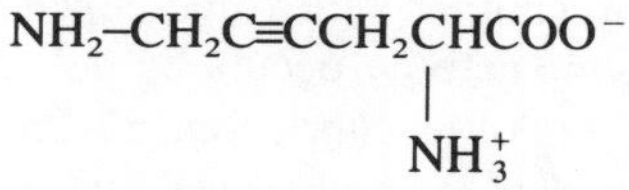

$$NH_2-CH_2C\equiv CCH_2CHCOO^-$$
$$|$$
$$NH_3^+$$

Scheme 37.   Structural Formula 7.24.

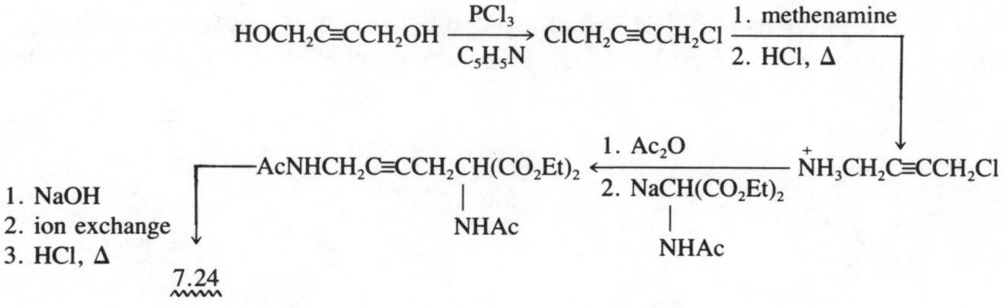

$$HOCH_2C\equiv CCH_2OH \xrightarrow[C_5H_5N]{PCl_3} ClCH_2C\equiv CCH_2Cl \xrightarrow[2.\ HCl,\ \Delta]{1.\ methenamine}$$

AcNHCH$_2$C≡CCH$_2$CH(CO$_2$Et)$_2$ $\xleftarrow[2.\ NaCH(CO_2Et)_2]{1.\ Ac_2O}$ $\overset{+}{N}H_3CH_2C\equiv CCH_2Cl$

1. NaOH
2. ion exchange
3. HCl, Δ   NHAc   NHAc

7.24

Scheme 38.

6. α-Ethynylhistamine

α-Ethynylhistamine (Structural Formula 7.25) was synthesized by Holbert and Metcalf[64] (Scheme 39), and was shown to be an irreversible inactivator of hamster placental histidine decarboxylase on the basis of the principle of microscopic reversibility; pseudo first-order kinetics were not obeyed.

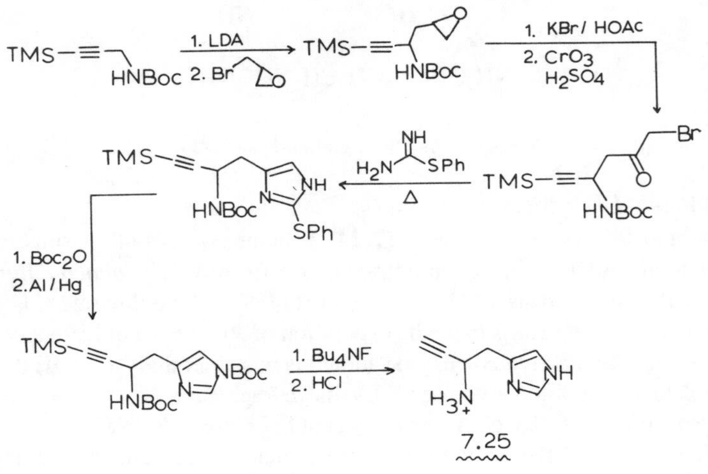

Scheme 39.   Containing Structural Formula 7.25.

7. *2-Butynylamine and 2-Propynylamine*

2-Butynylamine[65] and 2-propynylamine[65,66] are time-dependent inactivators of beef plasma amine oxidase; no return of enzyme activity occurs by gel filtration or dialysis. No inactivation occurs if the enzyme is devoid of its cupric ion.[66] 2-Propynyl[3-³H]amine inactivation results in the incorporation of 2 mol of inactivator per mole of enzyme after dialysis, which is stable to base, acid, and urea treatment.[65] Since the cofactor for this enzyme at the time was not known (it is now believed to be pyrroloquinone quinone[66a]), but was thought to be some sort of carbonyl group, an isomerization mechanism was proposed by Hevey et al.[65] (Scheme 40, pathway $\underline{a}$). An alternative mechanism is shown in pathway $\underline{b}$. 2-Butynylamine can be synthesized by a Gabriel amine synthesis[67] from 1-bromo-2-butene.[68]

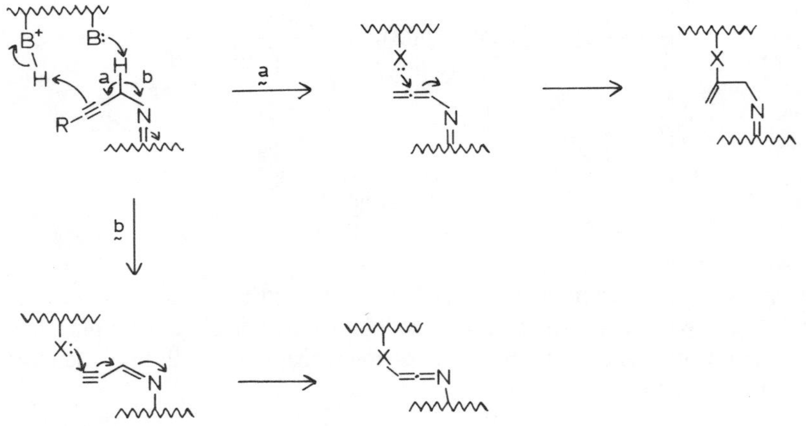

Scheme 40.

B. Allene-Containing Inactivators

1. *γ-Allenyl GABA*

γ-Allenyl GABA (4-aminohept-5,6-dienoic acid; Structural Formula 7.26) has been synthesized by several different routes. Castelhano and Krantz[69] prepared **7.26** (R = R′ = H) by a route that involves an aza-Cope rearrangement (Scheme 41). Hiemstra et al.[70] used the approach shown in Scheme 42. Casara et al.[71] devised a general synthesis of allenic amines and allenic amino acids; the route to **7.26** is shown in Scheme 43 (R = $CH_2CH_2CO_2Me$).

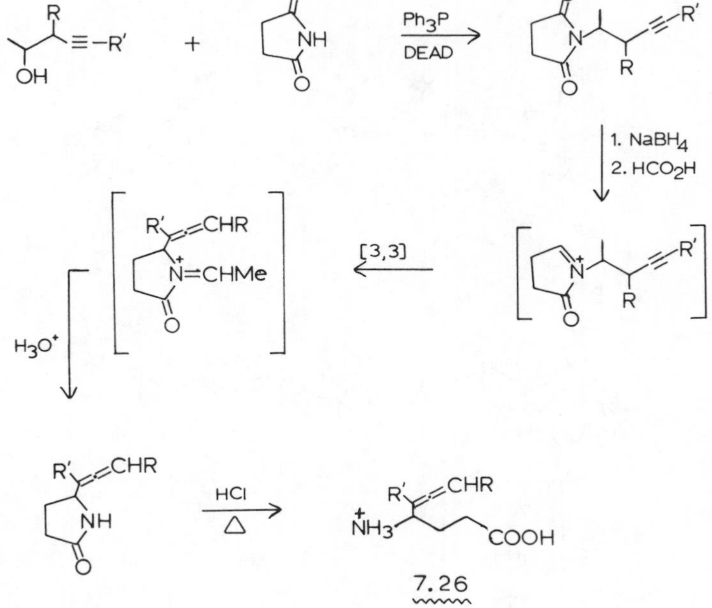

Scheme 41.   Containing Structural Formula 7.26.

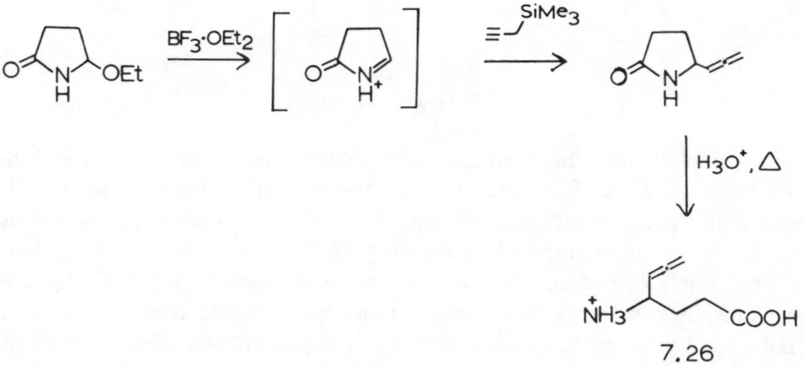

Scheme 42.

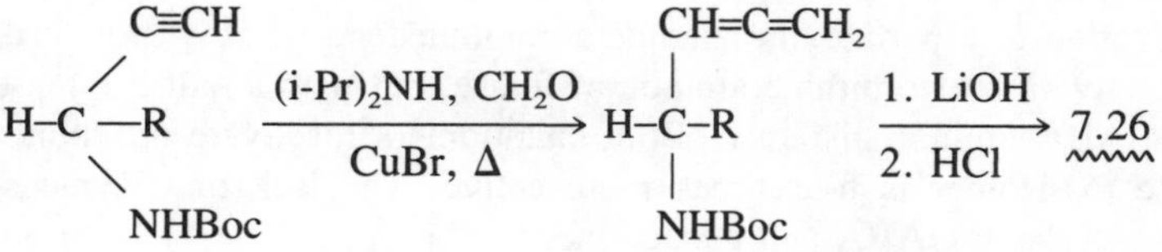

Scheme 43.

(S)-γ-Allenyl GABA is a mechanism-based inactivator of mammalian GABA aminotransferase.[71] The mechanisms proposed[71] for inactivation of PLP-dependent enzymes by allenic amines and amino acids are shown in pathways a and b and of Scheme 44.

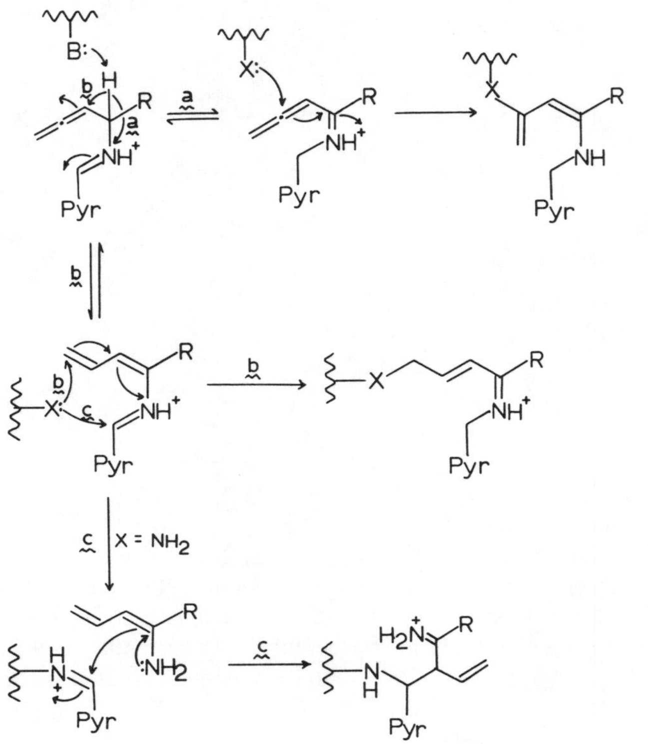

Scheme 44.

Pathway c is adapted from the enamine mechanism[42,43] of inactivation of PLP-dependent enzymes by L-serine *(O)*-sulfate described in Chapter 6 of Volume I (Section II.G.3.a.). The enamine could add to the enzyme-bound PLP with the β-carbon (as shown in Scheme 44) or with the δ-carbon. A more detailed study of **7.26** was carried out by Jung et al.[72] *(R,S)*-γ-Allenyl GABA produces time-dependent inactivation of pig brain GABA aminotransferase. This inactivator is two to three times more potent than γ-vinyl GABA (see Section III.C.1.); no return of enzyme activity is apparent after dialysis. The *(S)*-isomer inactivates the enzyme at the same rate as double the concentration of the racemate and the *(R)*-isomer has no effect at 10 times the concentration of the *(S)*-enantiomer. These results are consistent with those previously obtained with γ-acetylenic GABA[46] (see Section III.A.3.). γ-Allenic GABA does not inhibit glutamate decarboxylase nor does it produce time-dependent inactivation of aspartate and alanine aminotransferases. However, it does produce irreversible inactivation of ornithine aminotransferase, but with a half-life ten times longer than that for GABA aminotransferase. Both enantiomers inactivate ornithine aminotransferase, but the *(S)*-isomer is four times more active. The lack of enzyme selectivity is common to γ-acetylenic GABA and gabaculine.[48]

2. 5,6-Heptadiene-1,4-diamine (α-Allenylputrescine)

(R)-α-Allenylputrescine (Structural Formula 7.27, Scheme 45) inactivates *E. coli*[71,73] and rat liver[73] ornithine decarboxylase; a high concentration of the *(S)*-isomer does inactivate

the rat liver enzyme, but not the *E. coli* enzyme.[73] The stereospecificity of the *E. coli* enzyme for the *(R)*-isomer is consistent with the stereochemistry of the enzyme that decarboxylates L-ornithine with retention of configuration.[74] The inactivation mechanisms in Scheme 44 (Section III.B.1.) are reasonable possibilities for **7.27** as well.

$$CH=C=CH_2$$
$$|$$
$$NH_2CHCH_2CH_2CH_2NH_2$$

Scheme 45.   Structural Formula 7.27.

α-Allenylagmatine (Structural Formula 7.27a, Scheme 45a) is a mechanism-based inactivator of arginine decarboxylase from *E. coli,* oats, and barley.[62a]

$$CH=C=CH_2 \qquad NH$$
$$| \qquad\qquad ||$$
$$NH_2CHCH_2CH_2CH_2NHCNH_2$$

Scheme 45A.   Structural Formula 7.27a.

C. Alkene-Containing Inactivators
1. 4-Aminohex-5-enoic Acid (γ-Vinyl GABA) and Related

Incubation of rat brain GABA aminotransferase with 4-aminohex-5-enoic acid (**7.28**, R = $CH_2CH_2CO_2H$; Scheme 46) results in time-dependent pseudo first-order loss of enzyme activity; dialysis does not restore enzyme activity.[75] Dithiothreitol and β-mercaptoethanol have no effect on the inactivation rate. No inhibition of GABA aminotransferase from *Pseudomonas fluorescens* and no irreversible inactivation of rat brain glutamate decarboxylase or asparatate aminotransferase was observed. At 10 mM concentration, however, slow time-dependent inactivation of rat brain alanine aminotransferase occurs (ca. 10^3 times slower than for GABA aminotransferase). Inactivation of GABA aminotransferase by 4-[4-^{2}H]aminohex-5-enoic acid results in a primary kinetic isotope effect, suggesting deprotonation is involved in the rate-determining step. The mechanism proposed by Lippert et al.[75] is a typical isomerization/addition pathway (Scheme 46). An alternative pathway, however, would be based on the enamine mechanism[42,43] of inactivation of aspartate aminotransferase and glutamate decarboxylase by L-serine *O*-sulfate discussed in Chapter 6 of Volume I (Section II.G.3.a.) (Scheme 47). Consistent with either mechanism, γ-deuterio-γ-vinyl GABA exhibits a primary kinetic isotope effect of 3.88 on the rate of inactivation of GABA aminotransferase; the *pro-4S* proton is abstracted from (+)-γ-vinyl GABA.[44]

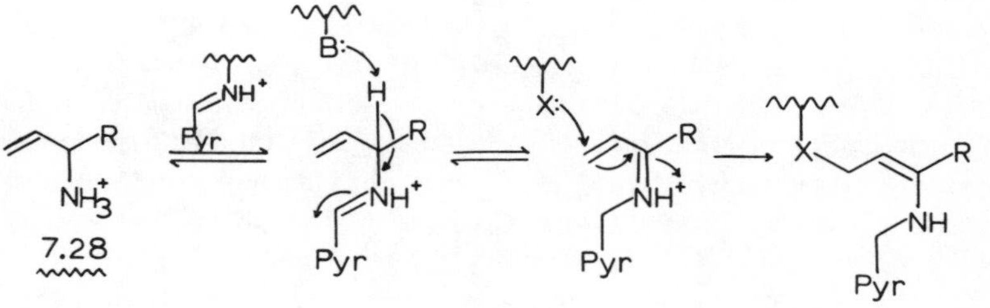

Scheme 46.   Containing Structural Formula 7.28.

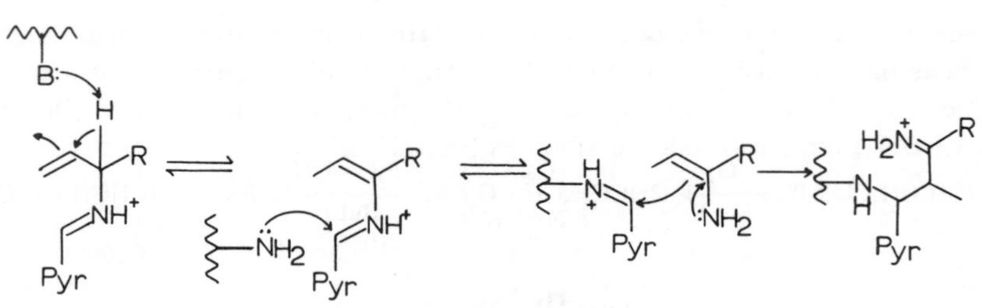

Scheme 47.

Compound **7.28** (R = CH$_2$CH$_2$COOH), synthesized[75] by the reduction of γ-ethynyl GABA with lithium (liquid NH$_3$/[NH$_4$]$_2$SO$_4$), is specific for inactivation of GABA aminotransferase; it does not inactivate rat liver ornithine aminotransferase,[50] the α-subform of pig heart cytoplasmic aspartate aminotransferase,[50] or glutamate decarboxylase.[44,45]

A series of analogues of β-alanine, γ-aminobutyrate, and δ-aminopentanoate containing a β-, γ-, and δ-vinyl group, respectively, was shown by Bey et al.[52] to inactivate pig brain GABA aminotransferase. Unlike the corresponding fluoromethyl compounds (see Chapter 6 of Volume I, Section II.A.2.), the potency of the vinyl-substituted inactivators increases with increasing chain length.

Analogues of γ-vinyl GABA in which 1-3 vinyl hydrogens are replaced by fluorine (Structural Formula 7.28a, Scheme 47A) were synthesized by Kolb et al.[75a] and shown to be time-dependent irreversible inactivators of pig brain GABA aminotransferase. The least active inactivator is **7.28a** (X=X'=X''=F); the most active compound is **7.28a** (X=X'=H; X''=F). Dialysis does not regenerate enzyme activity. Two of the compounds, **7.28a** (X=F, X'=X''=H) and **7.28a** (X=X'=X''=F) are not competitive with GABA. Little or no transamination was observed with any of the compounds. Compounds **7.28a** (X=F, X'=X''=H) and **7.28a** (X=X'=H, X''=F) were synthesized as shown in Scheme 47B.

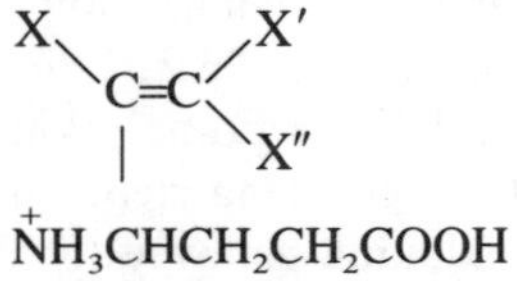

Scheme 47A.   Structural Formula 7.28a.

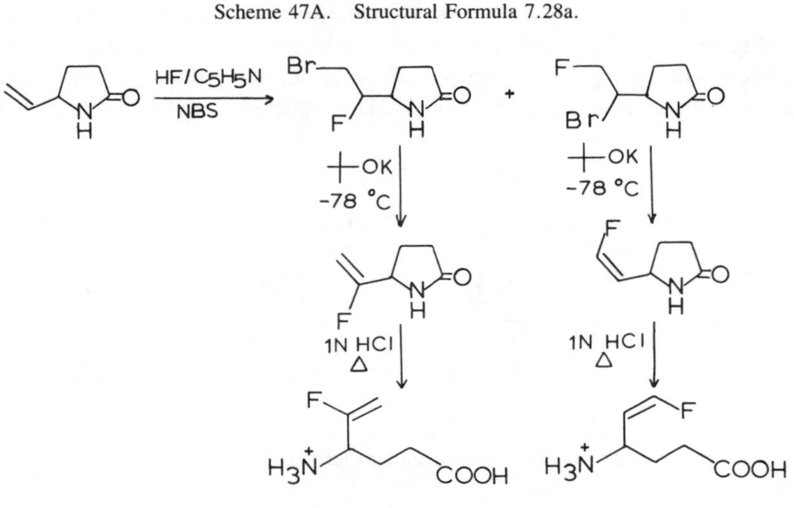

Scheme 47B.

Compounds **7.28a** (X=H, X'=X''=F) and **7.28a** (X=X'=X''=F) were synthesized by the route in Scheme 47C.

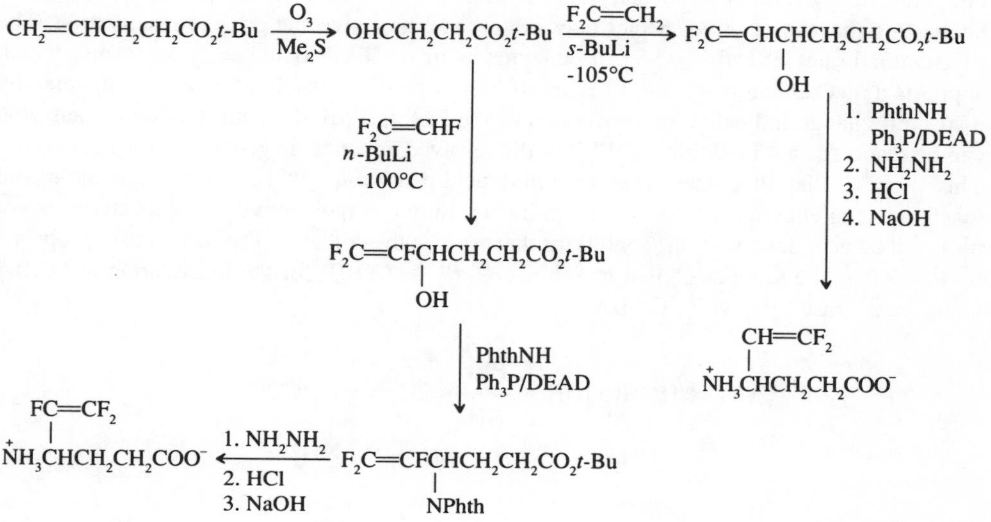

$$CH_2\text{=}CHCH_2CH_2CO_2t\text{-Bu} \xrightarrow[\text{Me}_2\text{S}]{\text{O}_3} OHCCH_2CH_2CO_2t\text{-Bu} \xrightarrow[\substack{s\text{-BuLi}\\-105°C}]{F_2C\text{=}CH_2} F_2C\text{=}CHCHCH_2CH_2CO_2t\text{-Bu}$$

with OH on the carbinol carbon of the last compound.

$$F_2C\text{=}CHF,\ n\text{-BuLi},\ -100°C \downarrow$$

$$F_2C\text{=}CFCHCH_2CH_2CO_2t\text{-Bu}$$
(OH)

1. PhthNH Ph₃P/DEAD
2. NH₂NH₂
3. HCl
4. NaOH

$$\underset{\text{CH=CF}_2}{\overset{+}{N}H_3CHCH_2CH_2COO^-}$$

$$\text{PhthNH},\ \text{Ph}_3\text{P/DEAD} \downarrow$$

$$F_2C\text{=}CFCHCH_2CH_2CO_2t\text{-Bu}$$
(NPhth)

$$\underset{\text{FC=CF}_2}{\overset{+}{N}H_3CHCH_2CH_2COO^-} \xleftarrow[\substack{2.\ \text{HCl}\\3.\ \text{NaOH}}]{1.\ \text{NH}_2\text{NH}_2}$$

Scheme 47C.

(*E*)-4-Amino-2,5-hexadienoic acid (Structural Formula 7.28, Scheme 47D) is a time-dependent inactivator of pig brain GABA aminotransferase; the rate constant for **7.28b** is about one sixth that for γ-vinyl GABA.[75b]

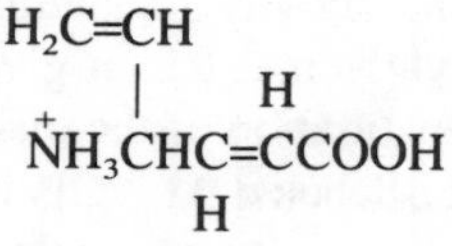

$$\underset{\text{H}}{\overset{\text{H}_2\text{C=CH}}{\overset{+}{N}H_3CHC\text{=}CCOOH}}\ \ \text{H}$$

Scheme 47D.   Structural Formula 7.28b.

Compound **7.28b** was synthesized by Bey et al.[75b] as shown in Scheme 47E.

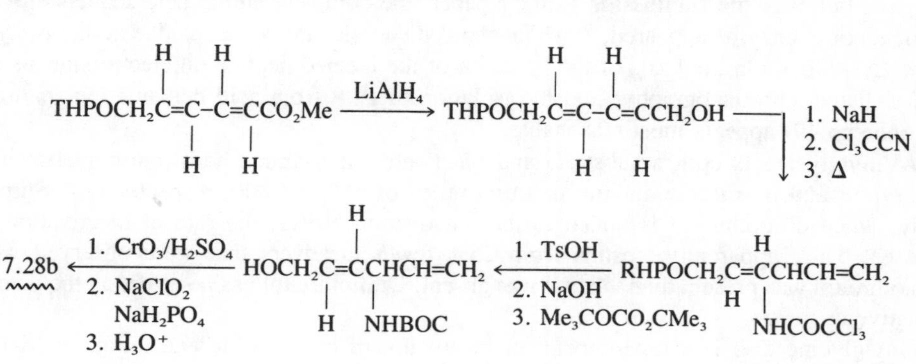

$$THPOCH_2C\text{=}C\text{-}C\text{=}CCO_2Me \xrightarrow{\text{LiAlH}_4} THPOCH_2C\text{=}C\text{-}C\text{=}CCH_2OH$$
(with H H / H H substituents)

1. NaH
2. Cl₃CCN
3. Δ

$$RHPOCH_2C\text{=}CCHCH\text{=}CH_2$$
(H | NHCOCCl₃)

$$\xrightarrow[\substack{2.\ \text{NaOH}\\3.\ \text{Me}_3\text{COCO}_2\text{CMe}_3}]{1.\ \text{TsOH}} HOCH_2C\text{=}CCHCH\text{=}CH_2$$
(H NHBOC)

$$\xrightarrow[\substack{2.\ \text{NaClO}_2\\ \text{NaH}_2\text{PO}_4\\3.\ \text{H}_3\text{O}^+}]{1.\ \text{CrO}_3/\text{H}_2\text{SO}_4} \text{7.28b}$$

Scheme 47E.

2. 2-Amino-3-butenoic Acid (Vinylglycine)

Vinylglycine (**7.29**) can be synthesized[76] from 2-hydroxy-3-butenoic acid[77] (Scheme 48). Cytoplasmic pig heart aspartate aminotransferase is irreversibly inactivated by **7.29**; no enzyme activity returns upon dialysis.[78] β-Mercaptoethanol does not protect the enzyme, suggesting that the reactive species does not escape the active site prior to inactivation. The inactivator is not effective when the enzyme is in the PMP or apoenzyme forms, which supports a mechanism involving PLP Schiff base formation. Inactivation is accompanied by a spectral change indicative of conversion of the PLP to PMP. The inactivated haloenzyme can be resolved, but readdition of PLP to the resolved apoenzyme gives no enzyme activity. This suggests that attachment is to an amino acid residue. [1-¹⁴C]2-Amino-3-butenoic acid inactivates the enzyme with concomitant incorporation of radioactivity. Radioactivity is not removed by acid denaturation, conditions that remove the cofactor. The mechanism proposed by Rando[78] is the same as shown in Scheme 46 (R = CO$_2$H) for the inactivation of GABA aminotransferase by γ-vinyl GABA.

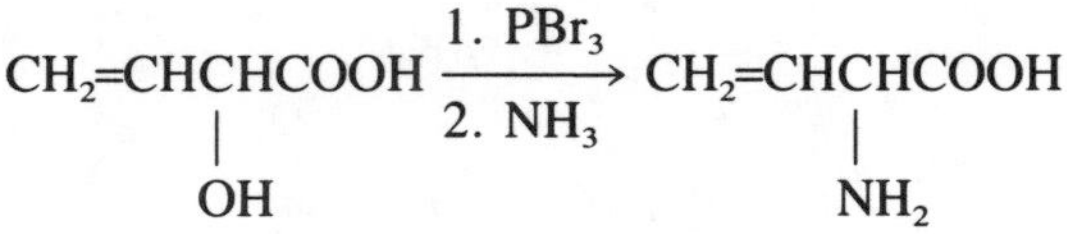

$$\text{CH}_2\text{=CHCHCOOH} \xrightarrow[\text{2. NH}_3]{\text{1. PBr}_3} \text{CH}_2\text{=CHCHCOOH}$$
$$\quad\quad\quad | \quad\quad\quad\quad\quad\quad\quad\quad\quad\quad\quad\quad | $$
$$\quad\quad\quad \text{OH} \quad\quad\quad\quad\quad\quad\quad\quad\quad\quad\quad \text{NH}_2$$

7.29

Scheme 48.   Containing Structural Formula 7.29.

Both the cytosolic (pig heart) and mitochondrial (chicken heart) aspartate aminotransferase isozymes also are irreversibly inactivated by vinylglycine; inactivation only occurs when the cofactor is in the PLP form.[79] In the absence of α-ketoglutarate, inactivation proceeds to 90%, indicating a partition ratio near zero. When α-ketoglutarate is added, complete inactivation results. The 10% enzyme activity remaining in the absence of α-ketoglutarate results from transamination of vinylglycine, which gives the PMP form. This protects the enzyme from inactivation. Acid denaturation of the inactivated enzyme results in release of PLP, suggesting that an amino acid is labeled. [1-¹⁴C]Vinylglycine inactivates both isozymes with incorporation of 1 mol of radioactivity per mole of subunit. Cyanogen bromide and chymotrypsin cleavages of the radioactively labeled isozymes led to the identification of the labeled amino acid in both cases as the lysine that secures the PLP at the active site (Gehring et al.[79] refer to this lysine residue as Lys-258 in both cases; it is Lys-258 for the cytosolic enzyme, but after the submission of their paper, the complete amino acid sequence of the mitochondrial enzyme appeared,[79a] which showed the relevant lysine residue in that isozyme to be Lys-250.). On the basis of the behavior of the labeled peptide on electrophoresis and ion exchange chromatography, and the isolation of PLP from acid denaturation, pathway a (Scheme 49) appears most reasonable.

D-Vinylglycine is both a substrate and inactivator of D-amino acid transaminase of *B. sphaericus* and *B. subtilis*, having partition ratios of 450 and 800, respectively.[80] Surprisingly, when dithiothreitol is added to the inactivation buffer, the rate of inactivation **increases**! This, apparently, results from 25% fewer turnovers that were observed when dithiothreitol was present than when it was absent. Dithiothreitol has no effect on the activity of native enzyme.

Vinylglycine also is a time-dependent inactivator of cytoplasmic and mitochondrial rat brain aromatic aminotransferase; inactivation occurs at the same rate for both isozymes.[81]

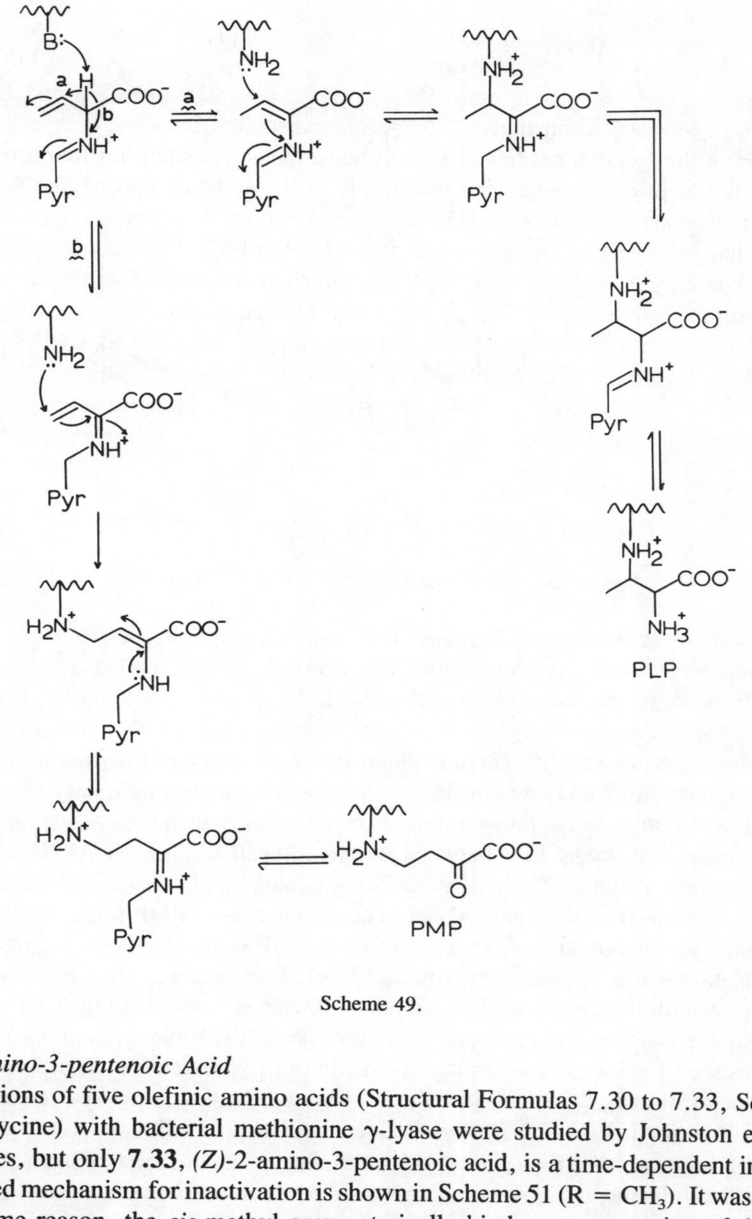

Scheme 49.

3. (Z)-2-Amino-3-pentenoic Acid

The reactions of five olefinic amino acids (Structural Formulas 7.30 to 7.33, Scheme 50, and vinylglycine) with bacterial methionine γ-lyase were studied by Johnston et al.[82] All are substrates, but only **7.33**, (Z)-2-amino-3-pentenoic acid, is a time-dependent inactivator. The proposed mechanism for inactivation is shown in Scheme 51 (R = CH₃). It was suggested that, for some reason, the *cis*-methyl group sterically hinders reprotonation of the quinoid intermediate **7.34** so that it accumulates and can be attacked by an active site nucleophile. Consistent with this explanation is the observation of a 550-nm absorbance, attributed to **7.34**, which only occurs when **7.33**, is the substrate.

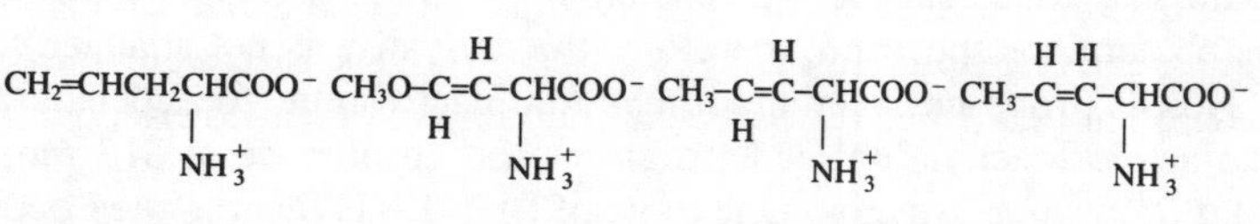

Scheme 50.   Structural Formulas 7.30 to 7.33.

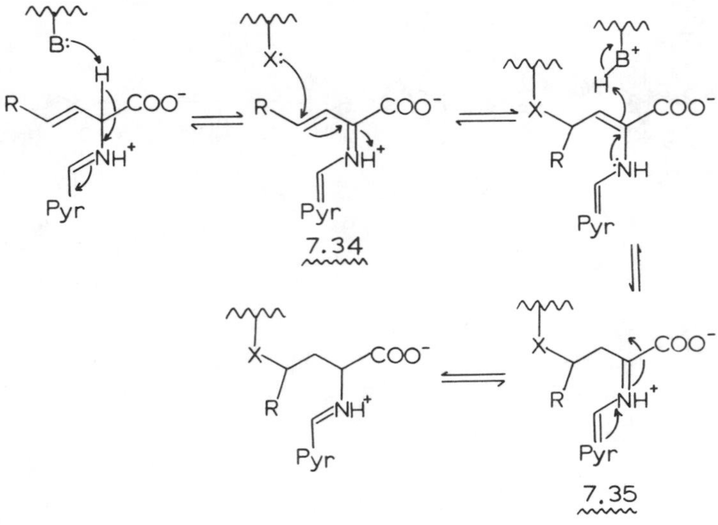

Scheme 51.   Containing Structural Formulas 7.34 and 7.35.

4. L-2-Amino-4-methoxy-trans-3-butenoic Acid (4-Methoxyvinylglycine)

L-2-Amino-4-methoxy-*trans*-3-butenoic acid (**7.31**) is a bacterial toxin that was isolated by Scannell et al.[83] from fermentation broths of *Pseudomnas aeruginosa*. It irreversibly inactivates L-aspartate aminotransferase; neither haloenzyme in the PMP form nor apoenzyme is affected by this compound.[84] There is a concomitant change in the PLP spectrum during inactivation; inactivated enzyme can be reactivated with fresh PLP. Therefore, an amino acid residue was implicated in the inactivation mechanism, which was suggested by Rando[84] to be the same as that shown in Scheme 51 (see Section III.C.3.) for **7.33**. The penultimate PMP-quinone intermediate (**7.35**, R = OMe) was suggested as the final adduct. Further studies by Rando et al.[85] were carried out. The enzyme in its PMP form is not inactivated by **7.31**, but it is inactivated by 2-keto-4-methoxy-*trans*-3-butenoic acid, suggesting that an intermediate in the inactivation is the imine (**7.36**). Furthermore, after inactivation of the native enzyme with the amino acid (**7.31**), the cofactor is converted into PMP. In a model study in which pyridoxal and the inactivator were heated, the loss of these two compounds can be accounted for by the concomitant formation of pyridoxamine. The proposed mechanism of inactivation is shown in Scheme 52. This mechanism is related to pathway b in Scheme 49 (see Section III.C.2.); vinylglycine is believed to follow pathway a in Scheme 49. This difference may be the result of the stabilizing effect of methoxyl on the β,γ-double bond, which diminishes isomerization to the α,β-unsaturated intermediate (pathway a, Scheme 49).

L-2-Amino-4-methoxy-*trans*-3-butenoic acid is both a substrate and inactivator of the $\alpha_2\beta_2$ complex and only an inactivator of the β_2 subunit of *E. coli* tryptophan synthase; no inactivation occurs in the absence of PLP.[86] β-Mercaptoethanol has no effect on the inactivation. Treatment of the inactivated enzyme with sodium borohydride or sodium cyanoborohydride does not alter the absorbance spectrum; therefore, the derivative is not attached to the PLP in a Schiff base. No SH group has been modified after inactivation. Acid denaturation does not release the cofactor. Inactivation leads to an intense absorbance at 317 nm, which is consistent with that of cyanine systems. One mole of [^{3}H]PLP is incorporated per monomer and is not removed by dialysis, urea, or H_2SO_4 precipitation. These results are consistent with the mechanism shown in Scheme 52 where X = NH_2.

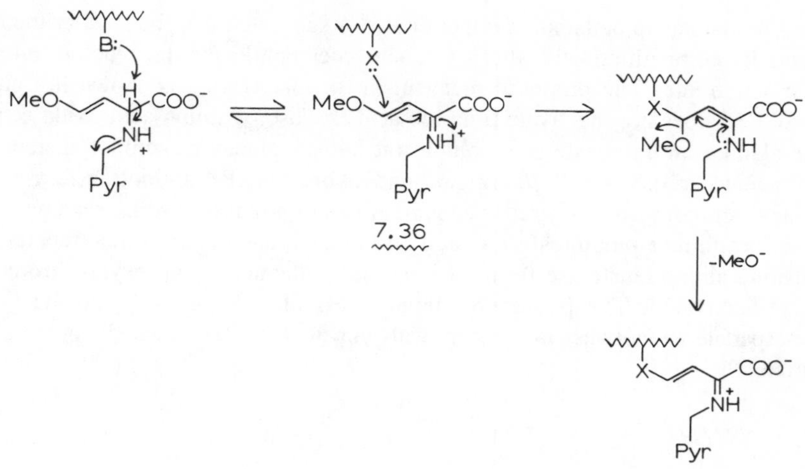

Scheme 52.   Containing Structural Formula 7.36.

5. *2-Amino-4-(2-amino-3-hydroxypropoxy)-trans-but-3-enoic Acid (Rhizobitoxine)*

The structure of rhizobitoxine was identified by Owens et al.[87] as 2-amino-4-(2-amino-3-hydroxypropoxy)-*trans*-but-3-enoic acid (Structural Formula 7.37, Scheme 53). Spinach β-cystathionase is inactivated by rhizobitoxine; gel filtration does not regenerate enzyme activity.[88] The inactivated enzyme, however, can be reactivated by incubation with PLP, suggesting the modification of the cofactor is responsible for inactivation.

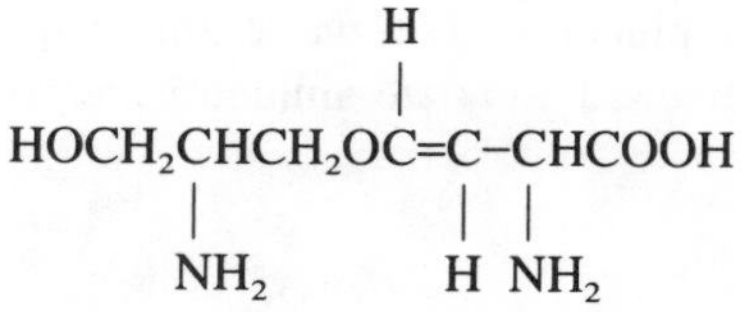

$$HOCH_2\underset{\underset{NH_2}{|}}{C}HCH_2OC=\underset{\underset{H}{|}}{C}-\underset{\underset{NH_2}{|}}{C}HCOOH$$

with H above the C

Scheme 53.   Structural Formula 7.37.

6. *β-Methylene-D,L-aspartic Acid and Related*

β-Methylene-D,L-aspartic acid (Structural Formula 7.38, Scheme 54) inactivates soluble pig heart glutamate-aspartate aminotransferase in a time-dependent reaction; dialysis does not restore enzyme activity.[89,90]

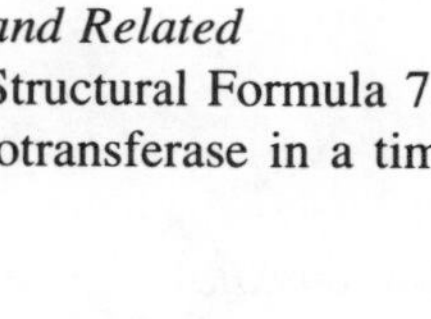

$$CH_2=\underset{\underset{NH_3^+}{|}}{C}-\overset{\overset{CO_2^-}{|}}{C}HCOO^-$$

Scheme 54.   Structural Formula 7.38.

The PLP spectrum changes to that of PMP. No inactivation occurs if the enzyme is in the PMP form prior to addition of inactivator. No transamination of the inactivator occurs without inactivation, i.e., the partition ratio is zero. The rate of inactivation is essentially the same in the presence or absence of glutathione, but is diminished by dithiothreitol and greatly diminished by α-mercaptoethanol. Cooper et al.[89,90] suggested that the smaller the thiol, the greater was its competition with the active site nucleophile for the reactive intermediate within the active site. The proposed mechanism for inactivation is shown in Scheme 55. Compound **7.38** does not inactivate D-amino acid oxidase, L-amino acid oxidase, pig heart glutamate-alanine aminotransferase, soluble rat kidney glutamine aminotransferase K, *E. coli* glutamate decarboxylase, *P. fluorescens* and rat brain GABA aminotransferase, bacterial D-amino acid aminotransferase, rat liver glutamine aminotransferase L, asparagine aminotransferase, ornithine aminotransferase, and rat kidney branch-chain aminotransferase. Glutamate-alanine aminotransferase from rat liver and glutamate decarboxylase from rat brain are slowly inactivated. The β,γ-double bond is essential for activity; no inactivation of glutamate-aspartate aminotransferase occurs with γ-methylene- D,L-glutamate or D,L-β-methyl-D,L-aspartate.[90]

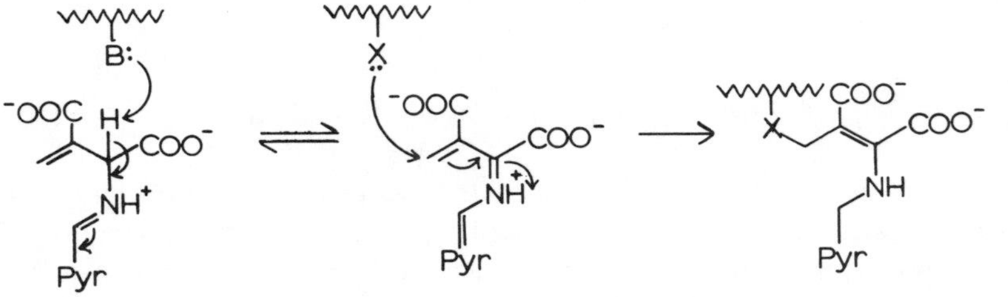

Scheme 55.

(±)-β-Methyleneaspartic acid was synthesized by Dowd and Kaufman[91] by the route in Scheme 56. β-Methylene-D,L-glutamate is a weak time-dependent inactivator (K_I = 200 mM) of soluble pig heart glutamate-aspartate aminotransferase; dialysis did not regenerate enzyme activity.[92]

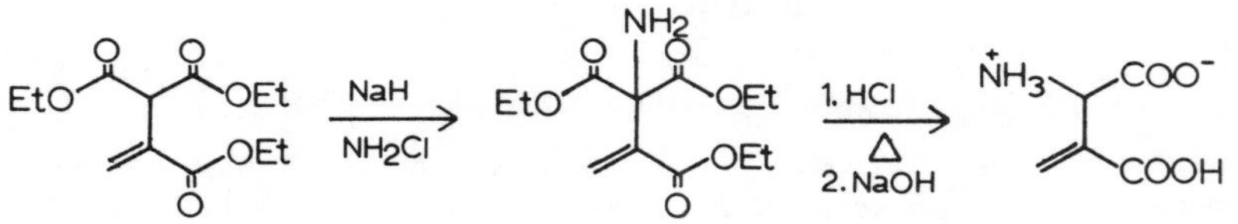

Scheme 56.

7. D,L-*cis*- and *trans*-2,6-Diamino-4-hexenoate

D,L-*cis*-2,6-Diamino-4-hexenoate (D,L-*cis*-lysene) (**7.39**) and D,L-*trans*-2,6-diamino-4-hexenoate (D,L-*trans*-lysene) (**7.40**) are both (see Scheme 57) substrates and time-dependent inactivators of L-lysine ε-aminotransferase from *Achromobacter liquidum*.[63] The partition ratio for *trans*-lysene is 160 and for *cis*-lysene is 1700 as measured by conversion of [^{14}C]α-ketoglutarate to [^{14}C]glutamate. The *cis*-isomer (**7.39**) was shown by Shannon et al.[63] to produce α-picolinate, the product of cyclization of the initially formed 6-aldehyde and auto-oxidation (Scheme 58); the *trans*-isomer also produces some α-picolinate. The ready cyclization of the *cis*-isomer, however, probably accounts for the high partition ratio. The inactivated enzyme adduct with either **7.39** or **7.40** is unstable; dialysis at pH 8.5 is sufficient to cause complete reactivation overnight. Lysine α-racemase is not inactivated by these compounds, but it catalyzes exchange of the α-proton. This indicates the specificity of these compounds for PLP enzymes that act at the ε-position. The *trans*-alkene, but not the *cis*-alkene, causes time-dependent inactivation of rat liver ornithine δ-aminotransferase, but no inactivation of GABA aminotransferase. The typical isomerization/addition mechanism for olefinic amine inactivators of PLP-dependent enzyme (see Scheme 46, Section II.C.1.) was suggested. These compounds were synthesized[63] by the same route used for the corresponding alkyne (Scheme 38, Section III.A.5.), starting from *cis-* and *trans*-2-butene-1,4-diol, respectively.

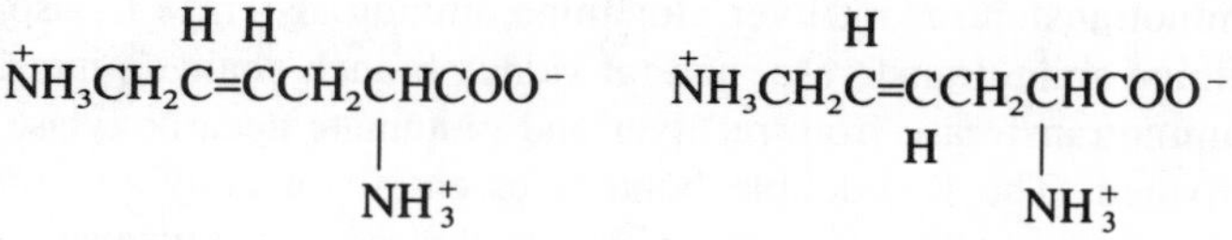

Scheme 57.   Structural Formulas 7.39 and 7.40.

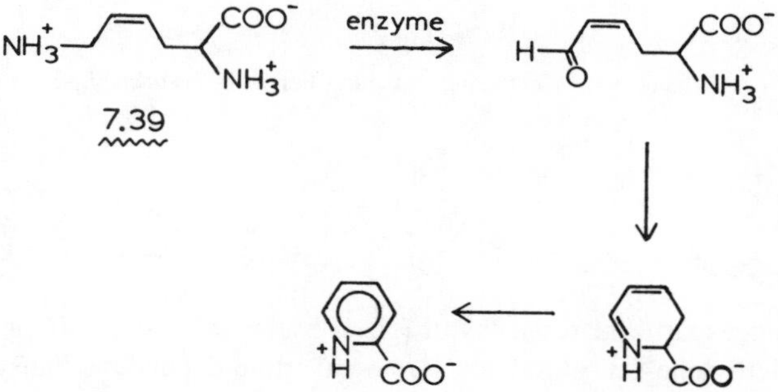

Scheme 58.

IV. ISOMERIZATION/SUBSTITUTION

A. 2-Amino-3-halobutyric Acids

The mechanism for the conversion of O-succinyl-L-homoserine to α-ketobutyrate by the PLP-dependent enzyme, cystathionase γ-synthetase, is believed to be that shown in Scheme 59 (R = $^-$OOCCH$_2$CH$_2$CO). On the basis of intermediates **7.41** and **7.42**, it would be expected that vinylglycine and 2-amino-3-halobutyric acids would be substrates for the enzyme, since they would produce **7.41** and **7.42**, respectively, following enzyme-catalyzed α-proton removal. Both are, indeed, substrates, however, (2R,3R)-2-amino-3-chloro(or fluoro)butyrate and (2R,3S)-2-amino-3-fluoro-butyrate also act as inactivators.[93] In order to accomodate these observations, an S$_N$2 displacement of halide by an enzyme nucleophile was proposed (Scheme 60).

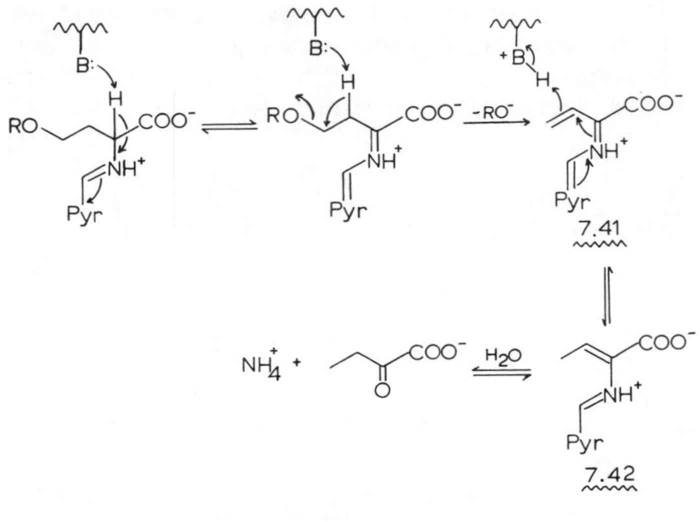

Scheme 59.   Containing Structural Formulas 7.41 and 7.42.

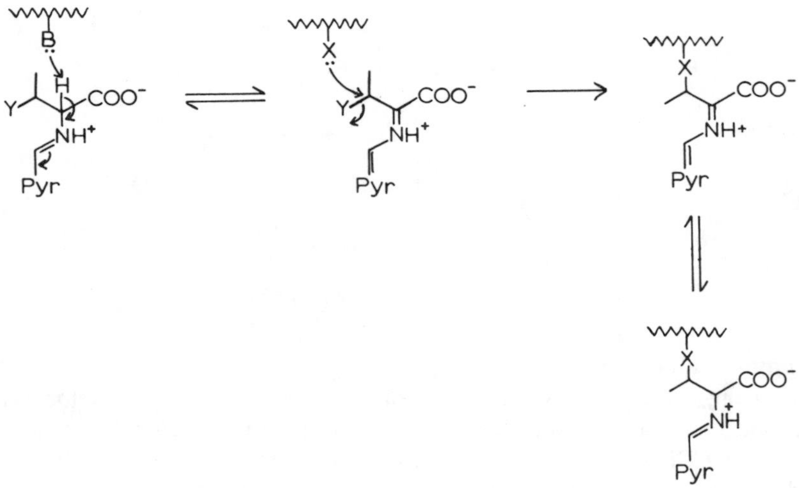

Scheme 60.

B. β-Substituted Alanines

The D-isomers of three amino acids with good leaving groups substituted at the β-position (β-chloroalanine, β-fluoroalanine, and O-acetylserine) are inactivators of the PLP-dependent 1-aminocyclopropane-1-carboxylate deaminase from *Pseudomonas* sp.[94] The partition ratios for the three inactivators are 68, 190, and 300 for β-Cl, β-F, and β-OAc substituted alanines, respectively. This suggests that the inactivating species is different for all three and eliminates the aminoacrylate-PLP intermediate (see Chapter 6 of Volume I, Section I, Scheme 2) as that responsible for inactivation. The mechanism favored by Walsh et al.[94] is an isomerization/substitution pathway similar to that shown in Scheme 60 (see Section IV.A.).

V. ISOMERIZATION/ACYLATION

A. Aminoacetonitrile

Maycock et al.[95] showed that aminoacetonitrile (Structural Formula 7.43) is an inactivator of beef plasma amine oxidase; [1-^{14}C]-labeled inactivator results in the incorporation of [^{14}C] into the enzyme. The mechanism proposed is shown in Scheme 61.

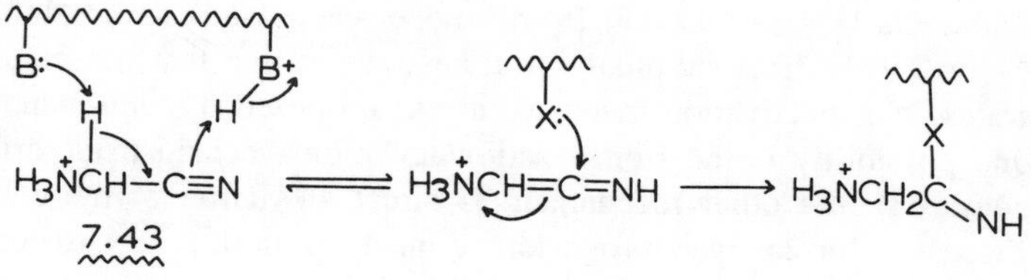

Scheme 61.   Containing Structural Formula 7.43.

B. α-Cyanoglycine

α-Cyanoglycine (Structural Formula 7.44, Scheme 62) inactivates both the β_2 subunit and the $\alpha_2\beta_2$ complex of *E. coli* tryptophan synthase.[96] Upon dialysis for an extended period of time, the β_2 subunit regains enzyme activity ($t_{1/2}$ = 24 hr). This reactivation, however, can be prevented by the addition of α subunit. This modification of the β_2 subunit increases its affinity for the α subunit. Modification of the $\alpha_2\beta_2$ complex increases its stability to heat, urea, and low pH. The mechanism proposed by Miles[96] is shown in Scheme 63.

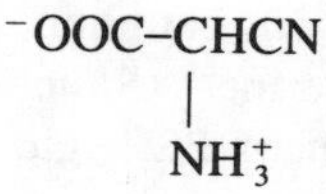

Scheme 62.   Structural Formula 7.44.

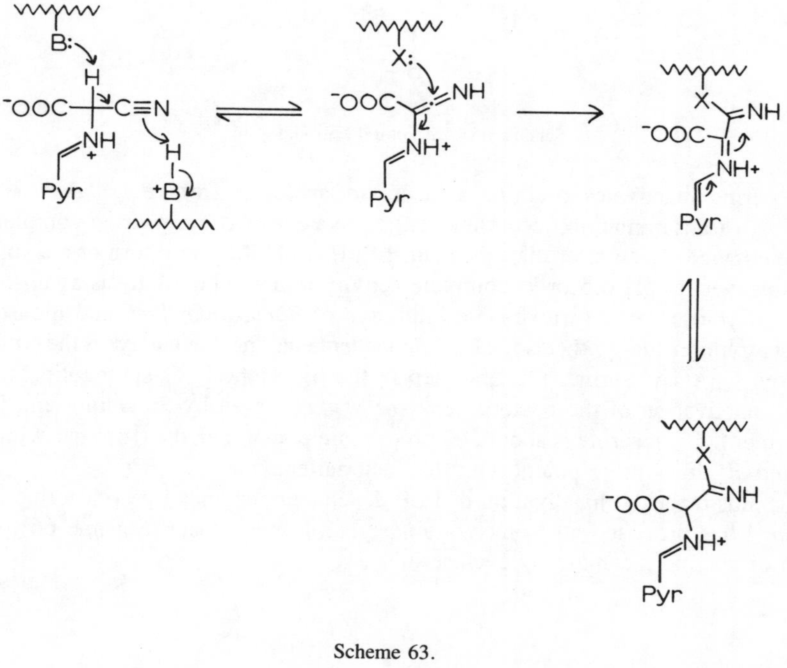

Scheme 63.

C. 2α-Cyano-17β-hydroxy-4,4,17-trimethylandrost-5-en-3-one

2α-Cyano-17β-hydroxy-4,4,17-trimethylandrost-5-en-3-one (Structural Formula 7.45, Scheme 64) titrates stoichiometrically 3β-hydroxysteroid dehydrogenase from testosterone-induced *Pseudomonas testosteroni* and from bovine and rat adrenals.[97] Inhibition occurs with either 3β- or 17β-hydroxysteroidal substrates, suggesting that the same enzyme acts on both substrates. No inactivation mechanism was proposed by Goldman,[97] but two are reasonable. One possibility is the isomerization/acylation mechanism described for cyanoglycine (Scheme 63); the other mechanism is direct alkylation of the enzyme (affinity labeling), as discussed for 2α-cyanoprogesterone inactivation of Δ^5-3-oxosteroid isomerase (see Section X.B.).

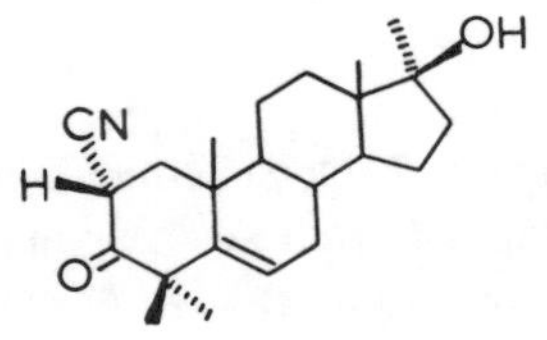

Scheme 64.   Structural Formula 7.45.

D. Cycloserine (4-Amino-3-isoxazolidinone)

GABA aminotransferases from *E. coli,* cat brain, and monkey brain are inactivated in a time-dependent fashion by D,L-cycloserine[98] (Structural Formula 7.46, Scheme 65). Inactivation is not reversed upon addition of PLP, glutathione, cysteine, Mg(II), Mn(II), or Zn(II).

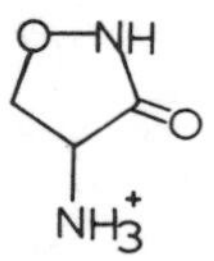

Scheme 65.   Structural Formula 7.46.

D-Cycloserine inactivates D-amino acid aminotransferase from *B. sphaericus* in a 1:1 titration.[99] No transamination occurs and 1.0 equivalent of D-cycloserine completely inactivates the enzyme. Even after dialysis against PLP at pH 8.5, no return of enzyme activity results. However, at pH 6.5 or 7, complete activity returns upon dialysis against PLP.

D- and L-Cycloserine are irreversible inhibitors of *Bacteriodes levii* and mouse brain 3-ketodihydrosphingosine synthetase, a PLP-dependent enzyme that catalyzes the condensation of palmitoyl CoA and serine, the first step in the biosynthesis of sphingolipids.[100] Time-dependent inactivation of the bacterial enzyme was observed by Sundaram and Lev[100] for L-cycloserine. L-Cycloserine is about 100 times more potent than the D-isomer with the brain enzyme and 50 times more potent with the bacterial enzyme.

No mechanism for the inactivation of PLP-dependent enzymes by cycloserine is known, but Rando[101] has proposed an isomerization/acylation mechanism (Scheme 66) for the inactivation of alanine racemase by cycloserine.

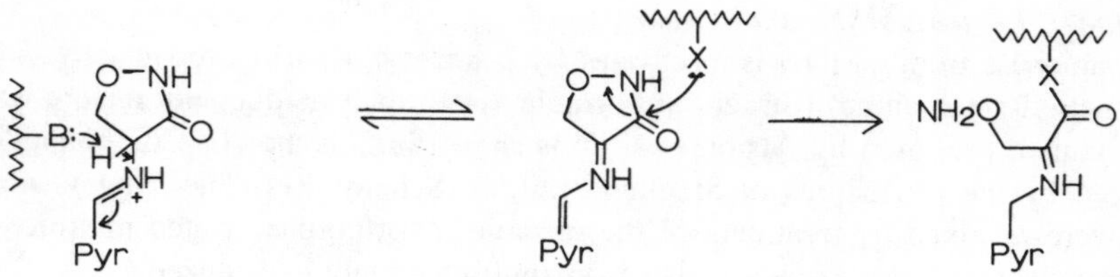

Scheme 66.

VI. ISOMERIZATION/ELIMINATION

A. Isomerization/Elimination/Addition

1. 2-Amino-4-chloro-4-pentenoic Acid

2-Amino-4-chloro-4-pentenoic acid (Structural Formula 7.47, Scheme 67) irreversibly inactivates rat liver γ-cystathionase; the mechanism suggested by Washtien and Abeles[102] is shown in Scheme 68.

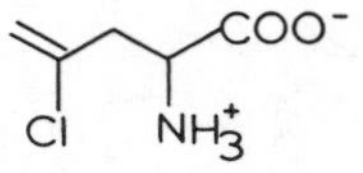

Scheme 67.   Structural Formula 7.47.

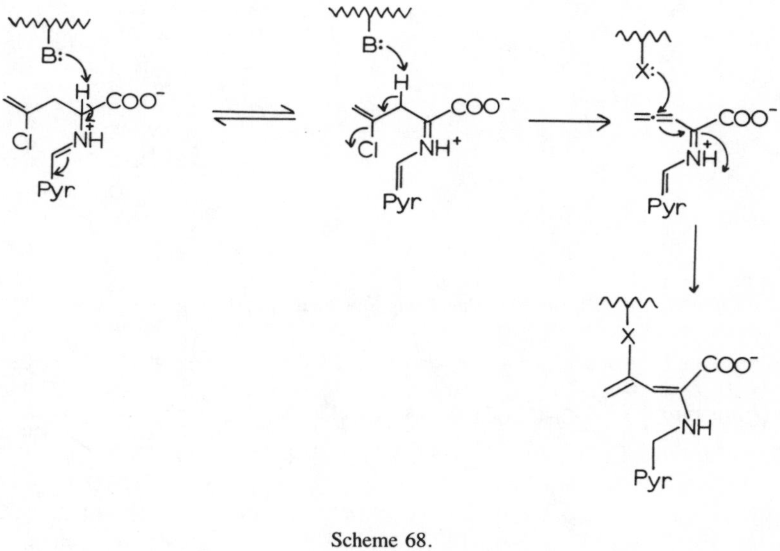

Scheme 68.

2. 2-Amino-4-chlorobutyric Acid

2-Amino-4-chlorobutyric acid (Structural Formula 7.48, Scheme 69) slowly inactivates PLP-dependent rat liver cystathionine synthase at high concentrations.[103] No mechanism was offered, but a mechanism similar to that shown in Scheme 68 could be drawn.

$$ClCH_2CH_2CHCOO^-$$
$$|$$
$$NH_3^+$$

Scheme 69.   Structural Formula 7.48.

3. 2-Amino-4,4,4-trifluorobutanoic Acid

γ-Cystathionase from rat liver is inactivated by 2-amino-4,4,4-trifluorobutanoic acid (β,β,β-trifluoromethyl-D,L-alanine) (Structural Formula 7.49); dialysis does not restore activity.[104] The mechanism proposed by Alston et al.[104] is shown in Scheme 70. Compound **7.49** was synthesized by the procedures of Steglich et al.[105] (Scheme 71). The R-(+)- and S-(−)-isomers were resolved by treatment of the racemic N-trifluoroacetylated mixture with hog kidney aminoacylase; the R-(+)-isomer is enantioselectively hydrolyzed.[105a]

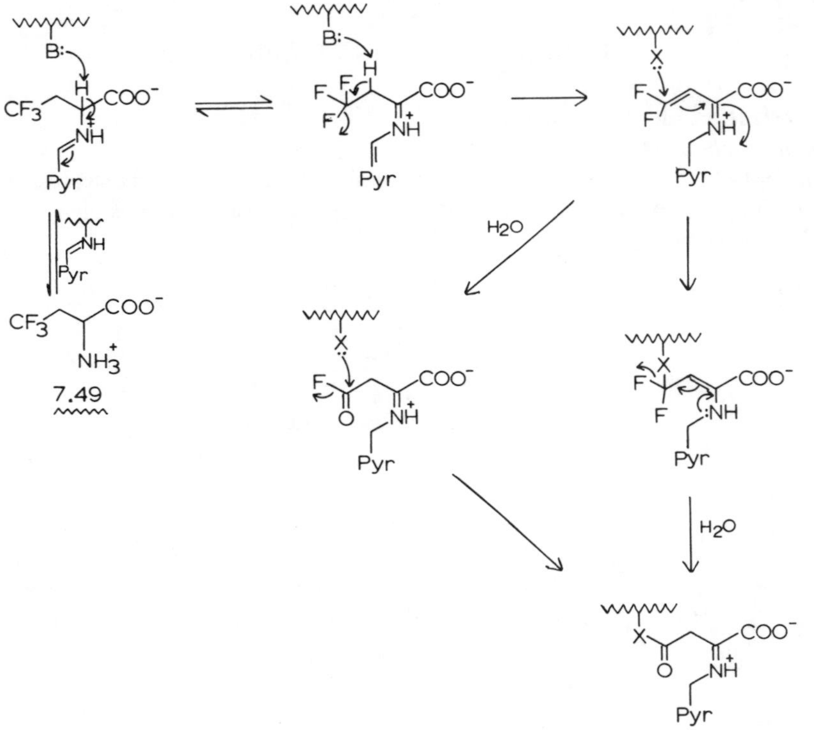

Scheme 70.   Containing Structural Formula 7.49

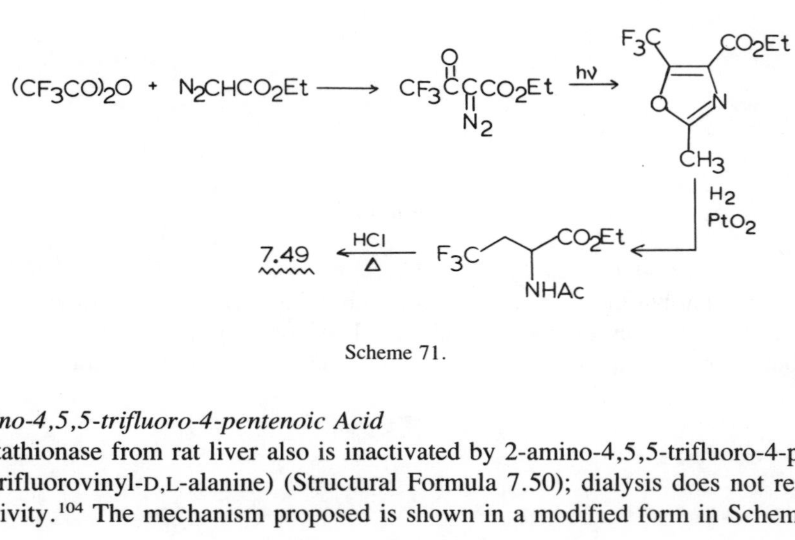

Scheme 71.

4. 2-Amino-4,5,5-trifluoro-4-pentenoic Acid

γ-Cystathionase from rat liver also is inactivated by 2-amino-4,5,5-trifluoro-4-pentenoic acid (β-trifluorovinyl-D,L-alanine) (Structural Formula 7.50); dialysis does not restore enzyme activity.[104] The mechanism proposed is shown in a modified form in Scheme 72.

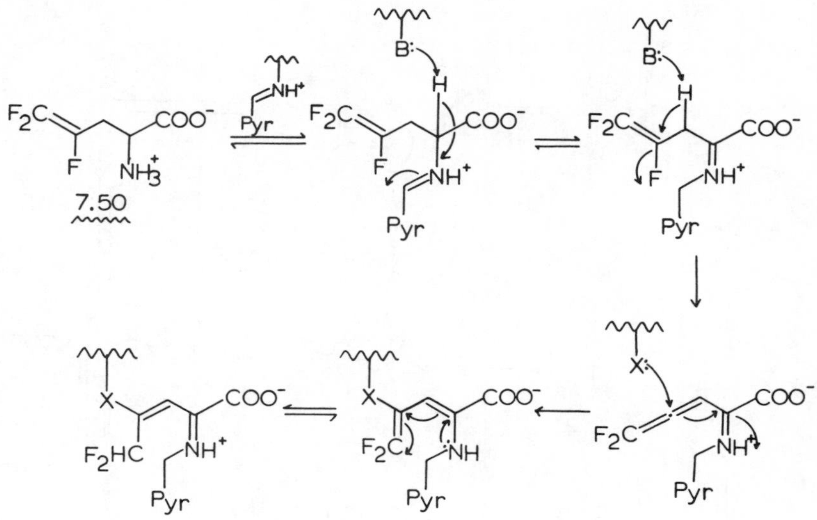

Scheme 72.   Containing Structural Formula 7.50.

Compound **7.50** was synthesized by Muramatsu and Ueda[106] (Scheme 73).

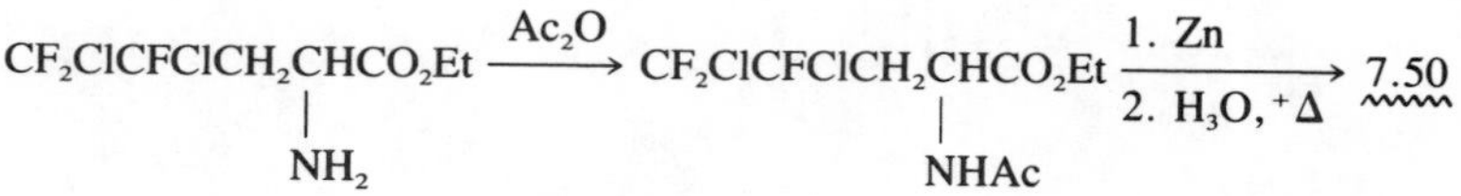

$$CF_2ClCFClCH_2\underset{\underset{NH_2}{|}}{C}HCO_2Et \xrightarrow{Ac_2O} CF_2ClCFClCH_2\underset{\underset{NHAc}{|}}{C}HCO_2Et \xrightarrow[2.\ H_3O,^+\Delta]{1.\ Zn} 7.50$$

Scheme 73.

B. Isomerization/Elimination/Rearrangement/Substitution

1. 2-Amino-4-chloro-5-(arylsulfinyl)pentanoic Acid

A novel approach to the inactivation of enzymes was suggested by Johnston et al.[107] The basis for the approach arises from a potential allyl sulfoxide to allyl sulfenate rearrangement ([2,3]-sigmatropic rearrangement), where the allyl sulfoxide would be generated by enzyme-catalyzed elimination of HX from a β-haloalkyl sulfoxide. This idea was tested with cystathionine γ-synthase from *Salmonella typhimurium* meA and methionine γ-lyase from *Pseudomonas ovalis*, two PLP-dependent enzymes previously shown to catalyze proton removal and elimination of halide from substrates. 2-Amino-4-chloro-5-(arylsulfinyl)pentanoic acid (**7.51** Ar = *p*-nitrophenyl; Scheme 74) inactivates both enzymes, but **7.51** (Ar = *p*-tolyl) does not.

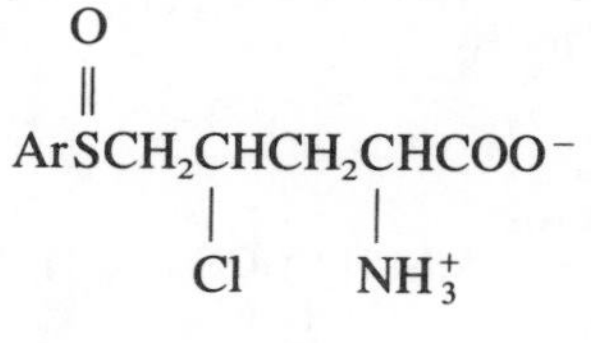

$$\overset{\overset{\displaystyle O}{\|}}{ArSCH_2}\underset{\underset{Cl}{|}}{C}HCH_2\underset{\underset{NH_3^+}{|}}{C}HCOO^-$$

Scheme 74.   Structural Formula 7.51.

This is consistent with the rearrangement mechanism shown in Scheme 75, since [2,3]-sigmatropic rearrangements are known to be much more sluggish when Ar = *p*-tolyl than *p*-nitrophenyl. Also, the compound in which the ArSO- and Cl groups are interchanged does not inactivate either enzyme.

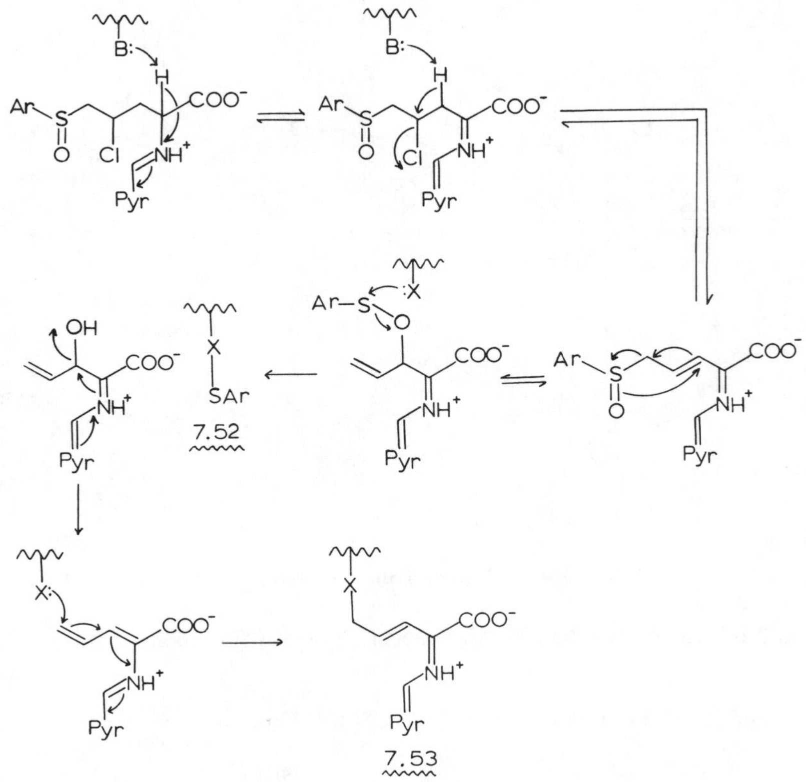

Scheme 75. Containing Structural Formulas 7.52 and 7.53.

Both enzymes are inactivated by the aryl ring-labeled inactivator with a stoichiometry of about 1 equivalent per monomer, but when [5-^{3}H]-inactivator was used, only 0.1 equivalent of [^{3}H] per monomer was incorporated. Adduct **7.52** could account for incorporation of ring-labeled inactivator into the enzyme; adduct **7.53** could be the cause for [5-^{3}H]-labeled inactivator incorporation. Inactivated methionine γ-lyase was fully reactivated by incubation with mono- or dithiols, but inactivated cystathionine γ-synthetase was only 25% reactivated and only by a dithiol (DTT). During reactivation, *p*-nitrophenylthiol is generated at about the same rate as reactivation. Partial reactivation of inactivated cystathionine γ-synthetase may be the result of attack by two different nucleophiles, cysteine attacking only one out of four times. Compound **7.51** can be synthesized[107] by the route shown in Scheme 76.

$CH_2{=}CHCH_2CH(NH_2)COOH$ —(1. Boc–ON=C(Ph)CN; 2. Ph_2CN_2)→ $CH_2{=}CHCH_2CH(NHBoc)CO_2CHPh_2$

—(ArSCl)→ $ArSCH_2CH(Cl)CH_2CH(NHBoc)CO_2CHPh_2$ —(1. m-CPBA; 2. TFA)→ 7.51

Scheme 76.

VII. ISOMERIZATION/ISOMERIZATION

A. Isomerization/Isomerization

1. Gabaculine

Gabaculine (5-amino-1,3-cyclohexadienylcarboxylic acid (Structural Formula 7.54, Scheme 77), a natural product isolated from *Streptomyces toyacaensis,* is an irreversible inactivator of mouse[108] and pig[108a] brain GABA aminotransferase. Gel filtration does not reactivate the enzyme; β-mercaptoethanol does not protect the enzyme. Synthetic *d,l*-gabaculine has one-half the activity of the natural *l*-isomer. The pH vs. rate of inactivation profile for gabaculine is identical to that for GABA transamination except that the pH rate maximum is shifted 0.5 units down from that with substrate. 4,5-Dideuteriogabaculine inactivates the enzyme at a rate 2.1 times slower than does the protio compound, indicating α–C–H bond cleavage is involved in the rate determining step. Two possible mechanisms were suggested by Rando and Bangerter[108] (Scheme 78). [2-³H]Gabaculine inactivates GABA aminotransferase in a time-dependent manner that corresponds to the incorporation of radioactivity into the enzyme.[109] Denaturation of labeled enzyme leads to release of the radioactivity, suggesting that gabaculine reacts with the cofactor not with an active site residue. The released radioactivity was identified by Rando[109] as *m*-carboxyphenylpyridoxamine phosphate (**7.55**).

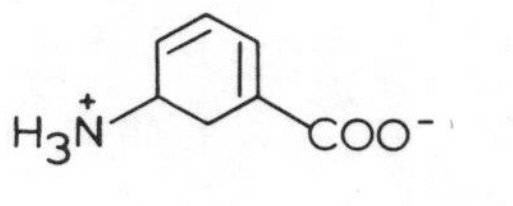

Scheme 77.   Structural Formula 7.54.

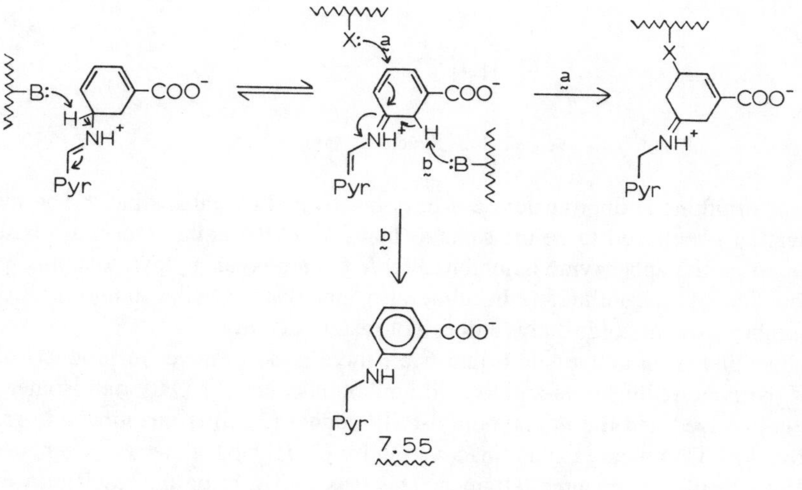

Scheme 78.   Containing Structural Formula 7.55.

The nonenzymatic reaction of gabaculine with PLP was studied by Rando and Bangerter[110] as a model for the mechanism of GABA aminotransferase by gabaculine (Scheme 79). A deuterium isotope effect of 4.3 on the reaction of 4,5-dideuteriogabaculine with PLP was observed relative to nondeuterated gabaculine. Consistent with the isotope effect, there is an increase in the reaction rate upon increasing the buffer concentration. The activation energy is 24.8 kcal/mol. These studies suggest that the mechanism for GABA aminotransferase inactivation by gabaculine is the isomerization/isomerization mechanism leading to **7.55** (Scheme 78, pathway b). Consistent with this mechanism is the observation that 1,2-

dehydrogabaculine (Structural Formula 7.56, Scheme 80) is not an inactivator. If a Michael addition mechanism were important, the 1,2-dehydro analogue also should be an inactivator. Gabaculine does not inactivate glutamate decarboxylase, ornithine decarboxylase, aspartate aminotransferase, and alanine aminotransferase.[110,111] Contrary to the report of Rando and Bangerter,[111] Wood et al.[112] found that gabaculine is not specific for GABA aminotransferase. Mouse brain and liver alanine aminotransferase also is inactivated, albeit to a much less extent. Aspartate aminotransferase is not inhibited. Not only is GABA aminotransferase inactivated by gabaculine, but also is D-amino acid aminotransferase,[99,113] L-alanine aminotransferase,[113] and L-aspartate aminotransferase;[113] alanine racemase and tryptophanase are not inactivated by gabaculine.[113] Soper and Manning[113] observed that all of the enzymes that are inactivated by gabaculine also are capable of catalyzing the removal of the β-proton from their substrates, whereas those that are not inactivated by gabaculine cannot. These results could account for the toxicity of gabaculine.

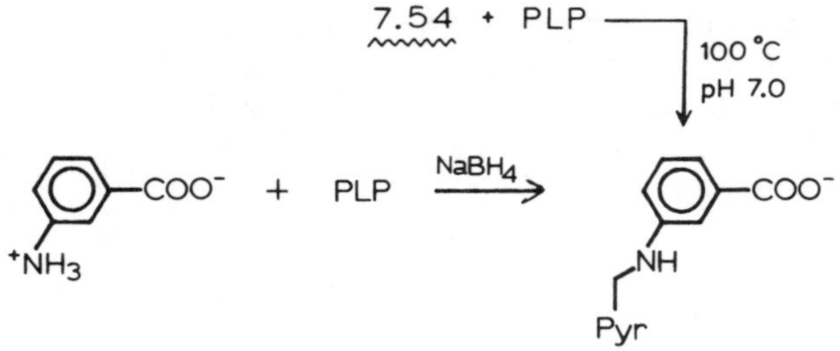

Scheme 79.

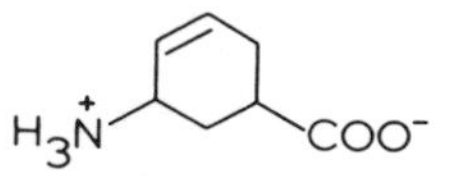

Scheme 80.   Structural Formula 7.56.

Rat liver ornithine aminotransferase also is inactivated by gabaculine.[48] The mechanism of inactivation is believed to be the same as that for GABA aminotransferase based on the observation that the apoenzyme is inhibited by N-m-carboxyphenylpyridoxamine phosphate (**7.55**), but not by gabaculine. Gabaculine also inactivates GABA aminotransferase from chicken embryo spinal cord; dialysis does not restore activity.[114]

DL-Gabaculine is an active site titrate (i.e., there is no turnover to product) of a *Pseudomonas* pyruvate-requiring ω-amino acid aminotransferase.[115] Only one isomer (presumably, L) is involved and the *m*-anthranilyl-PMP adduct (**7.55**) is quantitatively responsible for inactivation. Complete enzyme inactivation by [2-³H]gabaculine, however, occurs with only 0.45 molecule per enzyme tetramer. This was shown to be the result of a low extent of loading of the functional coenzyme per tetramer (only 0.45 equivalent of PLP per tetramer). When the enzyme is activated at 60°C in the presence of PLP, a phenylhydrazine assay indicates 1 PLP/subunit. However, inactivation of this activated enzyme only results in the incorporation of 1.5 molecules of gabaculine per tetramer after complete inactivation. Negative cooperativity among subunits was offered as an explanation by Burnett et al.[115] for this "quarter-site reactivity."

DL-Gabaculine also causes time-dependent inactivation of L-ornithine:α-ketoglutarate δ-aminotransferase from *Bacillus sphaericus* IFO 3525; apoenzyme and enzyme in the PMP forms are not affected.[116] Gel filtration and dialysis do not restore enzyme activity. When α-ketoglutarate is added, there is no return of enzyme activity, suggesting that inactivation occurs every turnover. Acid precipitation of the enzyme releases the cofactor which is shown to be converted to *m*-carboxyphenylpyridoxamine phosphate (**7.55**).

Gabaculine has been synthesized by three different routes shown in Schemes 81,[117] 82,[118] and 83.[119]

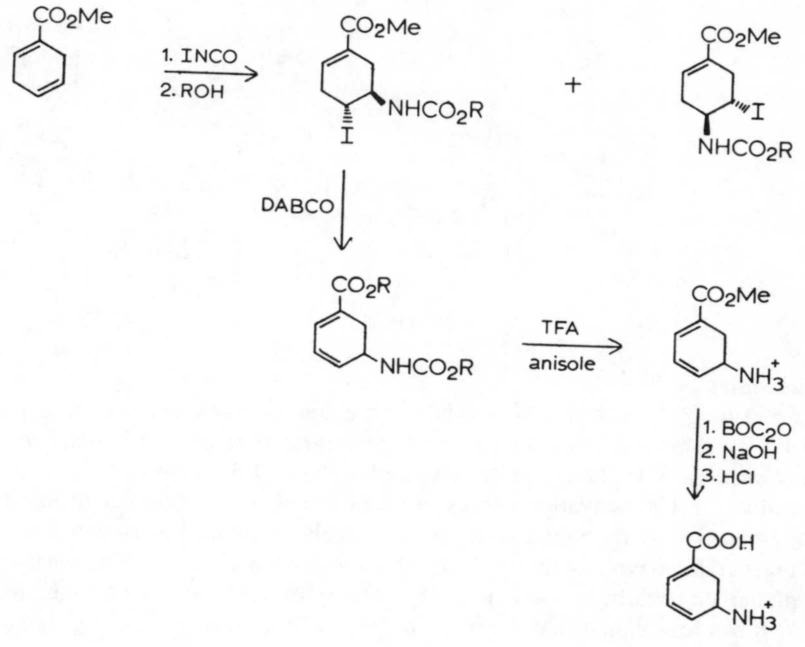

Scheme 81.

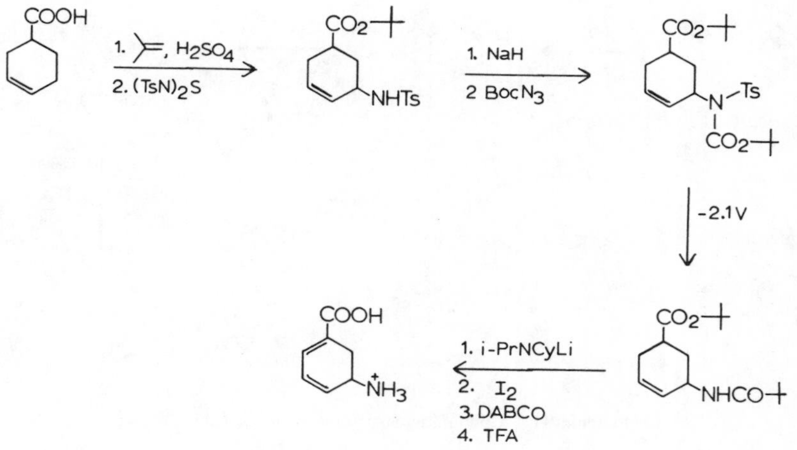

Scheme 82.

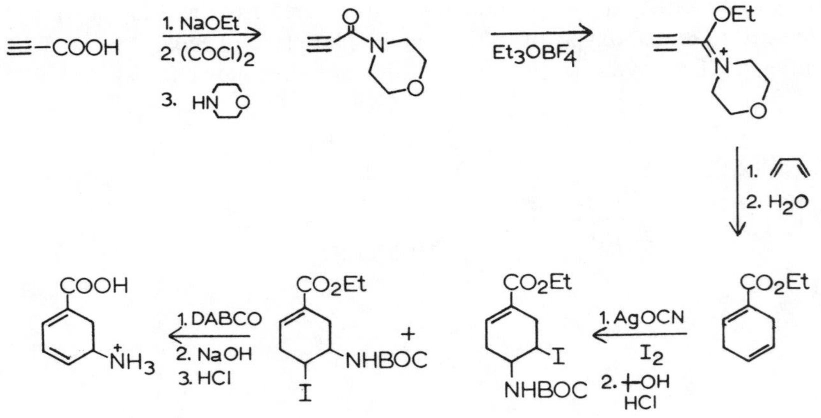

Scheme 83.

2. Isogabaculines

Metcalf and Jung[120] found that the product of the nonenzymatic reaction of isogabaculine (3-amino-1,5-cyclohexadienylcarboxylic acid; Structural Formula 7.57, Scheme 84) with PLP was *N-meta*-carboxyphenylpyridoxamine phosphate (**7.55**), the same as that obtained with gabaculine.[110] The activation energy, 17.2 kcal/mol, is very similar to that for gabaculine. Pig brain GABA aminotransferase is irreversibly inactivated by isogabauline; dialysis does not regenerate enzyme activity.[120] Ornithine aminotransferase also is inactivated, but rat brain glutamate decarboxylase is not. The affinity of **7.57** for GABA aminotransferase is 10 to 20 times lower than that reported for gabaculine; however, its k_{inact} is larger than that for gabaculine. This isogabaculine was synthesized[120] by the route shown in Scheme 84.

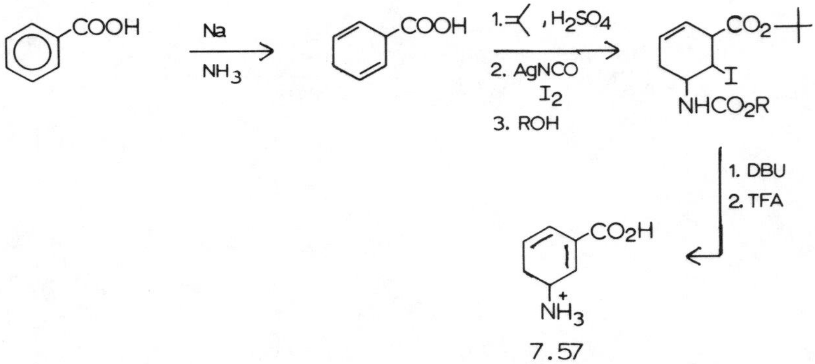

Scheme 84.   Containing Structural Formula 7.57.

The other possible isogabaculine (**7.58**) was synthesized by Danishefsky and Hershenson[121] (Scheme 85).

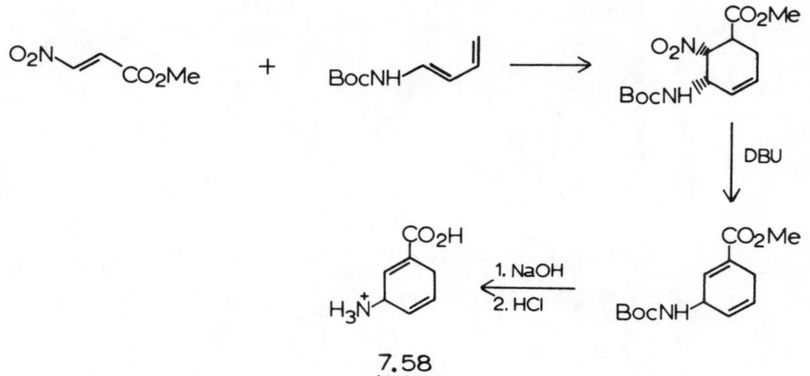

Scheme 85. Containing Structural Formula 7.58.

3. 4-Amino-4,5-dihydrofuran-2-carboxylic Acid and 4-Amino-4,5-dihydrothiophene-2-carboxylic Acid

On the basis of the mechanism of inactivation of GABA aminotransferase by gabaculine (see Section VII.A.1.), heteroaromatic analogues of gabaculine were synthesized.[122,123] *(R,S)*-4-Amino-4,5-dihydrofuran-2-carboxylic acid (**7.59**) was synthesized by Burkhart et al.[122] by the route shown in Scheme 86. A stereospecific synthesis, starting from *(S)*-glutamic acid, γ-methyl ester also was reported[122] (Scheme 87). The synthesis of racemic 4-amino-4,5-dihydrothiophene-2-carboxylic acid (**7.60**) is shown in Scheme 88.[123] Analogues of this compound having C3-substitution were made by phosphorylation of the intermediate alcohol followed by nucleophilic displacement of the phosphate ester. Enantiomerically pure *(R)*- and *(S)*-4-amino-4,5-dihydrothiophene-2-carboxylic acid were synthesized from *(R)*- and *(S)*-cysteine ethyl ester with some modifications of Scheme 88 to avoid racemization.[123] Compounds **7.59** and **7.60** are time-dependent inactivators of GABA aminotransferase.[124]

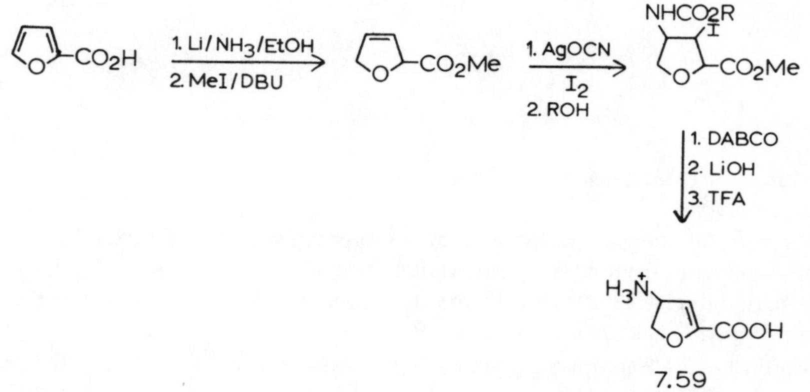

Scheme 86. Containing Structural Formula 7.59.

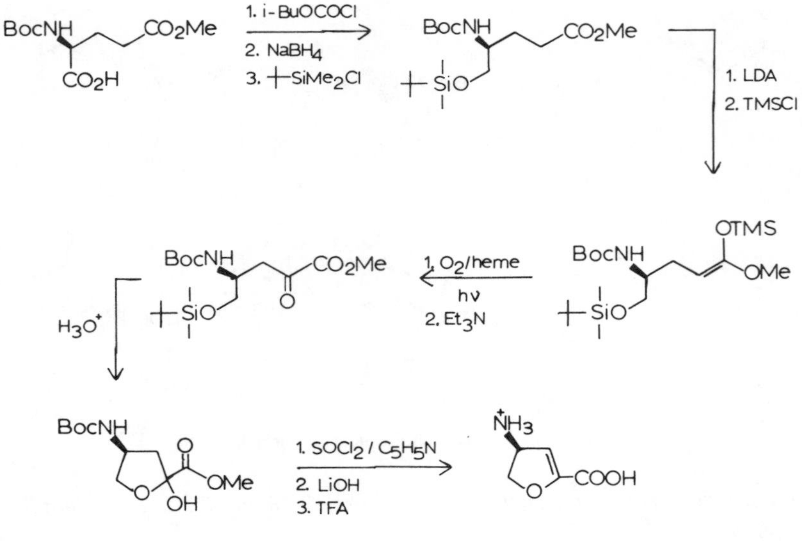

Scheme 87.

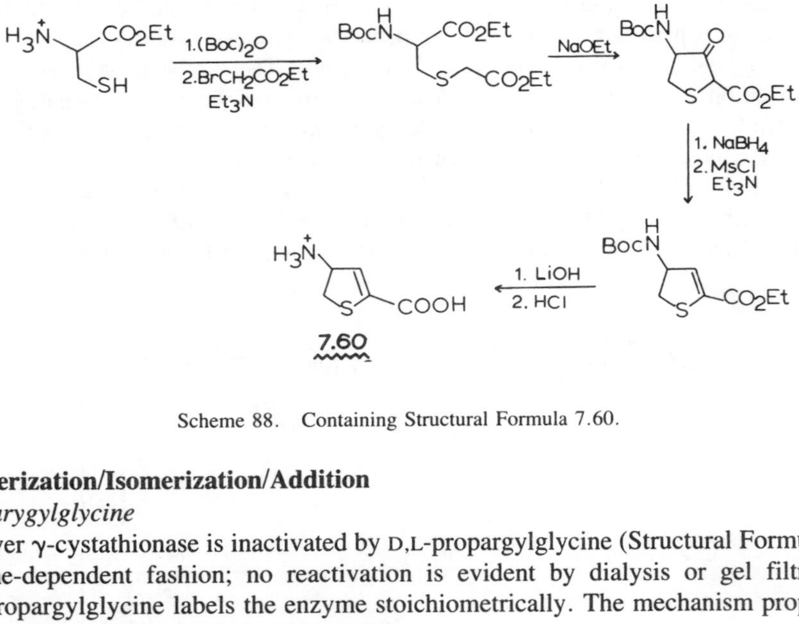

Scheme 88.   Containing Structural Formula 7.60.

B. Isomerization/Isomerization/Addition

1. Proparygylglycine

Rat liver γ-cystathionase is inactivated by D,L-propargylglycine (Structural Formula 7.61) in a time-dependent fashion; no reactivation is evident by dialysis or gel filtration.[125] [2-^{14}C]Propargylglycine labels the enzyme stoichiometrically. The mechanism proposed by Abeles and Walsh[125] is shown in Scheme 89.

Inactivation by [α-^{2}H]propargylglycine shows an isotope effect of 2.2.[102] Hydrolysis of enzyme inactivated with [2-^{14}C]propargylglycine produces the L-isomer of [2-^{14}C]2-amino-4-ketopentanoic acid (Structural Formula 7.62), indicating that reprotonation is enzyme catalyzed (i.e., stereospecific) rather than occurring after release from the enzyme. In order for this product to be generated, the α-proton must have been replaced during inactivation. Therefore, a modified mechanism of inactivation was suggested by Washtien and Abeles[102] in which reprotonation at the α-carbon occurs. This mechanism is shown in Scheme 90, beginning with the α,β,γ-allenic immonium intermediate from Scheme 89. Controlled hydrolyses of the labeled enzyme resulted in release of **7.62** at a rate comparable to that for a vinyl sulfide or a vinyl phenyl ether, suggesting that attachment is to a cysteine or tyrosine residue.

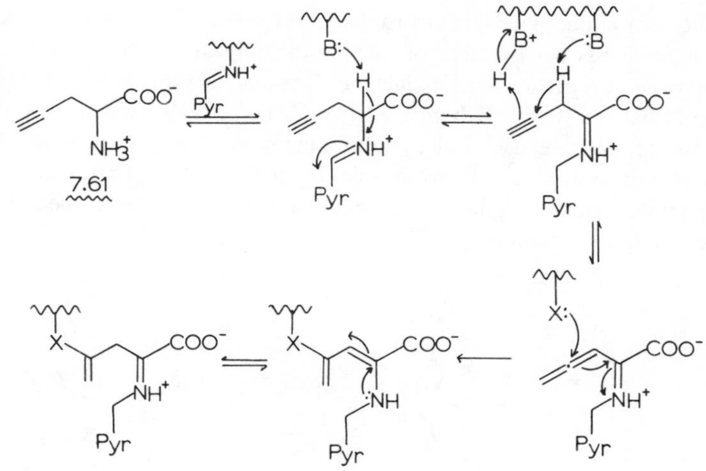

Scheme 89.   Containing Structural Formula 7.61.

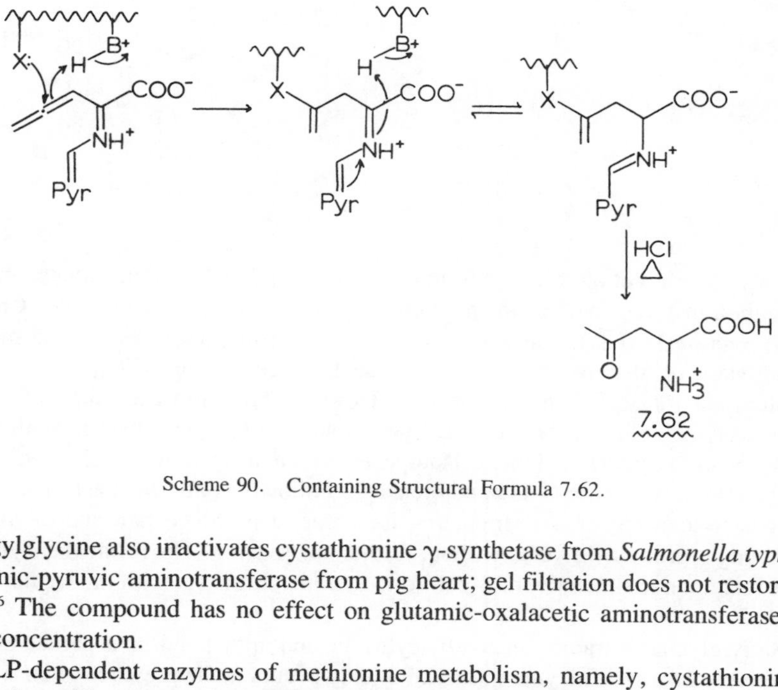

Scheme 90.   Containing Structural Formula 7.62.

Propargylglycine also inactivates cystathionine γ-synthetase from *Salmonella typhimurium* and glutamic-pyruvic aminotransferase from pig heart; gel filtration does not restore enzyme activity.[126] The compound has no effect on glutamic-oxalacetic aminotransferase, even at 100 m*M* concentration.

Two PLP-dependent enzymes of methionine metabolism, namely, cystathionine γ-synthase from *S. typhimurium* and methionine γ-lyase from *Pseudomonas ovalis,* are inactivated by the natural amino acid, L-propargylglycine.[127] The kinetics of inactivation are pseudo first order, but biphasic with a break in the initial phase of inactivation occurring after 80 to 90% inactivation. Two explanations for this phenomenon were afforded by Johnston et al.[127] One, previously suggested by Beeler and Churchich,[128] attributes this to nonequivalent binding of L-propargylglycine to the PLP sites of the oligomeric protein, especially if there is heterogeneity of subunit composition. Another possibility is kinetic cooperativity between subunits such that rapid modification of one subunit alters the susceptibility of an adjacent subunit to react at the same rate. The partition ratios for inactivation of cystathionine γ-synthase and methionine γ-lyase are 4 and 8, respectively. With [2-14C]propargylglycine, four equivalents of radioactivity are incorporated per tetramer of cystathionase γ-synthase, but only two equivalents per tetramer of methionine γ-lyase. In the latter case only one α and one β subunit per $\alpha_2\beta_2$ tetramer is labeled. The inactivation mechanism proposed is a modification of that shown in Scheme 90 (see Scheme 91). Because of the ability of propargylglycine to inactive bacterial cystathionine γ-synthase, Cheung et al.[129] prepared propargylglycine-containing depeptides in order to evaluate their prodrug antibacterial properties. The dipeptides containing the L-inactivator have up to 4000 times the in vitro antibiotic activity of the inactivator alone.

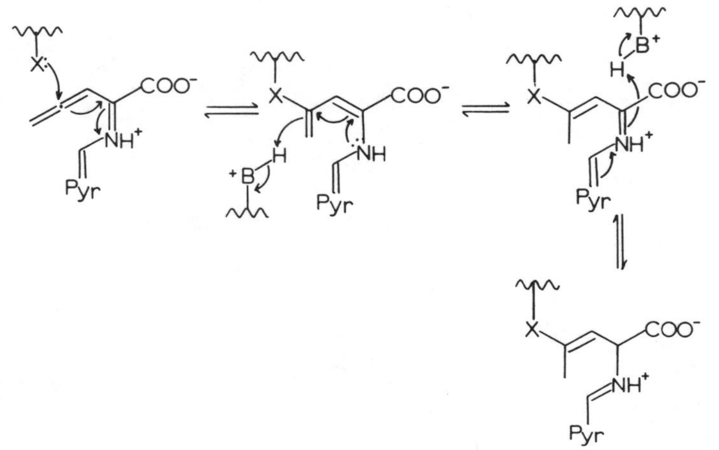

Scheme 91.

Previously, Morino et al.[130] reported that β-chloro-L-[14C]alanine inactivated pig heart L-alanine aminotransferase with a stoichiometry of about 2 labels per dimer (see Chapter 6 of Volume I, Section II.D.1.). Burnett et al.[131] found that propargylglycine is a mechanism-based inactivator of this enzyme, but only 1 label per dimer from [14C]propargylglycine is incorporated, leaving < 3% enzyme activity. The partition ratio varies, depending upon the age of the enzyme solution, between 2.7 per subunit to 6.5 per subunit. With the use of DL-[2-2H]propargylglycine, a kinetic isotope effect on inactivation of 3.5 was obtained, indicating a rate-determining C–H bond breakage. Following half-site inactivation, the modified enzyme is capable of transaminating L-alanine at a reduce rate and of binding and transaminating L-propargylglycine without further inactivation. The residual enzyme activity (< 3%) is lost upon treatment with β-chloro-L-alanine, consistent with the results of Morino et al.[130]

L-Propargylglycine is transaminated by cytosolic and mitochondrial aspartate aminotransferases from pig heart.[132] Enzyme inactivation only occurs in the presence of an α-keto acid; otherwise, conversion of PLP to PMP prevents inactivation. The presence of formate ion accelerates the inactivation rate. With L-[2-14C]propargylglycine, 1.1 equivalent of radioactivity becomes covalently bound to both isozymes. The same mechanism proposed by Abeles and Walsh[125] for γ-cystathionase inactivation (Scheme 89) was suggested by Tanase and Morino[132] in this case.

Propargylglycine has been synthesized by two similar routes (Scheme 92, R = CHO[133] or Ac[134]). The L-isomer can be prepared from racemic N-acetylpropargylglycine with hog kidney acylase.[134]

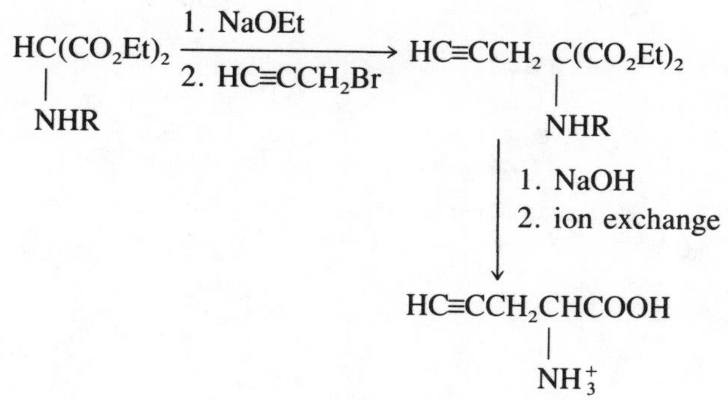

$$HC(CO_2Et)_2 \xrightarrow[\text{2. } HC{\equiv}CCH_2Br]{\text{1. NaOEt}} HC{\equiv}CCH_2\,C(CO_2Et)_2$$

(with NHR below the left structure and NHR below the right structure)

1. NaOH
2. ion exchange

$$HC{\equiv}CCH_2CHCOOH$$

$$NH_3^+$$

2. Allylglycine

Allylglycine does not inactivate glutamic-pyruvic aminotransferase, but does inactivate glutamic-oxalacetic aminotransferase.[126] Therefore, by changing the glycyl substituent from propargyl (see Section VII.B.1.) to allyl, specificity can be changed from the former aminotransferase to the latter. Guinea pig brain glutamate decarboxylase also undergoes time-dependent inactivation in the presence of PLP by **7.63**; dialysis only partially reactivates the enzyme.[135] Although no mechanism was suggested, that shown in Scheme 93 may be responsible for inactivation.

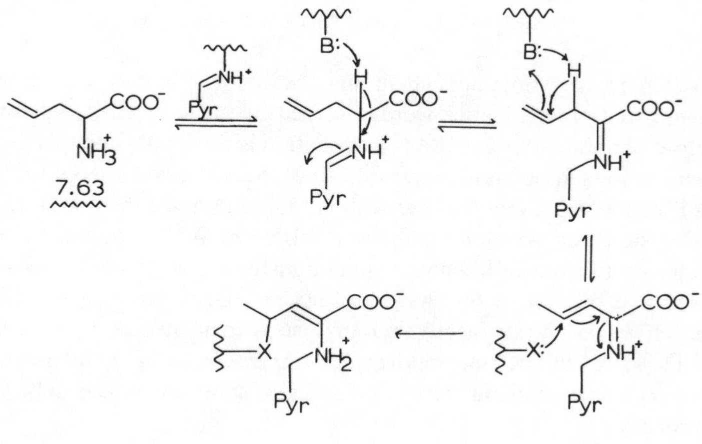

Scheme 93.   Containing Structural Formula 7.63.

C. Isomerization/Isomerization/Acylation

1. β-Cyanoalanine

β-Cyano-L-alanine (Structural Formula 7.64) inactivates pig heart alanine aminotransferase; enzyme activity is restored upon dialysis.[136] β-Merceptoethanol does not affect the rate of inactivation. The mechanism proposed by Alston et al.[136] is shown in Scheme 94.

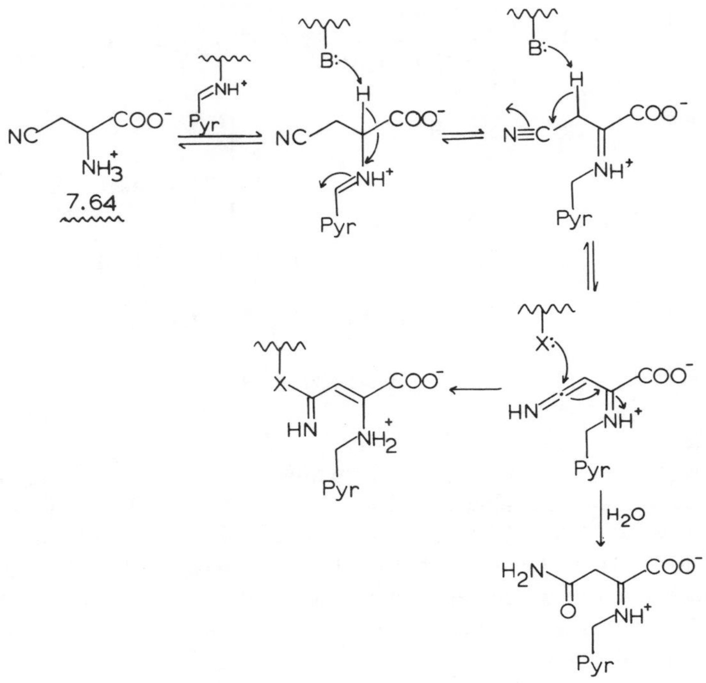

Scheme 94.   Containing Structural Formula 7.64.

β-Cyano-D-alanine is a time-dependent inactivator of D-amino acid aminotransferase; however, pseudo first-order kinetics were observed by Ueno et al.[137] only at temperatures of 10°C or below. At higher temperatures inactivation is reversible and this process becomes more significant as the temperature increases. Since the activation energy for transamination (10.4 kcal/mol) differs from that for inactivation (5.1 kcal/mol), there must be two different rate-determining steps for these two pathways. Alston et al.[136] concluded that dilution was the cause of the reactivation of L-alanine aminotransferase by β-cyano-L-alanine, but Ueno et al.[137] showed that dilution at 6°C has no effect on inactivated enzyme; raising the temperature does. However, when inactivated enzyme is incubated at 37°C in the absence of substrate, then assayed at 6°C, no recovery of enzyme activity is apparent. Therefore, a combination of increased temperature and presence of substrate appear to be involved in the reactivation process.

2. β-Aminopropionitrile

β-Aminopropionitrile (Structural Formula 7.65, Scheme 95) is a potent lathyrogen that irreversibly inactivates lysyl oxidase in embryonic chick bone[138] and cartilage.[139] Inactivation by [1-¹⁴C]- or [³H]-inactivator followed by dialysis and chromatography on agarose and DEAE-cellulose results in radioactively labeled protein that elutes at the same place as does lysyl oxidase in the control reaction.[139]

$$NH_2CH_2CH_2CN$$

Scheme 95.   Structural Formula 7.65.

β-Aminopropionitrile, N²-acetyl-L-lysine methyl ester, *n*-butylamine, and 1,4-diamino-butane are time-dependent inactivators of bovine aortic lysyl oxidase.[140] All but β-amino-propionitrile are oxidized by the enzyme during inactivation; biphasic inactivation kinetics were observed by Trackman and Kagan.[140] Dialysis results in partial return of enzyme activity, suggesting both reversible and irreversible inhibition pathways. Concomitant with inactivation, [1,2-¹⁴C]- and [3-¹⁴C]-β-aminopropionitrile bind covalently to purified aortic lysyl oxidase to equivalent extents.[141] This indicates that the nitrile moiety is not eliminated during inactivation. However, only 0.07 to 0.10 equivalent of radioactivity is bound. The copper content of the enzyme remains unchanged after inactivation; therefore, inactivation without labeling does not result from removal of copper. Tang et al.[141] suggest that the low incorporation of radioactivity may reflect the low concentration of functional active sites in the enzyme preparation. Since carbonyl-modifying reagents reduce the amount of radioac-tivity incorporated, Schiff base formation of the amine with an active site carbonyl was suggested as the first step in the inactivation. It is now known [141a] that the active-site carbonyl is a pyrroloquinoline quinone cofactor. The mechanism proposed is shown in Scheme 96.

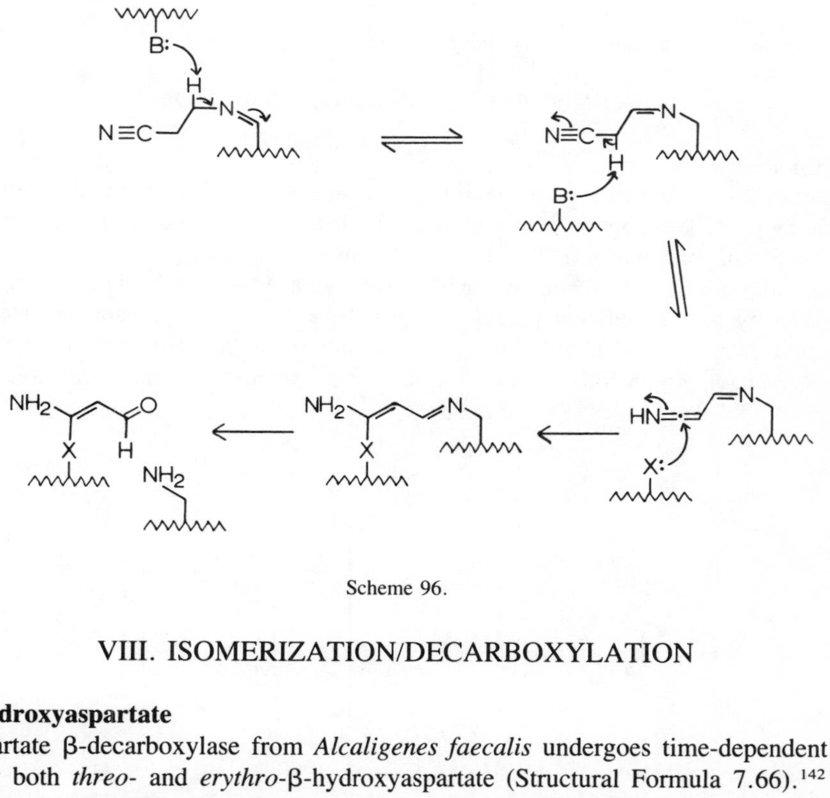

Scheme 96.

VIII. ISOMERIZATION/DECARBOXYLATION

A. β-Hydroxyaspartate

L-Aspartate β-decarboxylase from *Alcaligenes faecalis* undergoes time-dependent inhi-bition by both *threo-* and *erythro*-β-hydroxyaspartate (Structural Formula 7.66).[142] Both compounds are decarboxylated at a rate faster than serine formation. [1,2-¹⁴C]- and [3,4-¹⁴C]-β-hydroxyaspartate were used to show that carbon-4 is lost during inactivation. Acid precipitation produces [¹⁴C]serine, but NaBH₄ does not reduce the enzyme. One form of the altered enzyme that is consistent with these results is an oxazolidine (Structrual Formula 7.67, Scheme 97).

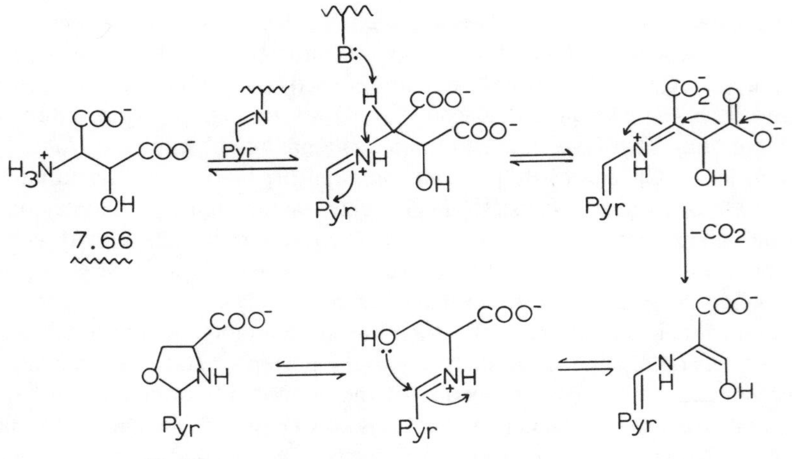

Scheme 97.   Containing Structural Formulas 7.66 and 7.67.

IX. NONCOVALENT INACTIVATION

A. Aminomalonate

Aspartate β-decarboxylase is irreversibly inactivated by aminomalonic acid.[143] When either [2-^{14}C]- or [1-^{14}C]aminomalonate is used as the inactivator, 1 mol of [^{14}C] is bound per 60,000 g of enzyme (mol wt 720,000; 12 subunits). No loss of [^{14}C] occurs after dialysis or ammonium sulfate precipitation of the labeled enzyme. Most of the radioactivity, however, is released by acid or ethanol precipitation of the enzyme. These data indicate that the inactivated enzyme contains all three carbon atoms of aminomalonate. Treatment of the inactivated enzyme with NaB^{3}H$_4$ results in the uptake of tritium. A noncovalent adduct was proposed by Palekar et al.[143] (Scheme 98).

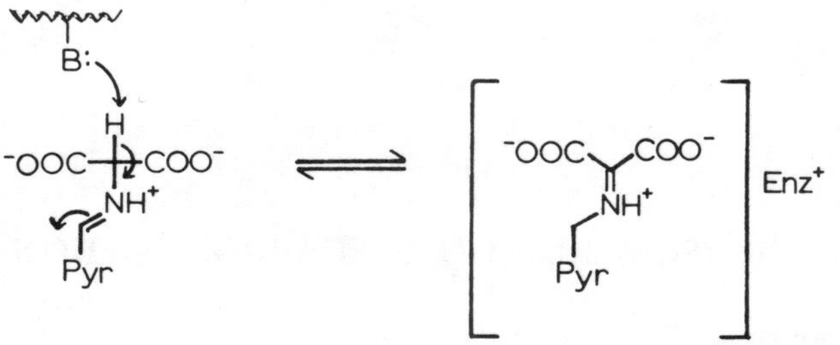

Scheme 98.

X. NOT MECHANISM-BASED INACTIVATION

A. Alk-3-ynoyl CoA

Pig heart cytoplasmic thiolase is inactivated by pent-3-ynoyl CoA (**7.68**, R = CH$_3$) and but-3-ynoyl CoA (**7.68**, R = H) (see Scheme 99), but addition of dithiothreitol[144,145] or CoASH[145] prevents inactivation; pseudo first-order kinetics were observed by Holland et al.[145] 2,3-Butadienoyl CoA also inactivates the enzyme which suggests an isomerization/addition mechanism. However, since thiols protect the enzyme, the reactive species is, presumably, escaping the active site prior to inactivation.

$$RC\equiv CCH_2COSCoA$$

Scheme 99.   Structural Formula 7.68.

A series of alk-3-ynoyl CoA esters was shown by Bloxham[146] to be irreversible inactivators of rat liver cytoplasmic thiolase. The effect of added thiol was not demonstrated, so it is not clear if the activated species is released from the active site as in the case with pig heart thiolase.[143,144]

B. 2α-Cyanoprogesterone

2α-Cyanoprogesterone (Structural Formula 7.69, Scheme 100) is a time-dependent pseudo first-order inactivator of Δ^5-3-oxosteroid isomerase from *Pseudomonas testosteroni*; dialysis does not restore enzyme activity.[147] The mechanism proposed by Penning[147] was affinity labeling involving direct S_N2 displacement of the cyano group by an active site nucleophile rather than isomerization/acylation (Scheme 63 in Section V.B., for example) for the following reasons: (1) no evidence was found for ketenimine in a nonenzymatic reaction with base; (2) there is no change in the UV spectrum during inactivation; and (3) the presence of β-mercaptoethanol protects the enzyme completely from inactivation. If this is the case, then **7.69** is an affinity-labeling agent. Regardless of the interpretation of their results, the fact that the enzyme is protected by β-mercaptoethanol suggests that the reactive species responsible for inactivation is outside of the active site and, therefore, this is not a mechanism-based inactivation.

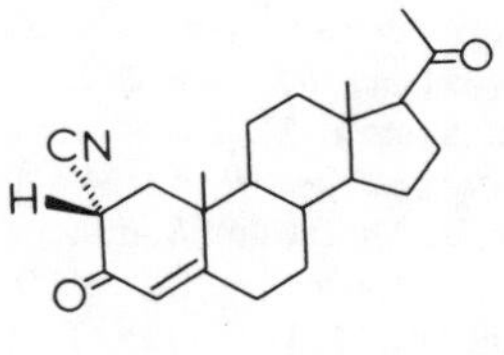

Scheme 100.   Structural Formula 7.69.

C. (E)-β-Fluoromethyleneglutamic Acid

(E)-β-Fluoromethyleneglutamic acid (Structural Formula 7.70, Scheme 101) was synthesized by McDonald et al.[148] as a potential dual enzyme-activated irreversible inhibitor of γ-aminobutyric acid aminotransferase. It was reasoned that L-glutamate decarboxyalse would catalyze its decarboxylation to *(E)*-β-fluoromethylene GABA and this would inactivate GABA aminotransferase. Unfortunately, **7.70** is not a substrate for L-glutamate decarboxylase.

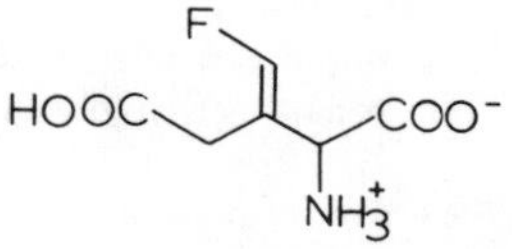

Scheme 101.   Structural Formula 7.70.

D. Inactivators that Utilize Abnormal Catalytic Mechanisms

3,5-Dioxocyclohexanecarboxylic acid (**7.2**, Scheme 4)[4] and β-D-galactopyranosyl-methyl(p-nitrophenyl)triazene (**7.3**, Scheme 6)[6,7] inactivate GABA aminotransferase and β-galactosidase, respectively, but the proposed inactivation mechanisms are not initiated by the normal catalytic mechanisms. Therefore, these are not true mechanism-based enzyme inactivators.

REFERENCES

1. **Pankaskie, M. and Abdel-Monem, M. M.,** Inhibitors of polyamine biosynthesis. VII. Evaluation of pyruvate derivatives as inhibitors of S-adenosyl-L-methionine decarboxylase, *J. Pharm. Sci.,* 69, 1000, 1980.

2. **Pankaskie, M. and Abdel-Monem, M. M.,** Inhibitors of polyamine biosynthesis. 8. Irreversible inhibition of mammalian S-adenosyl-L-methionine decarboxylase by substrate analogues, *J. Med. Chem.,* 23, 121, 1980.

3. **Kolb, M., Danzin, C., Barth, J., and Claverie, N.,** Synthesis and biochemical properties of chemically stable product analogues of the reaction catalyzed by S-adenosyl-L-methionine decarboxylase, *J. Med. Chem.,* 25, 550, 1982.

4. **Alston, T. A., Porter, D. J. T., Wheeler, D. M. S., and Bright, H. J.,** Mechanism-based inactivation of GABA aminotransferase by 3,5-dioxocyclohexanecarboxylic acid, *Biochem. Pharmacol.,* 31, 4081, 1982.

5. **Tobin, P. S., Basu, S. K., Grosserode, R. S., and Wheeler, D. M. S.,** Reactions of umpolung reagents with enol ethers of β-diketones: sterically induced selectivity of the dithiane anion, *J. Org. Chem.,* 45, 1250, 1980.

5a. **Porter, T. G. and Martin, D. L.,** Evidence for feedback regulation of glutamate decarboxylase by γ-aminobutyric acid, *J. Neurochem.,* 43, 1464, 1984.

5b. **Spink, D. C., Porter, T. G., Wu, S. J., and Martin, D. L.,** Characterization of three kinetically distinct forms of glutamate decarboxylase from pig brain, *Biochem. J.,* 231, 695, 1985.

5c. **Porter, T. G., Spink, D. C., Martin, S. B., and Martin, D. L.,** Transaminations catalysed by brain glutamate decarboxylation, *Biochem. J.,* 231, 705, 1985.

6. **Sinnott, M. L. and Smith, P. J.,** Affinity labelling with a deaminatively generated carbonium ion. Kinetics and stoichiometry of the alkylation of methionine-500 of the *lac*Z β-galactosidase of *Escherichia coli* by β-D-galactopyranosylmethyl *p*-nitrophenyltriazene, *Biochem. J.,* 175, 525, 1978.

7. **Sinnott, M. L. and Smith, P. J.,** Active-site-directed irreversible inhibition of *E. coli* β-galactosidase by the "hot" carbonium ion precursor, β-D-galactopyranosylmethyl-*p*-nitrophenyltriazene, *J. Chem. Soc. Chem. Commun.,* p. 223, 1976.

8. **Isaacs, N. S. and Rannala, E.,** Solvent effects upon the rates of acid-induced decomposition of 3-methyl-1-*p*-tolyltriazene and diphenyldiazomethane, *J. Chem. Soc. Perkin Trans. II,* p. 899, 1974.

9. **Kelly, M. A., Murray, M., and Sinnott, M. L.,** Substituent effects on tautomerisation constants of alkylaryltriazenes, *J. Chem. Soc. Perkin Trans. II,* p. 1649, 1982.

10. **Jones, C. C., Kelly, M. A., Sinnott, M. L., and Smith, P. J.,** Unimolecular heterolysis of a nitrogen-nitrogen bond, *J. Chem. Soc. Chem. Commun.,* p. 322, 1980.

11. **Jones, C. C., Kelly, M. A., Sinnott, M. L., Smith, P. J., and Tzotzos, G. T.,** Pathways for the decomposition of alkylaryltriazenes in aqueous solution, *J. Chem. Soc. Perkin Trans. II,* p. 1655, 1982.

12. **Sinnott, M. L., Tzotzos, G. T., and Marshall, S. E.,** Effect of aryl substituents on the kinetics of inactivation of glycosidases by glycosylmethylaryltriazenes: examination of the "suicide" nature of these inactivations, *J. Chem. Soc. Perkin Trans. II,* p. 1665, 1982.

13. **Fowler, A. V., Zabin, I., Sinnott, M. L., and Smith, P. J.,** Methionine 500, the site of covalent attachment of an active site-directed reagent of β-galactosidase, *J. Biol. Chem.,* 253, 5283, 1978.

14. **Van Diggelen, O. P., Galjaard, H., Sinnott, M. L., and Smith, P. J.,** Specific inactivation of lysosomal glycosidases in living fibroblasts by the corresponding glycosylmethyl-*p*-nitrophenyltriazenes, *Biochem. J.,* 188, 337, 1980.

15. **Van Diggelen, O. P., Schram, A. W., Sinnott, M. L., Smith, P. J., Robinson, D., and Galjaard, H.,** Turnover of β-galactosidase in fibroblasts from patients with genetically different types of β-galactosidase deficiency, *Biochem. J.,* 200, 143, 1981.

16. **Marshall, P. J., Sinnott, M. L., Smith, P. J., and Widdows, D.,** Active-site-directed irreversible inhibition of glycosidases by the corresponding glycosylmethyl-(*p*-nitrophenyl)triazenes, *J. Chem. Soc. Perkin Trans. I,* 366, 1981.

17. **Fowler, A. V. and Smith, P. J.,** The active site regions of *lac*Z and *ebg* β-galactosidases are homologous, *J. Biol. Chem.,* 258, 10204, 1983.

18. **Kass, L. R. and Bloch, K.,** On the enzymatic synthesis of unsaturated fatty acids in *Escherichia coli, Proc. Natl. Acad. Sci. U.S.A.,* 58, 1168, 1967.

19. **Helmkamp, G. M., Jr., Rando, R. R., Brock, D. J. H., and Bloch, K.,** β-Hydroxydecanoyl thioester dehydrase. Specificity of substrates and acetylenic inhibitors, *J. Biol. Chem.,* 243, 3229, 1968.

20. **Helmkamp, G. M., Jr., and Block, K.,** β-Hydroxydecanoyl thioester dehydrase. Studies on molecular structure and active site, *J. Biol. Chem.,* 244, 6014, 1969.

21. **Endo, K., Helmkamp, G. M., Jr., and Bloch, K.,** Mode of inhibition of β-hydroxydecanoyl thioester dehydrase by 3-decynoyl-*N*-acetylcysteamine, *J. Biol. Chem.,* 245, 4293, 1970.

22. **Morisaki, M. and Bloch, K.,** Inhibition of β-hydroxydecanoyl thioester dehydrase by some allenic acids and their thioesters, *Bioorg. Chem.,* 1, 188, 1971.

23. **Schwab, J. M. and Klassen, J. B.**, Steric course of the allylic rearrangement catalyzed by β-hydroxy-decanoylthioester dehydrase. Mechanistic implications, *J. Am. Chem. Soc.*, 106, 7217, 1984.

24. **Schwab, J. M., Lin, D. C. T., He, C., and Clardy, J.**, Absolute configuration of an allenic enzyme inactivator, *Tetrahedron Lett.*, 25, 4909, 1984.

25. **Morisaki, M. and Bloch, K.**, On the mode of interaction of β-hydroxydecanoyl thioester dehydrase with allenic acid derivatives, *Biochemistry*, 11, 309, 1972.

26. **Schwab, J. M., Li, W., Ho, C.-K., Townsend, C. A., and Salituro, G. M.**, Direct observation by carbon-13 NMR spectroscopy of the regioselectivity and stoichiometry of "suicide" enzyme inactivation, *J. Am. Chem. Soc.*, 106, 7293, 1984.

26a. **Schwab, J. M., Ho, C.-K., Li, W.-B, Townsend, C. A., and Salituro, G. M.**, β-Hydroxydecanoyl thioester dehydrase. Complete characterization of the fate of the "suicide" substrate 3-decynoyl-NAC, *J. Am. Chem. Soc.*, 108, 5309, 1986.

27. **Miesowicz, F. M. and Bloch, K. E.**, Acetylene-allene acyl thioester isomerase from hog liver, *Biochem. Biophys. Res. Commun.*, 65, 331, 1975.

28. **Miesowicz, F. M. and Bloch, K.**, Purification of hog liver isomerase. Mechanism of isomerization of 3-alkenyl and 3-alkynyl thioesters, *J. Biol. Chem.*, 254, 5868, 1979.

29. **Gomes, B., Fendrich, G. and Abeles, R. H.**, Mechanism of action of glutaryl-CoA and butyryl-CoA dehydrogenases. Purification of glutaryl-CoA dehydrogenase, *Biochemistry*, 20, 1481, 1981.

30. **Fendrich, G. and Abeles, R. H.**, Mechanism of action of butyryl-CoA dehydrogenase; reactions with acetylenic, olefinic, and fluorinated substrate analogues, *Biochemistry*, 21, 6685, 1982.

31. **Frerman, F. E., Miziorko, H. M., and Beckman, J. D.**, Enzyme-activated inhibitors, alternate substrates, and a dead end inhibitor of the general acyl-CoA dehydrogenase, *J. Biol. Chem.*, 255, 11192, 1980.

32. **Batzold, F. H. and Robinson, C. H**, Irreversible inhibition of Δ^5-3-ketosteroid isomerase by 5,10-secosteroids, *J. Am. Chem. Soc.*, 97, 2576, 1975.

33. **Penning, T. M., Westbrook, E. M., and Talalay, P.**, On the number of steroid-binding sites of Δ^5-3-oxosteroid isomerase, *Eur. J. Biochem.*, 105, 461, 1980.

34. **Penning, T. M., Covey, D. F., and Talalay, P.**, Irreversible inactivation of Δ^5-3-ketosteroid isomerase of *Pseudomonas testosteroni* by acetylenic suicide substrates. Mechanism of formation and properties of the steroid-enzyme adduct, *J. Biol. Chem.*, 256, 6842, 1981.

35. **Penning, T. M. and Talalay, P.**, Linkage of an acetylenic secosteroid suicide substrate to an active site of Δ^5-3-ketosteroid isomerase. Isolation and characterization of a tetrapeptide, *J. Biol. Chem.*, 256, 6851, 1981.

36. **Penning, T. M., Heller, D. N., Balasubramanian, T. M., Fenselau, C. C., and Talalay, P.**, Mass spectrometric studies of a modified active-site tetrapeptide from Δ^5-3-ketosteroid isomerase of *Pseudomonas testosteroni*, *J. Biol. Chem.*, 257, 12589, 1982.

37. **Batzold, F. H. and Robinson, C. H.**, Synthesis of β,γ-acetylenic 3-oxo steroids of the 5,10-seco series, *J. Org. Chem.*, 41, 313, 1976.

38. **Covey, D. F. and Robinson, C. H.**, Conjugated allenic 3-oxo-5,10-secosteroids. Irreversible inhibitors of Δ^5-3-ketosteroid isomerase, *J. Am. Chem. Soc.*, 98, 5038, 1976.

39. **Covey, D. F., Albert, K. A., and Robinson, C. H.**, Model studies with enzyme inhibitors. Addition of nucleophiles to conjugated allenic 3-oxo-5,10-secosteroids, *J. Chem. Soc. Chem. Commun.*, p. 795, 1979.

40. **Carrell, H. L., Glusker, J. P., Covey, D. F., Batzold, F. H., and Robinson, C. H.**, Molecular structures of substrates and inhibitors of Δ^5-3-keto steroid isomerase and their relevance to the enzymatic mechanism, *J. Am. Chem. Soc.*, 100, 4282, 1978.

41. **Jung, M. J. and Metcalf, B. W.**, Catalytic inhibition of γ-aminobutyric acid-α-ketoglutarate transaminase of bacterial origin by 4-aminohex-5-ynoic acid, a substrate analog, *Biochem. Biophys. Res. Commun.*, 67, 301, 1975.

42. **Likos, J. J., Ueno, H., Feldhaus, R. W., and Metzler, D. E.**, A novel reaction of the coenzyme of glutamate decarboxylase with L-serine *O*-sulfate, *Biochemistry*, 21, 4377, 1982.

43. **Ueno, H., Likos, J. J., and Metzler, D. E.**, Chemistry of the inactivation of cytosolic aspartate aminotransferase by serine *O*-sulfate, *Biochemistry*, 21, 4387, 1982.

44. **Lippert, B., Jung, M. J., and Metcalf, B. W.**, Biochemical consequences of reactions catalyzed by GAD and GABA-T, *Brain Res. Bull.*, 5 (Suppl. 2), 375, 1980.

45. **Jung, M. J., Metcalf, B. W., Lippert, B., and Casara, P.**, Mechanism of the sterospecific irreversible inhibition of bacterial glutamic acid decarboxylase by (R)-(−)-4-aminohex-5-ynoic acid, an analogue of 4-aminobutyric acid, *Biochemistry*, 17, 2628, 1978.

46. **Bouclier, M., Jung, M. J., and Lippert, B.**, Stereochemistry of reactions catalyzed by mammalian brain L-glutamate 1-carboxylase and 4-aminobutyrate: 2-oxoglutarate aminotransferase, *Eur. J. Biochem.*, 98, 363, 1979.

47. **Danzin, C., Claverie, N., and Jung, M. J.**, Stereochemistry of the inactivation of 4-aminobutyrate:2-oxoglutarate aminotransferase and L-glutamate 1-carboxylase by 4-aminohex-5-ynoic acid enantiomers, *Biochem. Pharmacol.*, 33, 1741, 1984.

48. **Jung, M. J. and Seiler, N.**, Enzyme-activated irreversible inhibitors of L-ornithine:2-oxoacid aminotransferase. Demonstration of mechanistic features of the inhibition of ornithine aminotransferase by 4-aminohex-5-ynoic acid and gabaculine and correlation with in vivo activity, *J. Biol. Chem.*, 253, 7431, 1978.

49. **Rando, R. R.**, The chemical labeling of glutamate decarboxylase *in vivo*, *J. Biol. Chem.*, 256, 1111, 1981.

50. **John, R. A., Jones, E. D., Fowler, L. J.**, Enzyme-induced inactivation of transaminases by acetylenic and vinyl analogues of 4-aminobutyrate, *Biochem. J.*, 177, 721, 1979.

51. **Jones, E. D., Basford, J. M., and John, R. A.**, An investigation of the properties of ornithine aminotransferase after inactivation by the "suicide" inhibitor aminohexynoate and use of the compound as a probe of intracellular protein turnover, *Biochem. J.*, 209, 243, 1983.

52. **Bey, P., Jung, M. J., Gerhart, F., Schirlin, D., Van Dorsselaer, V., and Casara, P.**, ω-Fluoromethyl analogues of ω-amino acids as irreversible inhibitors of 4-aminobutyrate:2-oxoglutarate aminotransferase, *J. Neurochem.*, 37, 1341, 1981.

53. **Metcalf, B. W. and Casara, P.**, Regiospecific 1,4 addition of a propargylic anion. A general synthon for 2-substituted propargylamines as potential catalytic irreversible enzyme inhibitors, *Tetrahedron Lett.*, p. 3337, 1975.

54. **Metcalf, B. W. and Casara, P.**, Synthetic access to α-substituted prop-2-ynylamines and α-acetylenic amino acids via the *t*-butyl *N*-trimethylsilylprop-2-ynylcarbamate dianion, *J. Chem. Soc. Chem. Commun.*, p. 119, 1979.

55. **Metcalf, B. W., Bey, P., Danzin, C., Jung, M. J., Casara, P., and Vevert, J. P.**, Catalytic irreversible inhibition of mammalian ornithine decarboxylase (E.C. 4.1.1.17) by substrate and product analogues, *J. Am. Chem. Soc.*, 100, 2551, 1978.

56. **Casara, P., Danzin, C., Metcalf, B. W., and Jung, M. J.**, Stereospecific irreversible inhibition of mammalian (S)-ornithine decarboxylase by (R)-(−)-hex-5-yne-1,4-diamine, *J. Chem. Soc. Chem. Commun.*, p. 1190, 1982.

57. **Kallio, A., McCann, P. P., and Bey, P.**, DL-α-Monofluoromethylputrescine is a potent irreversible inhibitor of *Escherichia coli* ornithine decarboxylase, *Biochem. J.*, 204, 771, 1982.

58. **Danzin, C., Jung, M. J., Seiler, N., and Metcalf, B. W.**, Effect of 5-hexyne-1,4-diamine on brain 4-aminobutyric acid metabolism in rats and mice, *Biochem. Pharmacol.*, 28, 633, 1979.

59. **Danzin, C., Casara, P., Claverie, N., and Grove, J.**, Effects of enantiomers of 5-hexyne-1,4-diamine on ODC, GAD, and GABA-T activities in the rat, *Biochem. Pharmacol.*, 32, 941, 1983.

60. **Danzin, C., Casara, P., Calverie, N., Metcalf, B., and Jung, M. J.**, (2R,5R)-6-Heptyne-2,5-diamine, an extremely potent inhibitor of mammalian ornithine decarboxylase, *Biochem. Biophys. Res. Commun.*, 116, 237, 1983.

61. **Mamont, P. S., Siat, M., Joder-Ohlenbusch, A. M., Bernhardt, A., and Casara, P.**, Effects of (2R,5R)-6-heptyne-2,5-diamine, a potent inhibitor of L-ornithine decarboxylase, on rat hepatoma cells cultured in vitro, *Eur. J. Biochem.*, 142, 457, 1984.

61a. **Bitonti, A. J., Bacchi, C. J., McCann, P. P., and Sjoerdsma, A.**, Catalytic irreversible inhibition of *Trypanosoma brucei brucei* ornithine decarboxylase by substrate and product analogs and their effects on murine trypanosomiasis, *Biochem. Pharmacol.*, 34, 1773, 1985.

62. **Casara, P., Danzin, C., Metcalf, B., and Jung, M.**, Stereospecific synthesis of (2R,5R)-hept-6-yne-2,5-diamine: a potent and selective enzyme-activated irreversible inhibitor of ornithine decarboxylase, *J. Chem. Soc. Perkin Trans. 1*, p. 2201, 1985.

62a. **Bitonti, A. J., Casara, P. J., McCann, P. P., and Bey, P.**, Catalytic irreversible inhibition of bacterial and plant arginine decarboxylase activities by novel substrate and product analogues, *Biochem. J.*, 242, 69, 1987.

63. **Shannon, P., Marcotte, P., Coppersmith, S., and Walsh, C.**, Studies with mechanism-based inactivators of lysine ε-transaminase from *Achromobacter liquidum*, *Biochemistry*, 18, 3917, 1979.

64. **Holbert, G. W. and Metcalf, B. W.**, Synthesis of α-ethynylhistamine, an inactivator of histidine decarboxylase, *Tetrahedron*, 40, 1141, 1984.

65. **Hevey, R. C., Babson, J., Maycock, A. L., and Abeles, R. H.**, Highly specific enzyme inhibitors. Inhibition of plasma amine oxidase, *J. Am. Chem. Soc.*, 95, 6125, 1973.

66. **Rando, R. R. and de Mairena, J.**, Propargylamine-induced irreversible inhibition of non-flavin-linked amine oxidases, *Biochem. Pharmacol.*, 23, 463, 1974.

66a. **Knowles, P. F., Pandeya, K. B., Rius, F. X., Spencer, C. M., Moog, R. S., McGuirl, M. A., and Dooley, D. M.**, The organic cofactor in plasma amine oxidase: evidence for pyrroloquinoline quinone and against pyridoxal phosphate, *Biochem. J.*, 241, 603, 1987.

67. **Ing, H. R. and Manske, R. H. F.**, A modification of the Gabriel synthesis of amines, *J. Chem. Soc.*, p. 2348, 1926.

68. **Schulte, K. E. and Reiss, K. P.**, Zur kenntnis der Acetylenecarbonsäuren. II. Mitteil: Die Darstellung der Hexinsäure, Pentin-(4)-säure und Heptadiin-(1,6)-carbonsäure-(4), *Chem. Ber.*, 87, 964, 1954.

69. **Castellano, A. and Krantz, A.,** Allenic amino acids. 1. Syntheses of γ-allenic GABA by a novel aza-Cope rearrangement, *J. Am. Chem. Soc.,* 106, 1877, 1984.

70. **Hiemstra, H., Fortgens, H. P., and Speckamp, W. N.,** Lewis acid induced reactions of propargyl trimethyl silane with ω-ethoxy lactams. Synthesis of γ-allenyl GABA, *Tetrahedron Lett.,* 25, 3115, 1984.

71. **Casara, P., Jund, K., and Bey, P.,** General synthetic access to α-allenyl amines and α-allenyl-α-amino acids as potential enzyme activated irreversible inhibitors of PLP dependent enzymes, *Tetrahedron Lett.,* 25, 1891, 1984.

72. **Jung, M. J., Heydt, J. -G., and Casara, P.,** γ-Allenyl GABA, a new inhibitor of 4-aminobutyrate aminotransferase. Comparison with other inhibitors of this enzyme, *Biochem. Pharmacol.,* 33, 3717, 1984.

73. **Danzin, C. and Casara, P.,** α-Allenyl putrescine, an enzyme-activated irreversible inhibitor of bacterial and mammalian ornithine decarboxylases, *FEBS Lett.,* 174, 275, 1984.

74. **Asada, Y., Tanizawa, K., Nakamura, K., Moriguchi, M., and Soda, K.,** Stereochemistry of ornithine decarboxylase reaction, *J. Biochem. (Tokyo),* 95, 277, 1984.

75. **Lippert, B., Metcalf, B. W., Jung, M. J., and Casara, P.,** 4-Aminohex-5-enoic acid, a selective catalytic inhibitor of 4-aminobutyric acid aminotransferase in mammalian brain, *Eur. J. Biochem.,* 74, 441, 1977.

75a. **Kolb, M., Barth, J., Heydt, J.-G., and Jung, M. J.,** Synthesis and evaluation of mono-, di-, and trifluoroethenyl-GABA derivatives as GABA-T inhibitors, *J. Med. Chem.,* 30, 267, 1987.

75b. **Bey, P., Gerhart, F., and Jung, M.,** Synthesis of (E)-4-amino-2,5-hexadienoic acid and (E)-4-amino-5-fluoro-2-pentenoic acid. Irreversible inhibitors of 4-amino-butyrate-2-oxoglutarate aminotransferase, *J. Org. Chem.,* 51, 2835, 1986.

76. **Baldwin, J. E., Haber, S. B., Hoskins, C., and Kruse, L. I.,** Synthesis of β,γ-unsaturated amino acids, *J. Org. Chem.,* 42, 1239, 1977.

77. **Glattfeld, J. W. E. and Hoen, R. E.,** The trihydroxybutyric acids, *J. Am. Chem. Soc.,* 57, 1405, 1935.

78. **Rando, R. R.,** Irreversible inhibition of aspartate aminotransferase by 2-amino-3-butenoic acid, *Biochemistry,* 13, 3859, 1974.

79. **Gehring, H., Rando, R. R., and Christen, P.,** Active-site labeling of aspartate aminotransferases by the β,γ-unsaturated amino acid vinylglycine, *Biochemistry,* 16, 4832, 1977.

79a. **Kagamiyama, H., Sakakibara, R., Wada, H., Tanase, S., and Morino, Y.,** The complete amino acid sequence of mitochondrial aspartate aminotransferase from pig heart, *J. Biochem. (Tokyo),* 82, 291, 1977.

80. **Soper, T. S., Manning, J. M., Marcotte, P. A., and Walsh, C. T.,** Inactivation of bacterial D-amino acid transaminases by the olefinic amino acid D-vinylglycine, *J. Biol. Chem.,* 252, 1571, 1977.

81. **King, S. and Phillips, A. T.,** Aromatic aminotransferase activity of rat brain cytoplasmic and mitochondrial aspartate aminotransferases, *J. Neurochem.,* 30, 1399, 1978.

82. **Johnston, M., Raines, R., Chang, M., Esaki, N., Soda, K., and Walsh, C.,** Mechanistic studies on reactions of bacterial methionine γ-lyase with olefinic amino acids, *Biochemistry,* 20, 4325, 1981.

83. **Scannell, J. P., Preuss, D. L., Demny, T. C., Sello, L. H., Williams, T., and Stempel, A.,** Antimetabolites produced by microorganisms. V. L-2-amino-4-methoxy-*trans*-3-butenoic acid, *J. Antibiot.,* 25, 122, 1972.

84. **Rando, R. R.,** β,γ-Unsaturated amino acids as irreversible enzyme inhibitors, *Nature (London),* 250, 586, 1974.

85. **Rando, R. R., Relyea, N., and Cheng, L.,** Mechanism of the irreversible inhibition of aspartate aminotransferase by the bacterial toxin L-2-amino-4-methoxy-*trans*-3-butenoic acid, *J. Biol. Chem.,* 251, 3306, 1976.

86. **Miles, E. W.,** Inactivation of tryptophan synthetase by α-cyanoglycine and L-2-amino-4-methoxy-*trans*-3-butenoic acid, in *Enzyme-Activated Irreversible Inhibitors,* Seiler, N., Jung, M. J., and Koch-Weser, J., Eds., Elsevier/North Holland, Amsterdam, 1978, 73.

87. **Owens, L. D., Thompson, J. F., Pitcher, R. G., and Williams, T.,** Structure of rhizobitoxine, an antimetabolic enol-ether amino-acid from *Rhizobium japonicum, J. Chem. Soc. Chem. Commun.,* p. 714, 1972.

88. **Giovanelli, J., Owens, L. D., and Mudd, S. H.,** Mechanism of inhibition of spinach β-cystathionase by rhizobitoxine, *Biochim. Biophys. Acta,* 227, 671, 1971.

89. **Cooper, A. J. L., Fitzpatrick, S. M., Kaufman, C., and Dowd, P.,** β-Methylene-D,L-aspartic acid: a selective inhibitor of glutamate-aspartate transaminase, *J. Am. Chem. Soc.,* 104, 332, 1982.

90. **Cooper, A. J. L., Fitzpatrick, S. M., Ginos, J. Z., Kaufman, C., and Dowd, P.,** Inhibition of glutamate-aspartate transaminase by β-methylene-D,L-aspartate, *Biochem. Pharmacol.,* 32, 679, 1983.

91. **Dowd, P. and Kaufman, C.,** (±)-β-Methyleneaspartic acid, *J. Org. Chem.,* 44, 3956, 1979.

92. **Cooper, A. J. L., Haber, M. T., Ginos, J. Z., Kaufman, P., Kaufman, C., Paik, H., and Dowd, P.,** Interaction of soluble pig heart glutamate-aspartate transaminase with various β,γ-unsaturated amino acids, *Biochem. Biophys. Res. Commun.,* 129, 193, 1985.

93. **Johnston, M., Marcotte, P., Donovan, J., and Walsh, C.,** Mechanistic studies with vinylglycine and β-haloaminobutyrates as substrates for cystathionine γ-synthetase from *Salmonella typhimurium, Biochemistry,* 18, 1729, 1979.

94. **Walsh, C., Pascal, R. A., Jr., Johnston, M., Raines, R., Dikshit D., Krantz, A., and Honma, M.,** Mechanistic studies on the pyridoxal phosphate enzyme 1-aminocyclopropane-1-carboxylate deaminase from *Pseudomonas* sp., *Biochemistry,* 20, 7509, 1981.

95. **Maycock, A. L., Suva, R. H., and Abeles, R. H.,** Novel inactivators of plasma amine oxidase, *J. Am. Chem. Soc.,* 97, 5613, 1975.

96. **Miles, E. W.,** Effects of modification of the β_2 subunit and of the $\alpha_2\beta_2$ complex of tryptophan synthase by α-cyanoglycine, a substrate analog, *Biochem. Biophys. Res. Commun.,* 64, 248, 1975.

97. **Goldman, A. S.,** Stoichiometric inhibition of various 3β-hydroxysteroid dehydrogenases by a substrate analogue, *J. Clin. Endocrinol.,* 27, 325, 1967.

98. **Dann, O. T. and Carter, C. E.,** Cycloserine inhibition of gamma-aminobutyric-alpha-ketoglutaric transaminase, *Biochem. Pharmacol.,* 13, 677, 1964.

99. **Soper, T. S. and Manning, J. M.,** Different modes of action of inhibitors of bacterial D-amino acid transaminase, a target enzyme for the design of new antibacterial agents, *J. Biol. Chem.,* 256, 4263, 1981.

100. **Sundaram, K. S. and Lev, M.,** Inhibition of sphingolipid synthesis by cycloserine *in vitro* and *in vivo,* *J. Neurochem.,* 42, 577, 1984.

101. **Rando, R. R.,** On the mechanism of action of antibiotics which act as irreversible enzyme inhibitors, *Biochem. Pharmacol.,* 24, 1153, 1975.

102. **Washtien, W. and Abeles, R. H.,** Mechanism of inactivation of γ-cystathionase by the acetylenic substrate analogue propargylglycine, *Biochemistry,* 16, 2485, 1977.

103. **Borcsok, E. and Abeles, R. H.,** Mechanism of action of cystathionine synthase, *Arch. Biochem. Biophys.,* 213, 695, 1982.

104. **Alston, T. A., Muramatsu, H., Ueda, T., and Bright, H. J.,** Inactivation of γ-cystathionase by γ-fluorinated amino acids, *FEBS Lett.,* 128, 293, 1981.

105. **Steglich, W., Heininger, H. U., Dowrschak, H., and Weygand, F.,** A general method for the preparation of β-perfluoroalkylalanines, *Angew. Chem. Int. Ed. Engl.,* 6, 807, 1967.

105a. **Keller, J. W. and Hamilton, B. J.,** Enzymatic resolution of 2-trifluoromethylalanine, *Tetrahedron Lett.,* 27, 1249, 1986.

106. Agency of Industrial Science and Technology, β-(Trifluorovinyl)alanine, *Chem. Abstr.,* 96, 20465r, 1982.

107. **Johnston, M., Raines, R., Walsh, C., and Firestone, R. A.,** Mechanism-based enzyme inactivation using an allyl sulfoxide-allyl sulfenate ester rearrangement, *J. Am. Chem. Soc.,* 102, 4241, 1980.

108. **Rando, R. R. and Bangerter, F. W.,** The irreversible inhibition of mouse brain γ-aminobutyric acid (GABA)-α-ketoglutaric acid transaminase by gabaculine, *J. Am. Chem. Soc.,* 98, 6762, 1976.

108a. **Churchich, J. E. and Moses, U.,** 4-Aminobutyrate aminotransferase. The presence of nonequivalent binding sites, *J. Biol. Chem.,* 256, 1101, 1981.

109. **Rando, R. R.,** Mechanism of the irreversible inhibition of γ-aminobutyric acid-α-ketoglutaric acid transaminase by the neurotoxin gabaculine, *Biochemistry,* 16, 4604, 1977.

110. **Rando, R. R. and Bangerter, F. W.,** Reaction of the neurotoxin gabaculine with pyridoxal phosphate, *J. Am. Chem. Soc.,* 99, 5141, 1977.

111. **Rando, R. R. and Bangerter, F. W.,** The in vivo inhibition of GABA transaminase by gabaculine, *Biochem. Biophys. Res. Commun.,* 76, 1276, 1977.

112. **Wood, J. D., Kurylo, E., and Tsui, D. S., K.,** Inhibition of aminotransferase enzyme systems by gabaculine, *Neurosci. Lett.,* 14, 327, 1979.

113. **Soper, T. S. and Manning, J. M.,** Inactivation of pyridoxal phosphate enzymes by gabaculine. Correlation with enzymic exchange of β-protons, *J. Biol. Chem.,* 257, 13930, 1982.

114. **Rando, R. R., Bangerter, F. W., and Farb, D. H.,** The inactivation of γ-aminobutyric acid transaminase in dissociated neuronal cultures from spinal cord, *J. Neurochem.,* 36, 985, 1981.

115. **Burnett, G., Yonaha, K., Toyama, S., Soda, K., and Walsh, C.,** Studies on the kinetics of inactivation of *Pseudomonas* ω-amino acid:pyruvate transaminase by gabaculine, *J. Biol. Chem.,* 255, 428, 1980.

116. **Yasuda, M., Toyama, S., Rando, R. R., Esaki, N., Tanizawa, K., and Soda, K.,** Irreversible inactivation of L-ornithine:α-ketoglutarate δ-aminotransferase by gabaculine, *Agric. Biol. Chem.,* 44, 3005, 1980.

117. **Kobayashi, K., Miyazawa, S., Terahara, A., Mishima, H., and Kurihara, H.,** Gabaculine: γ-aminobutyrate aminotransferase inhibitor of microbial origin, *Tetrahedron Lett.,* p. 537, 1976.

118. **Singer, S. P. and Sharpless, K. B.,** Synthesis of *dl*-gabaculine utilizing direct allylic amination as the key step, *J. Org. Chem.,* 43, 1448, 1978.

119. **François, J.-P. and Gittos, M. W.,** A preparative synthesis of *d,l*-gabaculine, *Syn. Commun.,* 9, 741, 1979.

120. **Metcalf, B. W. and Jung, M. J.,** Molecular basis for the irreversible inhibition of 4-aminobutyric acid:2-oxoglutarate and L-ornithine:2-oxoacid aminotransferases by 3-amino-1,5-cyclohexadienyl carboxylic acid (isogabaculine), *Mol. Pharmacol.,* 16, 539, 1979.

121. **Danishefsky, S. and Hershenson, F. M.,** Regiospecific synthesis of isogabaculine, *J. Org. Chem.,* 44, 1180, 1979.

122. **Burkhart, J. P., Holbert, G. W., and Metcalf, B. W.,** Enantiospecific synthesis of *(S)*-4-amino-4,5-dihydro-2-furancarboxylic acid, a new suicide inhibitor of GABA-transaminase, *Tetrahedron Lett.,* 25, 5267, 1984.

123. **Adams, J. L., Chen, T.-M., and Metcalf, B. W.,** 4-Amino-4,5-dihydrothiophene-2-carboxylic acid, *J. Org. Chem.,* 50, 2730, 1985.

124. **Metcalf, B. W. and Lippert, B.,** personal communication, 1985.

125. **Abeles, R. H. and Walsh, C. T.,** Acetylenic enzyme inactivators. Inactivation of γ-cystathionase, in vitro and in vivo, by propargylglycine, *J. Am. Chem. Soc.,* 95, 6124, 1973.

126. **Marcotte, P. and Walsh, C.,** Active site-directed inactivation of cystathionine γ-synthetase and glutamic pyruvic transaminase by propargylglycine, *Biochem. Biophys. Res. Commun.,* 62, 677, 1975.

127. **Johnston, M., Jankowski, D., Marcotte, P., Tanaka, H., Esaki, N., Soda, K., and Walsh, C.,** Suicide inactivation of bacterial cystathionine γ-synthase and methionine γ-lyase during processing of L-propargylglycine, *Biochemistry,* 18, 4690, 1979.

128. **Beeler, T. and Churchich, J. E.,** Reactivity of the phosphopyridoxal groups of cystathionase, *J. Biol. Chem.,* 251, 5267, 1976.

129. **Cheung, K.-S., Wasserman, S. A., Dudek, E., Lerner, S. A., and Johnston, M.,** Chloroalanyl and propargylglycyl dipeptides. Suicide substrate containing antibacterials, *J. Med. Chem.,* 26, 1733, 1983.

130. **Morino, Y., Kojima, H., and Tanase, S.,** Affinity labeling of alanine aminotransferase by 3-chloro-L-alanine, *J. Biol. Chem.,* 254, 279, 1979.

131. **Burnett, G., Marcotte, P., and Walsh, C.,** Mechanism-based inactivation of pig heart L-alanine transaminase by L-propargylglycine, half-site reactivity, *J. Biol. Chem.,* 255, 3487, 1980.

132. **Tanase, S. and Morino, Y.,** Irreversible inactivation of aspartate aminotransferases during transamination with L-propargylglycine, *Biochem. Biophys. Res. Commun.,* 68, 1301, 1976.

133. **Gershon, H., Meek, J. S., and Dittmer, K.,** Propargylglycine: an acetylenic amino acid antagonist, *J. Am. Chem. Soc.,* 71, 3573, 1949.

134. **Jansen, A. C. A., Weustink, R. J. M., Kerling, K. E. T., and Havinga, E.,** Studies on polypeptides. VII. Synthesis of D,L-, L-, and D-2-amino-6-dimethylamino-4-hexynoic acid, *Rec. Trav. Chim. Pays-Bas,* 88, 819, 1969.

135. **Fisher, S. K. and Davies, W. E.,** Some properties of guinea pig brain glutamate decarboxylase and its inhibition by the convulsant allylglycine (2-amino-4-pentenoic acid), *J. Neurochem.,* 23, 427, 1974.

136. **Alston, T. A., Porter, D. J. T., Mela, L., and Bright, H. J.,** Inactivation of alanine aminotransferase by the neurotoxin β-cyano-L-alanine, *Biochem. Biophys. Res. Commun.,* 92, 299, 1980.

137. **Ueno, H., Soper, T. S., and Manning, J. M.,** Enzyme-activated inhibition of bacterial D-amino acid transaminase by β-cyano-L-alanine, *Biochem. Biophys. Res. Commun.,* 122, 485, 1984.

138. **Pinnell, S. R. and Martin, G. R.,** The cross-linking of collagen and elastin: enzymatic conversion of lysine in peptide linkage to α-aminoadipic-δ-semialdehyde (allysine) by an extract from bone, *Proc. Natl. Acad. Sci. U.S.A.,* 61, 708, 1968.

139. **Narayanan, A. S., Siegel, R. C., and Martin, G. R.,** On the inhibition of lysyl oxidase by β-aminopropionitrile, *Biochem. Biophys. Res. Commun.,* 46, 745, 1972.

140. **Trackman, P. C. and Kagan, H. M.,** Nonpeptidyl amine inhibitors are substrates of lysyl oxidase, *J. Biol. Chem.,* 254, 7831, 1979.

141. **Tang, S.-S., Trackman, P. C., and Kagan, H. M.,** Reaction of aortic lysyl oxidase with β-aminopropionitrile, *J. Biol. Chem.,* 258, 4331, 1983.

141a. **Williamson, P. R. and Kagan, H. M.,** α-Proton abstraction and carbanion formation in the mechanism of action of lysyl oxidase, *J. Biol. Chem.,* 262, 8196, 1987.

142. **Miles, E. W. and Meister, A.,** The mechanism of the reaction of β-hydroxyaspartate with L-aspartate β-decarboxylase. A new type of pyridoxal 5'-phosphate-enzyme inhibition, *Biochemistry,* 6, 1734, 1967.

143. **Palekar, A. G., Tate, S. S., and Meister, A.,** Inhibition of aspartate β-decarboxylase by aminomalonate. Stereospecific decarboxylation of aminomalonate to glycine, *Biochemistry,* 9, 2130, 1970.

144. **Bloxham, D. P., Clark, M. G., and Holland, P. C.,** Inactivation of thiolase by alk-3-ynoyl-coenzyme A and bromoacyl coenzyme A esters: evidence for a basic group at the active site, *Biochem. Soc. Trans.,* 1, 1272, 1973.

145. **Holland, P. C., Clark, M. G., and Bloxham, D. P.,** Inactivation of pig heart thiolase by 3-butynoyl coenzyme A, 3-pentynoyl coenzyme A, and 4-bromocrotonyl coenzyme A, *Biochemistry,* 12, 3309, 1973.

146. **Bloxham, D. P.,** Selective inhibition of cholesterol synthesis by alkylating inhibitors of hepatic cytoplasmic thiolase, *Biochem. Soc. Trans.,* 2, 925, 1974.

147. **Penning, T. M.,** Irreversible inhibition of Δ⁵-3-oxosteroid isomerase by 2-substituted progesterones, *Biochem. J.,* 226, 469, 1985.

148. **McDonald, I. A., Palfreyman, M. G., Jung, M., and Bey, P.,** Synthesis of *(E)*-β-fluoromethyleneglutamic acid, *Tetrahedron Lett.,* 26, 4091, 1985.

Chapter 8

DECARBOXYLATION REACTIONS*

I. INTRODUCTION

This chapter includes those mechanism-based enzyme inactivations that are initiated by decarboxylation. The anion generated by decarboxylation, then, can undergo two principal processes: elimination of a leaving group, or "isomerization" (actually, it is just a resonance structure, but since it becomes protonated, it is classified as if a proton is removed from the point of decarboxylation and replaced at the isomerized position).

II. DECARBOXYLATION/ELIMINATION/ADDITION

Several general syntheses of mono-, di-, and trihalomethyl-α-amino acids have been reported by Bey and co-workers[1-4] and Kollonitsch and co-workers.[5-9] The route to mono-fluoro-, difluoro-, trifluoro-, monochloro-, and bromodifluoromethyl-α-amino acids[1,2] is shown in Scheme 1 (X = Cl or Br; Y = CH$_2$F, CHF$_2$,

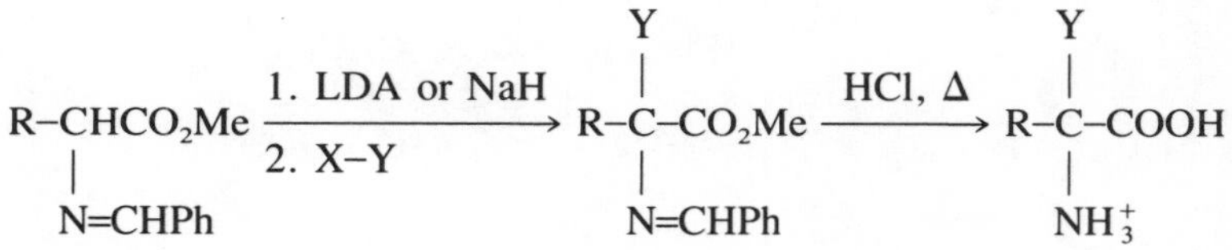

$$\begin{array}{ccccc}
 & & Y & & Y \\
 & & | & & | \\
R\text{--}CHCO_2Me & \xrightarrow[\text{2. X--Y}]{\text{1. LDA or NaH}} & R\text{--}C\text{--}CO_2Me & \xrightarrow{\text{HCl, }\Delta} & R\text{--}C\text{--}COOH \\
| & & | & & | \\
N\text{=}CHPh & & N\text{=}CHPh & & NH_3^+
\end{array}$$

Scheme 1.

CF$_3$, CH$_2$Cl, CF$_2$Br). The reactions appear to involve S$_N$2 mechanisms except for chloro-difluoromethane alkylation, which reacts via a difluorocarbene-mediated chain process (Scheme 2). A general synthesis of difluoromethyl amino

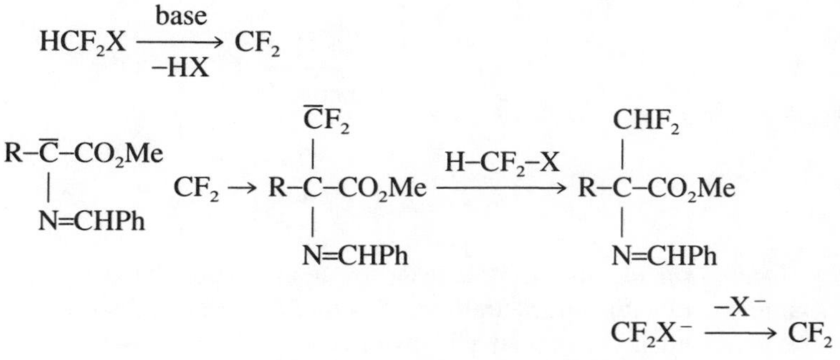

$$HCF_2X \xrightarrow[-HX]{\text{base}} CF_2$$

$$\begin{array}{ccccc}
R\text{--}\overline{C}\text{--}CO_2Me & & \overline{C}F_2 & & CHF_2 \\
| & & | & & | \\
N\text{=}CHPh & CF_2 \rightarrow & R\text{--}C\text{--}CO_2Me & \xrightarrow{\text{H--CF}_2\text{--X}} & R\text{--}C\text{--}CO_2Me \\
 & & | & & | \\
 & & N\text{=}CHPh & & N\text{=}CHPh
\end{array}$$

$$CF_2X^- \xrightarrow{-X^-} CF_2$$

Scheme 2.

acids[3] is shown in Scheme 3. α-Chlorofluoromethyl- and α-fluoromethyl α-amino

---

* A list of abbreviations and shorthand notations can be found prior to Chapter 7.

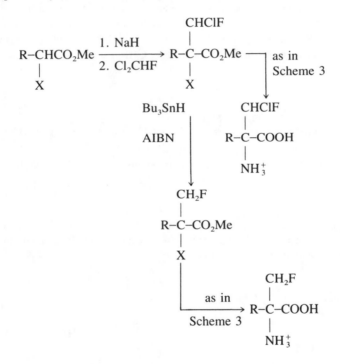

Scheme 3.

acids were prepared[4] by the routes shown in Scheme 4 (X = CO₂tBu or N=CHPh).

Scheme 4.

α-Fluoroalkyl α-amino acids also can be made by treatment of the corresponding α-hydroxyalkyl α-amino acid with sulfur tetrafluoride in liquid hydrogen fluoride[5,6] (Scheme 5). Treatment of α-mercaptoalkyl α-amino acids with HF

Scheme 5.

and fluoroxytrifluoromethane, chlorine, *N*-chlorosuccinimide, or fluorine gives the corresponding α-fluoroalkyl α-amino acid[7] (Scheme 6). Photofluorination

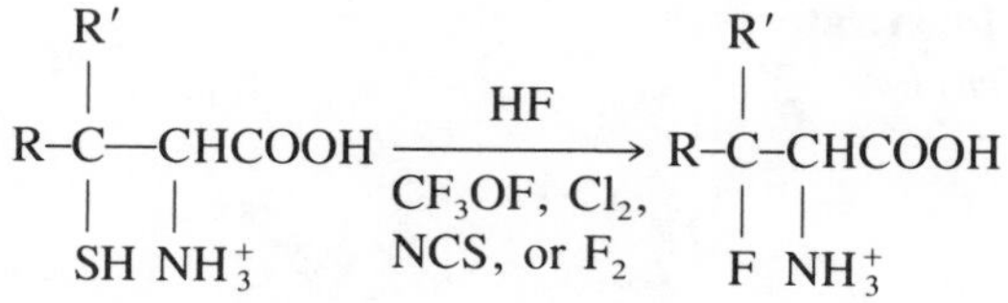

$$
\begin{array}{ccc}
\text{R}' & & \text{R}' \\
| & & | \\
\text{R–C–CHCOOH} & \xrightarrow[\text{CF}_3\text{OF, Cl}_2,]{\text{HF}} & \text{R–C–CHCOOH} \\
|\;\;| & \text{NCS, or F}_2 & |\;\;| \\
\text{SH NH}_3^+ & & \text{F NH}_3^+
\end{array}
$$

Scheme 6.

of alanine[8] and other α-methyl amino acids[9] with fluoroxytrifluoromethane gives the corresponding α-fluoromethyl amino acid (Scheme 7).

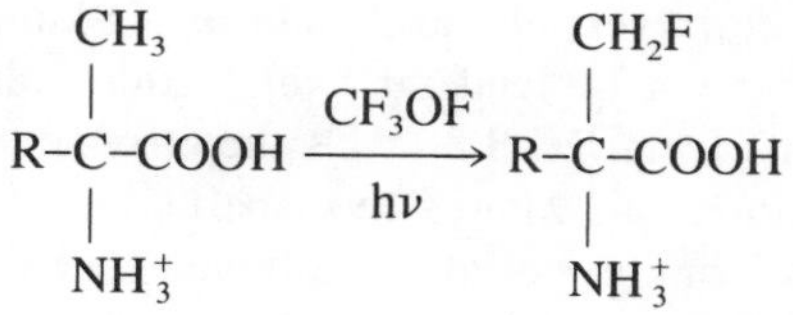

$$
\begin{array}{ccc}
\text{CH}_3 & & \text{CH}_2\text{F} \\
| & & | \\
\text{R–C–COOH} & \xrightarrow[h\nu]{\text{CF}_3\text{OF}} & \text{R–C–COOH} \\
| & & | \\
\text{NH}_3^+ & & \text{NH}_3^+
\end{array}
$$

Scheme 7.

A common proposed mechanism for inactivation of PLP-dependent decarboxylases by α-halomethyl (or dihalo- or trihalomethyl) α-amino acids is shown in pathway $\underset{\sim}{a}$ of Scheme 8. However, a mechanism of inactivation related to that described by Metzler and co-workers[10,11] for the inactivation of PLP-dependent enzymes by *L*-serine *O*-sulfate (see Chapter 6 of Volume I, Section II.G.3.a.) may be applicable (pathway $\underset{\sim}{b}$, Scheme 8). Scheme 8 will be referred to throughout this section.

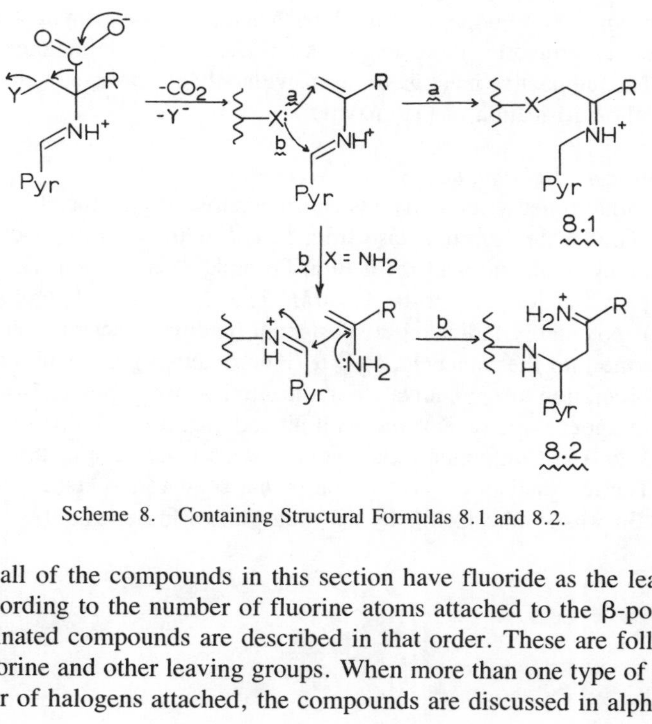

Scheme 8.   Containing Structural Formulas 8.1 and 8.2.

Since almost all of the compounds in this section have fluoride as the leaving group, it is organized according to the number of fluorine atoms attached to the β-position. Mono-, di-, and trifluorinated compounds are described in that order. These are followed by compounds with chlorine and other leaving groups. When more than one type of inactivator has the same number of halogens attached, the compounds are discussed in alphabetical order, e.g., (monofluoromethyl)dopa comes before (monofluoromethyl)glutamate (except α-(mon-ofluoromethyl)argenine, which was added in proof).

A. Monofluorinated Inactivators

1. α-(Monofluoromethyl)dopa

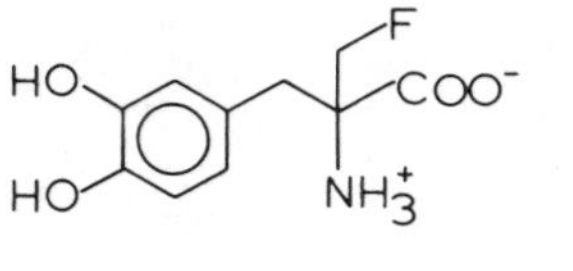

Scheme 9.   Structural Formula 8.3.

α-(Monofluoromethyl)dopa (2-(fluoromethyl)-3-(3,4-dihydroxyphenyl)alanine (Structural Formula 8.3, Scheme 9) is an irreversible inactivator or pig kidney dopa (L-aromatic amino acid) decarboxylase,[9,12,13] but not of fetal rat liver histidine decarboxylase.[9] No enzyme activity returns upon dialysis. [ring-^{3}H]-**8.3** labels the enzyme stoichiometrically with con-comitant release of one fluoride ion, whereas no label is incorporated with [carboxyl-^{14}C]-**8.3**.[9] The inactivation mechanism proposed by Kollonitsch et al.[9] is that shown in pathway a of Scheme 8. Further studies were carried out by Maycock et al.[14] Biphasic time-dependent inactivation was observed for (S)-**8.3**. The cause for the biphasic nature of the inactivation is not known, but multiple forms of the enzyme were suspected. A stoichiometric amount of [^{3}H]-inactivator is incorporated into the enzyme after inactivation. When [1-^{14}C]-labeled inactivator is used, essentially no radioactivity is incorporated; one equivalent of [^{14}CO$_2$], however, is released. During inactivation, 1 equivalent of fluoride ion is liberated. These stoichiometries suggest that only 1 molecule of inactivator is turned over during inactivation, or, in other words, a partition ratio of 0 is observed. The K_i for α-(fluoromethyl)dopa (0.04 μM) is much lower than the K_m for dopa (200 μM). The mechanism proposed is the same as in Scheme 8 (pathway a).

[^{3}H]α-(Monofluoromethyl)dopa was used by Maneckjee and Baylin[15] to label human pheochromacytoma L-aromatic amino acid decarboxylase and to define the subunit structure of the enzyme. The radioactive inactivator reacts with only one protein in crude homogenates and is useful in the purification of the enzyme.

2. α-(Monofluoromethyl)glutamate

(R,S)-α-(Monofluoromethyl)glutamate is an irreversible inactivator of rat brain glutamate decarboxylase.[9] Glutamate decarboxylase from *E. coli* is irreversibly inactivated by (R,S)-α-(monofluoromethyl)glutamic acid (Structural Formula 8.3a, to Scheme 9A); the K_i was determined by Kuo and Rando[15a] to be 1.4 μM. The K_i for α-methylglutamic acid is 10^4 higher (15 mM), potentially indicating a substantial binding effect as a result of the incor-poration of fluorine into the moleucle. [^{3}H]-α-(Fluoromethyl)glutamate inactivates the en-zyme with incorporation of 1 equivalent radioactivity per subunit. No radioactivity is incorporated into apoenzyme or enzyme with altered cofactor. The partition ratio is quite low, approximately 0.2. Compound **8.3a** is a very weak inactivator of the mouse and chick brain enzyme. The mechanism of inactivation is that shown in Scheme 8 (pathway a).

Compound **8.3a** was synthesized[15a] by the route shown in Scheme 9A.

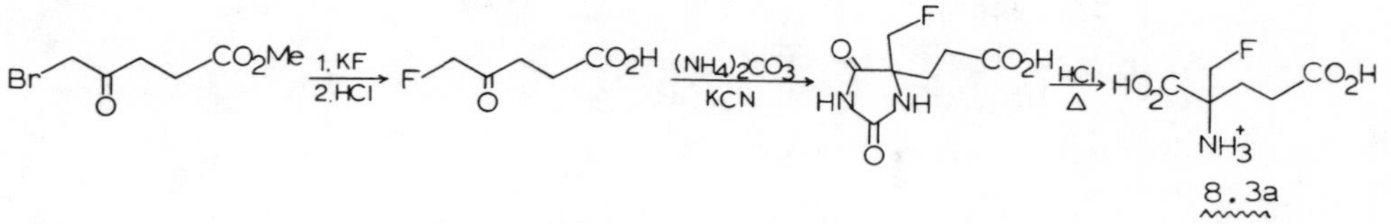

Scheme 9A.   Containing Structural Formula 8.3a.

3. α-(Monofluoromethyl)histidine

The (+)-isomer of α-fluoromethylhistidine (Structural Formula 8.4, Scheme 10) is a pseudo first-order time-dependent inactivator of rat hypothalamus[16] and rat gastric mucosa[17] histidine decarboxylase; dialysis does not reactivate the enzyme. The (−)-isomer is ineffective. At 2 μM (+)-**8.4**, hypothalamus histidine decarboxylase is inactivated > 90% in 30 min; at 5 mM (+)-**8.4**, dopa decarboxylase[9,16,17] and glutamate decarboxylase[17] are unaffected. *Lactobacillus* 30A histidine decarboxylase also is not inactivated.[9] α-(Fluoromethyl)histidine inactivates fetal rat brain, and stomach histidine decarboxylase irreversibly.[18] The use of [³H]α-(fluoromethyl)histidine results in radioactivity associated with the enzyme after gel filtration.

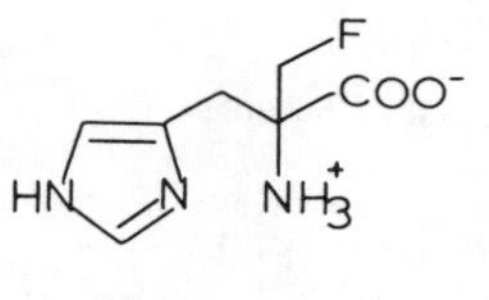

Scheme 10.   Structural Formula 8.4.

Histidine decarboxylase from fetal rat liver also is inactivated by α-(fluoromethyl)histidine;[9,19] only the (*S*)-enantiomer is active.[19] The use of *S*-α-[4-³H](fluoromethyl)histidine by Duggan et al.[19] showed that radioactivity remains bound to the enzyme following gel filtration and no α-(fluoromethyl)histamine is generated. Compound **8.4** does not inactivate dopa decarboxylase (hog kidney), glutamate decarboxylase (rat liver), histidase, diamine oxidase (hog kidney), or histamine *N*-methyltransferase (guinea pig brain). Kubota et al.[20] also found that α-fluoromethylhistidine inactivates fetal rat liver L-histidine decarboxylase irreversibly (dialysis does not restore activity); 2 mol of [*ring* 4-³H]-**8.4** are incorporated per mole of dimeric enzymes. No inactivation of apoenzyme occurs. L-Histidine, but not D-histidine, protects the enzyme. [*Carboxyl*-¹⁴C]-**8.4** inactivates the enzyme without incorporation of radioactivity; however, 2.8 molecules of ¹⁴CO₂ are generated for each molecule of inactivator incorporated into the enzyme. This indicates that almost twice as many activated molecules are released for each one that leads to inactivation. No α-(fluoromethyl)histamine was obtained as a reaction product; however, three unidentified products were detected by TLC. One could arise by decarboxylation and transamination (Scheme 11, pathway a̰) and one could arise from decarboxylation and elimination (Scheme 11, pathway b̰). No transamination experiments were

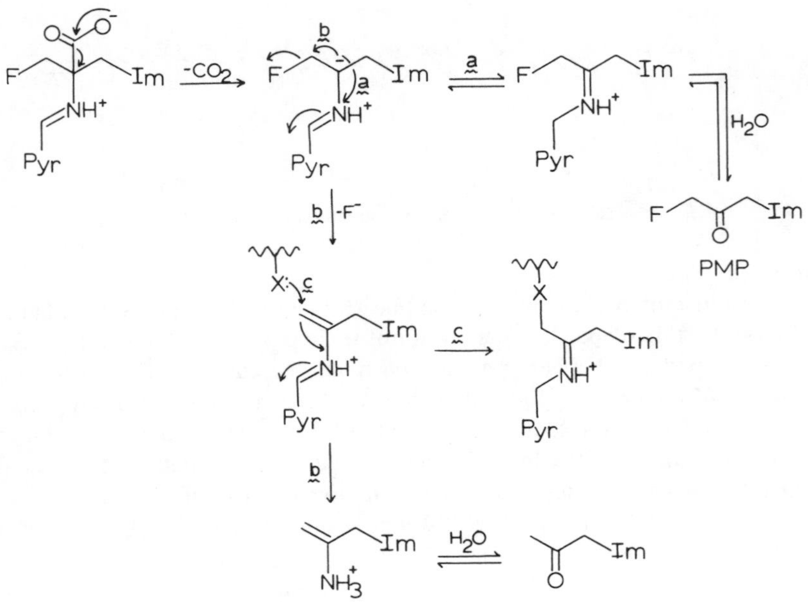

Scheme 11.

carried out to determine the feasibility of pathway <u>a</u>. Although the decarboxylation/elimination/addition mechanism (Scheme 11, pathway <u>c</u>) was suggested, the enamine mechanism[10,11] (see Scheme 8, pathway <u>b</u>) could not be excluded. The following results of Taguchi et al.[21] support the enamine mechanism. When [ring-4-^{3}H]-**8.4** was used to inactivate the enzyme and then the inactivated enzyme was subjected to NaDodSO$_4$-polyacrylamide gel electrophoresis, no radioactivity was bound to the protein. However, if the inactivated enzyme was reduced with NaBH$_4$ prior to electrophoresis, radioactivity was bound to a band that corresponded to that of a the monomer (**8.2**, Scheme 8). If NaBH$_4$ treatment occurred 20 hr after inactivation, again, no radioactivity was bound. This could be explained as elimination (Scheme 12) to **8.5** in the 20 hr prior to NaBH$_4$ treatment. Not enough

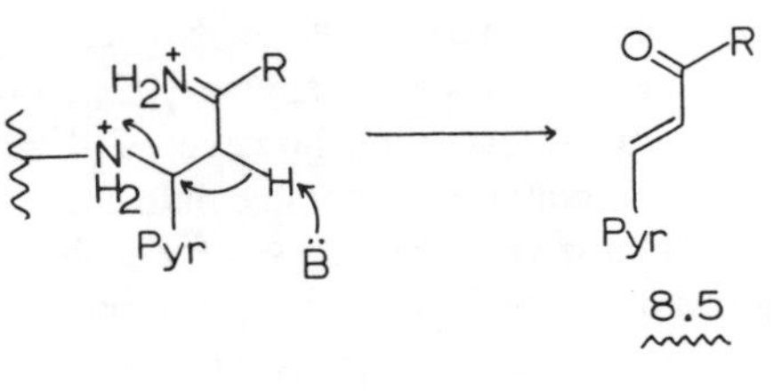

Scheme 12.   Containing Structural Formula 8.5.

enzyme was available, however, to identify the enzyme adduct. Histidine decarboxylase from human peripheral blood leukocytes is inactivated by (S)-α-fluoromethylhistidine.[21a]

Because of the lability and limited quantity of histidine decarboxylase from all sources, the structure of the inactivated enzyme could not be determined. However, the enzyme from *Morganella morganii* AM-15 was purified, characterized, and shown to be inactivated by (S)-α-fluoromethylhistidine.[21b] Inactivation with S-α-fluoromethyl[*ring* 4 − ^{3}H]histidine occurs with attachment of one molecule of inactivator per active site with a partition ratio of zero.[21c] Dialysis of labeled enzyme does not release the radioactivity, but urea or heat denaturation results in release of 90% of the tritium. The released product has spectral absorbances characteristic of the product expected from the enamine mechanism.[10,11] If the labeled enzyme is treated with $NaBH_4$ (but not $NaCNBH_3$), the tritium remains attached to the protein after denaturation. A tryptic peptide, containing 85% of the radioactivity, was isolated and shown to have the same amino acid analysis as the corresponding native enzyme peptide except for one serine residue. The peptide sequence was determined to be Asn–Ser–Ile–Thr–Val–Val–Phe–Pro–Cys–Pro–Ser–Glu–Arg, where × is the inactivator-

    |
    ×

modified coenzyme adduct. This suggests that a ternary complex is formed between the protein, the cofactor, and the inactivator. The ternary complex is somewhat unstable, even after $NaBH_4$ treatment. These results support the mechanism proposed by Hayashi et al.[21c] (Scheme 12A). The gene that encodes this enzyme has been cloned and sequenced and the corresponding amino acid sequence determined by Vaaler et al.[21d] The lysine that binds the PLP is residue 232 and the serine that binds the α-fluoromethylhistidine-pyridoxal phosphate adduct is residue 322.

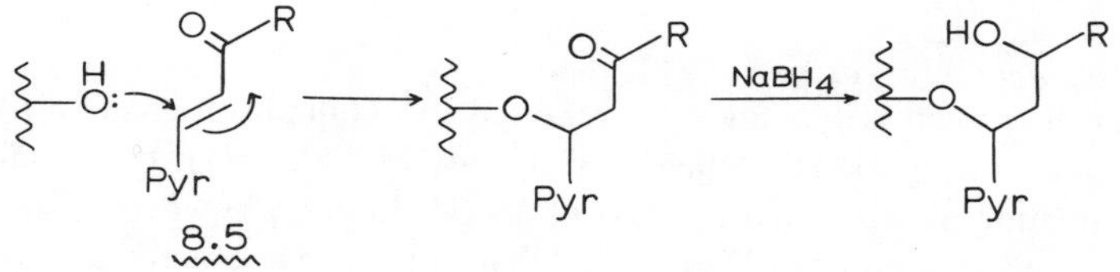

Scheme 12A.

4. α-(Monofluoromethyl)ornithine

Monofluoromethylornithine inactivates *Escherichia coli*,[22] *Pseudomonas aeruginosa*,[22] *Trypanosoma brucei brucei*[23] and thioacetamide-treated rat liver[9] ornithine decarboxylases in a time-dependent fashion; no enzyme activity is restored upon dialysis.

5. α-(Monofluoromethyl)dehydroornithine

α-(Monofluoromethyl-dehydroornithine (Structural Formula 8.6, Scheme 13) is a time-dependent inactivator of *Trypanosoma brucei brucei* ornithine decarboxylase[23] and ornithine decarboxylase from livers of thioacetamide-treated rats;[24] dialysis does not regenerate enzyme activity. In comparison with the saturated analogue (see Section II.A.4.), **8.6** has increased inactivator potency; the K_I is 30 times lower than the saturated compound with a fivefold increase in the rate of inactivation.[24]

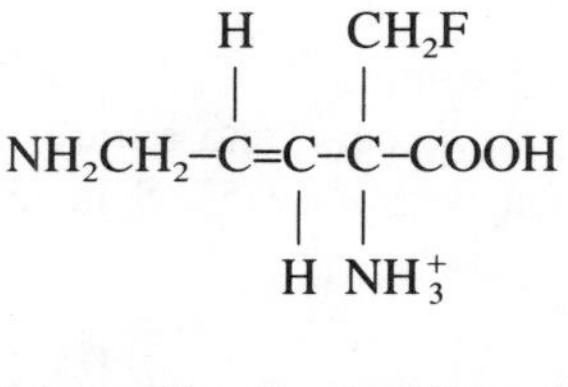

Scheme 13. Structural Formula 8.6.

Compound **8.6** was synthesized by Bey et al.[24] as shown in Scheme 14.

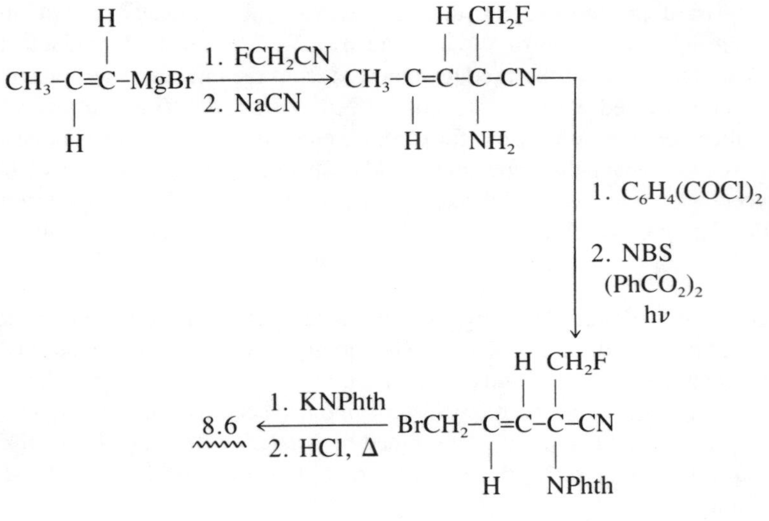

$$CH_3-\underset{\underset{H}{|}}{\overset{\overset{H}{|}}{C}}=C-MgBr \xrightarrow[\text{2. NaCN}]{\text{1. FCH}_2\text{CN}} CH_3-\underset{\underset{H}{|}}{\overset{\overset{H}{|}}{C}}=\overset{\overset{CH_2F}{|}}{C}-\underset{\underset{NH_2}{|}}{C}-CN$$

1. $C_6H_4(COCl)_2$

2. NBS
   $(PhCO_2)_2$
   $h\nu$

$$8.6 \xleftarrow[\text{2. HCl, }\Delta]{\text{1. KNPhth}} BrCH_2-\underset{\underset{H}{|}}{\overset{\overset{H}{|}}{C}}=\overset{\overset{CH_2F}{|}}{C}-\underset{\underset{NPhth}{|}}{C}-CN$$

Scheme 14.

6. α-(Monofluoromethyl)tyrosine

(S)-α-(Monofluoromethyl)tyrosine was reported by Kollonitsch et al.[9] to be a time-dependent inactivator of pig kidney dopa decarboxylase ($K_i > 2$ mM); Jung et al.,[13] however, found no inactivation up to a concentration of 1 mM. In vivo, however, it is an inactivator of dopa decarboxylase. Jung et al.[13] suggested that a dual enzyme-activated inactivation (see Chapter 1 of Volume I, Section III.D.) occurs, because hog adrenals tyrosine hydroxylase converts α-(monofluoromethyl)tyrosine to α-(monofluoromethyl)dopa, a known inactivator of dopa decarboxylase (see Section II.A.1.).

7. α-(Monofluoromethyl)arginine

α-(Monofluoromethyl)arginine (Structural Formula 8.6′, Scheme 14 was found by Bitonti et al.[24a] to be a mechanism-based inactivator of arginine decarboxylase from *E. coli*, oats, and barley that is more potent than α-(difluoromethyl)arginine.

$$\underset{\underset{NH_2}{|}}{HOOCC}\overset{\overset{FCH_2}{|}}{}CH_2CH_2CH_2NHC\overset{\overset{NH}{\|}}{}NH_2$$

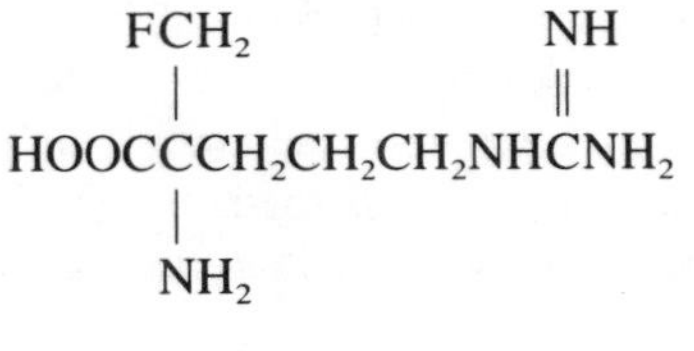

Scheme 14A.   Structural Formula 8.6′.

8. (E-α-(Monofluoromethyl)dehydroarginine

Arginine decarboxylase from *E. coli*, oats, and barley is inactive by (E)-α-(monofluoromethyl)dehydroarginine (Structural Formula 8.5″, Scheme 14B),[24a] which is more potent than α-(difluoromethyl)arginine.

$$\underset{\underset{NH_2}{|}}{HOOCC}\overset{\overset{FCH_2}{|}}{}\underset{\underset{H}{}}{\overset{\overset{H}{|}}{C}}=CCH_2\underset{\underset{H}{}}{N}C\overset{\overset{NH}{\|}}{}NH_2$$

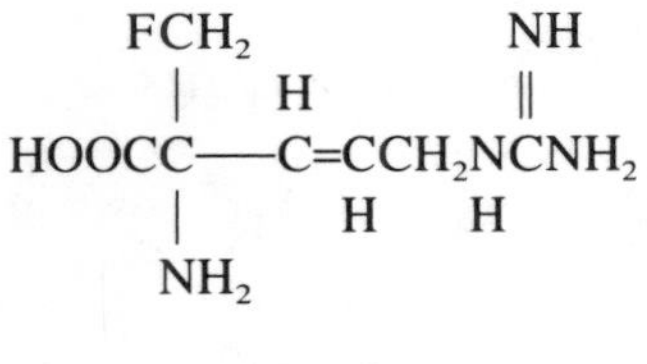

Scheme 14B.   Structural Formula 8.6″.

B. Difluorinated Inactivators

1. α-(Difluoromethyl)arginine

D,L-α-(Difluoromethyl)arginine (**8.6a**) is an irreversible inactivator of both the biosynthetic and biodegradative arginine decarboxylases of *Escherichia coli* and of the biosynthetic arginine decarboxylases of *Pseudomonas aeruginosa* and *Klebsiella pneumoniae*.[25] Only a few percent of enzyme activity returns after dialysis, and dithiothreitol has no effect on the rate of inactivation. Compound **8.6a** was synthesized by Bey et al.[2] as shown in Scheme 15.

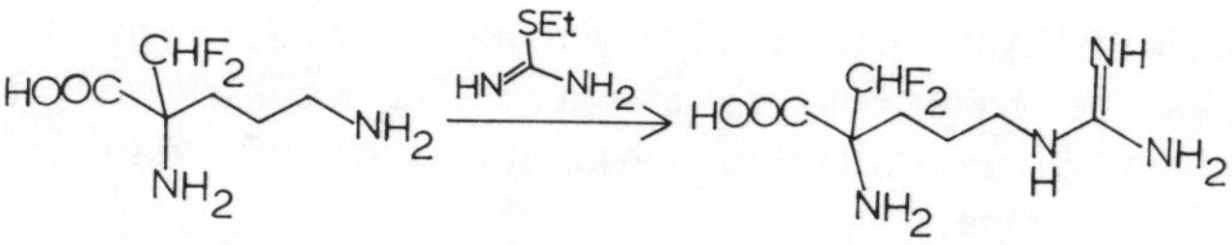

Scheme 15.   Containing Structural Formula 8.6a.

2. α-(Difluoromethyl)dopa

α-(Difluoromethyl)dopa is a mechanism-based inactivator of hog kidney aromatic L-amino acid decarboxylase.[9,26] Dialysis does not restore enzyme activity. The mechanism proposed by Palfreyman et al.[26] is the same as that proposed by Kollonitsch et al.[9] (see Scheme 8, pathway <u>a</u>). With ring-tritiated inactivator, 0.83 mol of inactivator was bound per mole of enzyme.[27] Only a slight modification in the PLP spectrum was observed after inactivation, suggesting that attachment is to an amino acid residue. Inactivation kinetics are not pseudo first order; however, no significant decrease in the inactivator concentration or formation of reaction product were observed. This phenomenon may be the result of the existence of isozymes. The inactivator was shown by Ribéreau-Gayon et al.[27] to react spontaneously with PLP to give **8.7** (Scheme 16). This compound, however, has no inhibitor effect on the enzyme, and was not detected in tissues or urine of rats treated with α-(difluoromethyl)dopa.

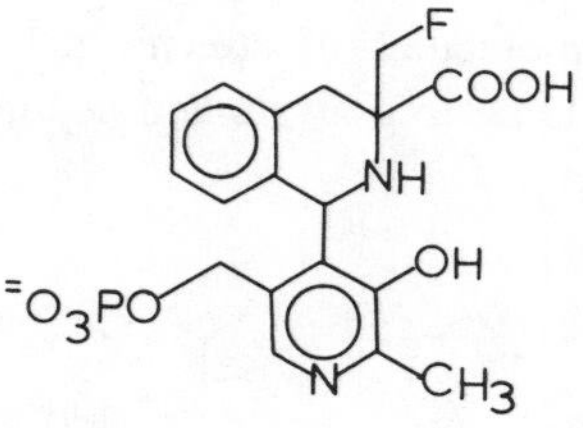

Scheme 16.   Structural Formula 8.7.

3. α-(Difluoromethyl)glutamic Acid

α-(Difluoromethyl)glutamic acid (Structural Formula 8.8, Scheme 17) has been synthesized by Tsushima et al;[28] however, enzyme inactivation studies with it have not been reported.

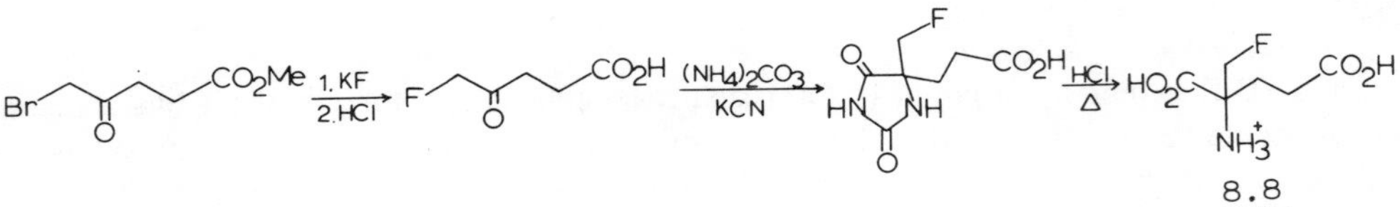

Scheme 17.   Containing Structural Formula 8.8.

4. α-(Difluoromethyl)lysine

D,L-α-(Difluoromethyl)lysine irreversibly inactivates lysine decarboxylase of *Mycoplasma dispar;* dialysis does not restore enzyme activity.[29] The specificity of this inactivation is evident by the fact that neither (difluoromethyl)ornithine nor (difluoromethyl)arginine causes any inhibition of this enzyme.

5. α-(Difluoromethyl)ornithine (DFMO)

D,L-α-(Difluoromethyl)ornithine (DFMO, **8.8a**) was synthesized by Bey et al.[2] (Scheme 18). DFMO irreversibly inactivates thioacetamide-treated rat liver ornithine decarboxylase; dialysis does not regenerate enzyme activity.[30] Dithiothreitol in the preincubation buffer does not prevent inactivation, indicating that the inactivating species is not released into solution prior to inactivation. Only the (−)-isomer is active. DFMO does not inactivate glutamate- or aromatic amino acid decarboxylase, GABA aminotransferase, or diamine oxidase. The mechanism proposed is that shown in Scheme 8 (pathway a). Partially purified rat liver ornithine decarboxylase is inactivated by DFMO.[31] [5-^{14}C]DFMO inactivation results in the incorporation of radioactivity concomitant with inactivation of the enzyme. Ornithine protects the enzyme and prevents radioactivity incorporation. Gel filtration of the labeled enzyme gives a single radioactive band that coelutes with enzyme activity (fresh enzyme was added just prior to gel filtration). More radioactivity is bound to protein in liver extracts from rats subjected to partial hepatectomy or treated with thioacetamide than from control rats. These conditions are known to increase ornithine decarboxylase activity. Enzyme purified 12 and 3500-fold has a similar ratio of radioactivity bound to enzyme activity (∼ 30 fmol DFMO per unit of enzyme). These results indicate the very low concentration of this enzyme in liver (only 200 molecules of enzyme per cell). The radioactive DFMO incorporated into the enzyme is stable to dialysis and heating to 90°C in 0.5 *M* perchloric acid.

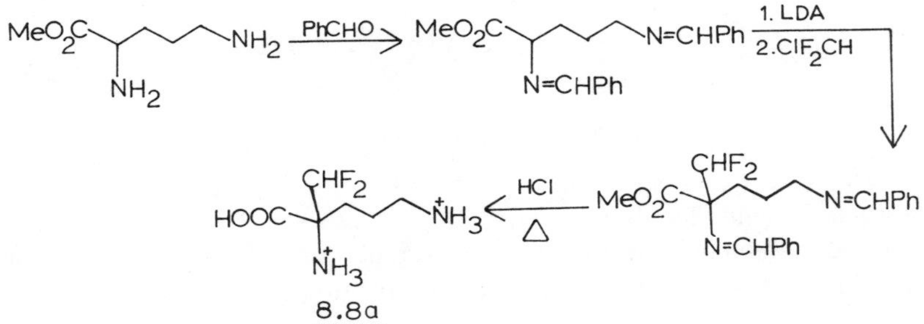

Scheme 18.   Containing Structural Formula 8.8a.

The inactivation of kidney ornithine decarboxylase from androgen-treated mice by α-difluoromethylornithine was studied by Pegg et al.[31a] [1-^{14}C]DFMO is a substrate and is decarboxylated to produce $^{14}CO_2$ concomitant with enzyme inactivation. Inactivation by [5-^{14}C]DFMO leads to incorporation of radioactivity into the protein (12.2 fmol/unit), proportional to the amount of $^{14}CO_2$ produced from [1-^{14}C]DFMO. The partition ratio was determined from these results to be 3.3. The same experiment with the enzymes from liver of thioacetamide-treated rats and from L1210 cells in culture gave partition ratios of 3.5.

DFMO has been used to determine the amount of ornithine decarboxylase present in rat liver under various physiological conditions and in mouse kidney after treatment with androgens.[32] Under these conditions, with [5-^{14}C]DFMO, only one protein is radiolabeled corresponding to òrnithine decarboxylase having a molecular weight of 100,000 and a subunit molecular weight of 55,000. Any treatment of the animal that affects the biosynthesis of this enzyme (either enhancement or inhibition), results in corresponding effect on the amount of radioactive DFMO incorporated into the liver. α-Difluoromethyl[5-^{3}H]ornithine also was used by Erwin et al.[32a] to titrate ornithine decarboxylase activity in mouse kidney extracts and purified mouse kidney ornithine decarboxylase.

Lysine- and ornithine decarboxylase from rat ventral prostate, kidney, liver, and regenerating liver are inactivated by α-(difluoromethyl)ornithine.[33] All of the data suggest that these two activities are derived from the same protein.

The purification of ornithine decarboxylase from kidneys of androgen-treated mice was shown by Seely et al.[34] to be homogeneous by treatment with [5-^{14}C]DFMO. All of the protein migrates in a single band with one equivalent of radioactivity bound as measured by native and NaDodSO$_4$ polyacrylamide gel electrophoresis and by thin-layer isoelectric focusing.

D,L-α-(Difluoromethyl)ornithine is a time-dependent inactivator of ornithine decarboxylase from *Pseudomonas aeruginosa,* but has no effect on the same enzyme from *Escherichia coli*[35,36] or *Klebsiella pneumoniae.*[35] Since D,L-α-(monofluoromethyl)putrescine is a potent inactivator of *E. coli* ornithine decarboxylase,[36] but has little effect on the enzyme from *P. aerugenosa* (see Chapter 6 of Volume I, Section II.A.5.), DFMO and α-(monofluoromethyl)putrescine have complementary activities. Both compounds, however, inactivate mammalian ornithine decarboxylase.

In the presence of limiting amounts of pyridoxal 5'-phosphate, DFMO is a selective time-dependent inactivator for the A-form of *Physarum* ornithine decarboxylase.[37] At higher PLP concentrations, the rates of inactivation of the A-(normal) and B-(modified) forms are comparable. The B-form does not bind PLP as well as does the A-form and this accounts for the selectivity of inactivation at low PLP concentrations.

Ornithine decarboxylase from the thermophilic bacterium *Clostridium thermohydrosulfuricum* is irreversibly inactivated by high concentrations (50 m*M*) of DFMO; enzyme activity is not regenerated by dialysis.[38] Radioactively labeled inactivator remains bound to the enzyme after dialysis. Arginine decarboxylase from the bacterium also is irreversibly inactivated by DFMO.

Polyamines induce the biosynthesis of a noncompetitive protein inhibitor of ornithine decarboxylase in various cells; these protein inhibitors have been termed antizymes.[39] α-(Difluoromethyl)ornithine was shown by Kyriakidis et al.[40] to have no effect on these antizymes and no radioactivity from α-[5-^{3}H]DFMO is incorporated. However, when ornithine decarboxylase-antizyme complexes are treated with α-[5-^{3}H]DFMO, radioactivity is incorporated into the ornithine decarboxylase. No matter what the ratio of enzyme to antizyme, nor the degree of inhibition of the enzyme by the antizyme, the same amount of [^{3}H] is incorporated, corresponding to the amount of enzyme present. The affinity of antizyme for DFMO-inactivated ornithine decarboxylase is less than with native enzyme. In contrast to the results of Kyriakidis et al.,[40] Kitani and Fujisawa[40a] observed no reaction of DFMO with the ornithine decarboxylase-antizyme complex, and suggested that the amounts of radioactive DFMO bound to the protein may not reflect the total amount of ornithine de-carboxylase in crude tissues.

6. α-(Difluoromethyl)dehydroornithine

α-(Difluoromethyl)dehydroornithine (Structural Formula 8.9, Scheme 19) inactivates or-nithine decarboxylase from livers of thioacetamide-treated rats; no enzyme activity returns upon dialysis.[24] In comparison with the saturated analogue, the double bond in the chain increases the inhibitory potency.

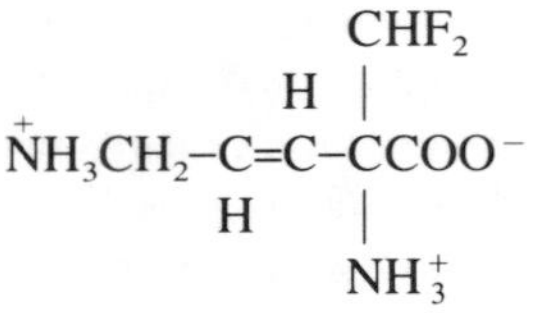

$$\overset{+}{N}H_3CH_2-\overset{\overset{\displaystyle H}{|}}{C}=\overset{\overset{\displaystyle CHF_2}{|}}{C}-\overset{\overset{\displaystyle |}{|}}{\underset{\underset{\displaystyle NH_3^+}{|}}{C}}COO^-$$

Scheme 19.   Structural Formula 8.9.

C. Trifluorinated Inactivators

1. 3,3,3-Trifluoro-2-aminoisobutyrate

α-Dialkylamino acid aminotransferase from *Pseudomonas cepacia* undergoes pseudo first-order time-dependent inactivation by 3,3,3-trifluoro-2-aminoisobutyrate (Structural Formula 8.9a, Scheme 20); dialysis does not restore enzyme activity.[41] [1-^{14}C]-Labeled inactivator results in no incorporation of radioactivity into inactivated enzyme, but ten equivalents of [14]CO$_2$ are liberated. The mechanism shown in Scheme 8 (pathway $\underline{a}$) was proposed by Keller and O'Leary.[41]

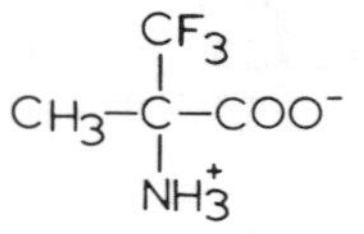

$$CH_3-\overset{\overset{\displaystyle CF_3}{|}}{\underset{\underset{\displaystyle NH_3^+}{|}}{C}}-COO^-$$

Scheme 20.   Structural Formula 8.9a

D. Monochlorinated Inactivators

1. α-(Monochloromethyl)histidine

Histidine decarboxylase from hamster placenta is irreversibly inactivated by α-(chloro-methyl)histidine.[42] No inactivation of aromatic amino acid decarboxylase, diamine oxidase, or bacterial histidine decarboxylase was observed; ornithine decarboxylase was inhibited 500 times slower than mammalian histidine decarboxylase.

2. α-(Monochloromethyl)ornithine

α-(Monochloromethyl)ornithine is a time-dependent inactivator of ornithine decarboxylase from liver of thioacetamide-treated rats; no saturation kinetics, however, were observed.[30] Dialysis does not regenerate enzyme activity.

E. Inactivators with Other Leaving Groups

1. α-(Cyanomethyl)ornithine

α-(Cyanomethyl)ornithine is a weak (K_I = 8.7 mM) time-dependent inactivator of or-nithine decarboxylase from liver of thioacetamide-treated rats; dialysis does not regenerate enzyme activity.[30]

III. DECARBOXYLATION/ISOMERIZATION

A. Decarboxylation/Isomerization

Various PLP-dependent decarboxylases have been found to be less than perfect in their reaction specificity. In addition to the normal catalysis that converts α-amino acids to amines (Scheme 21, pathway a), a small percentage of the time proton donation occurs at the PLP (Scheme 21, pathway b). The latter route is, in effect, decarboxylation followed by transamination. In the absence of an α-ketoacid, the cofactor (PLP) cannot be regenerated from the PMP formed in the side reaction. This, therefore inactivates the enzyme until either an α-ketoacid or fresh PLP is added. The side reaction, then, is responsible for mechanism-based inactivation. Bacterial α-dialkylamino acid aminotransferase (the bacterium was originally isolated as an *E. coli* contaminant and appeared to be a *Pseudomonad*), however, catalyzes a decarboxylation-dependent transamination of isovaline, aminoisobutyric acid, and cycloleucine **every turnover**.[43] In this case, decarboxylation/transamination appears to be the normal catalytic mechanism. A model study for the reactions catalyzed by PLP-dependent enzymes was carried out by Kalyankar and Snell.[44] When α-amino acids are heated with pyridoxal in the absence of metal ions in dilute aqueous solutions, two related reactions take place: decarboxylation and decarboxylation-dependent transamination. Chemical precedence for this inactivation reaction, therefore, exists.

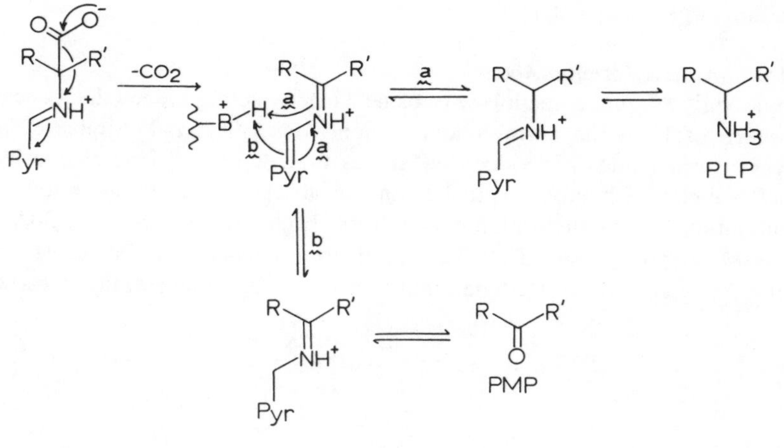

Scheme 21.

1. Amino Acids

L-Aspartate, L-glutamate, L-valine, and L-alanine inactivate aspartate β-decarboxylase from *Alcaligenes faecalis* in the absence of PLP and an α-ketoacid.[45,46] *E. coli* glutamate decarboxylase undergoes a rapid decarboxylation-dependent transamination, with release of PMP by L-glutamate and L-aspartate.[47] Rat brain glutamate decarboxylase is inactivated by glutamate in the absence of PLP; enzyme activity is restored by incubation with PLP.[47a] Glutamic acid also produces time-dependent inactivation of hog brain L-glutamate decarboxylase; a decarboxylative-transamination mechanism was suggested by Meeley and Martin.[47b] Addition of PLP reactivates the enzyme.[47c] Glutamate and GABA (see Chapter 7, Section I.C.) are time- and concentration-dependent inactivators of the three isozymic forms of pig brain glutamate decarboxylase with the order of inactivation being γ-form> β-form >·α-form.[47d] Although the inactivation rates by glutamate and GABA differ among the three forms, the kinetics of inactivation of each form by saturating glutamate and GABA are identical, suggesting a common intermediate. The mechanism of inactivation by glutamate is decarboxylation-dependent transamination, and by GABA is direct transamination with PLP to give PMP and succinic semialdehyde.[47e] Both reactions lead to the conversion of holoenzyme to apoenzyme. Rates of transamination of glutamate and GABA are similar, suggesting that the decarboxylation step of the former is not rate determining. Transamination is reversible; holoenzyme can be prepared from apoenzyme in the presence of PMP and succinic semialdehyde. Aspartate also is a time-dependent inactivator of pig brain glutamate decarboxylase; the rate of inactivation is 25 times greater than that for glutamate or GABA.[47f] Inactivation produces PMP and is reversed by PLP, suggesting that transamination is the cause. Production of β-alanine is two orders of magnitude faster than decarboxylative transamination. L-5-Hydroxytryptophan and L-3,4-dihydroxyphenylalanine (L-dopa) produce a time-dependent inhibition of hog kidney 5-hydroxytryptophan decarboxylase in the absence of PLP; addition of PLP reverses inhibition.[48] Hog kidney L-dopa decarboxylase is inactivated by L-dopa in the absence of PLP; L-dopa is converted into 3,4-dihydroxyphenylacetaldehyde.[49] Addition of PLP results in return of only 30 to 50% of the activity as a result of instability of the apoenzyme. Similar reactions occur with *m*-tyrosine; the corresponding aldehyde was isolated after inactivation.[50] The same reaction occurs with glutamic acid by a bacterial glutamate decarboxylase and with ornithine by bacterial ornithine decarboxylase.[50] With ornithine, a small amount of 4-aminobutanal is produced, resulting from decarboxylation and transamination;[51] transamination generally occurs in a few hundredths of a percent of the number of decarboxylations of normal substrates. The decarboxylation-dependent transamination is pH independent.

2. α-Methyl-Substituted Amino Acids

α-Methyldopa[49,50,52] and α-methyl-*m*-tyrosine[50] inactivate pig kidney dopa decarboxylase in the absence of PLP; the corresponding ketone was isolated. L-Aromatic amino acid decarboxylase form guinea pig kidney undergoes time-dependent inactivation by α-methyldopa and α-methyl-5-hydroxytryptophan in the absence of PLP; the addition of PLP reverses inhibition.[53] α-Methylornithine inactivates bacterial (*Lactobacillus* 30A) ornithine decarboxylase[50,51] in the absence of PLP; inactivation results from the formation of PMP and 5-amino-2-pentanone.[51] α-Methylarginine inactivates bacterial arginine decarboxylase.[50]

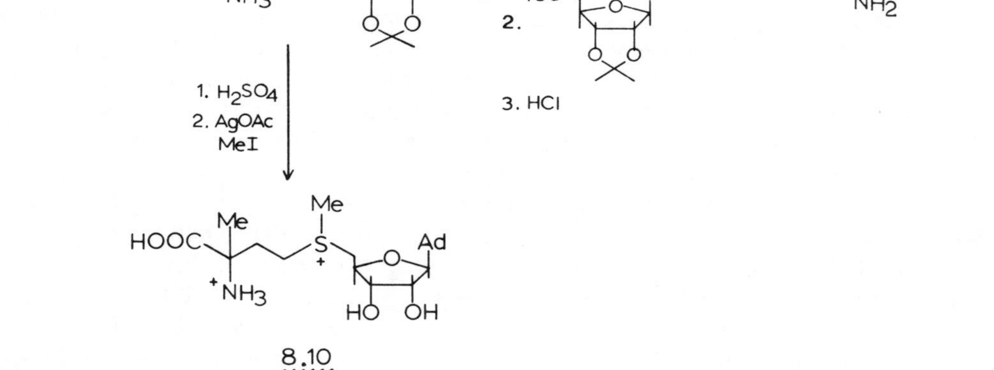

8.10

Scheme 22.   Containing Structural Formula 8.10.

Transamination generally occurs in a few percent of decarboxylations of α-methyl-substituted substrates.[50] α-Methylaspartate inactivates aspartate β-decarboxylase from *Alcaligenes faecalis* in the absence of PLP and an α-ketoacid.[45] *E. coli* glutamate decarboxylase is inactivated by α-methyl-D,L-glutamate with release of PMP.[47]

S-Adenosyl-L-methionine decarboxylase from rat liver, an enzyme with an essential pyruvyl residue as the prosthetic group, is inactivated by S-(5'-deoxy-5'-adenosyl)-(±)-2-methylmethionine dihydrogen sulfate (α-methyl S-adenosylmethionine; **8.10**); inactivation by this compound is not reversed by dialysis.[54] It is believed that decarboxylation-transamination occurs. The synthesis of **8.10** is shown in Scheme 22.

B. Decarboxylation/Isomerization/Addition

The inactivators discussed in this section are, α-acetylenic-, α-allenic-, and α-olefinic α-amino acids. A general synthesis of α-acetylenic α-amino acids was devised by Casara and Metcalf[55] with the use of trimethylsilylacetylene-*N*-carbethoxyglycinate dianion (Scheme 23). α-Allenic α-amino acids can be prepared by the method of Castelhano et al.[56] (Scheme 24) or by Casara et al.[57] (Scheme 25).

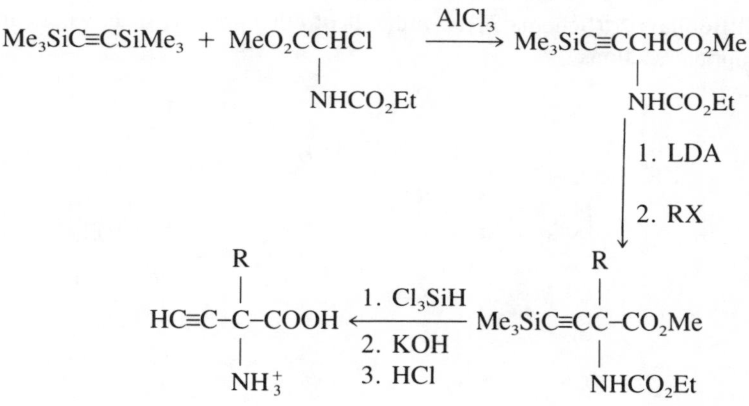

$$Me_3SiC{\equiv}CSiMe_3 + MeO_2CCHCl \xrightarrow{AlCl_3} Me_3SiC{\equiv}CCHCO_2Me$$

(with NHCO₂Et substituent)

1. LDA
2. RX

$$HC{\equiv}C{-}C({-}R)({-}NH_3^+){-}COOH \xleftarrow[\text{2. KOH}]{\text{1. Cl}_3\text{SiH} \quad \text{3. HCl}} Me_3SiC{\equiv}CC({-}R)({-}NHCO_2Et){-}CO_2Me$$

Scheme 23.

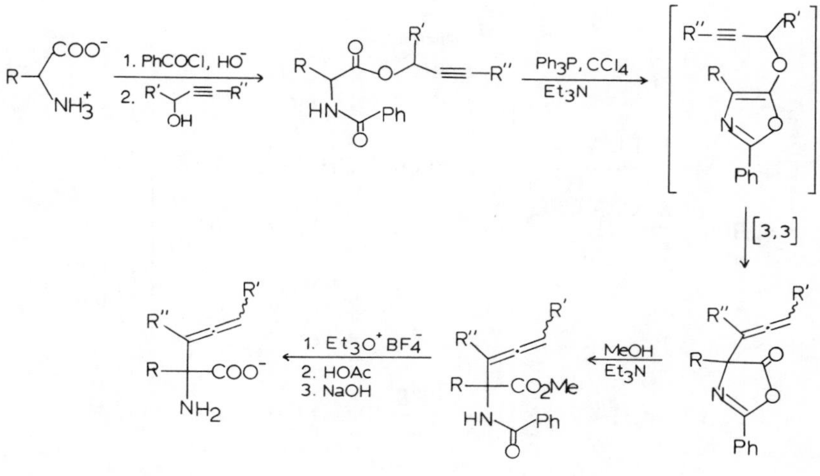

Scheme 24.

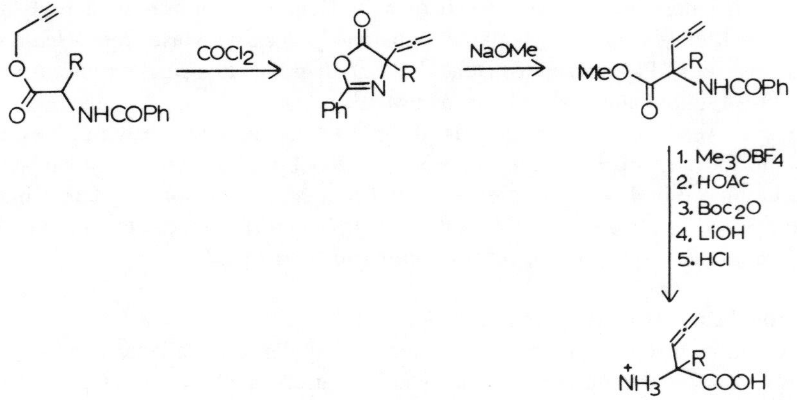

Scheme 25.

α-Olefinic α-amino acids were synthesized by Greenlee et al.[58] (Scheme 26) and by Metcalf and Bonilavri[59] (Scheme 27). Syntheses of other specific inactivators are described in the appropriate sections.

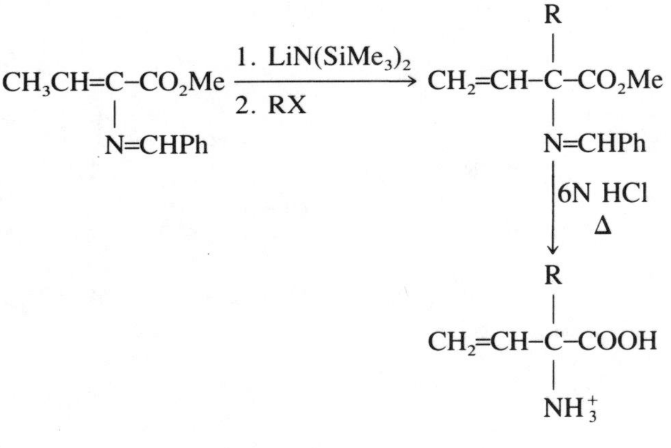

Scheme 26.

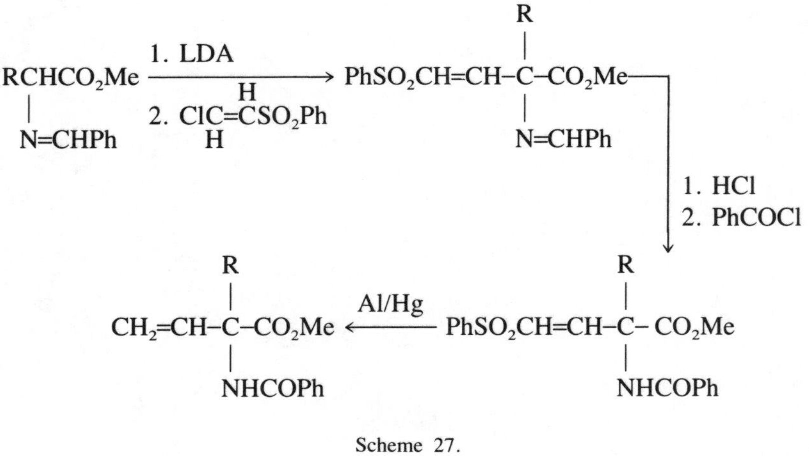

Scheme 27.

The typical mechanism for inactivation of PLP-dependent decarboxylases by α-acetylenic amino acids is shown in Scheme 28. Intermediate **8.11** is the same as that proposed for the inactivation of aminotransferases by acetylenic amines; one pathway, referred to as the enamine mechanism of inactivation[10,11] (see Chapter 7, Section III.A.3., Scheme 28), may be important in decarboxylase inactivation mechanisms as well (Scheme 28, pathway c). α-Allenic amino acids may inactivate these enzymes by the mechanisms shown in Scheme 29. Pathways a and b were proposed by Casara et al.[57] Intermediate **8.12**, however, also is susceptible to inactivation by an enamine mechanism[10,11] (see Chapter 7, Section III.B.1., Scheme 44). The corresponding α-olefinic amino acids may inactivate decarboxylases by the mechanisms shown in Scheme 30. Intermediate **8.13** is the same as that suggested as an intermediate in the inactivation of aminotransferases by olefinic amines and could proceed by the enamine mechanism[10,11] (see Chapter 7, Section III.C.1., Scheme 47).

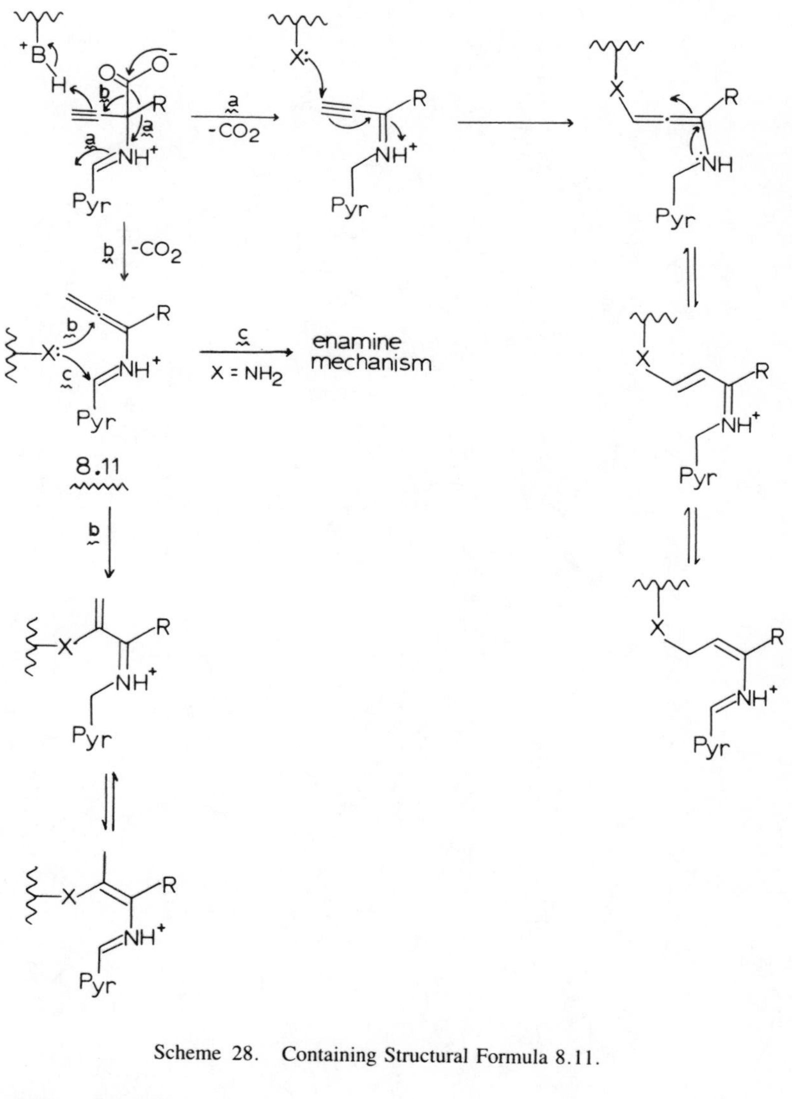

Scheme 28. Containing Structural Formula 8.11.

1. Acetylene-Containing Inactivators
See Scheme 28 for potential inactivation mechanisms.

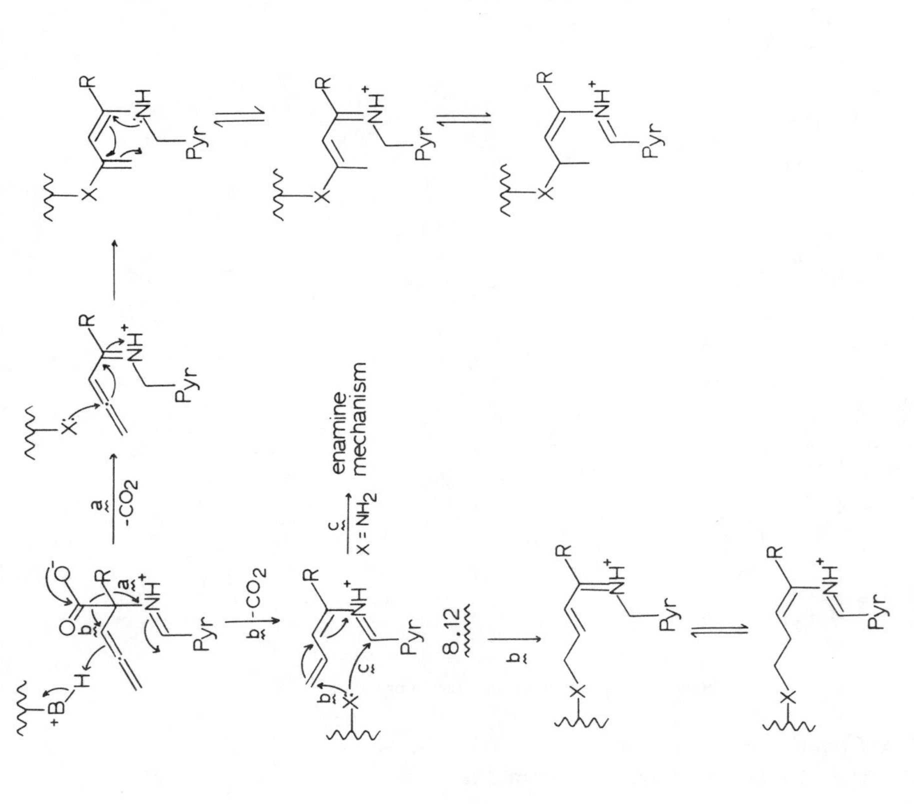

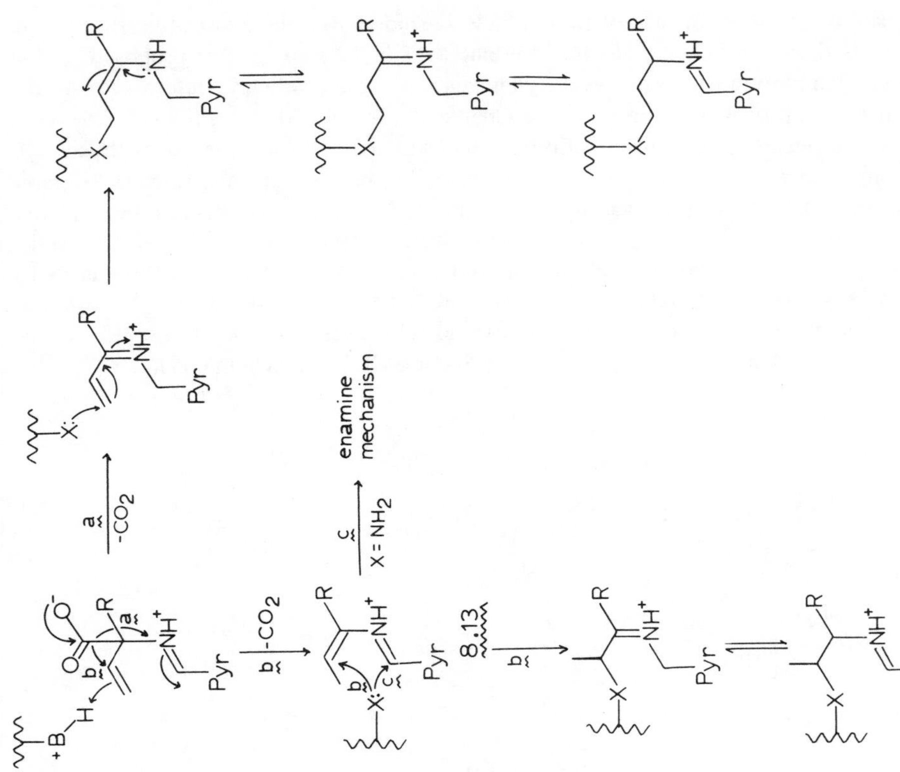

Scheme 30.   Containing Structural Formula 8.13.

Scheme 29.   Containing Structural Formula 8.12.

a. α-Acetylenic (Ethynyl) Dopa

α-Acetylenic dopa (Structural Formula 8.14, Scheme 31) produces weak nonpseudo first-order, time-dependent inactivation of hog kidney L-amino acid decarboxylase.[60] This flattening of the inactivation curves generally indicates either a consumption of the inactivator or product inhibition. The concentration of α-acetylenic dopa, however, did not change substantially (measured by HPLC) nor was a metabolite detected in the incubation medium. Dialysis does result in partial return of enzyme activity.

Compound **8.14** has been synthesized in a similar fashion by Taub and Patchett[61] (Scheme 31) and by Metcalf and Jund[62] (Scheme 32).

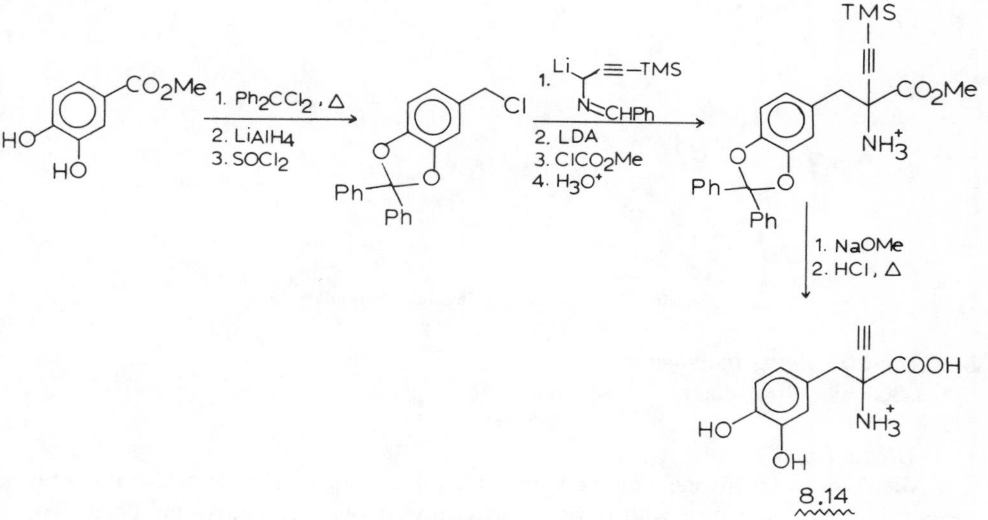

Scheme 31.   Containing Structural Formula 8.14.

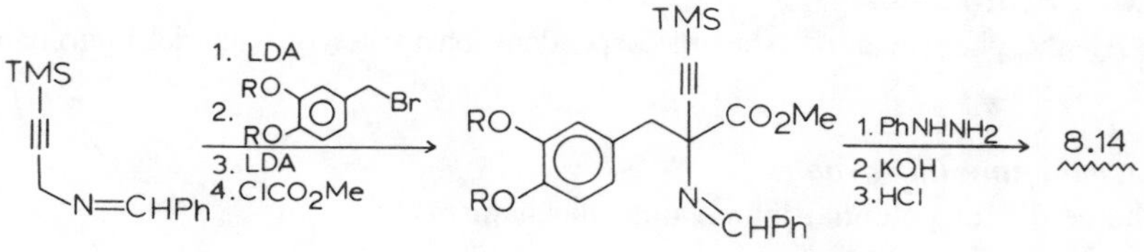

Scheme 32.

b. α-Ethynylornithine

α-Ethynylornithine (Structural Formula 8.15) is a time-dependent inactivator of the crude liver ornithine decarboxylase from thioacetamide-treated rats; about 15% of the enzyme activity returns after dialysis.[63] The synthesis of **8.15** is shown in Scheme 33.

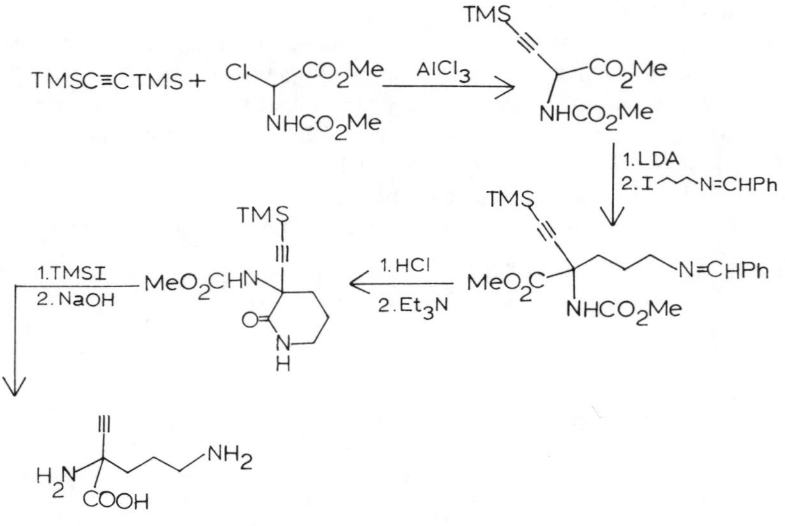

Scheme 33.   Containing Structural Formula 8.15.

2. Allene-Containing Inactivators

See Scheme 29 for potential inactivation mechanisms.

a. α-Allenic Dopa

α-Allenic dopa inactivates porcine kidney L-aromatic amino acid decarboxylase.[56] Inactivation is more rapid than with α-vinyl- or α-ethynyldopa and, unlike the latter two, α-allenic dopa is irreversible to dialysis. One diastereometric pair is at least an order of magnitude more potent than the other against mammalian aromatic amino acid decarboxylase.

b. α-Allenic Phenylalanine

(R,S)-α-Allenic phenylalanine is a time-dependent inactivator of bacterial L-aromatic amino acid decarboxylase.[57]

3. Olefin-Containing Inactivators

See Scheme 30 for potential inactivation mechanisms.

a. α-Vinyldopa

α-Vinyldopa (Structural Formula 8.16, Scheme 34) is a weak time-dependent, nonpseudo first-order inactivator of hog kidney L-amino acid decarboxylase; dialysis results in partial return of enzyme activity.[60]

Taub and Patchett[61] (Scheme 34) and Metcalf and Jund[62] (Scheme 35) synthesized **8.16** from protected α-ethynyldopa (see Section III.B.1.a.).

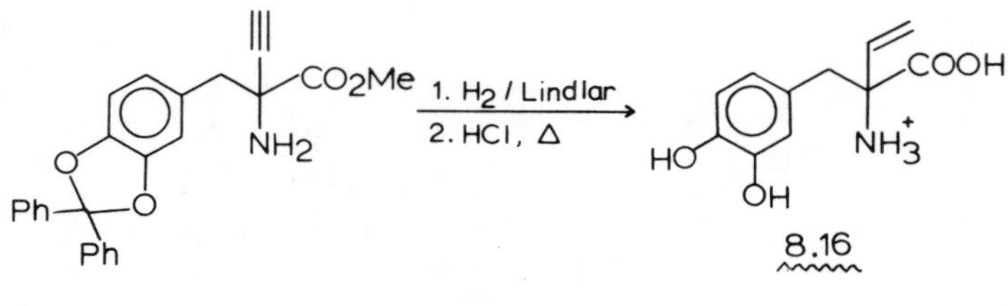

Scheme 34.   Containing Structural Formula 8.16.

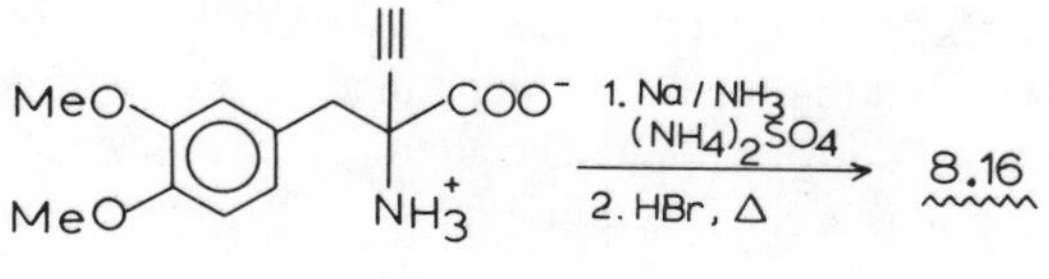

$$\text{MeO} \quad \overset{|||}{\underset{\overset{+}{N}H_3}{C}}\text{COO}^- \quad \xrightarrow[\text{2. HBr, }\Delta]{\begin{array}{c}\text{1. Na / NH}_3\\ \text{(NH}_4)_2\text{SO}_4\end{array}} \quad \text{8.16}$$

Scheme 35.

b. α-Vinylornithine

α-Vinylornithine (Structural Formula 8.17) is a time-dependent inactivator of ornithine decarboxylase from liver of thioacetamide-treated rats; about 15% of the enzyme activity returns after dialysis.[63] The synthesis[63] of **8.17** is shown in Scheme 36.

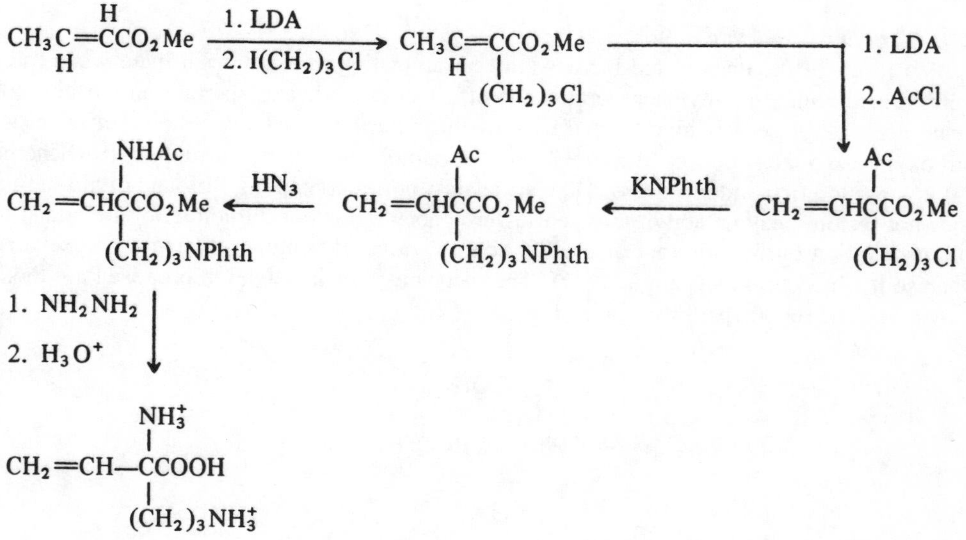

$$\underset{H}{\overset{H}{CH_3C}}=CCO_2Me \xrightarrow[\text{2. I(CH}_2)_3Cl]{\text{1. LDA}} \underset{H}{\overset{}{CH_3C}}=\underset{(CH_2)_3Cl}{\overset{|}{C}}CO_2Me \xrightarrow{\qquad} \begin{array}{c}\text{1. LDA}\\ \\ \text{2. AcCl}\end{array}$$

$$\underset{(CH_2)_3NPhth}{\overset{NHAc}{CH_2}=CHCCO_2Me} \xleftarrow{HN_3} \underset{(CH_2)_3NPhth}{\overset{Ac}{CH_2}=CHCCO_2Me} \xleftarrow{KNPhth} \underset{(CH_2)_3Cl}{\overset{Ac}{CH_2}=CHCCO_2Me}$$

$$\begin{array}{c}\text{1. NH}_2\text{NH}_2\\ \\ \text{2. H}_3\text{O}^+\end{array} \Bigg\downarrow$$

$$\underset{(\overset{}{CH_2})_3NH_3^+}{\overset{NH_3^+}{CH_2}=CH-CCOOH}$$

8.17

Scheme 36. Containing Structural Formula 8.17.

c. 2-(R,S)-2-Methyl-(E)-3,4-didehydroglutamic Acid

$$\underset{\overset{|}{H}}{\overset{H}{HOOCC}}=C-\underset{\overset{|}{NH_3^+}}{\overset{\overset{CH_3}{|}}{C}}-COOH$$

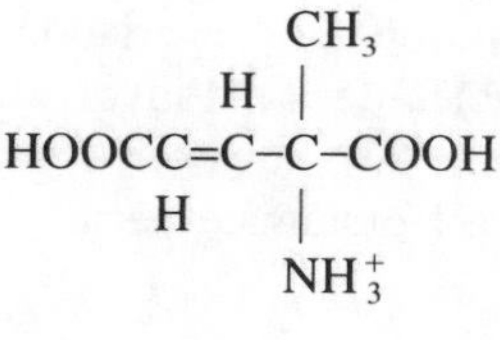

Scheme 37. Structural Formula 8.18.

(2RS,3E)-2-Methyl-3,4-didehydroglutamic acid (Structural Formula 8.18, Scheme 37) was shown by Chrystal et al.[64] to be a time-dependent irreversible inhibitor of L-glutamate-1-decarboxylase from chick embryo brain; dialysis does not cause reactivation. Reconstitution with PLP of apoenzyme treated with inactivator results in enzyme with complete activity, thus indicating the requirement of PLP in the inactivation process. Two mechanisms suggested[64] are those shown in Scheme 30 (pathways a and b). A stereospecific synthesis of the *E*-

isomer of 2-methyl-3,4-didehydroglutamic acid was reported by Bey and Vevert.[65] The alkylation reaction was shown to proceed with complete retention of configuration (Scheme 38; Y = CO_2H, CO_2Me, CO_2SiMe_3, CN). The attempted synthesis of the Z-isomer resulted in only cyclization to the unsaturated lactam.

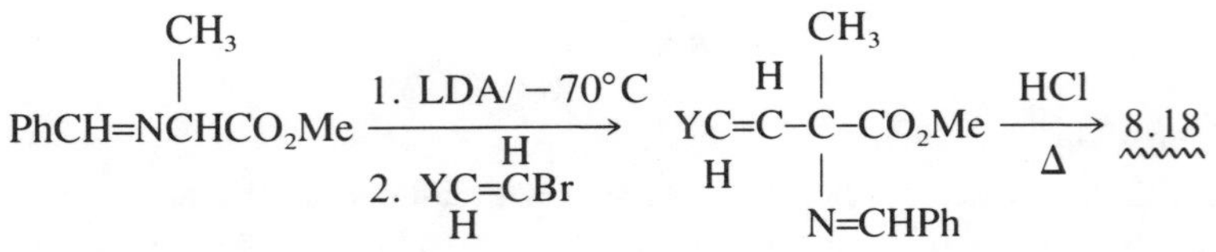

$$
\begin{array}{c}
CH_3 \\
| \\
PhCH{=}NCHCO_2Me
\end{array}
\quad
\xrightarrow[\substack{\text{H} \\ \text{2. YC=CBr} \\ \text{H}}]{\text{1. LDA/}-70°C}
\quad
\begin{array}{c}
CH_3 \\
\text{H} \quad | \\
YC{=}C{-}C{-}CO_2Me \\
\text{H} \quad | \\
N{=}CHPh
\end{array}
\quad
\xrightarrow[\Delta]{HCl}
\quad 8.18 \;\sim\!\sim\!\sim
$$

Scheme 38.

d. β-Methyleneaspartate

As a result of the unexpected increase in transamination of cysteinesulfinate when mice were treated with β-methyleneaspartate, the effect of β-methyeneaspartate on cysteinesulfinate decarboxylase was investigated by Griffith.[66] Purified rat liver cysteinesulfinate decarboxylase is inactivated by β-methylene-DL-aspartate (Structural Formula 8.19, Scheme 39) in a pseudo first-order process. Dialysis against buffer containing PLP and dithiothreitol does not restore enzyme activity. When lower concentrations of dithiothreitol are added to the inactivation buffer, the rate of inactivation increases (dithiothreitol cannot be excluded because it is needed to sustain activity). The mechanism of inactivation proposed are those shown in Scheme 30 (pathways a and b).

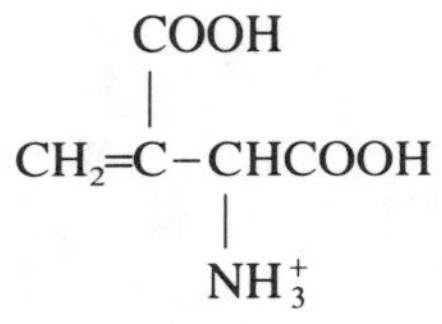

$$
\begin{array}{c}
COOH \\
| \\
CH_2{=}C{-}CHCOOH \\
| \\
NH_3^+
\end{array}
$$

Scheme 39.   Structural Formula 8.19.

C. Decarboxylation/Isomerization/Isomerization/Addition

1. Allyglycine

Guinea pig brain glutamate decarboxylase undergoes time-dependent inactivation by allylglycine (Structural Formula 8.20); dialysis only partially reactivates the enzyme.[67] No mechanism was given, but, since assays are done in the presence of PLP, the mechanism could be that shown in Scheme 40 (or one of the variations from **8.21** as described in Scheme 30). An alternative mechanism, which does not involve decarboxylation at all, is shown in Chapter 7, Section VII.B.2. (Scheme 93); however, if that is the case, then **8.20** would not be a mechanism-based inactivator of glutamate decarboxylase, because it does not utilize the normal catalytic mechanism.

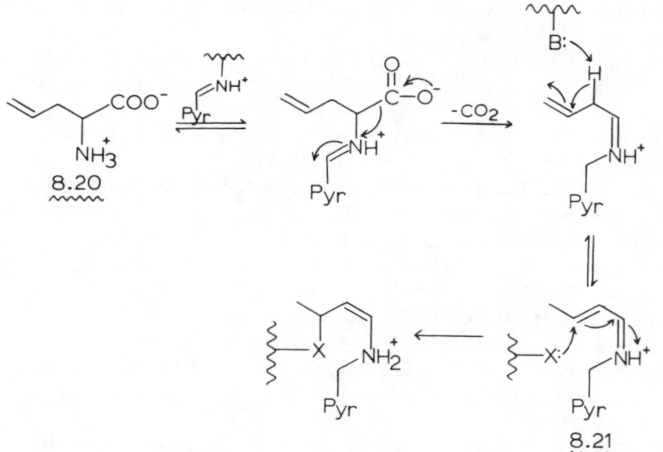

Scheme 40.   Containing Structural Formulas 8.20 and 8.21.

REFERENCES

1. **Bey, P. and Vevert, J. P.**, New approach to the synthesis of α-halogenomethyl-α-amino acids, *Tetrahedron Lett.*, p. 1215, 1978.
2. **Bey, P., Vevert, J. -P., Van Dorsselaer, V., and Kolb, M.**, Direct synthesis of α-halogenomethyl-α-amino acids from the parent α-amino acid, *J. Org. Chem.*, 44, 2732, 1979.
3. **Bey, P. and Schirlin, D.**, General approach to the synthesis of α-difluoromethyl amines as potential enzyme-activated irreversible inhibitors, *Tetrahedron Lett.*, p. 5225, 1978.
4. **Bey, P., Ducep, J. B., and Schirlin, D.**, Alkylation of malonates or Schiff base anions with dichloro-fluoromethane as a route to α-chlorofluoromethyl or α-fluoromethyl α-amino acids, *Tetrahedron Lett.*, 25, 5657, 1984.
5. **Kollonitsch, J., Marburg, S., and Perkins, L. M.**, Selective fluorination of hydroxy amines and hydroxy amino acids with sulfur tetrafluoride in liquid hydrogen fluoride, *J. Org. Chem.*, 40, 3808, 1975.
6. **Kollonitsch, J., Marburg, S., and Perkins, L. M.**, Fluorodehydroxylation, a novel method for synthesis of fluoroamines and fluoroamino acids, *J. Org. Chem.*, 44, 771, 1979.
7. **Kollonitsch, J., Marburg, S., and Perkins, L. M.**, Fluorodesulfurization. A new reaction for the formation of carbon-fluorine bonds, *J. Org. Chem.*, 41, 3107, 1976.
8. **Kollonitsch, J., and Barash, L.**, Organofluorine synthesis via photofluorination: 3-fluoro-D-alanine and 2-deuterio analogue, antibacterials related to the bacterial cell wall, *J. Am. Chem. Soc.*, 98, 5591, 1976.
9. **Kollonitsch, J., Patchett, A. A., Marburg, S., Maycock, A. L., Perkins, L. M., Doldouras, G. A., Duggan, D. E., and Aster, S. D.**, Selective inhibitors of biosynthesis of aminergic neurotransmitters, *Nature (London)*, 274, 906, 1978.
10. **Likos, J. J., Ueno, H., Feldhaus, R. W., and Metzler, D. E.**, A novel reaction of the coenzyme of glutamate decarboxylase with L-serine O-sulfate, *Biochemistry*, 21, 4377, 1982.
11. **Ueno, H., Likos, J. J., and Metzler, D. E.**, Chemistry of the inactivation of cytosolic aspartate aminotransferase by serine O-sulfate, *Biochemistry*, 21, 4387, 1982.
12. **Jung, M. J., Palfreyman, M. G., Wagner, J., Bey, P., Ribereau-Gayon, G., Zraïka, M., and Koch-Weser, J.**, Inhibition of monoamine synthesis by irreversible blockage of aromatic amino acid decarboxylase with α-monofluoromethyldopa, *Life Sci.*, 24, 1037, 1979.
13. **Jung, M. J., Hornsperger, J.-M., Gerhart, F., and Wagner, J.**, Inhibition of aromatic amino acid decarboxylase and depletion of biogenic amines in brain of rats treated with α-monofluoromethyl p-tyrosine: similitudes and differences with the effects of α-monofluoromethyldopa, *Biochem. Pharmacol.*, 33, 327, 1984.
14. **Maycock, A. L., Aster, S. D., and Patchett, A. A.**, Inactivation of 3-(3,4-dihydroxyphenyl)alanine decarboxylase by 2-(fluoromethyl)-3-(3,4-dihydroxyphenyl)alanine, *Biochemistry*, 19, 709, 1980.
15. **Maneckjee, R. and Baylin, S. B.**, Use of radiolabeled monofluoromethyl-dopa to define the subunit structure of human L-dopa decarboxylase, *Biochemistry*, 22, 6058, 1983.

15a. **Kuo, D. and Rando, R. R.,** Irreversible inhibition of glutamate decarboxylase by α-(fluoromethyl)glutamic acid, *Biochemistry,* 20, 506, 1981.

16. **Garbarg, M., Barbin, G., Rodergas, E., and Schwartz, J. C.,** Inhibition of histamine synthesis in brain by α-fluoromethylhistidine, a new irreversible inhibitor: *in vitro* and *in vivo* studies, *J. Neurochem.,* 35, 1045, 1980.

17. **Bouclier, M., Jung, M. J., and Gerhart, F.,** α-Fluoromethyl histidine. Inhibition of histidine decarboxylase in pylorus ligated rat, *Biochem. Pharmacol.,* 32, 1553, 1983.

18. **Watanabe, T., Yamada, M., Taguchi, Y., Kubota, H., Maeyama, K., Yamatodani, A., Fukui, H., Shiosaka, S., Tohyama, M., and Wada, H.,** Purification and properties of histidine decarboxylase isozymes and their pharmacological significance, *Adv. Biosci.,* 33, 93, 1982.

19. **Duggan, D. E., Hooke, K. F., and Maycock, A. L.,** Inhibition of histamine synthesis *in vitro* and *in vivo* S-α-fluoromethylhistidine, *Biochem. Pharmacol.,* 33, 4003, 1984.

20. **Kubota, H., Hayashi, H., Watanabe, T., Taguchi, Y., and Wada, H.,** Mechanism of inactivation of mammalian L-histidine decarboxylase by (S)-α-fluoromethylhistidine, *Biochem. Pharmacol.,* 33, 983, 1984.

21. **Taguchi, Y., Watanabe, T., Kubota, H., Hayashi, H., and Wada, H.,** Purification of histidine decarboxylase from the liver of fetal rats and its immunochemical and immunohistochemical characterization, *J. Biol. Chem.,* 259, 5214, 1984.

21a. **Tung, A. S., Blake, J. T., Roman, I. J., Vlasses, P. H., Ferguson, R. K., and Zweerink, H. J.,** *In vivo* and *in vitro* inhibition of human histidine decarboxylase by (S)-α-fluoromethylhistidine, *Biochem. Pharmacol.,* 34, 3509, 1985.

21b. **Tanase, S., Guirard, B. M., and Snell, E. E.,** Purification and properties of a pyridoxal 5'-phosphate-dependent histidine decarboxylase from *Morganella morganii* AM-15, *J. Biol. Chem.,* 260, 6738, 1985.

21c. **Hayashi, H., Tanase, S., and Snell, E. E.,** Pyridoxal 5'-phosphate-dependent histidine decarboxylase. Inactivation by α-fluoromethylhistidine and comparative sequences at the inhibitor- and coenzyme-binding sites, *J. Biol. Chem.,* 261, 11003, 1986.

21d. **Vaaler, G. L., Brasch, M. A., and Snell, E. E.,** Pyridoxal 5'-phosphate dependent histidine decarboxylase. Nucleotide sequence of the *hdc* gene and the corresponding amino acid sequence, *J. Biol. Chem.,* 261, 11010, 1986.

22. **Bitonti, A. J., McCann, P. P. and Sjoerdsma, A.,** Restriction of bacterial growth by inhibition of polyamine biosynthesis by using monofluoromethylornithine, difluoromethylarginine and dicyclohexylammonium sulphate, *Biochem. J.,* 208, 435, 1982.

23. **Bitonti, A. J., Bacchi, C. J., McCann, P. P., and Sjoerdsma, A.,** Catalytic irreversible inhibition of *Trypanosoma brucei brucei* ornithine decarboxylase by substrate and product analogs and their effects on murine trypanosomiasis, *Biochem. Pharmacol.,* 34, 1773, 1985.

24. **Bey, P., Gerhart, F., Van Dorsselaer, V., and Danzin, C.,** α-(Fluoromethyl)dehydroornithine and α-(fluoromethyl)dehydroputrescine analogues as irreversible inhibitors of ornithine decarboxylase, *J. Med. Chem.,* 26, 1551, 1983.

24a. **Bitonti, A. J., Casara, P. J., McCann, P. P., and Bey, P.,** Catalytic irreversible inhibition of bacterial and plant arginine decarboxylase activities by novel substrate and product analogues, *Biochem. J.,* 242, 69, 1987.

25. **Kallio, A., McCann, P. P., and Bey, P.,** DL-α-(Difluoromethyl)arginine: a potent enzyme-activated irreversible inhibitor of bacterial arginine decarboxylases, *Biochemistry,* 20, 3163, 1981.

26. **Palfreyman, M. G., Danzin, C., Bey, P., Jung, M. J., Ribereau-Gayon, G., Aubry, M., Vevert, J. P., and Sjoerdsma, A.,** α-Difluoromethyl dopa, a new enzyme-activated irreversible inhibitor of aromatic L-amino acid decarboxylase, *J. Neurochem.,* 31, 927, 1978.

27. **Ribéreau-Gayon, G., Palfreyman, M. G., Zraïka, M., Wagner, J., and Jung, M. J.,** Irreversible inhibition of aromatic L-amino acid decarboxylase by α-difluoromethyl dopa and metabolism of the inhibitor, *Biochem. Pharmacol.,* 29, 2465, 1980.

28. **Tsushima, T., Kawada, K., Shiratori, O., and Uchida, N.,** Fluorine-containing amino acids and their derivatives. V. Synthesis of novel fluorinated analogues of the antitumor agent methotrexate, *Heterocycles,* 23, 45, 1985.

29. **Pösö, H., McCann, P. P., Tanskanen, R., Bey, P., and Sjoerdsma, A.,** Inhbition of growth of *Mycoplasma dispar* by DL-α-difluoromethyllysine, a selective irreversible inhibitor of lysine decarboxylase, and reversal by cadaverine (1,5-diaminopentane), *Biochem. Biophys. Res. Commun.,* 125, 205, 1984.

30. **Metcalf, B. W., Bey, P., Danzin, C., Jung, M. J., Casara, P., and Vevert, J. P.,** Catalytic irreversible inhibition of mammalian ornithine decarboxylase (E.C. 4.1.1.17) by substrate and product analogues, *J. Am. Chem. Soc.,* 100, 2551, 1978.

31. **Pritchard, E. L., Seely, J. E., Pösö, H., Jefferson, L. S., and Pegg, A. E.,** Binding of radioactive α-difluoromethylornithine to rat liver ornithine decarboxylase, *Biochem. Biophys. Res. Commun.,* 100, 1597, 1981.

31a. **Pegg, A. E., McGovern, K. A., and Wiest, L.,** Decarboxylation of α-difluoromethylornithine by ornithine decarboxylase, *Biochem. J.,* 241, 305, 1987.

32. **Seely, J. E., Pösö, H., and Pegg, A. E.,** Measurement of the number of ornithine decarboxylase molecules in rat and mouse tissues under various physiological conditions by binding of radiolabelled α-difluoromethylornithine, *Biochem. J.,* 206, 311, 1982.

32a. **Erwin, B. G., Seely, J. E., and Pegg, A. E.,** Mechanism of stimulation of ornithine decarboxylase activity in transformed mouse fibroblasts, *Biochemistry,* 22, 3027, 1983.

33. **Pegg, A. and McGill, S.,** Decarboxylation of ornithine and lysine in rat tissues, *Biochim. Biophys. Acta,* 568, 416, 1979.

34. **Seely, J. E., Pösö, H., and Pegg, A.,** Purification of ornithine decarboxylase from kidneys of androgen-treated mice, *Biochemistry,* 21, 3394, 1982.

35. **Kallio, A. and McCann, P. P.,** Difluoromethylornithine irreversibly inactivates ornithine decarboxylase of *Pseudomonas aeruginosa,* but does not inhibit the enzymes of *Escherichia coli, Biochem. J.,* 200, 69, 1981.

36. **Kallio, A., McCann, P. P., and Bey, P.,** DL-α-Monofluoromethylputrescine is a potent irreversible inhibitor of *Escherichia coli* ornithine decarboxylase, *Biochem. J.,* 204, 771, 1982.

37. **Mitchell, J. L. A., Yingling, R. A., and Mitchell, G. K.,** Selective inactivation of *Physarum* A-form ornithine decarboxylase by DL-difluoromethylornithine, *FEBS Lett.,* 131, 305, 1981.

38. **Paulin, L. and Pösö, H.,** Ornithine decarboxylase activity from an extremely thermophilic bacterium, *Clostridium thermohydrosulfuricum, Biochim. Biophys. Acta,* 742, 197, 1983.

39. **Heller, J. S., Rostomily, R., Kyriakidis, D. A., and Canellakis, E. S.,** Regulation of polyamine biosynthesis in *Escherichia coli* by basic proteins, *Proc. Natl. Acad. Sci. U.S.A.,* 80, 5181, 1983.

40. **Kyriakidis, D. A., Flamigni, F., Pawlak, J. W., and Canellakis, E. S.,** Mode of interaction of ornithine decarboxylase with antizyme and α-difluoromethylornithine, *Biochem. Pharmacol.,* 33, 1575, 1984.

40a. **Kitani, T. and Fjuisawa, H.,** α-Difluoromethylornithine does not bind to ornithine decarboxylase-antizyme complex, *Biochem. Biophys. Res. Commun.,* 137, 1101, 1986.

41. **Keller, J. W. and O'Leary, M. H.,** 3,3-3-Trifluoro-2-aminoisobutyrate: a mechanism-based inhibitor of *Pseudomonas cepacia* α-dialkylamino acid transaminase, *Biochem. Biophys. Res. Commun.,* 90, 1104, 1979.

42. **Lippert, B., Bey, P., Van Dorsselaer, V., Vevert, J. P., Danzin, C., Ribereau-Gayon, G., and Jung, M. J.,** Selective irreversible inhibition of mammalian histidine decarboxylase by α-chloromethyl histidine, *Agents and Actions,* 9, 38, 1979.

43. **Bailey, G. B. and Dempsey, W. B.,** Purification and properties of an α-dialkyl amino acid transaminase, *Biochemistry,* 6, 1526, 1967.

44. **Kalyankar, G. D. and Snell, E. E.,** Pyridoxal-catalyzed decarboxylation of amino acids, *Biochemistry,* 1, 594, 1962.

45. **Novogrodsky, A., Nishimura, J. S., and Meister, A.,** Transamination and β-decarboxylation of aspartate catalyzed by the same pyridoxal phosphate-enzyme, *J. Biol. Chem.,* 238, PC1903, 1963.

46. **Novogrodsky, A. and Meister, A.,** Control of aspartate β-decarboxylase activity by transamination, *J. Biol. Chem.,* 239, 879, 1964.

47. **Sukhareva, V. S. and Braunshtein, A. E.,** Investigation of the nature of the interactions of glutamate decarboxylase from *Escherichia coli* with the substrate and its analogs, *Mol. Biol. (U.S.S.R.),* 5, 241, 1971.

47a. **Miller, L. P., Martin, D. L., Mazumder, A., and Walters, J. R.,** Studies on the regulation of GABA synthesis: substrate-promoted dissociation of pyridoxal-5′-phosphate from GAD, *J. Neurochem.,* 30, 361, 1978.

47b. **Meeley, M. P. and Martin, D. L.,** Inactivation of brain glutamate decarboxylase and the effects of adenosine 5′-triphosphate and inorganic phosphate, *Cell. Mol. Neurobiol.,* 3, 39, 1983.

47c. **Meeley, M. P. and Martin, D. L.,** Reactivation of substrate-inactivated brain glutamate decarboxylase, *Cell. Mol. Neurobiol.,* 3, 55, 1983.

47d. **Spink, D. C., Porter, T. G., Wu, S. J., and Martin, D. L.,** Characterization of three kinetically distinct forms of glutamate decarboxylase from pig brain, *Biochem. J.,* 231, 695, 1985.

47e. **Porter T. G., Spink., D. C., Martin, S. B., and Martin, D. L.,** Transaminations catalysed by brain glutamate decarboxylation, *Biochem. J.,* 231, 705, 1985.

47f. **Porter, T. G. and Martin, D. L.,** Rapid inactivation of brain glutamate decarboxylase by aspartate, *J. Neurochem.,* 48, 67, 1987.

48. **Yuwiler, A., Geller, E., and Eiduson, S.,** Studies of 5-hydroxytryptophan decarboxylase. I. *In vitro* inhibition and substrate interaction, *Arch. Biochem. Biophys.,* 80, 162, 1959.

49. **O'Leary, M. H. and Baughn, R. L.,** New pathway for metabolism of dopa, *Nature (London),* 253, 52, 1975.

50. **O'Leary, M. H. and Baughn, R. L.,** Decarboxylation-dependent transamination catalyzed by mammalian 3,4-dihydroxyphenylalanine decarboxylase, *J. Biol. Chem.,* 252, 7168, 1977.

51. **O'Leary, M. H. and Herreid, R. M.,** Mechanism of inactivation of ornithine decarboxylase by α-methylornithine, *Biochemistry,* 17, 1010, 1978.

52. **Borri Voltattorni, C., Minelli, A., and Turano, C.,** Spectral properties of the coenzyme bound to dopa decarboxylase from pig kidney, *FEBS Lett.,* 17, 231, 1971.
53. **Lovenberg, W., Barchas, J., Weissbach, H., and Udenfriend, S.,** Characteristics of the inhibition of aromatic L-amino acid decarboxylase by α-methylamino acids, *Arch. Biochem. Biophys.,* 103, 9, 1963.
54. **Pankaskie, M. and Abdel-Monem, M. M.,** Inhibitors of polyamine biosynthesis. 8. Irreversible inhibition of mammalian *S*-adenosyl-L-methionine decarboxylase by substrate analogues, *J. Med. Chem.,* 23, 121, 1980.
55. **Casara, P. and Metcalf, B. W.,** Trimethylsilylacetylene-*N*-carboethoxy glycinate dianion. A general synthon for α-acetylenic α-amino acids, *Tetrahedron Lett.,* p. 1581, 1978.
56. **Castelhano, A. L., Pliura, D. H., Taylor, G. J., Hsieh, K. C., and Krantz, A.,** Allenic suicide substrates. New inhibitors of vitamin B_6 linked decarboxylase, *J. Am. Chem. Soc.,* 106, 2734, 1984.
57. **Casara, P., Jund, K., and Bey, P.,** General synthetic access to α-allenyl amines and α-allenyl-α-amino acids as potential enzyme activated irreversible inhibitors of PLP dependent enzymes, *Tetrahedron Lett.,* 25, 1891, 1984.
58. **Greenlee, W. J., Taub, D., and Patchett, A. A.,** A general synthesis of α-vinyl-α-amino acids, *Tetrahedron Lett.,* p. 3999, 1978.
59. **Metcalf, B. W. and Bonilavri, E.,** Phenyl *trans*-2-chlorovinyl sulphone, a vinyl cation equivalent, *J. Chem. Soc. Chem. Commun.,* 914, 1978.
60. **Ribéreau-Gayon, G., Danzin, C., Palfreyman, M. G., Aubry, M., Wagner, J., Metcalf, B. W., and Jung, M. J.,** *In vitro* and *in vivo* effects of α-acetylenic dopa and α-vinyl dopa on L-amino acid decarboxylase, *Biochem. Pharmacol.,* 28, 1331, 1979.
61. **Taub, D. and Patchett, A. A.,** Syntheses of α-ethynyl-3,4-dihydroxyphenylalanine and α-vinyl-3,4-dihydroxyphenylalanine, *Tetrahedron Lett.,* p. 2745, 1977.
62. **Metcalf, B. W. and Jund, K.,** Synthesis of β,γ-unsaturated amino acids as potential catalytic irreversible enzyme inhibitors, *Tetrahedron Lett.,* p. 3689, 1977.
63. **Danzin, C., Casara, P., Claverie, N., and Metcalf, B. W.,** α-Ethynyl and α-vinyl analogues of ornithine as enzyme-activated inhibitors of mammalian ornithine decarboxylase, *J. Med. Chem.,* 24, 16, 1981.
64. **Chrystal, E., Bey, P., and Rando, R. R.,** The irreversible inhibition of brain L-glutamate-1-decarboxylase by (2RS, 3E)-2-methyl-3,4-didehydroglutamic acid, *J. Neurochem.,* 32, 1501, 1979.
65. **Bey, P. and Vevert, J. P.,** Stereospecific alkylation of the Schiff base ester of alanine with 2-substituted-(E)- and -(Z)-vinyl bromides. An efficient synthesis of 2-methyl-(E)-3,4-didehydroglutamic acid, a potent substrate-induced irreversible inhibitor of L-glutamate-1-decarboxylase, *J. Org. Chem.,* 45, 3249, 1980.
66. **Griffith, O. W.,** Cysteinesulfinate metabolism. Altered partitioning between transamination and decarboxylation following administration of β-methyleneaspartate, *J. Biol. Chem.,* 258, 1591, 1983.
67. **Fisher, S. K. and Davies, W. E.,** Some properties of guinea pig brain glutamate decarboxylase and its inhibition by the convulsant allylglycine (2-amino-4-pentenoic acid), *J. Neurochem.,* 23, 427, 1974.

Chapter 9

OXIDATION REACTIONS*

I. INTRODUCTION

Mechanism-based enzyme inactivation that is initiated by an oxidation reaction is the most prolific type. The oxidation can occur via single- or two-electron transfer processes. Only within the last decade have radical intermediates in enzyme-catalyzed reactions become popularized. There now appears to be strong support, principally based on inactivation studies, for one-electron pathways in reactions catalyzed by monoamine oxidase,[1,2] cytochrome P-450,[3-5] dopamine β-hydroxylase,[6] horseradish peroxidase,[7] alcohol oxidase,[8] methanol oxidase,[9] ribonucleotide reductase,[10] and 5-lipoxygenase.[11] It is likely, therefore, that many other enzymes catalyzing oxidation reactions also will be shown to utilize one-electron pathways.

Since it is often difficult to differentiate a two one-electron mechanism from a single two-electron mechanism many of the examples described in this chapter do not distinguish between these initiation pathways. If evidence has been presented to support a particular mechanism, then it is specified; otherwise, a generic "oxidation" is denoted. Consequently, most of the examples in this chapter are in the Oxidation/Addition section. As a means of organizing the examples in this section, the inactivation reactions are grouped according to what bond is oxidized or activated, in the following order: C–C, C–N, C–O, C–S, C–X, and N–N. Within each of these groups the compounds are organized according to what bond is cleaved in the process, starting with C–H, then as above. For example, the oxidation of 4-pentenoyl CoA (Structural Formula 9.1) to 2,4-pentadienoyl CoA (Structural Formula 9.2) involves oxidation of a C–C bond with cleavage of C–H bonds (Scheme 1), and, therefore, is described in Section II.A.1. (Oxidation/Addition; Carbon-Carbon Bond Oxidation; With Carbon-Hydrogen Bond Cleavage).

$$CH_2=CH-CH_2CH_2COSCoA \quad CH_2=CH-CH=CHCOSCoA$$

$$\underset{\sim\sim\sim\sim}{9.1} \qquad\qquad\qquad \underset{\sim\sim\sim\sim}{9.2}$$

Scheme 1.   Structural Formulas 9.1 and 9.2.

II. OXIDATION/ADDITION

A. Carbon-Carbon Bond Oxidation
1. With Carbon-Hydrogen Bond Cleavage
a. 3-Nitropropionate
3-Nitropropionate, which is isoelectronic with succinate, is a time-dependent inactivator of succinate dehydrogenase from rat liver mitochondria; centrifugation and washing of the pellet does not result in return of enzyme activity.[12] This compound and the mechanism proposed by Alston et al.[12] are described in Chapter 4 of Volume I, Section II.B.4., Scheme 28. The same enzyme from bovine heart mitochondria also is inactivated by 3-nitropropionate; Coles et al.[13] refuted the mechanism suggested by Alston et al.[12] (see Chapter 4 of Volume I Section II.B.4., Scheme 29). Since the enzyme-catalyzed oxidation of carbon-

---

* A list of abbreviations and shorthand notations can be found prior to Chapter 7.

carbon bonds is not known definitively, this inactivator also is included here; carbon-carbon bond oxidation (by whatever mechanism) initiates the inactivation.

b. Amino Acid-Derived Hydantoins

As in the case with 3-nitropropionate, amino acid-derived hydantoins are discussed in Chapter 4 of Volume I (Section II.B.5.) on the assumption that the mechanism of carbon-carbon bond oxidation by flavoenzymes involves an initial covalent complex between the substrate and the flavin. However, these compounds, which inactivate bacterial dihydroorotate dehydrogenase,[14] also are cited here, since the oxidation mechanism is not clearly defined. The inactivation mechanism is shown in Chapter 4 of Volume I (Scheme 31).

c. 3-Arylpropenes

2-X-3-(*p*-Hydroxyphenyl)-1-propenes (X = H, Br, Cl) (Structural Formulas 9.3, Scheme 2) are mechanism-based inactivators of the copper-containing enzyme dopamine β-hydroxylase from bovine adrenal medulla; no reactivation occurs after dialysis.[15] These compounds also are substrates, being oxidized to 2-X-3-hydroxy-3-(*p*-hydroxyphenyl)-1-propenes; partition ratios are in the range of 40 to 100. The syntheses of **9.3** are shown in Scheme 2. A more detailed study of inactivation of dopamine β-hydroxylase by 2-bromo-3-(*p*-hydroxyphenyl)-1-propene (**9.3**, X = Br) was carried out by Colombo et al.[16] Inactivation is dependent on O_2 and a reducing agent; the rate of inactivation increases with increasing O_2 concentration with a partition ratio of 36. The rate of inactivation at saturation increases to a maximum value when eight Cu(II) ions are added to the apoenzyme tetramer.[16,17] The product of oxidation, 2-bromo-3-hydroxy-3-(*p*-hydroxyphenyl)-1-propene (Structural Formula 9.4) also inactivates the enzyme, but only in the absence of ascorbate. A proposed mechanism for inactivation by **9.4** is shown in Scheme 3. However, treatment of inactivated enzyme with [^{3}H]sodium cyanoborohydride or [^{3}H]sodium borohydride leads to no incorporation of tritium, even under denaturing conditions. This, therefore, leaves doubt concerning the proposed inactivation mechanism. The inactivation by compounds **9.3** may involve an initial oxygenation reaction and, therefore, also is discussed in Chapter 10 (Section II.A.3.e.).

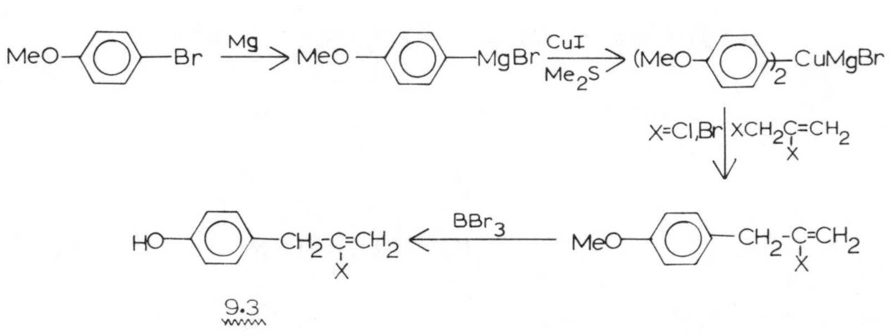

Scheme 2.   Containing Structural Formulas 9.3.

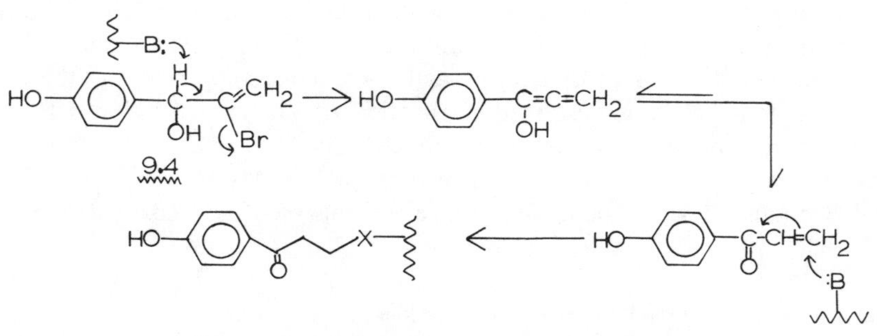

Scheme 3.   Containing Structural Formula 9.4.

$$ArCH_2CH=CH_2$$

Scheme 4.   Structural Formula 9.5.

A series of ring-substituted 3-phenylpropenes (Structural Formula 9.5, Scheme 4) was investigated by Fitzpatrick et al.[6] as mechanism-based inactivators for dopamine β-hydroxylase. All compounds require ascorbate and oxygen for inactivation. When the benzylic protons of 3-(*p*-hydroxyphenyl)propene were replaced by deuterium, a kinetic isotope effect of 2.0 on the k_{inact}/K_{O_2} was observed. However, there was no effect on the partition ratio (V_{max}/k_{inact}). This suggests that there is a stepwise mechanism for hydrogen abstraction and oxygen insertion. Values of V/K_{O_2} were calculated from the k_{inact}/K_{O_2} and partition ratio values and a linear free energy plot of V/K vs. σ^+ gives a good correlation with a ρ value of -1.2. Carbanion mechanisms generally give large ρ values; a radical mechanism is consistent with the small ρ value obtained and the correlation with σ^+. The rate of inactivation, however, is not affected by the radical trap mannitol, suggesting that, if a radical forms, it does not escape the active site. The negative ρ value is consistent with abstraction of a benzylic hydrogen atom by an electrophilic species, e.g., a high energy copper-oxygen species. 3-(*p*-Hydroxyphenyl)propane does not inactivate the enzyme, indicating the importance of the double bond for inactivation. These results indicate that C–H bond cleavage occurs prior to the intermediate that partitions between product formation and inactivation, and the radical mechanism shown in Scheme 5 was proposed. The inactivation mechanism for **9.5** also may be applicable to compounds **9.3**.

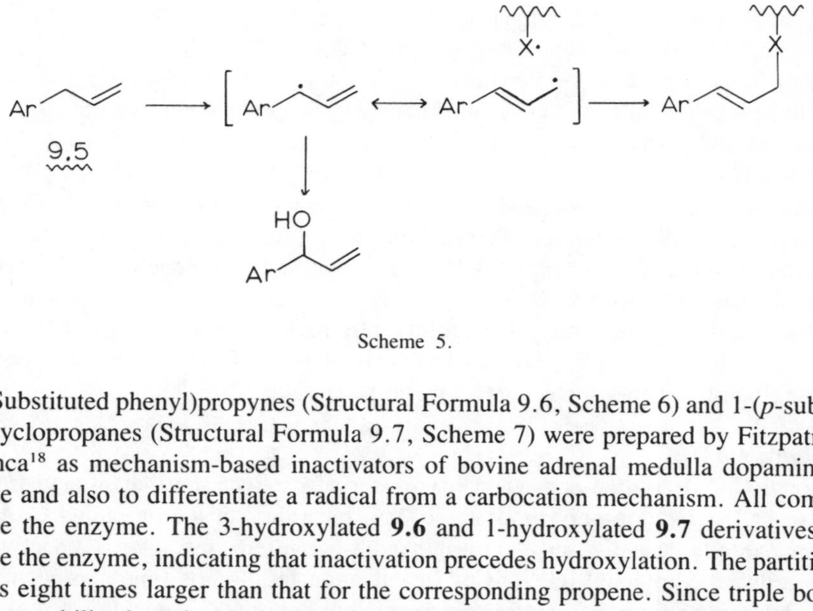

Scheme 5.

3-(*p*-Substituted phenyl)propynes (Structural Formula 9.6, Scheme 6) and 1-(*p*-substituted benzyl cyclopropanes (Structural Formula 9.7, Scheme 7) were prepared by Fitzpatrick and Villafranca[18] as mechanism-based inactivators of bovine adrenal medulla dopamine β-hydroxylase and also to differentiate a radical from a carbocation mechanism. All compounds inactivate the enzyme. The 3-hydroxylated **9.6** and 1-hydroxylated **9.7** derivatives do not inactivate the enzyme, indicating that inactivation precedes hydroxylation. The partition ratio for **9.6** is eight times larger than that for the corresponding propene. Since triple bonds are more electrophilic than double bonds, a mechanism involving a carbocation would have resulted in a **lower** partition ratio for the propyne, although there are considerable geometric differences in these compounds that could have accounted for the difference. Inactivation by cyclopropyl analogues[1,3,4,7-9] has been used as a probe for radical formation. These results in conjunction with those obtained with 3-arylpropenes[6] suggested[18] inactivation mechanisms shown in Schemes 6 and 7, respectively. Compound **9.6** was synthesized[18] by the route shown in Scheme 8 and compound **9.7** was prepared[18] as shown in Scheme 9.

$$ArCH_2C\equiv CH \rightarrow [Ar\dot{C}H\text{–}C\equiv CH \leftrightarrow ArCH=C=\dot{C}H] \rightarrow \text{inactivation}$$

9.6

Scheme 6.   Containing Structural Formula 9.6.

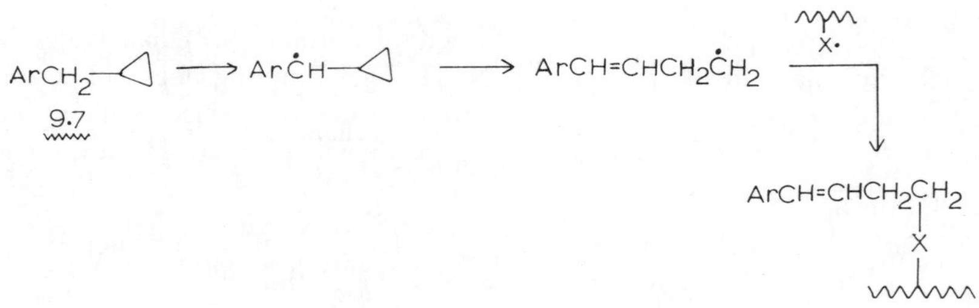

Scheme 7.   Containing Structural Formula 9.7.

$$ArMgBr \xrightarrow{\quad BrCH_2C\equiv CH \quad} 9.6$$

Scheme 8.

$$ArCH_2C=CH_2 \xrightarrow[\text{Cu}]{\quad CH_2I_2 \quad} 9.7$$

Scheme 9.

d. 4-Pentenoyl CoA

4-Pentenoic acid is a time-dependent irreversible inactivator of acetoacetyl CoA- and 3-ketodecanoyl CoA thiolases in coupled rat heart mitochondria;[19] presumably the CoA ester is the actual inactivator. Holland et al.[20] have shown that 4-pentenoic acid is metabolized to 2,4-pentadienoyl CoA (**9.2**, Scheme 1) in mitochondria, and this compound is a time-dependent inactivator of 3-oxoacyl CoA thiolase from pig heart. It is not clear if inactivation occurs only after release of **9.2** or prior to release as well. If inactivation only occurs after its release, then this is not a mechanism-based inactivator.

e. Propionyl CoA

Propionyl CoA is a time-dependent inactivator of ox liver short-chain acyl CoA dehydrogenase.[20a] Inactivation only proceeds to 78% regardless of the propionyl CoA concentration, apparently reaching an equilibrium with active enzyme. During inactivation the flavin is reduced. Propionyl [G-³H] CoA inactivation results in a time-dependent incorporation of [³H] into ammonium sulfate-precipitated protein with a stoichiometry of 1 mol of radioactivity per mole of inactivated enzyme. When the flavin is released by acid precipitation of the protein, a stable [³H]-containing flavin semiquinone is generated. This is suggestive of an N5-alkylflavin adduct. The partition ratio is 4.3. Although no mechanism was suggested by Shaw and Engel,[20a] an oxidation-Michael addition pathway (see Scheme 10) is consistent with these results.

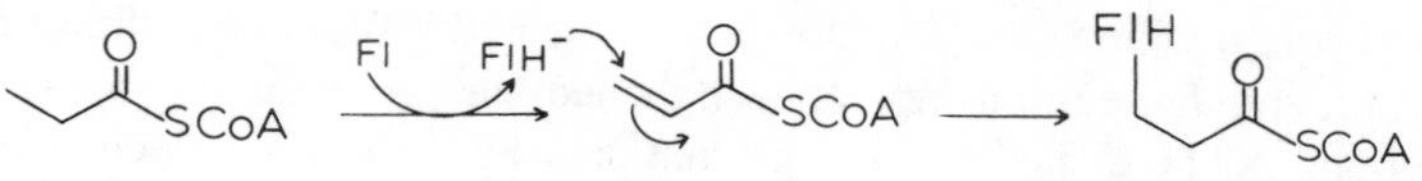

Scheme 10.

f. Cyclopropylmethanols

exo-Bicyclo[4.1.0]heptane-7-methanol (Structural Formula 9.8) inactivates horse liver alcohol dehydrogenase in the presence of NAD$^+$; gel filtration does not restore enzyme activity.[21] In the presence of NADH, 2-vinylcyclohexanol is produced without inactivation. Compound **9.8** was prepared by MacInnes et al.[21] as shown in Scheme 11. Several monocyclic and bicyclic cyclopropymethanols were prepared by MacInnes et al.[22] Biphasic time-dependent inactivation was observed and no reactivation occurred with gel filtration. Partition ratios varied from 126 for **9.8** to 3540 for (1-hydroxyethyl)cyclopropane. When the enzyme is in the NADH form, the product of hydrolysis of **9.8** is *trans*-2-vinylcyclohexanol, presumably derived from the reaction shown in Scheme 12. With the use of the α,α-dideuterio compound, an inverse isotope effect on inhibition was observed. This was suggested to indicate bonding of the enzyme to the inhibitor prior to formation of the carbonyl (Scheme 13). Cyclopropylglycolic acid inactivates lactate dehydrogenase in a time-dependent biphasic manner with a partition ratio of 274.

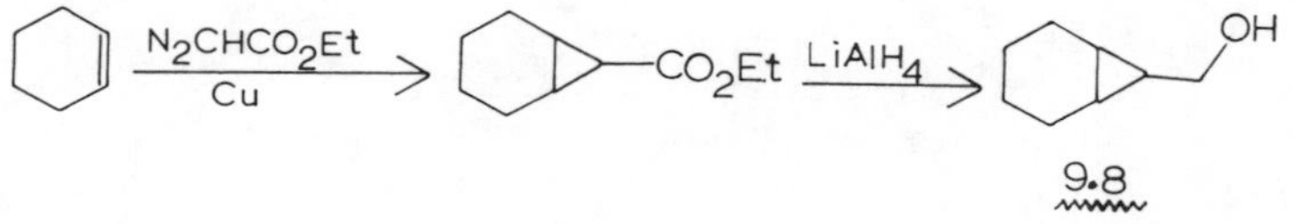

Scheme 11.   Containing Structural Formula 9.8.

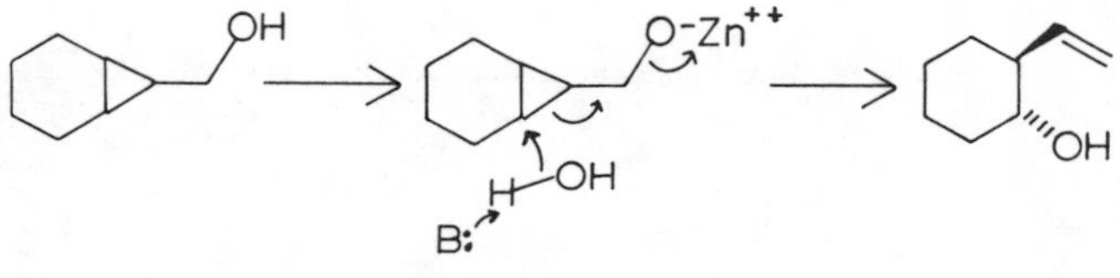

Scheme 12.

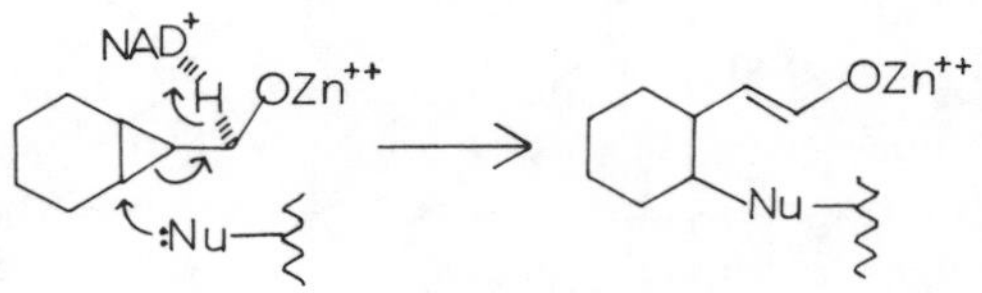

Scheme 13.

Molecular models for the active sites of carboxypeptidase A, horse liver alcohol dehydrogenase, and lactate dehydrogenase were constructed with the use of molecular graphics; the cyclopropane-containing compounds were inserted by superimposition on the bound substrate (or inhibitor)[22a] The substrate (or inhibitor) was deleted and the steric interactions of the cyclopropane inactivators with the active sites were determined. Inactivation mechanisms previously proposed[21] were shown by Breckenridge and Suckling[22a] to be competent with the active-site environments with respect to space and stereoelectronic requirements for cyclopropyl ring opening, the juxtaposition of the active-site nucleophiles relative to the cyclopropyl ring, and the relationship between the leaving group (H^-) in the dehydrogenases and the C–C bond of the cyclopropane ring that cleaves. Frontier molecular orbital considerations were shown to be useful in rationalizing the electrophilicity of the cyclopropane-containing compounds during enzyme inactivation.

2. With Carbon-Nitrogen Bond Cleavage
a. 2,3-Bis(carbethoxy)-2,3-diazabicyclo[2.2.0]hex-5-ene

2,3-Bis(carbethoxy)-2,3-diazabicyclo[2.2.0]hex-5-ene (Structural Formula 9.9) synthesized as shown in Scheme 14, inactivates phenobarbital (but not 3-methylcholanthrene-inducible cytochrome P-450.[23] Inactivation requires the cyclobutenyl π-bond and the diazobicyclo[2.2.0] skeleton, but not the carbamate groups. Enzyme is protected by CO. Inactivation leads to alkylation of the heme on the nitrogen of pyrrole ring D as is the case when terminal olefins are inactivators; however, in the case of terminal olefins an oxygen atom is incorporated into the adduct (see Chapter 10, Section II.A.3.b.). The adduct was shown by Stearns and Ortiz de Montellano[23] to be *N*-(2-cyclobutenyl)protoporphyrin IX (Structural Formula 9.10, Scheme 15). Except for *N*-aryl derivatives, this is the first adduct attached to the heme through a secondary rather than a primary carbon. Internal acetylenes inactivate P-450, but do not alkylate the heme. A one-electron mechanism was proposed (Scheme 15).

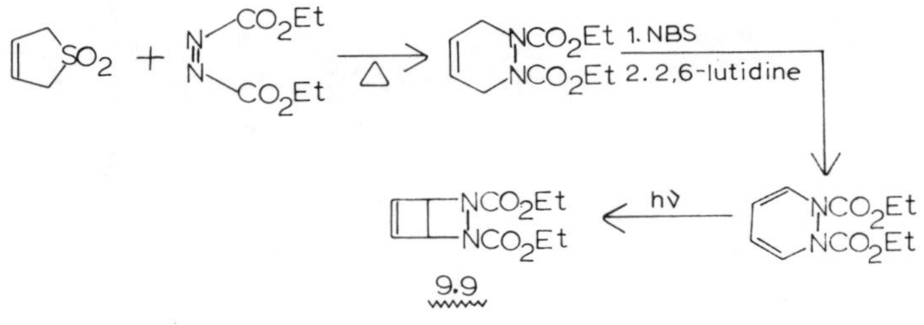

Scheme 14.   Containing Structural Formula 9.9.

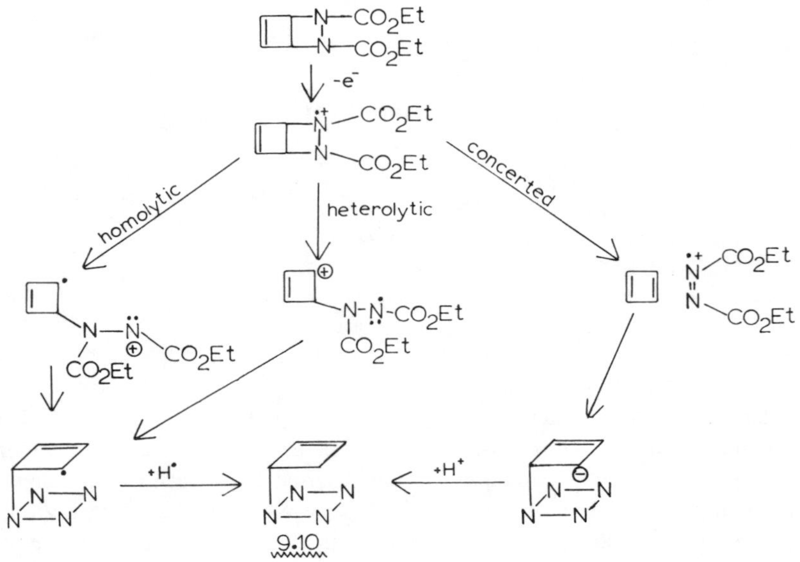

Scheme 15.   Containing Structural Formula 9.10.

3. With Carbon-Silicon Bond Cleavage

a. 20-(2-[Trimethylsilyl]ethyl)-5-pregnen-3β,20α-diol

On the basis of the known electrophilicity of silyl substituents that are beta to carbocations, a mechanism-based inactivator of cytochrome P-450$_{scc}$ was designed by Nagahisa et al.[24] Inactivation by 20-(2-[trimethylsilyl]ethyl)-5-pregnen-3β,20α-diol (Structural Formula 9.11, Scheme 16) requires NADPH and O$_2$ and is accompanied by the production of at least four oxidation products. No enzyme activity returns after gel filtration and no protection by mercaptans or fluoride ion was observed. The mechanism proposed (Scheme 16) involves a β-cation intermediate. 20-Vinyl-5-pregnen-3β,20α-diol was identified as one of the oxidation products. Compound **9.11** was synthesized[24] by the method of Wilson and Shedrinsky[25] (Scheme 17).

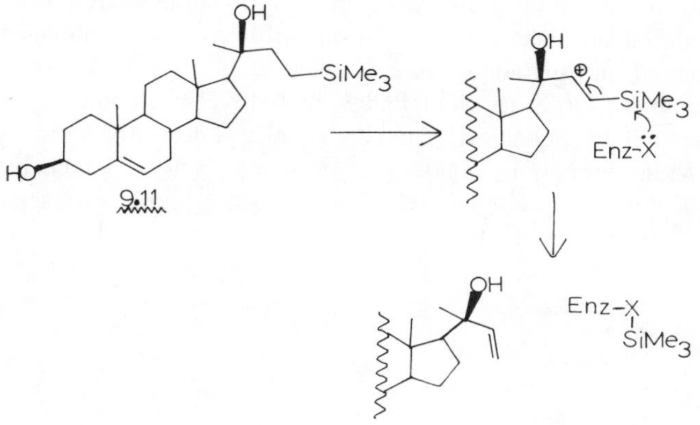

Scheme 16.   Containing Structural Formula 9.11.

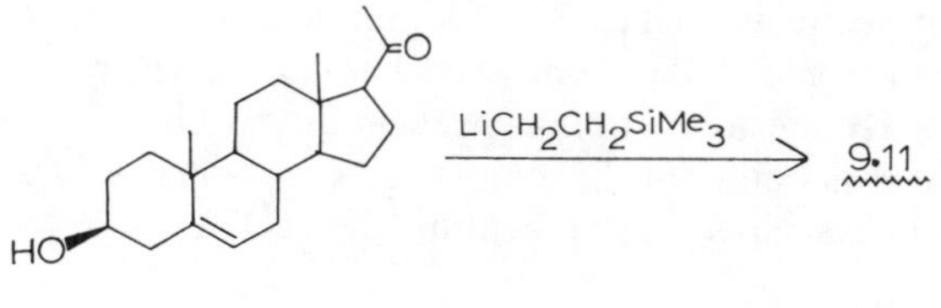

Scheme 17.

4. With Carbon-Carbon Bond Cleavage

a. 12-Methylidene-10(Z),13(Z)-nonadecadienoic Acid

12-Methylidene-10(Z),13(Z)-nonadecadienoic acid (Structural Formula 9.11a, Scheme 17A) is an irreversible inactivator of soybean lipoxygenase; enzyme activity is not restored by dialysis.[25a] An inactivation mechanism, proposed by Corey and d'Alarcao,[25a] is shown in Scheme 17A.

Compound **9.11a** was synthesized by the route shown in Scheme 17B.

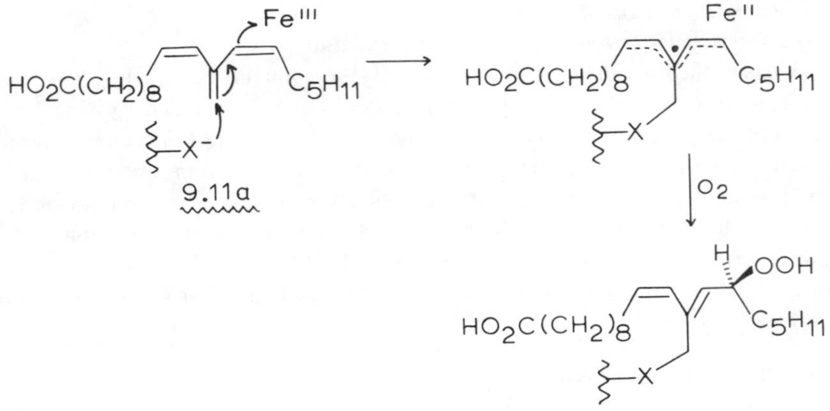

Scheme 17A.   Containing Structural Formula 9.11a.

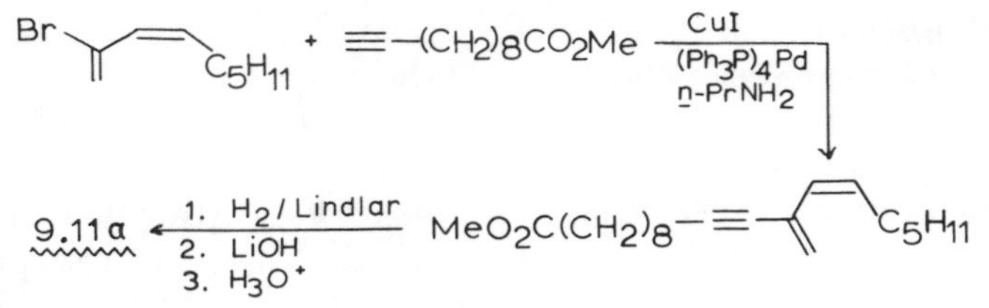

Scheme 17B.

b. Cyclopropane Derivatives

Bicyclo[2.1.0]pentane, methylcyclopropane, and nortricyclane (Structural Formulas **9.11b**, **9.11c**, and **9.11d**, respectively, Scheme 17C) were tested as radical clock substrates for liver microsomal cytochrome P-450 from phenobarbital-treated rats.[25] In the presence of NADPH **9.11b** and **9.11c** are weak, time-dependent inactivators of the enzyme. A possible inactivation mechanism is shown in Scheme 17D. Compound **9.11b** was synthesized by Ortiz de Montellano and Stearns[25b] as shown in Scheme 17E.

9.11b   9.11c   9.11d

Scheme 17C.   Structural Formulas 9.11b, 9.11c, and 9.11d.

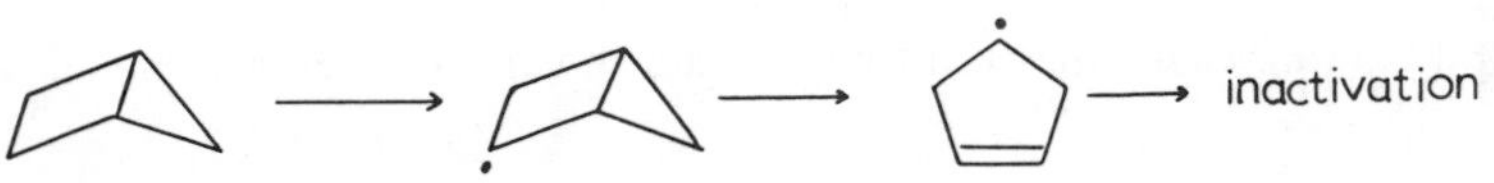

Scheme 17D.

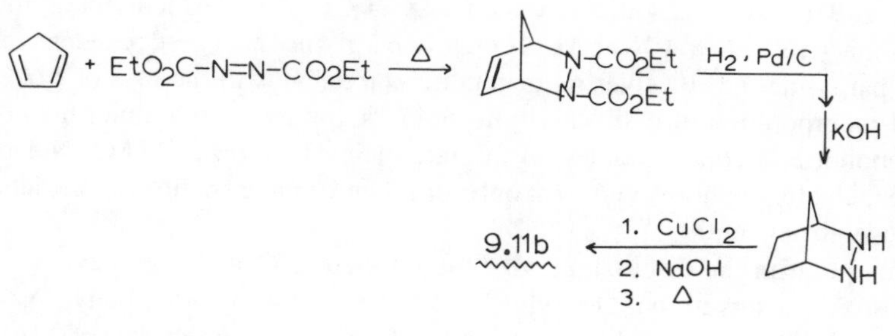

Scheme 17E.

B. Carbon-Nitrogen Bond Oxidation

1. With Carbon-Hydrogen Bond Cleavage

This section is divided according to those compounds containing triple bonds, allene groups, and olefins.

a. Acetylene-Containing Inactivators

(1) N-Benzyl-N-methyl-2-propynylamine (Pargyline) and Derivatives

In the first part of this section different inactivators of mitochondrial monoamine oxidase (MAO), an enzyme containing a covalently bound flavin, will be discussed. This is followed by studies on specific inactivation of the isozymes MAO A and MAO B. Then the questions of the structure of the enzyme adduct and the chemistry of inactivation will be dealt with. At the end of the section are acetylene-containing inactivators of sarcosine oxidase and cytochrome C reductase.

Both *N*-benzyl-*N*-methyl-2-propynylamine (pargyline; **9.12**, Ar = Ph, X = CH$_2$; see Scheme 18) and 2-propynylamine are time-dependent inactivators of MAO;[26,27] [^{3}H] pro-pynylamine inactivates MAO, and tritium is incorporated into the enzyme.[28] An isotope effect was observed when α,α-dideuteriopargyline inactivated rat liver MAO.[29,30]

$$\underset{\text{ArX}-\text{NCH}_2\text{C}\equiv\text{CH}}{\overset{\text{CH}_3}{|}}$$

Scheme 18.   Structural Formula 9.12.

$$\text{ArCH}_2\text{NHCH}_3 + \text{BrCH}_2\text{C}\equiv\text{CH} \rightarrow \underset{\text{ArCH}_2\text{N}-\text{CH}_2\text{C}\equiv\text{CH}}{\overset{\text{CH}_3}{|}}$$

Scheme 19.

A series of pargyline analogues was prepared by Swett et al.[31] (Scheme 19) and structure-activity relationships were made in vitro and in vivo. Clorgyline (**9.12**, Ar = 2,4-dichlo-rophenyl, X = O(CH$_2$)$_3$) is a time-dependent inactivator of rat brain monoamine oxidase.[32] A variety of analogues of propargylamine were prepared and studied by Williams and Lawson[33] as inhibitors of porcine brain MAO. Compounds **9.12** (Ar = phenyl, X = (CH$_2$)$_n$ where n = 1 to 4; Ar = 2,4-dichlorophenyl, X = O(CH$_2$)$_3$; Ar = 2,4-dichlorophenyl, X = CH$_2$; Ar = phenyl, X = CH$_2$) and *N*-4-phenylpiperidinyl-*N*-methylpropargylamine are time-dependent inactivators. *N*-3-Phenylpiperidylpropargylamine is inactive at 10 μ*M*. The most potent is **9.12** (Ar = 2,4-dichlorophenyl, X = CH_2). *N*-Desmethylpargyline is a much less potent inactivator of rat liver MAO than is pargyine.[34] Time-dependent incorporation of [ring-^{3}H]pargyline into rat liver mitochondria corresponds to the loss of MAO activity.[34] The rate of incorporation of radioactivity from [7-^{14}C]pargyline into outer membranes of rat liver mitochondria is concomitant with the rate of inactivation of MAO; $NaDodSO_4$-polyacrylamide gel electrophoresis shows only one band that is radioactively labeled with a molecular weight of 60,000.[35]

Up to this point in the discussion, MAO is referred to as a single enzyme. However, in 1968, Johnston[36] observed that clorgyline (**9.12**, Ar = 2,4-dichlorophenyl, X = $O(CH_2)_3$) inhibits rat and human brain MAO and a plot of percentage inhibition of MAO (tyramine as the substrate) vs. concentration of this compound does not show the usual sigmoidal curve. Instead, a pair of sigmoidal curves joined by a horizontal section where inhibition is invariant occurs. Johnston[36] suggested that MAO is actually a binary system of enzymes, each of which has a different sensitivity to this inhibitor. The form of the enzyme that is more sensitive to clorgyline was referred to as enzyme A (MAO A), and the less sensitive one as enzyme B (MAO B); the pI_{50} values were separated by 3.6 units (a factor of 4 × 10^3) in rat brain and by 3.9 units (a factor of 8 × 10^3) in human brain. Neither tranylcypromine nor iproniazid shows differential inhibition of the two forms of MAO. Maître[37] also noted that MAO may exist in more than one form. It was suggested by Youdim et al.[38] that the reason some MAO inhibitors are antidepressants, but others are not, is because of differential inhibition of MAO isozymes. Therefore, compounds specific for certain isozymes should be tested. These early results led to a vast amount of research involving the pharmacology of MAO inhibition. However, only the work having to do with the chemistry and enzymology is discussed here.

[^{14}C]Pargyline forms a 1:1 molar ratio with pig liver MAO; $NaDodSO_4$-PAGE, in the absence of β-mercaptoethanol, gives two bands of 55,000 and 63,000 mol wt.[39] Both bands have a 1:1 molar ratio of [^{14}C] label. In the presence of β-mercaptoethanol only one 60,000-mol wt band was observed by $NaDodSO_4$-PAGE. Therefore, it was concluded by Oreland et al.[39] that the two bands are not the two subunits, but rather two forms of the enzyme, presumably differing in the number of disulfide bonds.

A variety of analogues of pargyline has been tested for differential isozyme inactivation. Human brain MAO A undergoes time-dependent inactivation by clorgyline and MAO B is inactivated by L-deprenyl (**9.12**, Ar = Ph, X = $CH_2CH(CH_3)$).[40] The kinetics of inhibition of the A and B forms of rat liver MAO by clorgyline, L-deprenyl, and pargyline were studied by Fowler et al.[41] since this, presumably, is the only difference in the selectivity of these reactions. The results indicate a reversible preequilibrium mechanism followed by time-dependent formation of the covalent adduct. The K_i value for reversible interaction of clorgyline with MAO A is 1000 times lower than with MAO B; this difference accounts for almost all of the selectivity of clorgyline. The rate for the irreversible component of inactivation is 13 times faster for MAO A than B, so there also is considerable selectivity of clorgyline in its irreversibility. The K_I value for L-deprenyl with MAO B is only 40 times lower than with MAO A, but the rate of the irreversible component of the inactivation is greater than 7 times faster for MAO B than MAO A. Pargyline has a K_i value only 8 times lower for MAO B than for MAO A (although the K_I is 26 times lower), and the inactivation rates are the same for the two isozymes. These results account for the relative nonselectivity of pargyline.

Dupont et al.[41a] observed that clorgyline, pargyline, deprenyl, and iproniazid are NADPH- and time-dependent inactivators of rat liver cytochrome P-450.

A clonal line of rat hepatoma cells was treated with [7-^{3}H]pargyline by Castro Costa and Breakefield[42] and both MAO A and B activities were inactivated. $NaDodSO_4$-PAGE showed only one radioactive band of mol wt 57,000.

Following treatment with clorgyline or deprenyl to block MAO A or B binding sites selectively, [^{3}H]pargyline was used by Edwards and Pak[43] to label the remaining active sites of rat liver mitochondrial MAO in the intact membrane. After the labeled protein was solubilized, a ratio of 1:3.3 for MAO A:B was determined by this method. Subunits containing the labeled MAO A and B sites could not be separated by NaDodSO$_4$-PAGE.

MAO A is irreversibly inactivated by dinitranyl (**9.12**, Ar = 2,4-dinitrophenyl, X = NH(CH$_2$)$_3$) with essentially no effect on MAO B; dialysis does not regenerate enzyme activity.[44]

N-Methyl-N-propargyl-(1-indanyl)ammonium hydrochloride (**9.12**, ArX = 1-indanyl) is a time-dependent irreversible inactivator of rat brain[45] and human brain[46] MAO B and is more potent than L-deprenyl. This inactivator was used by Fowler et al.[46] to titrate MAO B from human brain. A variety of propargylamine derivatives were tested by Kalir et al.[47] as inactivators of rat liver mitochondrial MAO A and B. Structure-activity relationships regarding selectivity for MAO A and B were made. N-Methylated propargyl derivatives are most active. The distance between the aromatic ring and the nitrogen is crucial to its selectivity; for MAO A inactivation, a distance equivalent to at least three carbon atoms is required, and for MAO B inactivation, this distance can be one or two carbons. A series of α-methyl- and N-propargyl amine analogues was prepared by Tipton et al.[48] in order to determine what structural features lead to specificity for rat liver MAO A and B, and if the specificity is the same with reversible and irreversible inhibitors. In general, the α-methyl substituted primary and secondary amines show selectivity as reversible inhibitors towards MAO A. All of the N-propargylamine derivatives are irreversible inhibitors, but the structural features that confer selectivity are difficult to define. Substitution of an N-propargyl group into a selective reversible inhibitor of MAO does not necessarily result in retention of the selectivity in the newly formed irreversible inhibitor. Also, if the propargyl group in a MAO B-selective inactivator is replaced by an allyl group, the resulting compound can be a reversible inhibitor showing slight selectivity for MAO A. Ten different propynylamine derivatives were synthesized and tested by Williams[49] for specificity with MAO A and B. Modification in the general structure of clorgyline were made and the effect on inactivation of MAO A and B from crude rat liver mitochondria was determined.

5'-(N-Dansyl)cadaveryl-p-carboxymethylpargyline (Structural Formula 9.13), a fluorescent analogue of the MAO inactivator pargyline, was prepared by Rando[50] (Scheme 20) to study the binding site of the enzyme. Pig liver mitochondrial MAO is irreversibly inactivated by this compound with biphasic kinetics. The biphasic nature, which also was observed with pargyline, was attributed to oligomers of the enzyme or isozymes. Dialysis does not regenerate activity. The dansyl moiety appears to be bound in a polar environment based on the λ$_{max}$ of emission of the enzyme-bound inactivator. It also has rotational degrees of freedom independent of the macromolecule based on depolarization studies. Analogous compounds made with fluorescein and rhodamine are as potent as the dansyl compound.

A number of studies have been carried out to determine the structure of the adduct formed after the inactivation of MAO by pargyline and derivatives. [7-^{14}C]Pargyline inactivates bovine kidney MAO and 1 equivalent of radioactivity is bound after gel filtration, dialysis, and acid precipitation.[51] Upon inactivation, the flavin coenzyme becomes reduced as evidenced by the bleaching of the 445-nm chromophore and formation of a new 410-nm absorbance concomitant with incorporation of 1.1 equivalents of radioactivity into the enzyme.[52] Proteolysis of the labeled enzyme led to the isolation of a radioactive peptide fragment containing an altered flavin, and Chuang et al.[52] suggested that pargyline reacts with the flavin of MAO. This is not a spontaneous reaction, since pargyline does not react with other flavoenzymes, e.g., succinic dehydrogenase, D-amino acid oxidase, and lipoyl dehydrogenase. If the flavin is initially reduced with substrate anaerobically, pargyline does not react with the enzyme. Sulfhydryl analysis indicated that pargyline is not attached to a

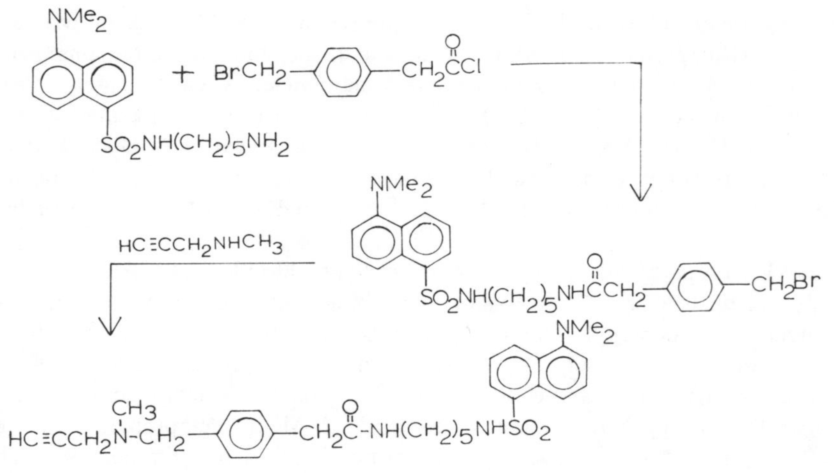

NMe₂ ... SO₂NH(CH₂)₅NH₂ + BrCH₂—⟨◯⟩—CH₂CCl (O)

HC≡CCH₂NHCH₃

NMe₂ ... SO₂NH(CH₂)₅NHCCH₂—⟨◯⟩—CH₂Br (O)

NMe₂

CH₃
HC≡CCH₂N–CH₂—⟨◯⟩—CH₂C–NH(CH₂)₅NHSO₂ (O)

9.13

Scheme 20.   Containing Structural Formula 9.13.

cysteine residue.[51] [7-^{14}C]Pargyline also inactivates pig liver MAO with incorporation of 1.1 equivalents of radioactivity.[53] Pronase digestion produces a [^{14}C]-labeled, flavin-containing pentapeptide that is composed of cysteine, aspartate, serine, and glycine in a 1:1:1:2 molar ratio. At each stage of purification of the radioactive peptide, Oreland et al.[53] found an equimolar ratio of radioactivity to flavin. Thin-layer chromatography in two acidic and one neutral solvent system confirms the attachment of the pargyline to the flavin-containing fragment. When a basic solvent system was used, however, the radioactivity was released from the flavin.

[^{3}H]Pargyline was used by Yu[54] to inactivate MAO A and B from several different tissues (rat heart and human placenta were MAO A sources; pig and beef liver were MAO B sources; and rat liver and brain represented mixed A and B sources). The labeled enzymes were proteolytically hydrolyzed and the tritiated peptides were isolated. With Pronase® P, a broad spectrum protease, only one peptide was isolated from all sources, having the structure **9.14** (Y is the pargyline-flavin adduct; Scheme 21). These results suggest that the active sites of MAO A and B around the pargyline binding sites are the same.

Ser–Gly–Gly–Cys–Tyr
|
Y

Scheme 21.   Sequence 9.14.

As a chemical model for the pargyline inactivation of MAO, 3-methyllumiflavin (Structural Formula 9.15) was photolyzed in the presence of *N,N*-dimethyl-2-propynylamine. UV and NMR spectral analyses indicated to Zeller et al.[55] that the two principal products were covalent cycloaddition products (Structural Formulas 9.16 and 9.17, Scheme 22). These results support the hypothesis that the C4a-N5 double bond in flavin is the primary redox center of the flavin nucleus. Further detailed NMR studies by Maycock[56] and Gärtner and Hemmerich[57] of the product formed from this photolysis reaction indicated that the structure proposed by Zeller et al.[55] is incorrect. The revised structure[56,57] is **9.18** (Scheme 23) having a flavocyanine structure. The mechanism proposed[58] for the photochemical reaction of propargylamine with flavin involves the flavoquinone triplet (Scheme 24; only the first step was proposed). The adduct formed by inactivation of bovine liver MAO (MAO B) with 3-dimethylamino-1-propyne was studied by Maycock et al.[59] When either 3-dimethylamino-[3-^{14}C]-1-propyne or 3-[*methyl*-^{14}C]-dimethylamino-1-propyne is used, 2.3 to 2.8 mol of inactivator per mole of enzyme are incorporated. The more than stoichiometric amounts of radioactivity was suggested to be nonspecific labeling. Tryptic-chymotryptic digestion of the labeled enzyme was followed by isolation of a [^{14}C]-labeled, flavin-containing penta-peptide. UV and NMR spectral studies, as well as chemical degradation reactions and comparison with model compounds, led to the conclusion that the adduct was related to that proposed for the product of photochemical reaction of 3-methyllumiflavin with 3-dimethylamino-1-propyne.[56,57] The most likely mechanism for its formation was suggested by Maycock et al.[59] as that shown in Scheme 25 (Fl is oxidized flavin; FlH$^-$ is reduced flavin); attachment to the N5-position of the flavin (**9.19**) was proposed. *N*-Benzyl-*N*-methyl-2-butynylamine also inactivates beef liver MAO;[60] spectral changes similar to those observed with 3-dimethylamino-1-propyne accompany inactivation. Dialysis, acid precipitation, and enzymatic digestion do not regenerate oxidized flavin, supporting a mechanism similar to that in Scheme 25.

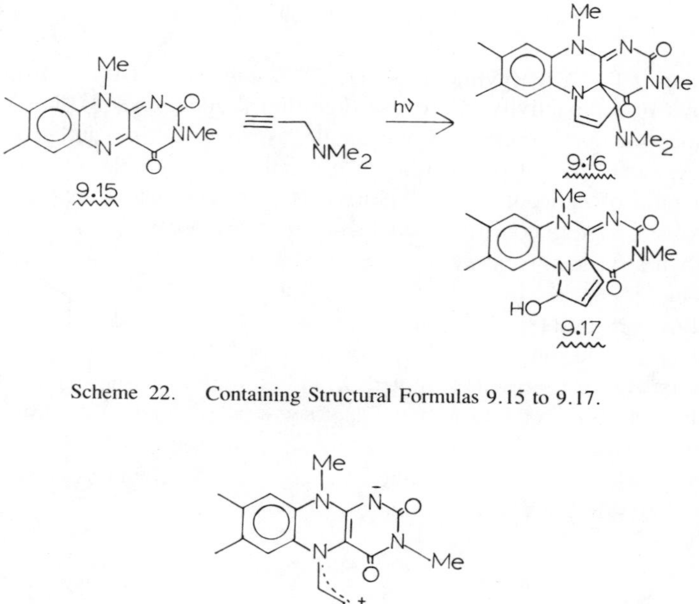

Scheme 22.   Containing Structural Formulas 9.15 to 9.17.

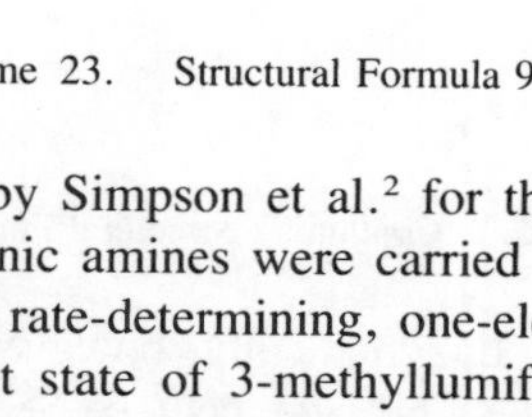

Scheme 23.   Structural Formula 9.18.

Photochemical model studies by Simpson et al.[2] for the inactivation of MAO by β,γ-acetylenic amines and β,γ,δ-allenic amines were carried out using 3-methyllumiflavin as the model flavin. Evidence for a rate-determining, one-electron mechanism of amine oxidation was provided. The singlet state of 3-methyllumiflavin is quenched by all of the secondary and tertiary amines tested at or near the diffusion-controlled limit. Likewise, the quantum yield and kinetic data for the reaction of triplet 3-methyllumiflavin shows little variation with amine structure. When β,γ-acetylenic- or allenic amines were used, addition products to the 3-methyllumiflavin were obtained. The reactions with acetylenic amines give complex product mixtures, including flavocyanines and C4a-N5 adducts; with allenic amines,

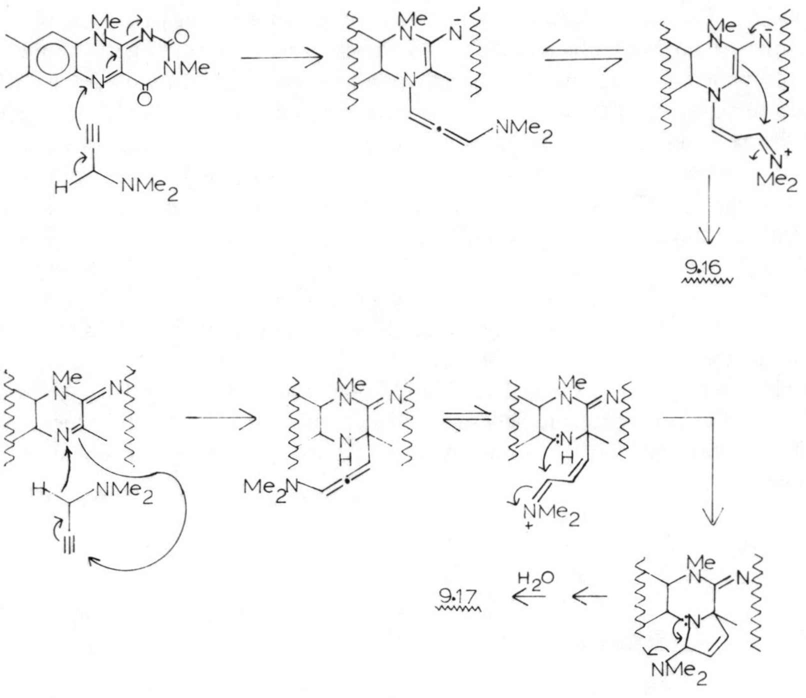

Scheme 24.

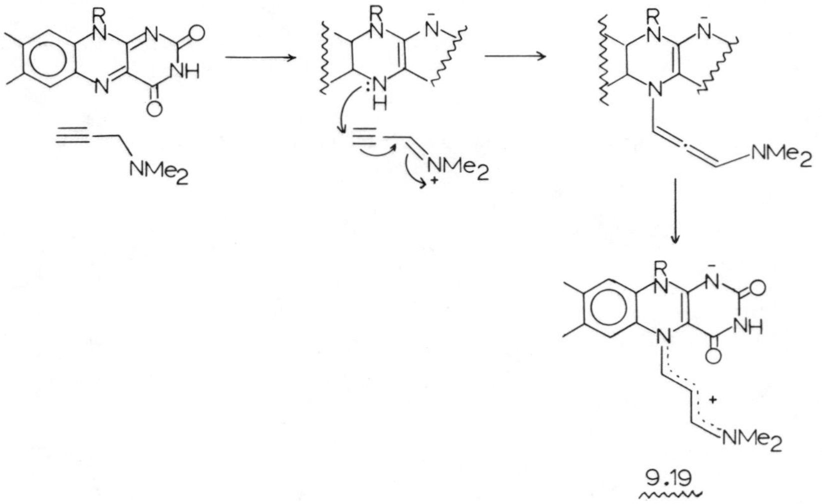

Scheme 25.   Containing Structural Formula 9.19.

only flavocyanines are obtained. Enzymatically, however, acetylenic amines give flavocyanines, and allenic amines give C4a-N5 adducts (see Section II.B.1.b.(2)). An initial one-electron transfer mechanism, however, is supported by reactions of cyclopropylamines with MAO (see Section II.B.2.a.(4)). In order for this type of mechanism to be relevant, the free base amine, not the protonated amine, must be the active form. McEwen et al.[26] studied the effect of pH on the kinetics of inactivation of human liver MAO by pargyline and 2-propynylamine and concluded that the unionized form of the amines interact with the enzyme.

If the one-electron mechanism of MAO is applied to propargylamine inactivators, the inactivation mechanism shown in Scheme 26 would be relevant.

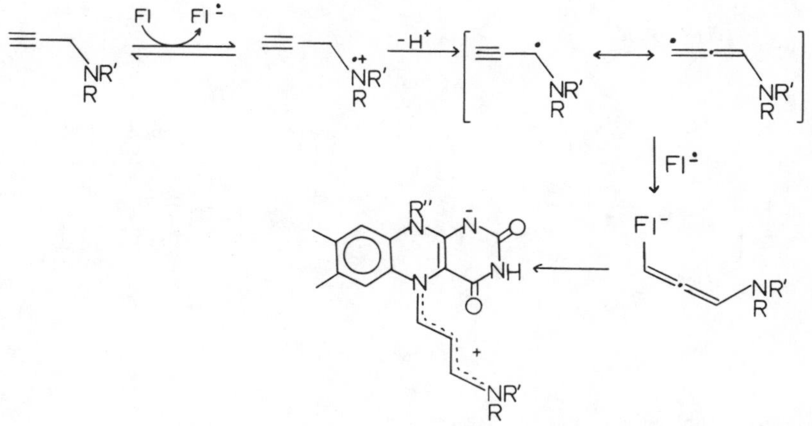

Scheme 26.

(2) *N*-Propargylglycine and Derivatives

N-Propargylglycine (Structural Formula 9.20, R = CH$_2$C≡CH; see Scheme 27) is a time-dependent inactivator of rat liver sarcosine (*N,N*-dimethylglycine) oxidase; dialysis does not regenerate enzyme activity.[61] A deuterium isotope effect of 2.0 was observed by Kraus and Belleau[61] for the corresponding α,α-dideuterio analogue. *N*-Propargyl substituted *N,N*-disubstituted glycines also are time-dependent inactivators of rat liver mitochondrial *N,N*-dimethylglycine oxidase.[29] On the basis of an isotope effect with 1,1-dideuterio-*N*-propargyl-*N*-methylglycine, it was proposed by Kraus et al.[29] that this compound is a mechanism-based inactivator for the enzyme. The inactivators were synthesized by the route shown in Scheme 27 (R = propargyl or γ-phenylpropargyl). Inactivation by *N*-methyl-*N*-propargyl-glycine is oxygen-dependent.[30] The inactivation of the enzyme by propiolaldehyde is negligible at a concentration that produced 50% inactivation with *N*-methyl-*N*-propargylglycine. Therefore, Kraus and Yaouanc[30] excluded a mechanism of inactivation involving hydrolysis of the oxidized inactivator to propiolaldehyde prior to inactivation. Scheme 28 depicts the proposed mechanism of inactivation. Rat liver *N,N*-dimethylglycine oxidase also is irreversibly inactivated by *N*-(1-methylpropargyl)- and *N*-propargylsarcosine in a time-dependent manner; the 1-methylpropargyl compound is 10^4 times less potent than the propargyl compound.[62] A steric effect was proposed by Yaouanc et al.[62] for the lower potency of the 1-methylpropargyl compound.

$$\text{R–Br} + \text{CH}_3\text{NHCH}_2\text{CO}_2\text{Et} \rightarrow \underset{\underset{\text{CH}_3}{|}}{\text{R–NCH}_2\text{CO}_2\text{Et}} \xrightarrow[\text{2. Dowex}^\circledR]{\text{1. KOH}} \underset{\underset{\text{CH}_3}{|}}{\text{RNCH}_2\text{COOH}}$$

9.20

Scheme 27.   Containing Structural Formula 9.20.

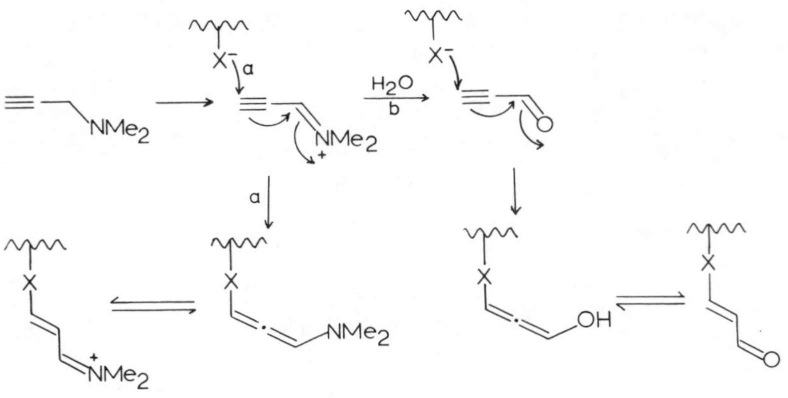

Scheme 28.

(3) Other Propargylamine-Containing Inactivators

N-Propargyl analogues of SKF-525A (**9.21**; R = Me or Et, R′ = CH$_2$C≡CH) and acetylmethadol (**9.22**; R = Me, R′ = CH$_2$C≡CH) were prepared (see Scheme 29) and shown to be time-dependent inactivators of rabbit and rat liver microsomal mixed-function oxidase (cytochrome C reductase). Dialysis does not regenerate enzyme activity. An interaction between hepatic flavoprotein and the acetylenic group of the inactivators was suggested by Kraus and Yaouanc[63] because spectral studies indicate partial reduction of the flavin after inactivation.

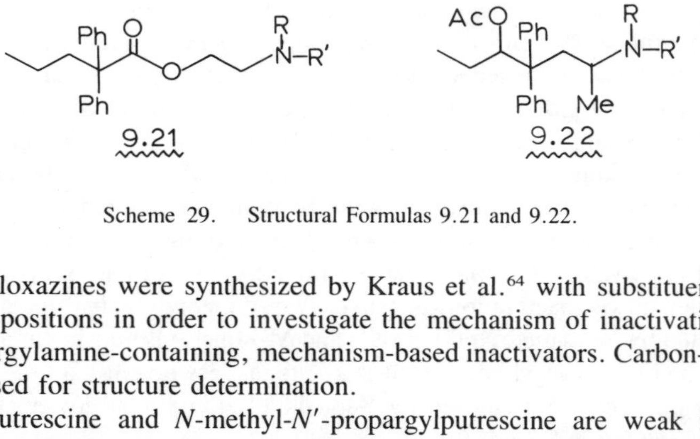

Scheme 29.   Structural Formulas 9.21 and 9.22.

Various isoalloxazines were synthesized by Kraus et al.[64] with substituents at the N3-, N10-, and C4a-positions in order to investigate the mechanism of inactivation of flavoenzymes by propargylamine-containing, mechanism-based inactivators. Carbon-13 NMR spectroscopy was used for structure determination.

N-Propargylputrescine and *N*-methyl-*N*′-propargylputrescine are weak inactivators of mammalian polyamine oxidase[65] (see Section II.B.1.a.(2) for synthetic methods).

b. Allene-Containing Inactivators
(1) α-Allenic Amines

Primary- (**9.23**) and *N*-methyl α-allenic amines (**9.24**) were synthesized by Claesson and co-workers[66,67] as shown in Scheme 30. Tertiary α-allenic amines (**9.25**) were prepared by Claesson and co-workers[67,68] (Scheme 31). A method used to synthesize primary, secondary, and tertiary allenic amines is shown in Scheme 32 (X = CH$_3$SO$_3$,[67] or Cl[69]). N-Benzyl-*N*-methylamine condenses with vinylacetylene to give *N*-2,3-butadienyl-*N*-benzyl-*N*-methylamine.[69]

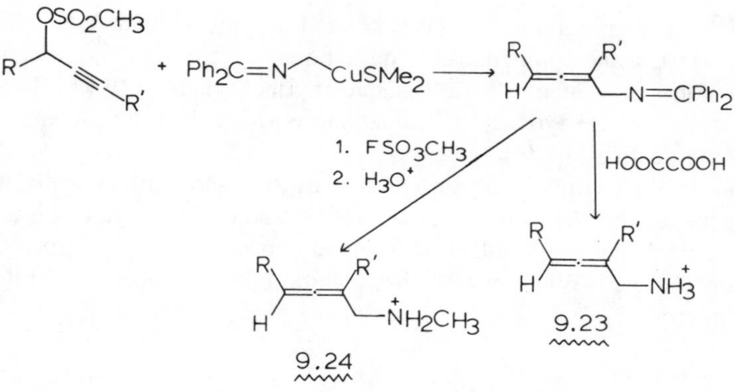

Scheme 30.   Containing Structural Formulas 9.23 and 9.24.

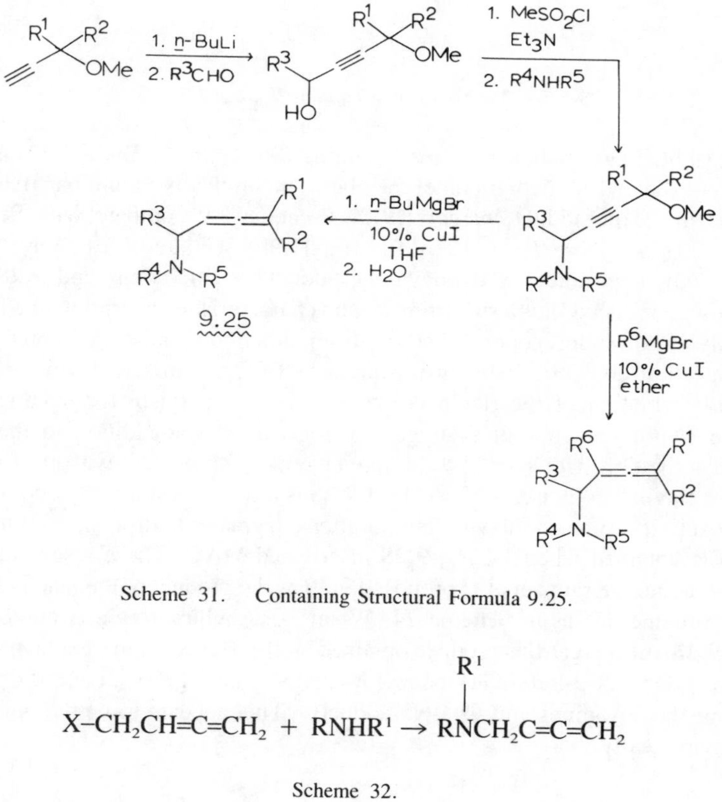

Scheme 31.   Containing Structural Formula 9.25.

$$R^1$$
$$|$$
$$X-CH_2CH=C=CH_2 \ + \ RNHR^1 \rightarrow RNCH_2C=C=CH_2$$

Scheme 32.

β-Substituted-α-allenyl primary amines can be synthesized in one step as shown in Scheme 32A.

$$RMgBr \ + \ MeOCH_2C\equiv CCH_2N(SiMe_3)_2 \ \xrightarrow[\text{2. HX}]{\substack{\text{1. CuBr} \cdot \text{SMe}_2 \quad \text{or} \\ \text{NiCl}_2(\text{Ph}_2\text{P}(\text{CH}_2)_3\text{PPh}_2)}} \ RCCH_2NH_3^+X^-$$

$$\underset{\overset{\|}{C}}{\overset{CH_2}{\|}}$$

Scheme 32A.

A series of 15 α-allenic amines was synthesized and tested as MAO inhibitors in vivo and in vitro.[67] Most of the compounds are quite potent inactivators and selective for MAO B. The *R*-(−)-form of *N*-methyl-*N*-(2,3-pentadienyl)benzylamine (**9.25**, R = R^3 = Me, R^1 = R^2 = H, R^4 = CH$_2$Ph) was 2.7 times more active than the *S*-(+)-isomer in vivo and 25 times more active in vitro.

Bovine liver MAO is irreversibly inactivated by two series of optically active allenic amines.[60] In general, the (*R*)-isomers (Structural Formula 9.26) were much more effective (larger k_{inact} values) than their (*S*)-allenic (Structural Formula 9.27) counterparts (see Scheme 33). It was concluded by White et al.[60] that a chiral allenic moiety is useful to probe the active site geometry.

Scheme 33.   Structural Formulas 9.26 and 9.27.

Since *N*-2,3-butadienyl-*N*-benzyl-*N*-methylamine (Structural Formula 9.28) is a tautomer of *N*-2-butynyl-2-benzyl-*N*-methylamine, they both, theoretically, could inactivate beef liver MAO to give the same adduct, namely, **9.29** (Scheme 34; compare with Scheme 26 in Section II.B.1.a.(1), where R = CH$_3$, R′ = CH$_2$Ph). Unlike *N*-2-butynyl-*N*-benzyl-*N*-methylamine, which produces a flavocyanine adduct (**9.29**), compound **9.28** inactivates MAO and shows a reduced flavin spectrum, but no characteristic absorption of a flavocyanine results.[70] Dialysis does not regenerate activity from either compound. Acid precipitation and enzymatic digestion does not result in regeneration of the oxidized flavin, indicating that covalent bond formation of the flavin is evident. The mechanism for inactivation by the allenic amine is not known, but a suggestion regarding cycloaddition to the flavin was mentioned. Inactivation with 1-[^{14}C]-**9.28** give enzyme with one equivalent of radioactivity attached to the flavin.[71] When *N*-[^{14}C-methyl]-**9.28** is the inactivator, only about 0.05 equivalent of radioactivity remains enzyme bound after enzymatic hydrolysis. A kinetic isotope effect of 5 was obtained when 1-[^{2}H$_2$]-**9.28** inactivated MAO. These results are consistent with a C4a-N5 adduct; Krantz et al.[71] proposed **9.30** as the product of the inactivation derived from **9.30a** (Scheme 35) as in Scheme 24. When 3-methyllumiflavin is photolyzed in the presence of **9.28**, however, the product obtained is the flavocyanine exclusively; the corresponding acetylene, *N*-2-butynyl-*N*-benzyl *N*-methylamine, gives a complex product mixture, including flavocyanines and C4a-N5 adducts.[2] This appears to be the opposite of that observed enzymatically.

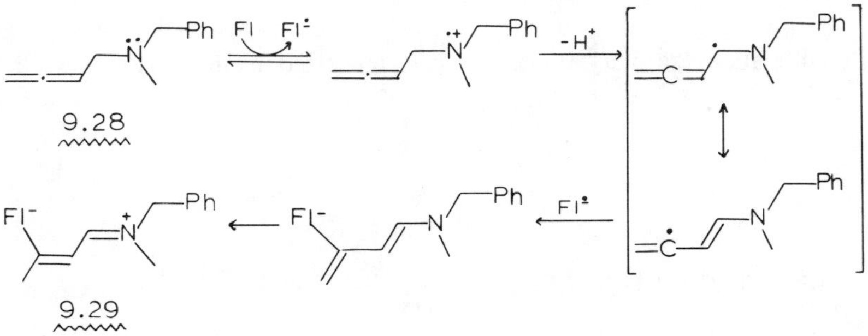

Scheme 34.   Containing Structural Formula 9.28 and 9.29.

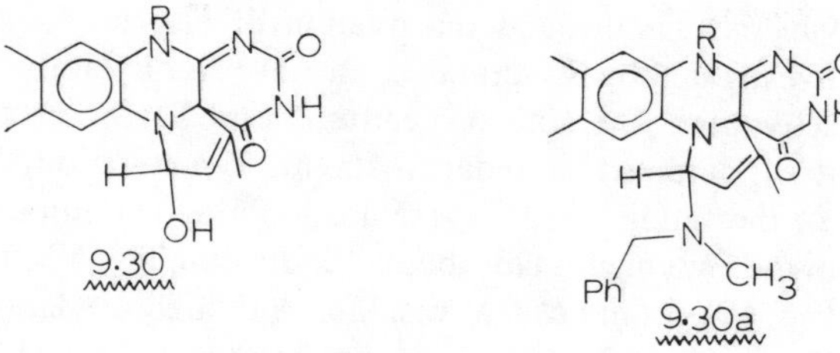

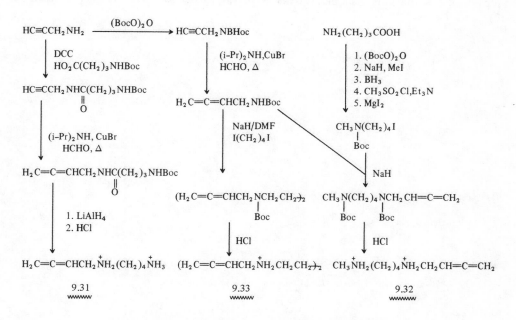

Scheme 35.   Structural Formulas 9.30 and 9.30a.

(2) *N*-2,3-Butadienyl-1,4-butanediamine and Derivatives

N-2,3-Butadienyl-1,4-butanediamine (**9.31**), *N*-2,3-butadienyl-*N'*-methyl-1,4-butanedi-amine (**9.32**), and bis-*N*,*N'*-2,3-butadienyl-1,4-butanediamine (**9.33**) are pseudo first-order, time-dependent, irreversible inactivators of pig liver polyamine oxidase.[65] None of these compounds inhibits MAO, diamine oxidase, or any other enzyme in polyamine metabolism; these are the first specific irreversible inactivators of polyamine oxidase reported. The inactivators were synthesized[65] as shown in Scheme 36.

Scheme 36.   Containing Structural Formulas 9.31 to 9.33.

c. Olefin-Containing Inactivators
(1) Allylamine and Derivatives

Pig liver monoamine oxidase is inactivated by allylamine; activity is not reversed upon dialysis.[72] [1-³H]Allylamine inactivation results in tritium incorporation concomitant with enzyme inactivation. A deuterium isotope effect of 2.35 on the rate of inactivation was observed for [1-²H₂]allylamine. Since inactivation is accompanied by reduction of the flavin, it was suggested that attachment of allylamine is to the flavin. Treatment of [1-¹⁴C]allylamine-inactivated enzyme with benzylamine led to complete reactivation and release of a radioactive compound which, after silver oxide oxidation, was identified as *N*-benzyl-β-alanine (Structural Formula 9.34). On the basis of these results, the inactivation and reactivation mechanisms shown in Scheme 37 were proposed by Rando and Eigner.[72] The mechanisms for inactivation of MAO by allylamine and reactivation of allylamine-inactivated MAO by benzylamine were reinvestigated by Silverman et al.[73] [1-³H]Allylamine is incorporated into the enzyme during inactivation. The optical spectrum of the FAD coenzyme changes during inactivation from that of oxidized to reduced flavin. Denaturation, however, results in complete reoxidation of the flavin. Under these denaturating conditions, 96% of the radio-activity associated with the enzyme remains bound, indicating that allylamine is not attached to the flavin but, rather, to an amino acid residue. Although allylamine and *N*-cyclopro-pylbenzylamine (see Section II.B.2.a.(4)) appear to be oxidized by MAO to give 3-(amino acid residue)propanal adducts, two different amino acid residues seem to be involved; reactivation by benzylamine of each of these adducts occurs at different rates. There also are differences in rates of adduct release by base and acid treatment. The inactivation mechanism proposed by Silverman et al.[73] is shown in Scheme 38. Rando and Eigner[72] suggest an S_N2 mechanism for reactivation (see Scheme 37) without experimental support. On the basis of results obtained for reactivation of *N*-cyclopropylbenzylamine-inactivated MAO (see Section II.B.2.a.(4)), an elmination mechanism for reactivation of allylamine-inactivated MAO was proposed by Silverman et al.[73] (Scheme 39). Michael addition of benzylamine to the released benzyl immonium ion of acrolein could give 3-benzylamino-propanal, the presumed product that is oxidized by silver oxide to *N*-benzyl-β-alanine (Scheme 37).

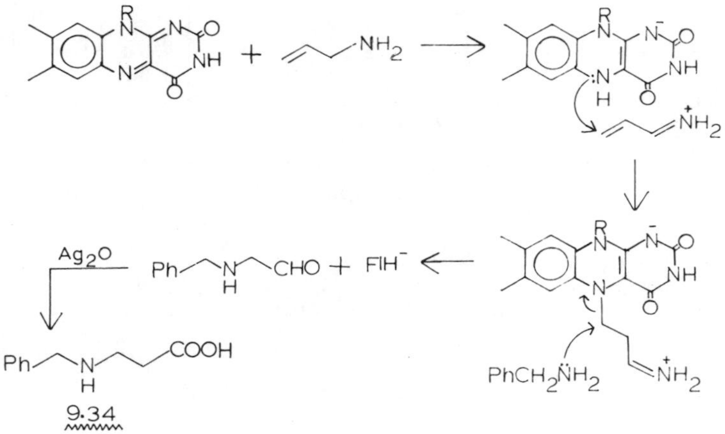

Scheme 37.   Containing Structural Formula 9.34.

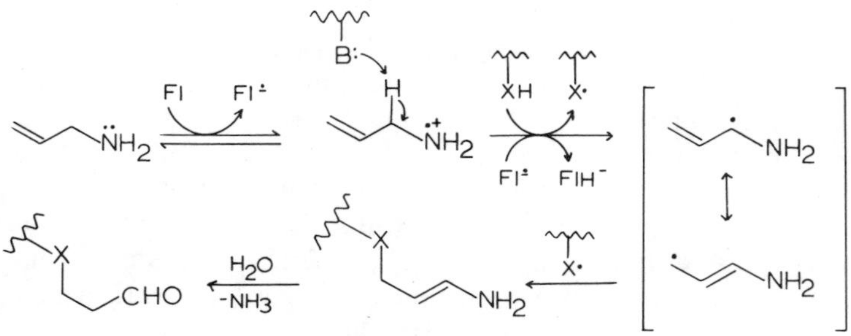

Scheme 38.

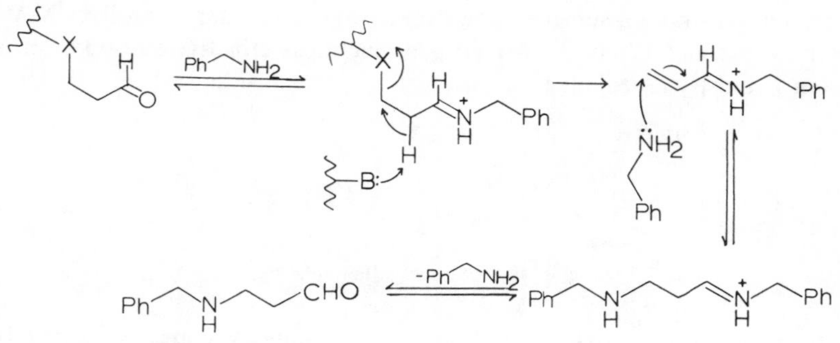

Scheme 39.

N-Benzylallylamine also inactivates MAO; in this case, however, the enzyme activity does not return upon benzylamine treatment.[72] 2-Amino-3-butene was reported[72] to inactivate MAO partially. Silverman and Yamasaki[74] found that 2-amino-3-butene completely inactivates the enzyme and forms an adduct that is stable to benzylamine treatment.

(2) 1-Methyl-4-phenyl-1,2,3,6-tetrahydropyridine (MPTP)

Symptoms of Parkinson's disease were observed by administration of 1-methyl-4-phenyl-1,2,3,6-tetrahydropyridine (MPTP; Structural Formula 9.35, Scheme 40) to man[75] and animals.[76] Pretreatment of animals with the MAO inactivators, deprenyl and pargyline, but not clorgyline, protects the animal from the neurotoxicity of MPTP.[76] This indicates that MAO is involved in the activation of MPTP. Both MAO A and B oxidize MPTP to 1-methyl-4-phenyl-2,3-dihydropyridinium ion (MPDP$^+$), although oxidation by MAO B is more rapid.[77] Time-dependent, pseudo first-order inactivation of beef liver MAO B was observed by Salach et al.[77] with 5 mM MPTP, but biphasic kinetics were observed at 1 mM MPTP, suggesting that more than one species of inactivator may be involved. With the use of fresh beef liver MAO B, its inactivation by MPTP or MPDP$^+$ was observed by Singer et al.[78] to be time-dependent and pseudo first order, even at 1 mM concentration of **9.35**. The rates of inactivation for MPTP and MPDP$^+$ are the same. 1-Methyl-4-phenylpyridinium ion (MPP$^+$) does not inactivate the enzyme. Inactivation of MAO B by [methyl-^{3}H]MPTP is accompanied by a concomitant bleaching of the flavin spectrum and incorporation of [^{3}H] into the enzyme. Upon precipitation of the inactivated enzyme and resuspension in Na-DodSO$_4$, the oxidized flavin spectrum is obtained, yet radioactivity remains bound. This suggests attachment is to an active-site amino acid residue, not to the flavin. Incorporation of radioactivity into the enzyme continues even after inactivation is complete, indicating alkylation probably can occur at more than one site. [methyl-^{3}H]MPTP inactivation of MAO B leads to the incorporation of 5 mol of radioactivity per mole of enzyme, which is stable to gel filtration or denaturation of the enzyme; flavin peptides isolated after proteolysis indicate that attachment is not to the flavin.[78a] In contrast to the report of Singer et al.,[78] MPTP was found by Fuller and Hemrick-Luecke[79] to be a competitive, reversible inhibitor of MAO A and a noncompetitive, time-dependent, partially reversible inactivator of MAO B. The results of Salach et al.[77] were disputed by Buckman and Eiduson.[80] They found that bovine liver MAO converts MPTP into a metabolite of unknown structure with a mass of 184 that absorbs at 345 nm. The metabolite could not be isolated nor characterized; however, it is not MPDP$^+$ or MPP$^+$, but is converted to MPTP with NaBH$_4$. It was shown that in the dark MAO B is **not** inactivated by MPTP, but in the light, it is. The 345-nm metabolite was shown to be responsible for the photo-induced inactivation reaction. Inactivation by this metabolite is specific for MAO B; no inactivation of placenta MAO A was observed by this metabolite in the dark or light. MPTP is a time-dependent inactivator of rat brain

MAO B[80a] and acts as both a substrate and irreversible inactivator of rat liver MAO B with a partition ratio of about 17,000.[80b] MPTP is a much less effective substrate for MAO A, and inhibition is mostly reversible.

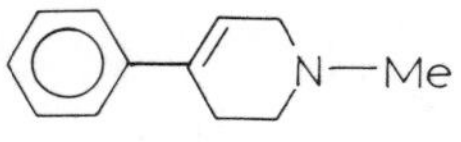

Scheme 40. Structural Formula 9.35.

Human liver monoamine oxidase B: monoclonal antibody complex (MAO B-IC2) is irreversibly inactivated in a time-dependent manner by MPTP, MPDP$^+$, and MPP$^+$.[81] The order of potency of MPDP$^+$ > MPTP > MPP$^+$. No enzyme activity is recovered upon dialysis. [^{3}H]MPTP inactivation results in incorporation of tritium into the protein. 4-Phenyl-1,2,3,6-tetrahydropyridine is not an inactivator, suggesting the importance of the *N*-methyl substitution to inactivation.

(3) *N*-Allyl-1,4-butanediamine and Related

N-Allyl-1,4-butanediamine (**9.36**, R = H) and *N*-allyl-*N*'-methyl-1,4-butanediamine (**9.36**, R = Me) (see Scheme 41) inactivate pig liver polyamine oxidase, but not with pseudo first-order kinetics.[65] Inactivation could be reversed by incubation of inactivated enzyme with substrate. These compounds also are oxidized by the enzyme to putrescine and *N*-methyl-putrescine, respectively.

$$CH_2=CHCH_2NH(CH_2)_4NHR$$

Scheme 41. Structural Formula 9.36.

(4) *N*-Allylglycine

N-Allylglycine (Structural Formula 9.37) is a time-dependent inactivator of rat liver *N*,*N*-dimethylglycine (sarcosine) oxidase; dialysis does not regenerate enzyme activity.[61] A deuterium isotope effect of 2.3 was observed for the corresponding α,α-dideuterio analogue. The mechanism proposed is shown in Scheme 42.

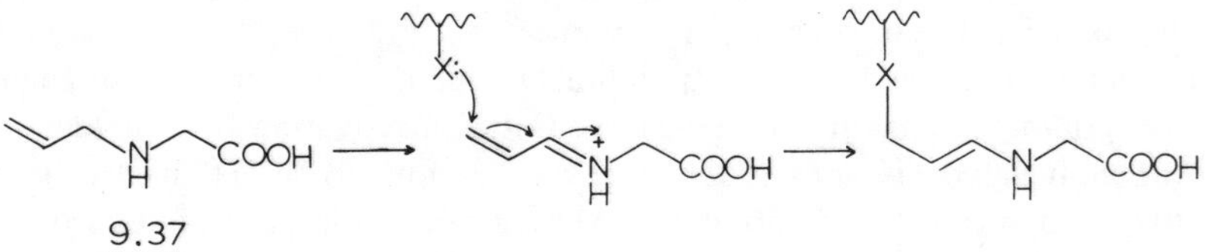

Scheme 42. Containing Structural Formula 9.37.

N-Allyl-substituted *N*,*N*-disubstituted glycines are time-dependent inactivators of rat liver mitochondrial *N*,*N*-dimethylglycine oxidase.[29,62] On the basis of an isotope effect with 1,1-dideuterio-*N*-allyl-*N*-methylglycine, it was proposed by Kraus et al.[29] that these compounds are mechanism-based inactivators of the enzyme. The synthesis of these compounds was shown in Section II.B.1.a.(2) (Scheme 27, R = allyl).

(5) Other Allylamine-Containing Inactivators

N-Allyl-*N*-methylacetylmethadol (Structural Formula 9.38, Scheme 43) is a time-dependent inactivator of rabbit and rat liver microsomal mixed-function oxidase (cytochrome

C reductase); dialysis does not regenerate enzyme activity.[63] Spectral studies indicate partial reduction of the flavin after inactivation.

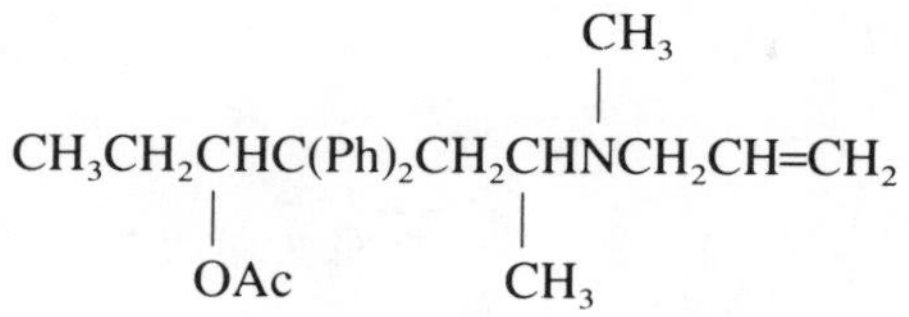

$$CH_3CH_2CHC(Ph)_2CH_2\overset{\overset{\displaystyle CH_3}{|}}{C}HNCH_2CH=CH_2$$
$$\underset{OAc}{|}\qquad\underset{CH_3}{|}$$

Scheme 43.   Structural Formula 9.38.

(6) 2-Amino-3-butenoic Acid (Vinylglycine)

D,L-Vinylglycine (Structural Formula 9.39, Scheme 44) is rapidly oxidized by the flavoenzymes, D-amino acid oxidase from pig kidney and L-amino acid oxidase from *Crotalus adamanteus* venom; however, only the L-amino acid oxidase is inactivated.[82] Turnover occurs about 2000 times prior to inactivation. Dithiothreitol does not protect the enzyme from inactivation which appears to result from attachment of vinylglycine to an active-site amino acid residue, not the flavin cofactor, because there is no bleaching of the flavin spectrum. D,L-Vinylglycine can be synthesized from 2-hydroxy-3-butenoic acid by treatment with PBr$_3$, then aqueous ammonia.[83] L-Vinylglycine was prepared from L-glutamic acid[84] and from L-methionine[85] (Scheme 44).

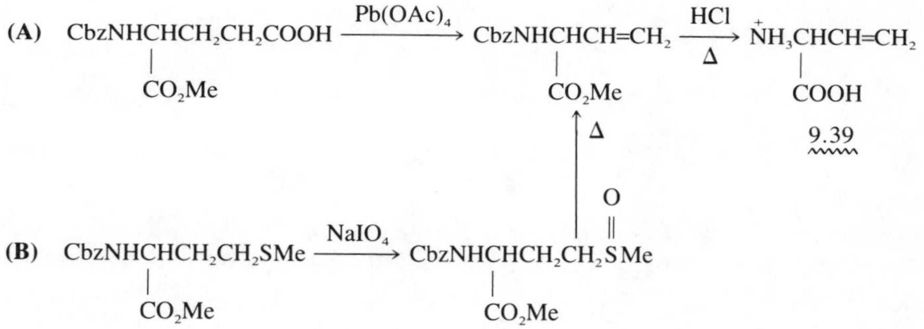

(A) $\underset{\underset{CO_2Me}{|}}{CbzNHCHCH_2CH_2COOH}\xrightarrow{Pb(OAc)_4}\underset{\underset{CO_2Me}{|}}{CbzNHCHCH=CH_2}\xrightarrow[\Delta]{HCl}\underset{\underset{COOH}{|}}{\overset{+}{N}H_3CHCH=CH_2}$

9.39

$\uparrow \Delta$

(B) $\underset{\underset{CO_2Me}{|}}{CbzNHCHCH_2CH_2SMe}\xrightarrow{NaIO_4}\underset{\underset{CO_2Me}{|}}{CbzNHCHCH_2CH_2\overset{\overset{\displaystyle O}{\|}}{S}Me}$

Scheme 44.   L-Vinylglycine (Structural Formula 9.39) prepared from (A) L-glutamic acid and (B) L-methionine.

(7) 3-Fluoroallylamine Derivatives

(*E*)-2-(3,4-Dimethoxyphenyl)-3-fluoroallylamine (Structural Formula 9.40, R = R′ = OMe; see Scheme 45) is a time-dependent inactivator of rat brain mitochondrial MAO; a much greater preference for inactivation of MAO B over MAO A was noted by Bey et al.[86] No enzyme activity was recovered by dialysis or treatment with benzylamine. Compound **9.40** (R = R′ = OMe) was synthesized[86] by the route shown in Scheme 46. Two other flavin-dependent enzymes, D-amino acid oxidase and polyamine oxidase, are not inhibited by **9.40**.[87] Unlike L-deprenyl, the standard for MAO B inactivators, this compound is not metabolized to amphetamine, has no indirect sympathetic activity, and does not inhibit monoamine uptake. The inactivation mechanism proposed by Zreika et al.[87] is shown in Scheme 47.

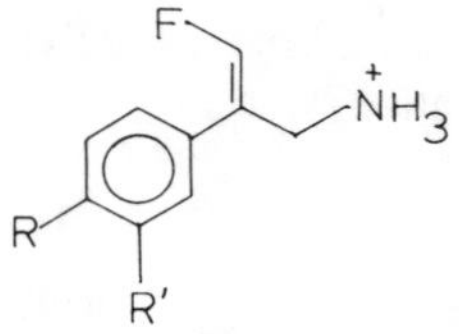

Scheme 45.   Structural Formula 9.40.

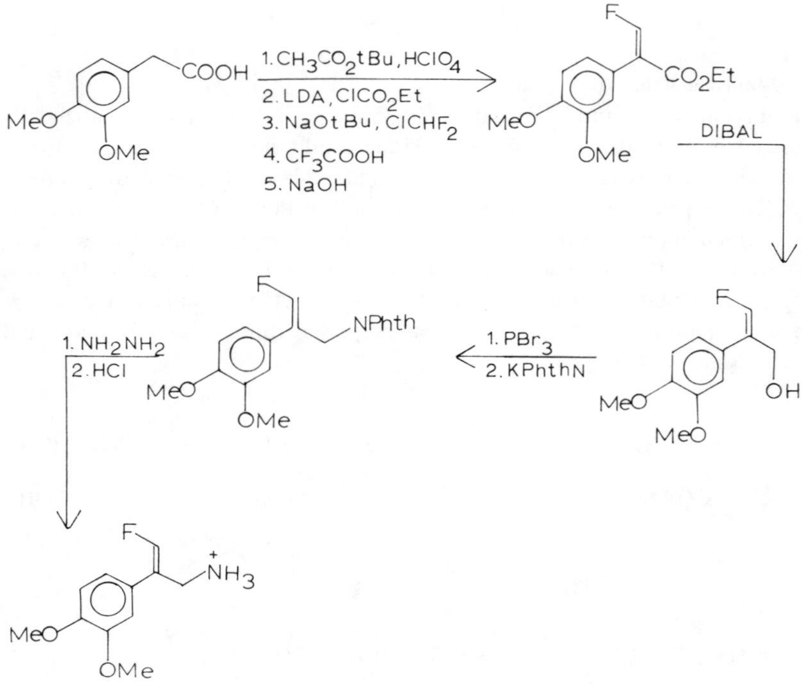

Scheme 46.

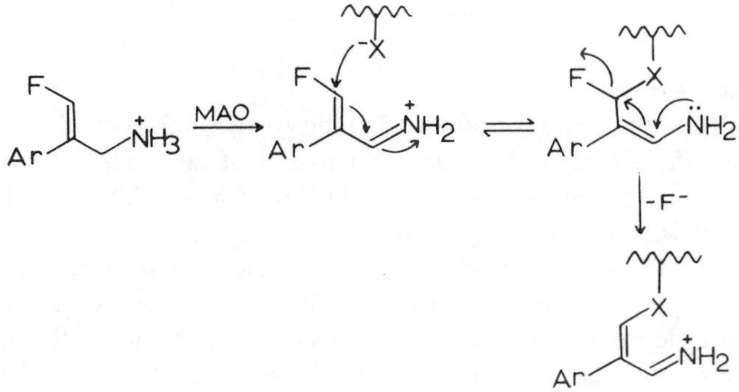

Scheme 47.

A series of (*E*)- and (*Z*)-2-aryl-3-fluoro- and 2-aryl-3,3-difluoroallylamines were synthesized by McDonald et al.[88] The synthetic route to the (*E*)-isomers is shown in Scheme 46. The (*Z*)-isomer of 2-phenyl-3-fluoroallylamine (Structural Formula 9.41) was prepared from the corresponding phthalimido-protected (*E*)-isomer (Scheme 48).

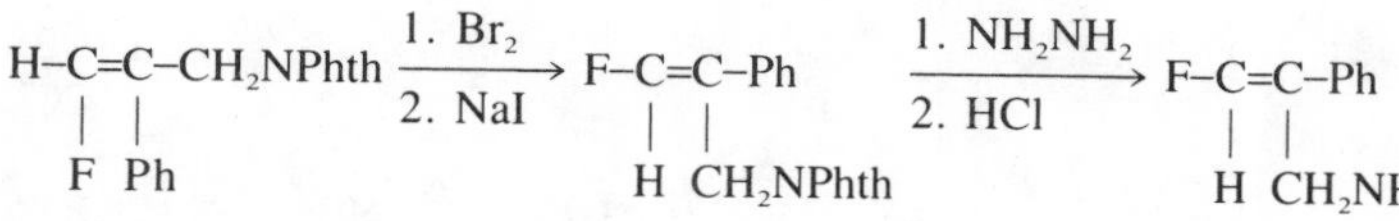

$$H-C=C-CH_2NPhth \xrightarrow[\text{2. NaI}]{\text{1. Br}_2} F-C=C-Ph \xrightarrow[\text{2. HCl}]{\text{1. NH}_2\text{NH}_2} F-C=C-Ph$$

with substituents: F, Ph below first structure; H, CH$_2$NPhth below second; H, CH$_2$NH$_3^+$ below third.

9.41

Scheme 48.   Containing Structural Formula 9.41.

2-Phenyl-3,3-difluoroallylamine (Structural Formula 9.42) was synthesized by the route in Scheme 49. All of the compounds are time-dependent inactivators of a rat brain mitochondrial preparation of MAO; pseudo first-order kinetics were observed for both MAO A and B.[88] Dialysis does not regenerate activity; β-mercaptoethanol does not inhibit the inactivation. The vinylic monofluorine derivative (chloro- and bromo analogues also were prepared; see Sections II.B.1.c.(8) and II.B.1.c.(9)) are the most potent of the nonsubstituted aromatic compounds tested. The 3,4-dimethoxy- and 4-methoxy analogues of **9.40** are as selective for MAO B as is L-deprenyl. The 3-hydroxy- and 3,4-dihydroxy analogues are less than 1/10 as selective for MAO A as is clorgyline, but still ten times more selective for MAO A than MAO B.

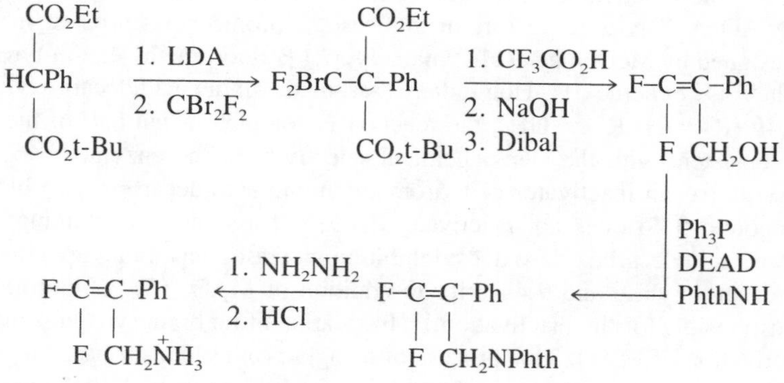

$$\text{HCPh with CO}_2\text{Et and CO}_2\text{t-Bu} \xrightarrow[\text{2. CBr}_2\text{F}_2]{\text{1. LDA}} \text{F}_2\text{BrC}-\text{C}-\text{Ph with CO}_2\text{Et and CO}_2\text{t-Bu} \xrightarrow[\substack{\text{2. NaOH}\\\text{3. Dibal}}]{\text{1. CF}_3\text{CO}_2\text{H}} F-C=C-Ph \text{ with F and CH}_2\text{OH}$$

$$F-C=C-Ph \text{ (F, CH}_2\text{NH}_3^+) \xleftarrow[\text{2. HCl}]{\text{1. NH}_2\text{NH}_2} F-C=C-Ph \text{ (F, CH}_2\text{NPhth)} \xleftarrow[\substack{\text{DEAD}\\\text{PhthNH}}]{\text{Ph}_3\text{P}}$$

9.42

Scheme 49.   Containing Structural Formula 9.42.

Several fluoroallylamines containing a β-substituted heteroatom (Structural Formula 9.43) were synthesized by McDonald and Bey[89] as shown in Scheme 50. When RX = PhS- or 2,4-dichlorophenoxy-, the products are inhibitors (they do not mention what kind) of MAO. When RX = NH$_2$, diamine oxidase inhibition was noted.

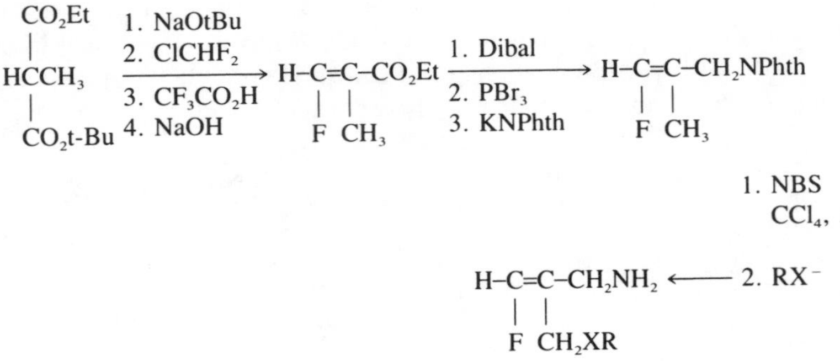

$$\begin{array}{c} \text{CO}_2\text{Et} \\ | \\ \text{HCCH}_3 \\ | \\ \text{CO}_2\text{t-Bu} \end{array} \quad \begin{array}{l} \text{1. NaOtBu} \\ \text{2. ClCHF}_2 \\ \text{3. CF}_3\text{CO}_2\text{H} \\ \text{4. NaOH} \end{array} \longrightarrow \begin{array}{c} \text{H–C=C–CO}_2\text{Et} \\ | \quad | \\ \text{F} \quad \text{CH}_3 \end{array} \quad \begin{array}{l} \text{1. Dibal} \\ \text{2. PBr}_3 \\ \text{3. KNPhth} \end{array} \longrightarrow \begin{array}{c} \text{H–C=C–CH}_2\text{NPhth} \\ | \quad | \\ \text{F} \quad \text{CH}_3 \end{array}$$

$$\begin{array}{l} \text{1. NBS} \\ \quad \text{CCl}_4, \ \Delta \\ \\ \begin{array}{c} \text{H–C=C–CH}_2\text{NH}_2 \\ | \quad | \\ \text{F} \quad \text{CH}_2\text{XR} \end{array} \longleftarrow \text{2. RX}^- \end{array}$$

9.43

Scheme 50.   Containing Structural Formula 9.43.

The type A selectivity of clorgyline has been attributed by Knoll et al.[90] to the four-atom linkage between the dichlorobenzene ring and the nitrogen atom. The fluoroallylamine analogue of clorgyline (Z)-2-(2,4-dichlorophenoxy)methyl-3-fluoroallylamine (**9.43**, RX = 2,4-dichlorophenoxy) was prepared by McDonald et al.[91] because the average aromatic ring-nitrogen distance for the two compounds appear to be similar. However, this compound is a potent, selective inactivator of rat brain MAO B, which indicates that the aromatic ring-nitrogen atom distance is not the dominant feature for selectivity.

(E)-2-(3-Hydroxyphenyl)-3-fluoroallylamine (**9.40**, R = H, R′ = OH) is a time-dependent irreversible inactivator of rat brain MAO with a several-fold preference for inactivation of MAO A.[92] A prodrug (or, in this case, a promechanism-based inactivator) of MAO was designed by McDonald et al.,[92] namely, (E)-β-fluoromethylene-m-tyrosine (Structural Formula 9.44, Scheme 51). Hog kidney L-aromatic amino acid decarboxylase converts **9.44** into **9.40** (R = H, R′ = OH); the reaction is complete when half of the substrate is consumed, consistent with the stereochemical selectivity of the enzyme.[92,92a] DL-α-(Monofluoromethyl)dopa, an inactivator of L-aromatic amino acid decarboxylase blocks this reaction. Compound **9.44** does not inactivate MAO, but a solution containing **9.44** and L-aromatic amino acid decarboxylase does; inhibition increases depending upon the incubation time of the decarboxylase and **9.44** prior to addition of MAO. (E)-β-Fluoromethylene-m-tyrosine is responsible for the inactivation.[92a] Incubation of rat brain synaptosomes with **9.44** leads to inactivation of MAO. This process of using a promechanism-based inactivator was termed dual enzyme-activated irreversible inhibition (see Chapter 1 of Volume I Section III.D.). The synthesis of **9.44** is shown in Scheme 51.[92,92b]

A series of β-substituted m-tyrosine derivatives (Structural Formula 9.44a, Scheme 52) was synthesized by McDonald et al.[92b] as potential dual enzyme-activated irreversible inhibitors of MAO. Compounds **9.44a** (X = F, Cl, or H and Y = R = H; Y = F and X = R = H; and X = F, R = Me, Y = H) are substrates for hog kidney L aromatic amino acid decarboxylase, producing the corresponding amines without inactivation of the enzyme. Compound **9.44a** (X = F, R = OMe, Y = H) is not a substrate. Aliquots removed for the incubation mixtures of those compounds that are substrates for L aromatic amino acid decarboxylase were used to inactivate MAO. The rate of inactivation of MAO is related to the length of time that the compounds are preincubated with the decarboxylase. None of these compounds inactivates MAO in the absence of preincubation with the decarboxylase. Since racemic mixtures of compounds were used, decarboxylation proceeded until 50% of

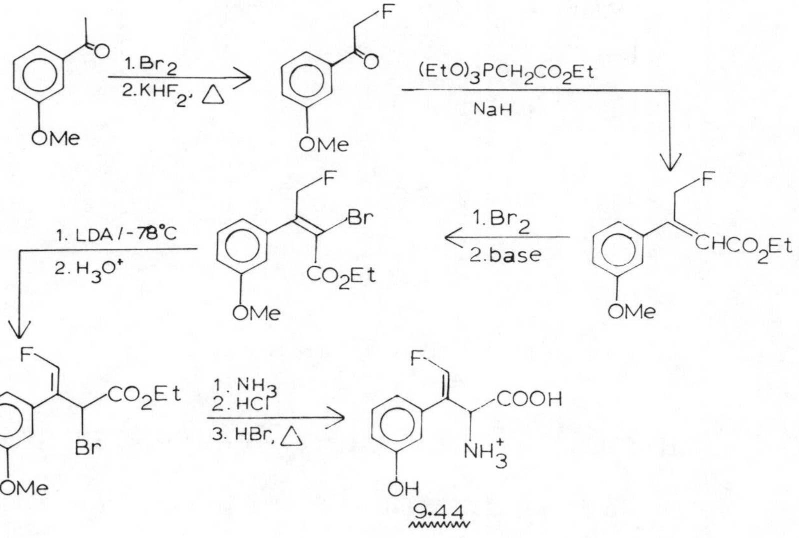

Scheme 51.   Containing Structural Formula 9.44.

the compounds were consumed. Compound **9.44a** (Y = F; X = R = H) is decarboxylated, but does not inactivate MAO. When the other compounds were tested for their selectivity with MAO A and B, all showed a preference for one isozyme. However, **9.44a** (X = F or H; Y = R = H) showed selectivity for MAO A and **9.44a** (X = Cl, Y = R = H and X = F, R = Me, Y = H) inactivated MAO B preferentially.

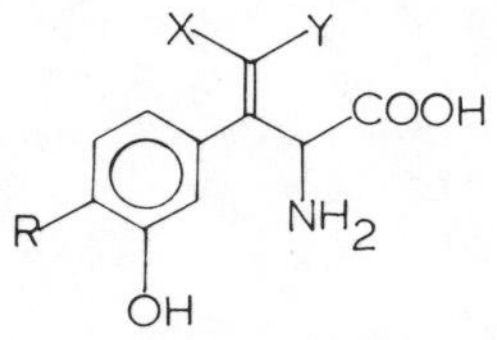

Scheme 52.   Structural Formula 9.44a.

(8) Chloroallylamine Derivatives

cis-3-Chloroallylamine (Structural Formula 9.45) is an irreversible inactivator of mono-amine oxidase; a flavin spectrum identical with that obtained when pargyline is the inactivator was observed by Rando and Eigner.[72] The *trans*-isomer is not an inactivator. The inactivation mechanism proposed is shown in Scheme 53. (Fl is oxidized flavin; FlH$^-$ is reduced flavin). Compound **9.45** can be synthesized from 1,3-dichloro-1-propene by treatment with hexamethylenetetramine[93] followed by acid hydrolysis.

(*E*)- (**9.46**, X = Cl) and (*Z*)-2-Phenyl-3-chloroallylamine (**9.47**, X = Cl) were synthesized by McDonald et al.[88] (Scheme 54). The (*E*)-isomer (**9.46**, X = Cl), but not the (*Z*)-isomer (**9.47**, X = Cl) is a time-dependent inactivator of rat brain MAO, and pseudo first-order kinetics were observed for both MAO A and MAO B.[88] Dialysis does not regenerate enyme activity and β-mercaptoethanol does not inhibit inactivation.

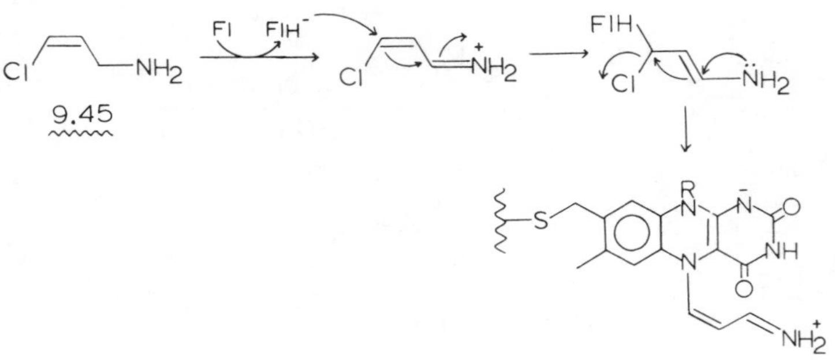

Scheme 53.   Containing Structural Formula 9.45.

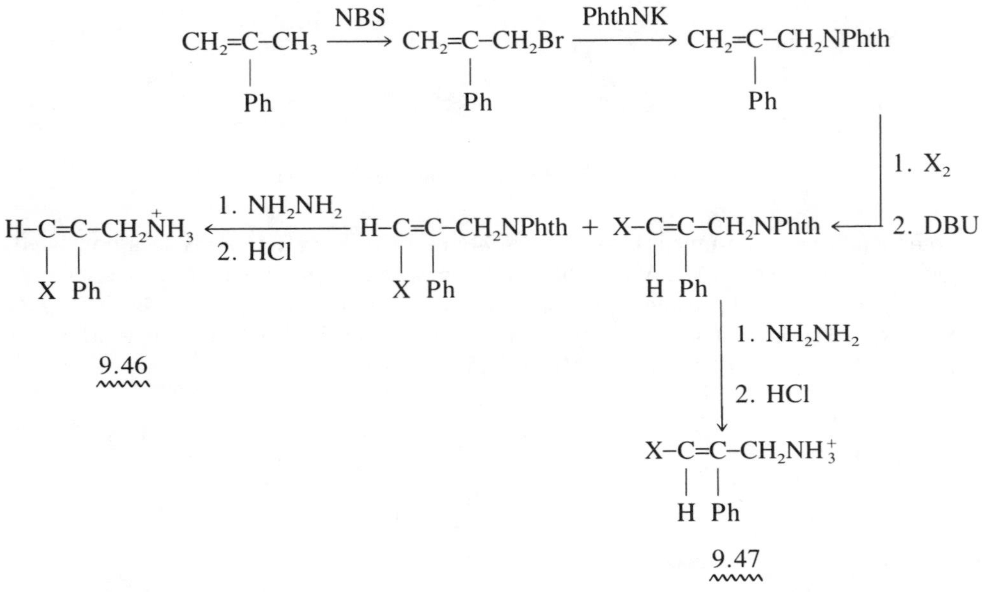

Scheme 54.   Containing Structural Formulas 9.46 and 9.47.

Rat liver N,N-dimethylglycine oxidase is irreversibly inactivated by N-(2-chloroallyl)sarcosine, but it is only half as potent as allylsarcosine.[62]

(9) 3-Bromoallylamine Derivatives

A mixture of *cis*- and *trans*-3-bromoallylamine is a time-dependent irreversible inactivator of rat liver mitochondrial monoamine oxidase; the mechanism proposed by Rando[94] (Scheme 55) was based on Hamilton's[95] mechanism for MAO, which has not had much support, since it is an apparent violation of the principle of microscopic reversibility. 1,3-Dibromo-1-propene can be treated with hexamethylenetetramine,[93] followed by hydrolysis, to give 3-bromoallylamine.

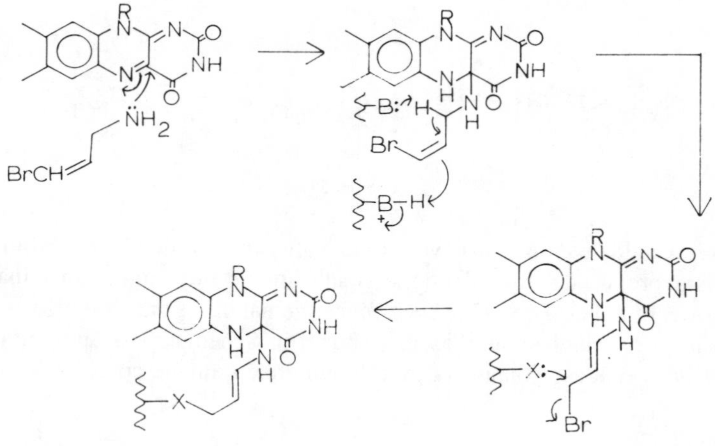

Scheme 55.

(*E*)-2-Phenyl-3-bromoallylamine (**9.46**, X = Br) was synthesized by McDonald et al.[88] as shown in Scheme 54. 2-Phenyl-3,3-dibromoallylamine (Structural Formula 9.48) was prepared by a similar route (Scheme 56). Both compounds are time-dependent inactivators of rat brain MAO A and B. Dialysis does not regenerate enzyme activity and β-mercaptoethanol does not inhibit inactivation.

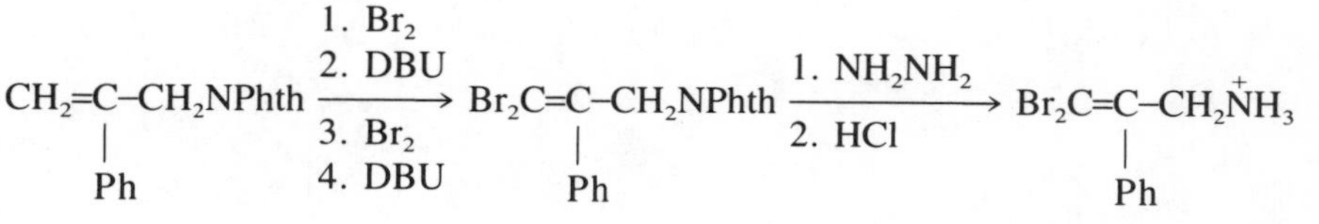

Scheme 56. Containing Structural Formula 9.48.

d. 2-Substituted Ethylamines

Beef plasma amine oxidase is irreversibly inactivated by 2-bromoethylamine with a partition ratio of 8.[96] An equivalent amount of oxygen is consumed during turnover. No inactivation occurs if the enzyme is in the reduced form. When 2-bromo-[U-^{14}C]ethylamine is used, 2 mol of [^{14}C] are incorporated per mole of enzyme dimer. 2-Chloroethylamine is oxidized, but does not inactivate the enzyme, presumably because chloride is a poorer leaving group than is bromide. The mechanism proposed by Neumann et al.[96] is shown in Scheme 57. It is not clear if alkylation occurs when the enzyme is in the oxidized or reduced form. Acid or Pronase hydrolysis of the labeled and sodium borohydride-reduced enzyme gives about 12 different radioactive products. One of these was identified by Suva and Abeles[97] as *S*-(2-hydroxyethyl)cysteine, suggesting that a cysteine residue (*inter alia*) is labeled during inactivation.

A series of β-substituted ethylamines was investigated by Tang et al.[93] as mechanism-based inactivators of bovine aorta lysyl oxidase. The substituents chosen were bromo, chloro, nitro, hydroxyl, and β,β,β-trifluoro. The first three are time-dependent inactivators which exhibit biphasic kinetics. [1,2-^{14}C]β-Bromoethylamine, as in the case of [1,2-^{14}C]β-aminopropionitrile (see Chapter 7, Section VII.C.2.) inactivates lysyl oxidase with incorporation of only 0.17 equivalent of radioactivity bound. Low active enzyme was suggested as the

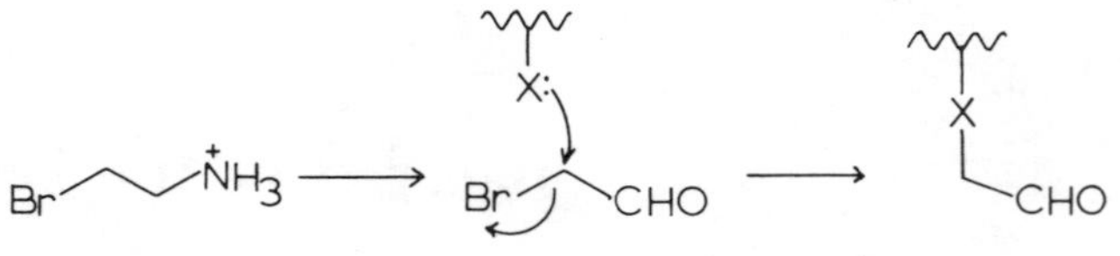

Scheme 57.

cause for this. Acetaldehyde is a turnover product after inactivation by the bromo compound. The mechanism proposed[98] for all of the β-substituted ethylamines that inactivate lysyl oxidase is shown in Scheme 58. However, since the nature of the oxidation is not clear, an oxidation/addition mechanism such as that shown in Scheme 57 is appropriate. It is now known[98a] that lysyl oxidase contains a pyrroloquinoline quinone cofactor.

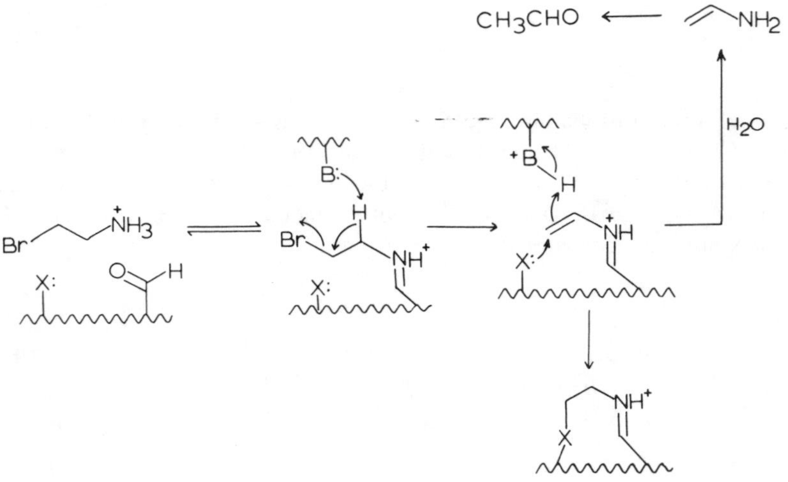

Scheme 58.

Weyler[98b] found that aerobic and anaerobic incubation of beef liver MAO B with 2-chloro-2-phenylethylamine leads to irreversible inactivation; enzyme activity is not restored by gel filtration or dialysis. Aerobic inactivation proceeds at about three times the rate of anaerobic inactivation. However, it was found that 2-chloro-2-phenylethylamine is rapidly oxidized to 2-chloro-2-phenyl-acetaldehyde, believed to be the actual inactivating species. A thiol protection experiment, however, was not carried out, so it is not clear if inactivation occurs after release of the aldehyde. If this is the case, then 2-chloro-2-phenylethylamine is not a mechanism-based inactivation. The anaerobic inactivation mechanism is not known.

e. 3-{4-[(3-Chlorophenyl)methoxy]phenyl}-5-[(methylamino)methyl]-2-oxazolidinone

Rat liver MAO is inactivated in a time-dependent manner by 3-{4-[(3-chlorophenyl)methoxy]phenyl}-5-[(methylamino)methyl]-2-oxazolidinone methanesulfonate (Structural Formula 9.49, Scheme 59).[99] The K_i value with MAO A is seven times greater than that with MAO B. This compound also is a substrate for both MAO A and B. The partition ratio for MAO B is about 530; the aldehyde was assumed by Tipton et al.[99] to be the product. The *R*- and *S*-enantiomers of **9.49** were separated by Dostert et al.[100] The two isomers are about equipotent inhibitors of MAO B, but the *S*-isomer is more active against MAO A. The mechanism proposed[100] for inactivation is shown in Scheme 60. It is not clear why the intermediate immonium ion in Scheme 60 should be any more reactive than other oxidized amines that do not inactivate the enzyme. The metabolism of the (*R*)- and (*S*)-enantiomers of **9.49** was studied by Strolin-Benedetti et al.[101] ex vivo and in vitro in rat liver and brain. The (*R*)-isomer and the (*R,S*)-racemate appear to be metabolized to the aldehyde in a similar way. The (*S*)-isomer appears to be the agent responsible for irreversible inactivation of MAO B; this isomer is selective for MAO B.

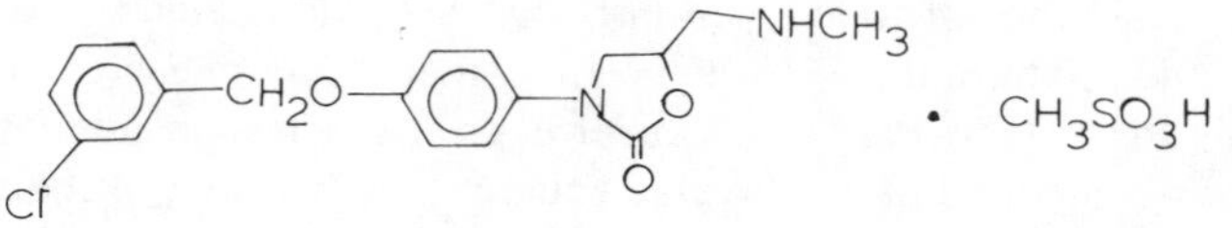

Scheme 59.   Structural Formula 9.49.

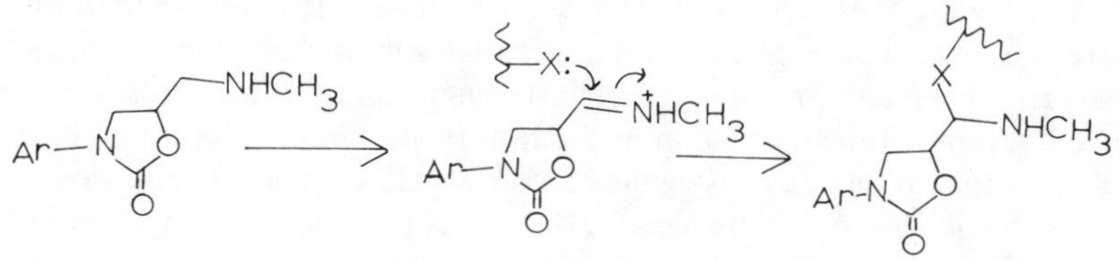

Scheme 60.

f. Cyclopropylamines

It was originally suggested by Silverman and Hoffman,[102] Paech et al.,[103] and Hanzlik et al.[104] that cyclopropylamines inactivate MAO[102,103] and cytochrome P-450[104] by a mechanism involving oxidation to a cyclopropanimmonium ion (Structural Formula 9.50, Scheme 61). This mechanism was proven to be incorrect (see Section II.B.2.a.).

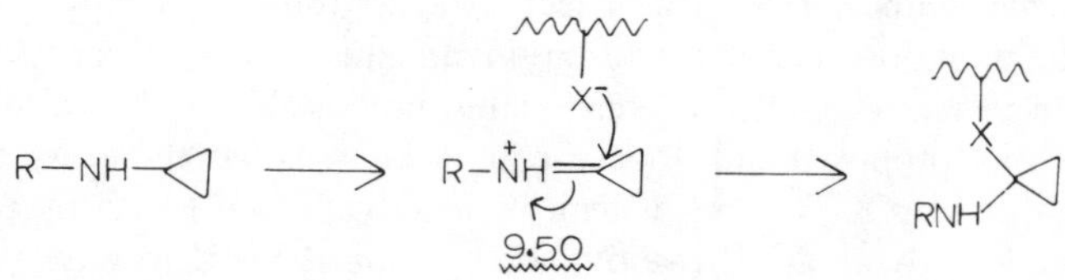

Scheme 61.   Containing Structural Formula 9.50.

2. With Carbon-Carbon Bond Cleavage
a. Cyclopropylamines
(1) *trans*-2-Phenylcyclopropylamine (Tranylcypromine) and Related

trans-2-Phenylcyclopropylamine (Structural Formula 9.51, Scheme 62) is a time-dependent inactivator of beef liver mitochondrial MAO;[105,106] oxygen is not required for inactivation.[105] Also, *trans*-2-methylcyclopropylamine and *cis*-2-phenylcyclopropylamine are potent inhibitors. Inactivation of MAO by tranylcypromine is reversed upon treatment with 4-phenyl-*n*-butylamine. Compound **9.51** also is a time-dependent inactivator of rat liver mitochondrial MAO; dialysis does not restore enzyme activity.[107] The inhibitory effect of *trans*-2-phenylcyclopropylamine was rationalized by Belleau and Moran[108] as resulting from its properties as a transition-state analogue. It was suggested that since the α-carbon of substrates develops sp² character in the transition state, tranylcypromine mimics this (cyclopropane has sp² character in the ground state). Also, π-electrons develop in the transition state, so the electron density of the cyclopropane ring resembles this as well. A comparative inhibition study was made by Burger and Nara[109] of (+)-, (−)-, and (±)-*trans*-2-phenylcyclopropylamine, (±)-*cis*-2-phenylcyclopropylamine, (±)-*trans*-N,N-dimethyl-N-(2-phenylcyclopropyl)amine, pargyline, and α-methylphenethylhydrazine with purified beef liver mitochondrial MAO. The (+)-tranylcypromine, (±)-*N,N*-dimethyl tranylcypromine, and pargyline are the most potent inactivators. Inactivation of human liver mitochondrial MAO by tranylcypromine and *cis*-2-phenylcyclopropylamine is pseudo first-order and time-dependent.[26] The (+)-isomer of tranylcypromine has a K_i value equal to one half of the K_i value of the racemic mixture; the K_i for the (−)-isomer is 50 times greater than that for the (±)-isomer. This suggests that the (+)-isomer is primarily responsible for the competitive inhibition. The *cis*-compound has a K_i value equal to one half that of the *trans*-compound. The rate data for inactivation of MAO by different 2-phenylcyclopropylamine compounds parallel the relative K_i data except that the *cis*-isomer is less effective than tranylcypromine. Hellerman and Erwin[51] found that when increasing amounts of (+)-tranylcypromine are allowed to react with MAO, a linear decrease of enzyme activity occurs until complete inhibition results with 1 mol of inactivator per mole of enzyme; inhibition, however, is partially reversed by prolonged dialysis against benzylamine. It also was found that tranylcypromine does inhibit the enzyme in its reduced flavin form. Contrary to these results,[51] Paech et al.[103] observed that dialysis against buffer containing benzylamine does not restore enzyme activity. The rates of inactivation for the D- and L-isomers of tranylcypromine differ. The apparent second-order rate constant for the D-isomer is 77 times greater than that for the L-isomer; the apparent K_i for the D-isomer (4 μM) is 83 times lower than that for the L-isomer. The D-isomer, therefore, is a much more efficient inactivator than the L-isomer. Concomitant with inactivation, the flavin spectrum is converted to its reduced form. Complete inactivation and bleaching of the flavin spectrum occurs with 1.1 to 1.2 mol of D-tranylcypromine per mole of enzyme. With [^{14}C]-labeled inactivator, 1 mol of radioactivity becomes incorporated into the enzyme. Also contrary to the report of Hellerman and Erwin,[51] the enzyme must be in its oxidized form in order for inactivation to occur. It was suggested previously[51] that arylcyclopropylamines combine with MAO at the same site as arylhydrazines and acetylenic amines. However, Paech et al.[103] found that tranylcypromine reacts with an active-site amino acid residue, but arylhydrazines[110] and propargylamines[59] form a bond to the flavin. The mechanism of inactivation proposed[103] is shown in Scheme 63. A cysteine residue was proposed only because it is known that there are free active site sulfhydryl groups in MAO. The mechanism of inactivation of MAO by tranylcypromine was reinvestigated by Silverman.[111] Inactivation by *trans*-2-[2-^{14}C]phenylcyclopropylamine results in the incorporation of 1.08 mol of [^{14}C]. Acid denaturation of labeled enzyme in the presence of 2,4-dinitrophenylhydrazine produces the 2,4-dinitrophenylhydrazone of cinnamaldehyde, not of 2-phenylcyclopropanone, as was previously reported by Paech et al.[103] Also, the product of diazotization, hydrolysis, oxidation, and 2,4-dinitrophenylhydrazine treatment of tranylcypromine is the 2,4-dinitrophenylhydrazone of cinnamaldehyde, not of 2-phenylcyclopropanone. Denaturation of labeled MAO results in the release of 94% of the radioactivity, but when the labeled enzyme is treated with sodium borohydride prior to denaturation, only 68% of the radioactivity is released. All of these results are consistent with the mechanism shown in Scheme 64.[111] Paech et al.[103] showed that X is not the flavin. The X· could arise from hydrogen atom abstraction of some amino acid residue (XH) by FlH·.

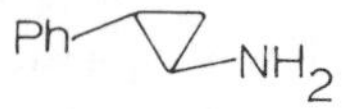

Scheme 62.   Structural Formula 9.51.

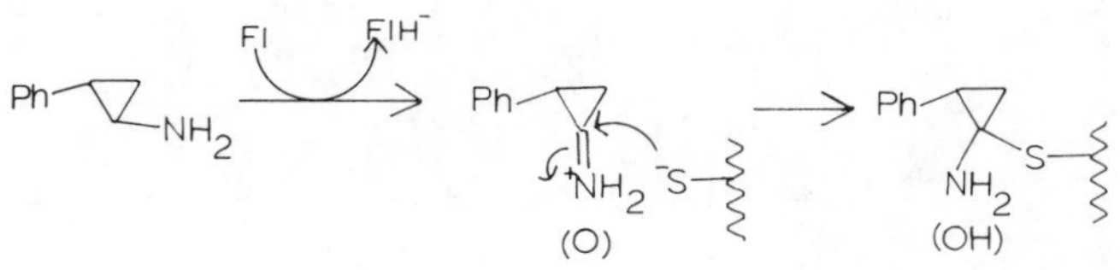

Scheme 63.

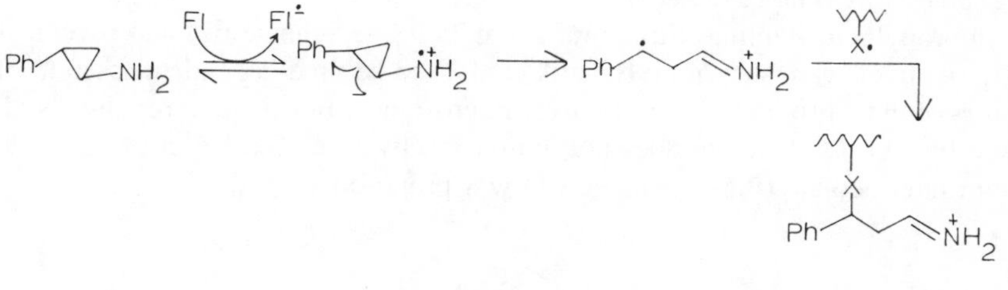

Scheme 64.

(2) *N*-(Phenoxyethyl)cyclopropylamines and Related

A series of phenyl-substituted *N*-(phenoxyethyl)cyclopropylamines (Structural Formula 9.52, Scheme 65) were shown by Fuller et al.[112] to be potent inhibitors of MAO. The type of inhibition, however, was not investigated. The inactivation of MAO by **9.52** (R = *o*-Cl) is time-dependent and irreversible to dialysis.[113] Other derivatives[114] showed similar behavior, but no mechanism was presented. Some selectivity for inhibition of serotonin oxidation was shown over that for phenethylamine oxidation. At the time this was published, however, the MAO A and MAO B terminology of Johnston[36] was just being published, so this compound was not yet termed an MAO A inhibitor.

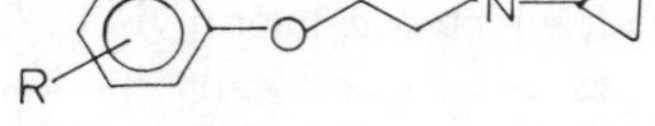

Scheme 65.   Structural Formula 9.52.

The simple substitution of iodine for chlorine to give **9.52** (R = *o*-I) produces a more potent and selective time-dependent irreversible inhibitor of MAO.[115] Whereas the chloro analogue has an IC$_{50}$ for MAO A of 1.2×10^{-9} *M*, the iodo compound has an IC$_{50}$ of 4×10^{-10} *M*. The IC$_{50}$ for MAO B by the chloro and iodo derivatives are 8×10^{-7} and 1×10^{-6} *M*, respectively. Therefore, the iodo compound binds more tightly to MAO A and less tightly to MAO B than does the chloro compound. Harmaline, a selective reversible inhibitor for MAO A protects the enzyme from inactivation. This potency and selectivity also is apparent in vivo.

(3) 5-Phenyl-3-(*N*-cyclopropyl)ethylamine-1,2,4-oxadiazole and Related

5-Phenyl-3-(*N*-cyclopropyl)ethylamine-1,2,4-oxadiazole (Structural Formula 9.53, Ar = Ph; see Scheme 66) is a time-dependent inactivator of two forms of rat liver MAO, the one which utilizes serotonin and the one which utilizes tyramine as substrate (MAO A); the activity that oxidizes benzylamine (MAO B) is much less sensitive to this compound.[116] The 5-(2-chlorophneyl) analogue shows some selectivity for MAO A, the 5-(3-chlorophenyl) analogue shows no selectivity, and the 5-(3-nitrophenyl)analogue is selective for MAO B.[117]

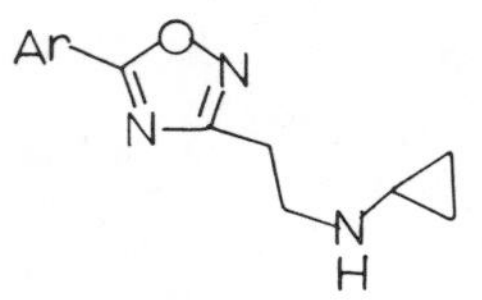

Scheme 66.   Structural Formula 9.53.

(4) *N*-Cyclopropylbenzylamine and Related

N-Cyclopropylbenzylamine (Structural Formula 9.54, Scheme 67) and several para-substituted derivatives were shown by Hanzlik et al.[104] to be time-dependent, pseudo first-order inactivators of cytochrome P-450 in rat liver microsomes. Inactivation requires NADPH and is blocked by CO. Glutathione does not inhibit inactivation. Oxidation to the cyclopropan-immonium intermediate (**9.50**, Scheme 61) was proposed.

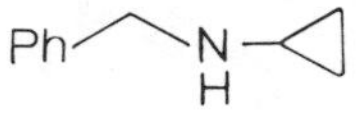

Scheme 67.   Structural Formula 9.54.

Pig liver MAO is irreversibly inactivated by *N*-cyclopropylbenzylamine (**9.54**) and *N*-cyclopropyltryptamine; dialysis does not regenerate enzyme activity.[102] Inactivation by [phenyl-^{14}C]-**9.54** leads to incorporation of 1.2 to 1.4 equivalents of radioactivity per mole of enzyme. An equivalent of nonamine tritium is released from *N*-[1-^{3}H]-**9.54** during inactivation. Silverman and Hoffman[102] wrongly assumed that the released tritium was [3]H$_2$O since *N*-[1-^{2}H]-**9.54** inactivates the enzyme with a deuterium isotope effect of 1.5. Neither *N*-isopropylbenzylamine nor *N*-cyclopropylmethylbenzylamine inactivate MAO, indicating the requirement of a cyclopropyl group attached directly to the nitrogen. The enzyme in its reduced form is not inactivated by **9.54**. Benzylamine reactivates **9.54**-inactivated MAO.[102,118] The inactivation mechanism proposed, which was later revised,[119] is shown in Scheme 61. Since amines can be oxidized to imines electrochemically, photochemically, and with chemical oxidants, and the mechanism of oxidation by all three of these chemical means is known to proceed via two one-electron transfer processes, it was proposed by Silverman et al.[1] that MAO also oxidizes amines by this mechanism. Compound **9.54** labeled with [^{14}C] in the benzyl group or at carbon-1 of the cyclopropyl group with [^{3}H] was allowed to be oxidized both by MAO and electrochemically and the product distribution was measured. The similarity in the results by both oxidation methods supported the mechanistic hypothesis shown in Scheme 68 which was later revised.[119]

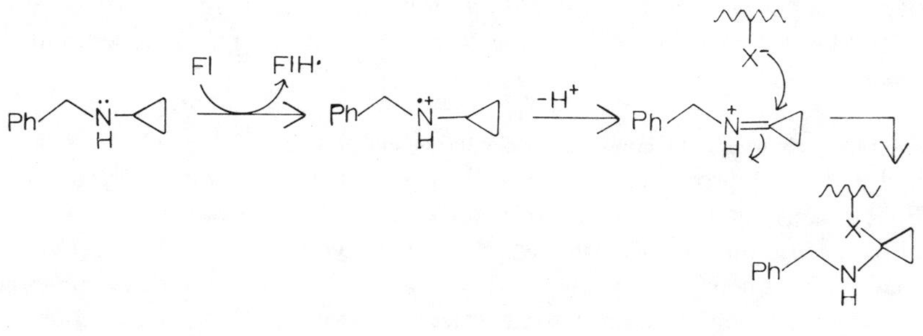

Scheme 68.

The earlier mechanisms (Schemes 61 and 68) proposed for the inactivation of MAO by **9.54**[102] were reinvestigated by Vazquez and Silverman.[119] The nonamine tritium released from *N*-[1-³H]-**9.54** during inactivation was identified as [³H]acrolein, not ³H₂O. Even after the enzyme is completely inactivated, [³H]acrolein continues to be released in a linear, time-dependent fashion. This is accounted for by a **9.54**-catalyzed regeneration of enzyme activity by a mechanism much the same as suggested for benzylamine reactivation of **9.54**-inactivated enzyme (see Scheme 72, Pathway a̱). The rate of inactivation, however, is nearly 500 times faster than the rate of reactivation; consequently, the enzyme appears to be inactive even though acrolein is being generated. One equivalent of [³H]acrolein is released by benzylamine from the tritiated enzyme formed from inactivation by *N*-[1-³H]-**9.54**. In accord with the incorporation of tritium into the enzyme during inactivation, the revised mechanism for inactivation of MAO by **9.54** is shown in Scheme 69 (R = R' = H).

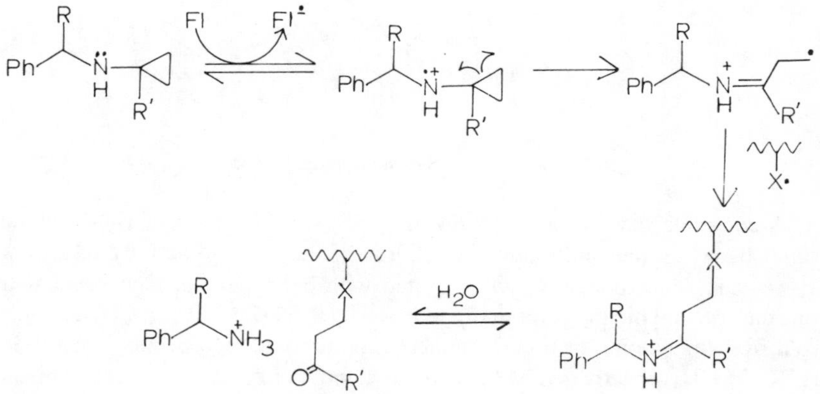

Scheme 69.

During inactivation of MAO by **9.54**, oxidation of the benzyl substituent of **9.54** competes with inactivation. The reactions leading to benzyl oxidation (pathway a̱) and inactivation (pathway ḇ) are shown in Scheme 70 (R = H). It is known that for electrochemical oxidation of secondary amines (a model for MAO-catalyzed oxidation) the ratio of α-proton removal from the two carbons adjacent to the amine radical cation intermediate depends upon their kinetic acidities. As a carbon atom becomes more alkyl substituted, the pK_a of a proton attached to the central carbon atom increases. Silverman[120] found that substitution of a methyl group at the benzyl methylene group of **9.54** (Scheme 70, R = Me), which decreases the kinetic acidity of the remaining benzyl methylene proton, results in α-methylbenzyl oxidation at a rate of only 1% of that for cyclopropyl ring opening. With **9.54**, benzyl oxidation and cyclopropyl oxidation occur equally. The K_i values for (*R*)-(+)- and (*S*)-(−)-α-methylbenzylamine were shown to be similar. This indicates that α,α-dimethylation of **9.54** should not interfere with binding. In fact, *N*-cyclopropyl-α,α-dimethylbenzylamine (Structural Formula 9.55, R = R' = CH₃; see Scheme 71) also is a mechanism-based inactivator.

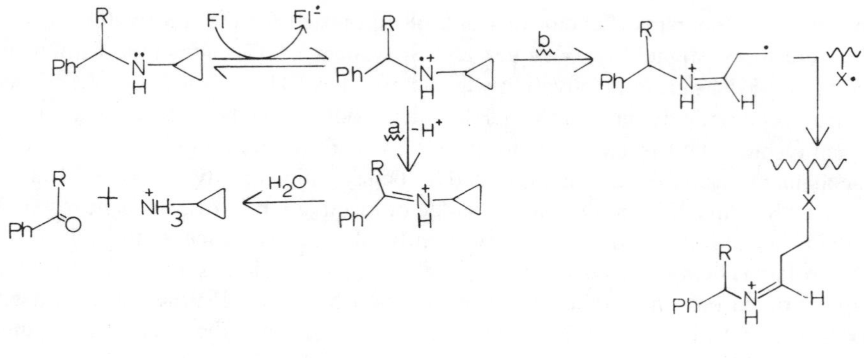

Scheme 70.

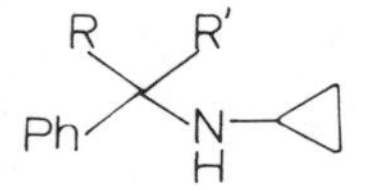

Scheme 71.   Structural Formula 9.55.

Whereas complete reactivation of **9.54**- or **9.55** (R = H, R′ = CH$_3$)-inactivated MAO occurs with benzylamine, only about two thirds of the enzyme activity returns when **9.55** (R = R′ = CH$_3$)-inactivated MAO is treated with benzylamine. Consistent with this observation, the flavin which is completely reduced by **9.55** (R = R′ = CH$_3$), is only partially reoxidized by benzylamine or urea; complete flavin reoxidation occurs when **9.54**- or **9.55** (R = H, R′ = CH$_3$)-inactivated MAO is treated with benzylamine or urea. This is evidence for two different adducts produced by **9.55** (R = R′ = CH$_3$), one attached to the flavin, and another to an amino acid residue.

When MAO is inactivated by *N*-[1-^{3}H]-**9.55** (R = H, R′ = CH$_3$), only 1 equivalent of tritium is incorporated into the enzyme instead of 3 equivalents with *N*-[1-^{3}H]-**9.54**. The adducts formed when **9.54**, **9.55** (R = H, R′ = CH$_3$), and *N*-methyl-**9.54** inactivate MAO appear to be identical based on rates of reactivation by benzylamine at pH 7.2 and 9.0. These three inactivators become attached to an amino acid residue.

The mechanism for reactivation of **9.54**-inactivated MAO by benzylamine was studied by Yamasaki and Silverman.[118] Eighteen different amines, two mercaptans, and two alcohols were tested as reactivators of the **9.54**-inactivated enzyme. All of the compounds that reactivate the enzyme produce a time-dependent, pseudo first-order return of enzyme activity and exhibit saturation kinetics. There is no direct correlation between the ability of a compound to serve as a substrate for native MAO and its ability to reactivate **9.54**-inactivated MAO. Amines containing an aromatic moiety, in general, are better reactivators than the aliphatic amines. The amine must be primary or secondary in order for reactivation to occur. The distance between the aromatic portion and the amino group is critical to the reactivation properties of the compound. The mercaptans and alcohols do not reactivate **9.54**-inactivated MAO nor do they interfere with reactivation by benzylamine. Three mechanisms for the reactivation reaction were considered (Scheme 72); the evidence supports mechanism <u>a</u>.

The synthesis of **9.54**[121] is outlined in Scheme 73. Compound **9.55** (R = H, R′ = CH$_3$) was prepared[120] by the same route with the substitution of acetophenone for benzaldehyde. The second methyl group for **9.55** (R = R′ = CH$_3$) was incorporated into Cbz-protected **9.55** (R = H, R′ = CH$_3$) with lithium cyclohexylisopropylamine and methyl iodide.[120]

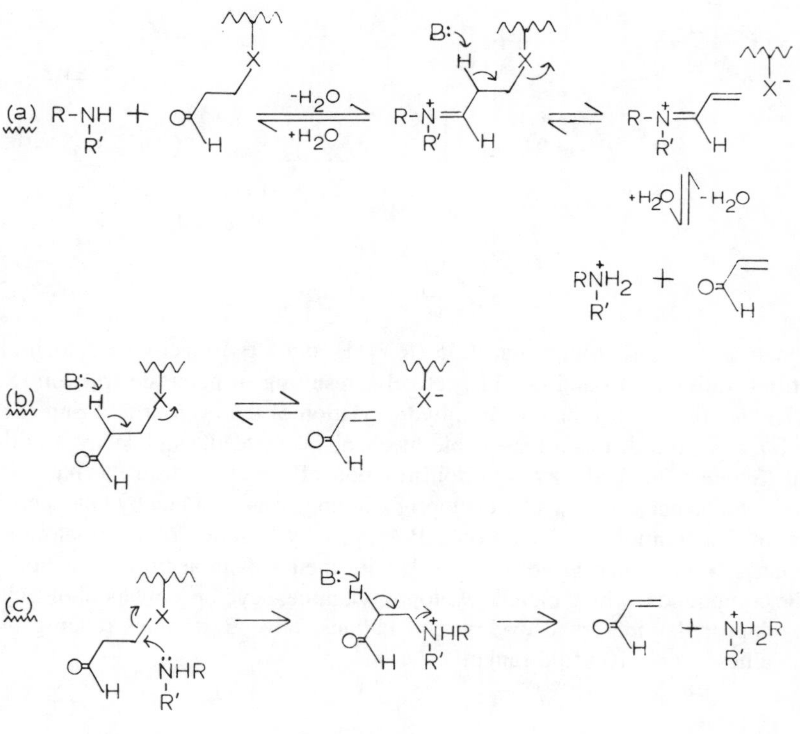

Scheme 72.

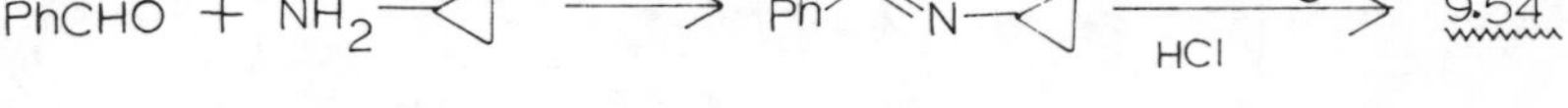

Scheme 73.

N-Cyclopropylbenzylamine (Structural Formula 9.56, R = H; see Scheme 74) labeled with either [³H] or [¹⁴C] at the benzylic position produces a time-dependent incorporation of radioactivity into rat liver microsomes.[3] When [7-¹⁴C, 7-³H]-**9.56** (R = H) is used, the [³H]/[¹⁴C] ratio remains constant during inactivation. Glutathione inhibits the total amount of radioactivity incorporated, but does not affect inactivation of P-450. *N*-(1-Methylcyclo-propyl)benzylamine (**9.56**, R = CH₃) also is a time-dependent inactivator, and [³H] from [7-³H]-**9.56** (R = CH₃) is covalently incorporated with inactivation. Two mechanisms were proposed by Hanzlik and Tullman[3] (Scheme 75). Because of the retention of the radioactive label with [³H]- or [¹⁴C]-**9.56** (R = H), pathway a̲ was favored. The cation-radical mechanism also was supported by lack of reaction of **9.56** (R = H) with strong oxidants.

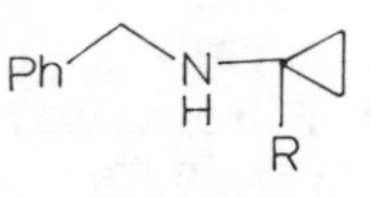

Scheme 74.   Structural Formula 9.56.

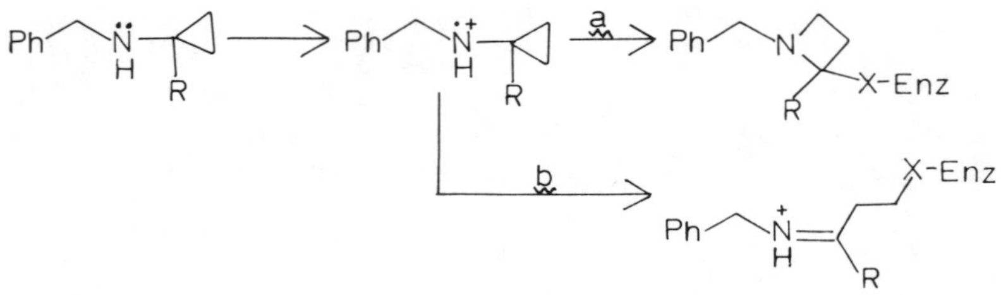

Scheme 75.

Macdonald et al.[4] also found that **9.56** (R = H and CH$_3$) inactivate cytochrome P-450 with partition ratios of 40 and 60, respectively, resulting in heme destruction. Since **9.56** (R = CH$_3$) also inactivates the enzyme, hydroxylation of the cyclopropyl ring (Scheme 76, pathway b) was excluded as a reasonable mechanism. Instead, one-electron oxidation was suggested (Scheme 76, pathway a). Confirmation of a one-electron mechanism for inactivation of cytochrome P-450$_{PB-B}$ by cyclopropylamines was obtained by Guengerich et al.[122] The rates of inactivation of cytochrome P-450$_{PB-B}$ by a series of heteroatom-substituted cyclopropanes were shown to correlate well with their single-electron oxidation potentials (E$_{1/2}$). The compounds, which include cyclopropylamines, cyclopropyl alcohols, cyclopropyl ethers, cyclopropylamides, and cyclopropyl halides, have E$_{1/2}$ values ranging over 1.5 V and k$_{inact}$ values over a 100-fold range.

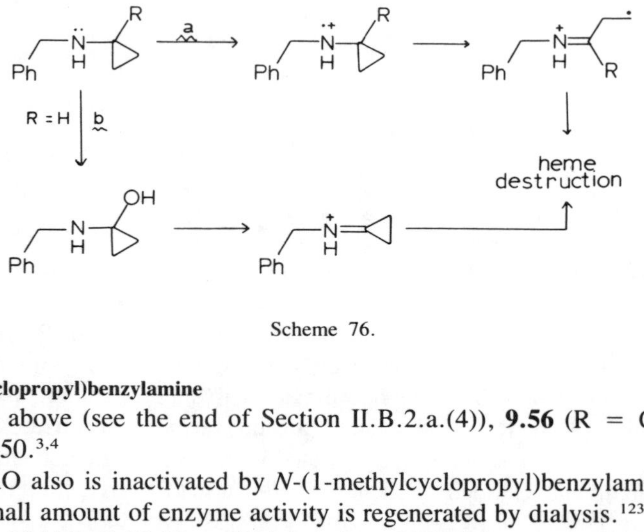

Scheme 76.

(5) N-(1-Methylcyclopropyl)benzylamine

As discussed above (see the end of Section II.B.2.a.(4)), **9.56** (R = CH$_3$) inactivates cytochrome P-450.[3,4]

Pig liver MAO also is inactivated by N-(1-methylcyclopropyl)benzylamine (**9.56**, R = CH$_3$); only a small amount of enzyme activity is regenerated by dialysis.[123] Unlike **9.56** (R = H)-inactivated MAO, **9.56** (R = CH$_3$)-inactivated MAO is not reactivated by benzylamine. This indicates that a more stable adduct is formed. Two possible mechanisms were suggested by Silverman and Hoffman[123] (Scheme 77). In order to elucidate the mechanism of inactivation of MAO by **9.56** (R = CH$_3$), the inactivator was synthesized with three different radioactive labels (Structural Formulas 9.57a, b, c; Scheme 78).[74] Inactivation of either beef or pig liver MAO by **9.57b** and **9.57c** leads to incorporation of about 1 mol of radioactivity per mole of enzyme after dialysis or urea denaturation; essentially no radioactivity from **9.57a** remains bound. Benzylamine treatment does not release the enzyme-bound radioactivity. The benzyl moiety from **9.57a** is released as benzylamine. In order to account for this distribution of labels, the mechanism shown in Scheme 79 was proposed by Silverman and Yamasaki.[74] Several organic reactions were carried out on the inactivated enzyme in order to identify the structure of the adduct. The benzylamine Schiff base product

(**9.58**) was supported by treatment of the **9.57a**-inactivated enzyme with sodium cyanoborohydride prior to dialysis. In this case, label in the benzyl group remains attached to the enzyme, as expected for reduction of the immonium ion to the amine. The structure of **9.59** was elucidated by three reactions. Sodium boro[³H]hydride treatment leads to incorporation of one tritium, as expected for reduction of the ketone. The iodoform reaction (KI₃/⁻OH) on the **9.57b**-inactivated enzyme produces [¹⁴C]iodoform in a 57% yield, thereby confirming a methyl ketone. Base was shown to release nonamine radioactivity (later identified as methyl vinyl ketone), but not if the adduct was pretreated with NaBH₄. The methyl vinyl ketone would be derived from a retro-Michael reaction (Scheme 80). The X group on the enzyme to which the adduct is attached was shown to be the flavin cofactor. The ring opening step was shown to occur with the amine radical cation, not the amine radical, since the rate of inactivation by *N*-methyl-**9.56** (R = CH₃) is the same as for **9.56** (R = CH₃). In the former case, deprotonation is not possible. A model reaction was carried out to test a mechanism involving concerted attack of oxidized flavin by the cyclopropyl ring (heterolytic cleavage). This mechanism was dismissed, since no reaction could be detected of **9.56** (R = CH₃) with 3-methyllumiflavin at 95°C for 12 hr.

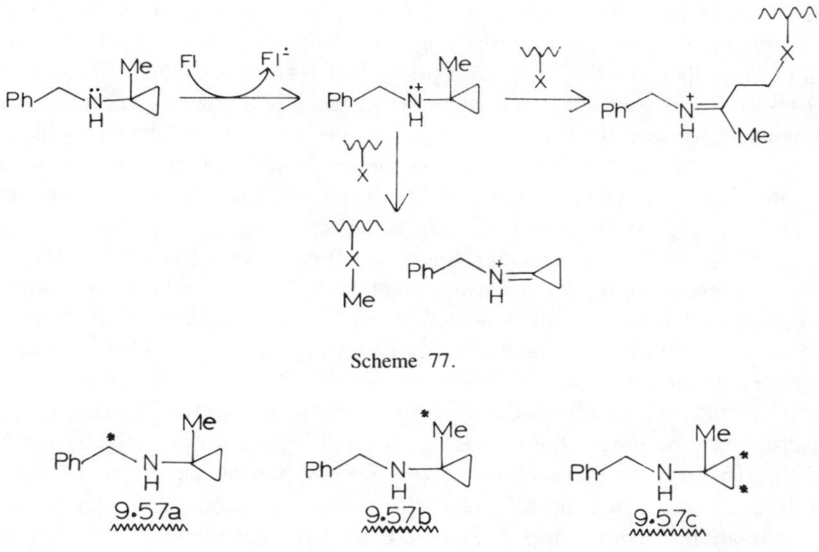

Scheme 77.

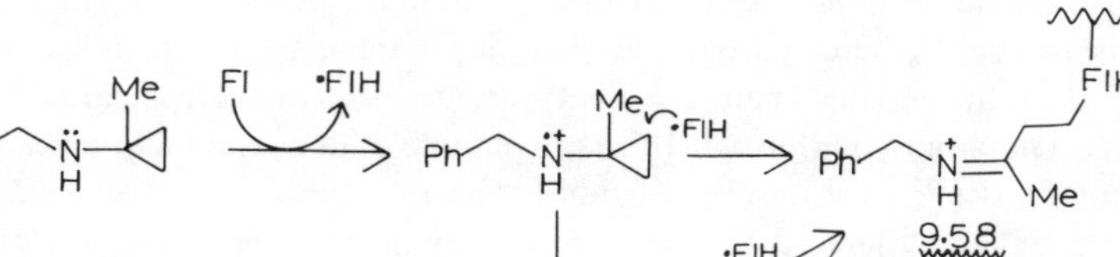

Scheme 78.　Structural Formulas 9.57a, b, and c.

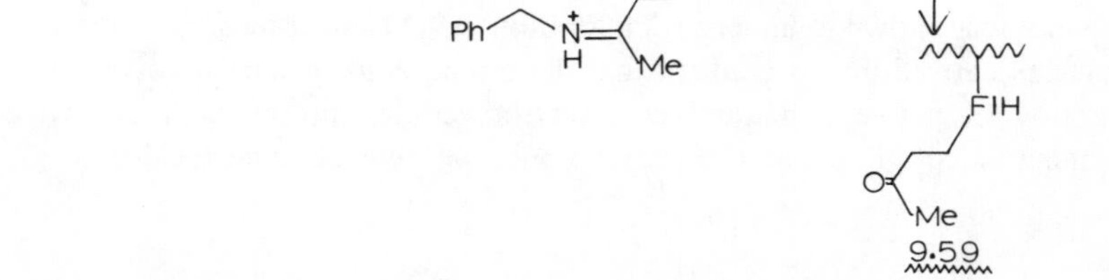

Scheme 79.　Containing Structural Formulas 9.58 and 9.59.

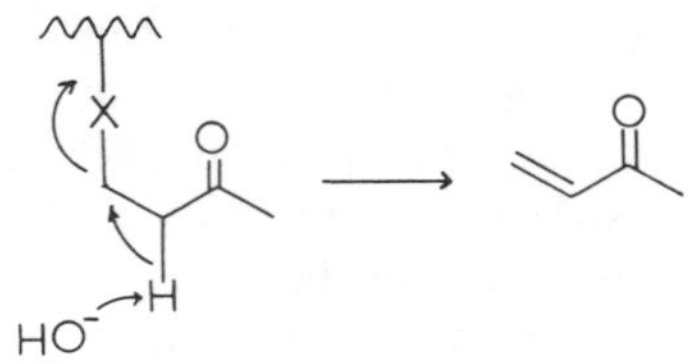

Scheme 80.

Compound **9.56** (R = CH$_3$) was synthesized by the route shown in Scheme 73, substituting 1-methylcyclopropylamine for cyclopropylamine.

(6) 1-Phenylcyclopropylamine

1-Phenylcyclopropylamine (Structural Formula 9.60, see Scheme 81) is a time-dependent inactivator of pig[123] and beef[105,124] liver MAO, but pseudo first-order kinetics are apparent for only three or four half-lives. The biphasic kinetics observed by Silverman and Zieske[124] were shown to be the result of **9.60** inactivation by two different pathways, both believed to be derived from a common intermediate. One pathway leads to irreversible inactivation and a 1:1 stoichiometry of radioactivity to enzyme when 1-[*phenyl*-^{14}C]-**9.60** is used as the inactivator. The other pathway results in a covalent reversible adduct. Since it requires 8 mol of **9.60** to irreversibly inactivate the enzyme, these two processes occur in a 1:7 ratio. The mechanism proposed for irreversible inactivation is shown in Scheme 81 (pathway a). Three organic reactions were carried out on the irreversibly labeled enzyme to determine the structure of the active-site adduct. Sodium boro[^{3}H]hydride reduction results in the incorporation of 0.73 equivalent of tritium, as expected for a ketone. Baeyer-Villiger oxidation (CF$_3$CO$_3$H) followed by saponification (NaOH) produces 0.8 equivalent of [^{14}C]phenol, indicating the presence of a phenyl ketone. Treatment of the labeled enzyme with hydroxide produces [^{14}C]acrylophenone, the expected product of a retro-Michael reaction on a β-X-propiophenone. The group to which the inactivator is attached was identified as the flavin and evidence was presented for N5-flavin attachment. The reversible pathway (Scheme 81, pathway b) occurs at a rate 7 times that of the irreversible pathway and produces 7 equivalents of [^{14}C]acrylophenone during the course of irreversible inactivation. The acrylophenone is believed to arise from an elimination reaction (−HX) on the same type of adduct as the irreversible one, except that the active site group (X) is an amino acid residue. This spontaneous elimination reaction can be prevented by NaBH$_4$ reduction of the ketone (or imine). Since an amino acid residue is involved (a good leaving group), it is reasonable that the retro-Michael elimination could occur spontaneously. An amino acid radical could be generated by hydrogen atom transfer from an amino acid to the purported flavin radical (Scheme 82). If this is the case, then hydrogen atom transfer must be seven times faster than the reaction of FlH· with the ring-opened radical intermediate (or the radical cation precursor) in Scheme 81. The amino acid to which the reversible adduct is attached was identified by Silverman and Zieske[125] as a cysteine residue. The reversible adduct was locked onto the enzyme by NaBH$_4$ treatment, then the labeled enzyme was treated with Raney nickel, a reducing agent for carbon-sulfur bonds. Only *trans*-β-methylstyrene (Structural Formula 9.61), the product of C–S bond reduction and dehydration (Scheme 83), was obtained (1-phenylpropanol was shown to undergo dehydration to **9.61** under these conditions). A second experiment was carried out to confirm that the amino acid residue involved is a cysteine. DTNB titration of native denatured enzyme showed the presence of six cysteines, but reversibly inactivated and denatured enzyme only has five cysteine residues.

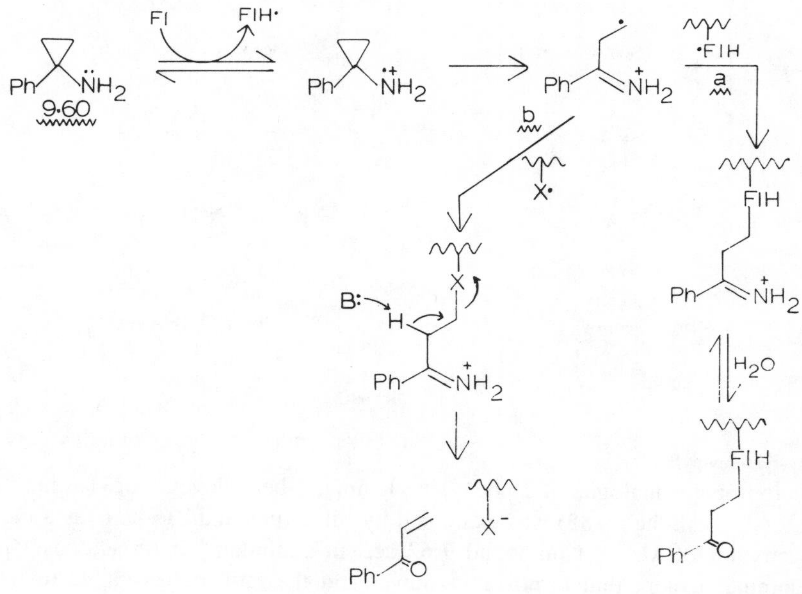

Scheme 81.   Containing Structural Formula 9.60.

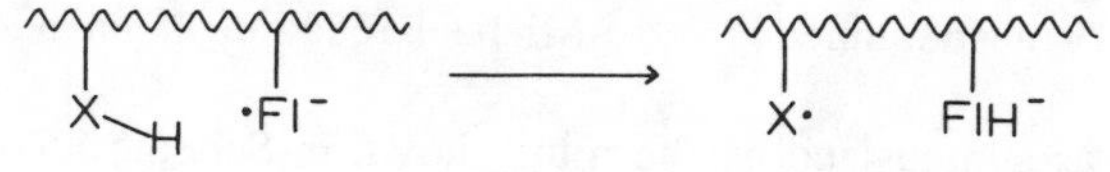

Scheme 82.

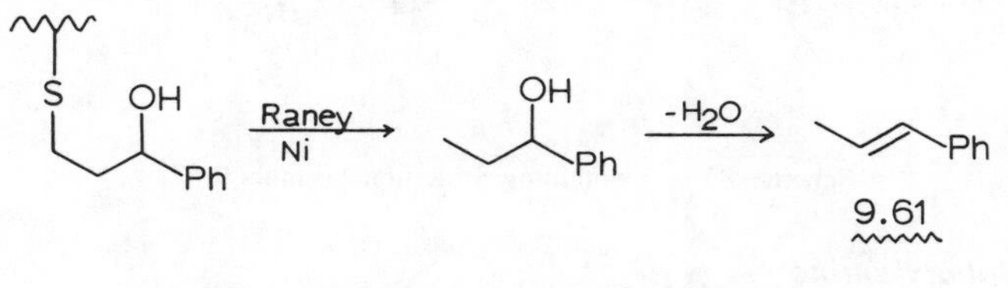

Scheme 83.   Containing Structural Formula 9.61.

A rationalization for why **9.56** (R = CH_3) only attaches to the flavin, tranylcypromine only attaches to an amino acid, but **9.60** attaches to both is that the intermediate radicals from each of these inactivators may be juxtaposed for these particular attachments (Scheme 84).

Compound **9.60** was synthesized[124] from 1-phenylcyclopropanecarboxylic acid by a Schmidt rearrangement (H_2SO_4/NaN_3).

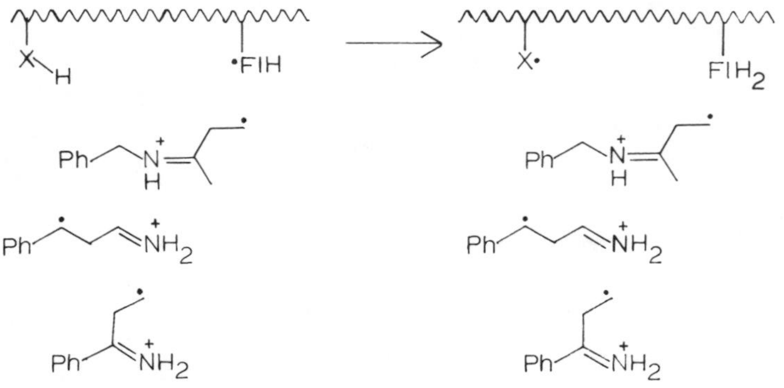

Scheme 84.

(7) 1-Benzylcyclopropylamine

The cyclopropane analogue of 2-phenylethylamine, 1-benzylcyclopropylamine (Structural Formula 9.62; see Scheme 85) was examined by Silverman and Zieske[126] as a mechanism-based inactivator of MAO. Compound **9.62** acts in a similar fashion as does 1-phenylcyclopropylamine, except that turnover is slower and the ratio of reversible to irreversible inactivation is about 2 instead of 8. Irreversible inactivation by **9.62** was shown to involve flavin attachment. The mechanism for irreversible inactivation is the same as was shown (Scheme 81) for **9.60**. The reversible component to the overall reaction yields benzyl vinyl ketone, the product of a spontaneous retro-Michael reaction of an adduct attached at an amino acid.

Compound **9.62** was synthesized by the route shown in Scheme 85.

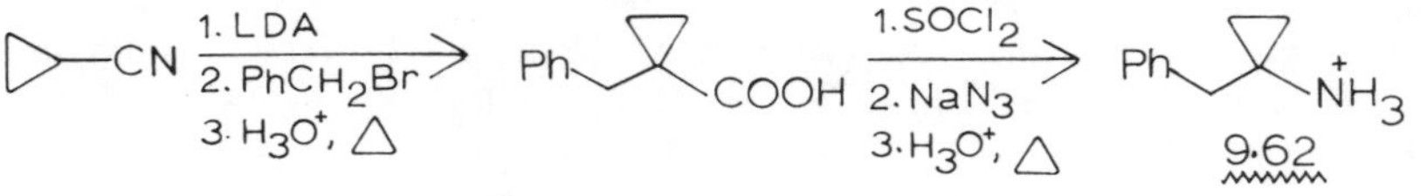

Scheme 85.   Containing Structural Formula 9.62.

b. 1-Phenylcyclobutylamine

1-Phenylcyclobutylamine (Structural Formula 9.63) was prepared by Silverman and Zieske[127] as a compound with the potential of forming an intermediate having a built-in radical trap (Structural Formula 9.64, Scheme 86). Compound **9.63** was shown to be both a substrate and a time-dependent irreversible inactivator of MAO; inactivation results in flavin attachment. For every inactivation event, 325 molecules of **9.63** are converted to product. The first metabolite formed was identified as 2-phenyl-1-pyrroline (Structural Formula 9.65), then, after a lag time, 3-benzoylpropanal (Structural Formula 9.66) and 3-benzoylpropanoic acid (Structural Formula 9.67) are produced (refer to Scheme 86). The carboxylic acid is a nonenzymatic oxidation product of **9.66**. Compound **9.65** is the expected product of cyclization and oxidation from the radical intermediate **9.64**. This is the first example of a cyclobutylamine-containing inactivator of MAO, and it provides strong support for a radical mechanism of action of MAO. Compound **9.63** was synthesized from 1-phenylcyclobutylcarbonitrile by saponification to the corresponding carboxylic acid followed by Schmidt rearrangement to the amine.

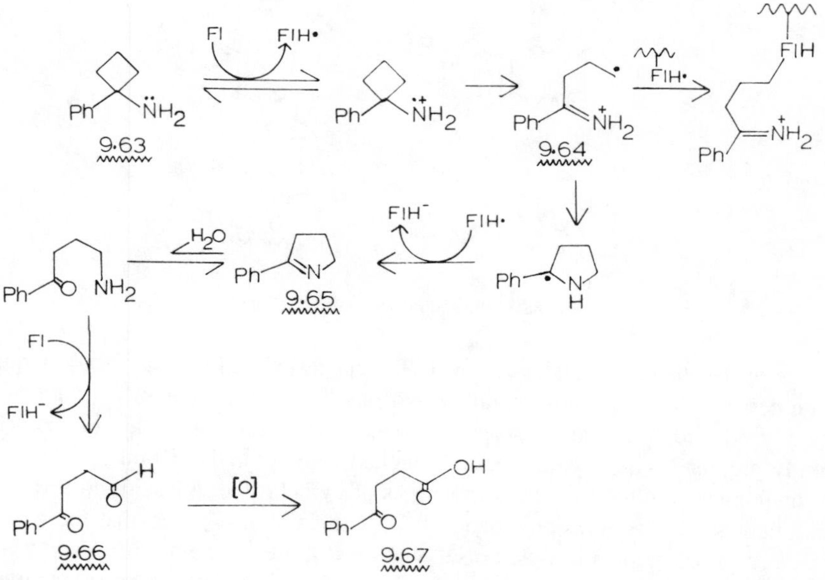

Scheme 86.   Containing Structural Formulas 9.63 to 9.67.

c. 3,5-Dicarbethoxy-1,4-dihydrocollidine and Related

In order to determine the source of the methyl group attached to protoporphyrin IX during inactivation of liver microsomal cytochrome P-450 from phenobarbital-treated rats by 3,5-dicarbethoxy-1,4-dihydrocollidine (Structural Fromula 9.68), two analogues were prepared by Ortiz de Montellano et al.,[5] namely, N-ethyl-3,5-dicarbethoxy-1,4-dihydrocollidine (Structural Formula 9.69) and 3,5-dicarbethoxy-2,6-dimethyl-4-ethyl-1,4-dihydropyridine (Structural Formula 9.70, see Scheme 87). N-Methylprotoporphyrin IX was isolated when either **9.68** or **9.69** was used, but not when **9.70** was the inactivator. N-Ethylprotoporphyrin IX is the green pigment isolated by treatment of phenobarbital-treated rats with **9.70**. All four isomers of N-ethylprotoporphyrin IX are formed. Compound **9.68** and **9.70**, but not **9.69**, inactivate liver microsomal cytochrome P-450 from phenobarbital-treated rats and destroy the heme; **9.70** is much more effective as an inactivator. NADPH and O_2 are required for inactivation. It was concluded, then, that the 4-alkyl group is transferred to the heme during inactivation. The reactivity difference of **9.68** and **9.70** suggest radical or cationic character in the migrating carbon during alkylation of the heme. Two possible intermediates suggested are **9.71** and **9.72** (Scheme 88). Transfer of R˙ to the heme from either of these intermediates would result in aromatization. Alternatively, R$^+$ could be transferred to the heme subsequent to second-electron transfer from one of the above intermediates to the heme.

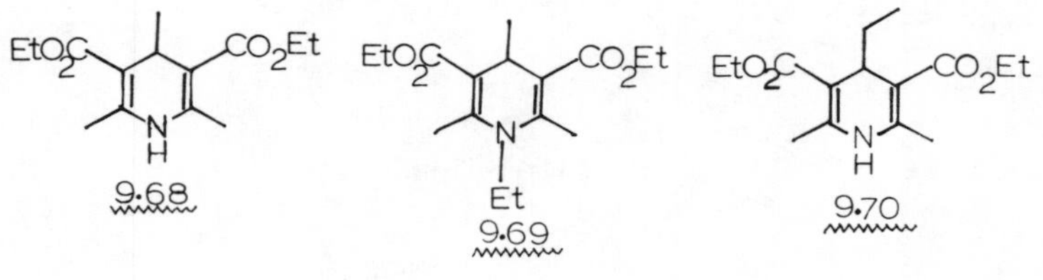

Scheme 87.   Structural Formulas 9.68 to 9.70.

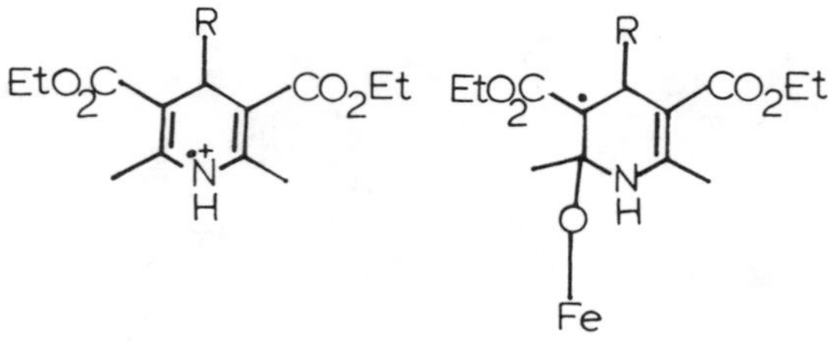

Scheme 88. Structural Formulas 9.71 and 9.72.

A series of 4-substituted analogues of **9.70** were tested as inactivators of cytochrome P-450 from hepatocytes of phenobarbital-treated rats.[128] Only the 4-methyl, ethyl, and propyl analogues produced N-alkylated porphyrins (increasing amount in this order) with correspondingly greater loss of cytochrome. Branched (4-isopropyl) or bulky (4-benzyl) 4-substituted analogues resulted in considerable loss of cytochrome, but with no formation of N-alkylated hemes. The conclusions made by DeMatteis et al.[128] were that the 4-alkyl group is transferred to the heme and this transalkylation is more efficient with 4-alkyl groups that give rise to stable carbocations. However, for N-alkylated porphyrins to accumulate, the size and shape of the 4-alkyl substituent is important. Perhaps steric constraints may hinder access to the pyrrole nitrogen, and instead lead to attack at other sites on the heme. A similar study of 4-substituted analogues of **9.70** was carried out by Augusto et al.[129] except attempts were made to trap any radicals formed with various spin traps. A qualitative correlation was found between the ability of the 4-substituted dihydropyridine to destroy cytochrome P-450 and the stability of the radical that would be formed by homolytic cleavage of the 4-substituent. The compounds having the 4-substituent of low radical stability (phenyl, methyl) caused little or no destruction of cytochrome P-450, those with substituents of intermediate stability (ethyl, propyl, i-butyl) are effective inactivators and produce alkylated hemes, and analogues with high stability substituents (benzyl, isopropyl) are effective destructive agents but do not give detectable heme adducts. In order to gain evidence for a radical intermediate, the inactivation by **9.70** was carried out in the presence of spin trapping agents. Of the three reagents tried, 5,5-dimethyl-1-pyrroline-N-oxide, α-phenyl N-t-butylnitrone, and α-(4-pyridyl-1-oxide)N-t-butylnitrone (Structural Formula 9.73, Scheme 89), only the latter was effective and resulted in the spin trapping of ethyl radicals; the radical product was identified by ESR spectroscopy. This radical-generating process is enzyme-catalyzed. The fact that **9.73** does not inhibit destruction of cytochrome P-450 by **9.70** suggests that only the ethyl radicals that escape the active site are trapped and that destruction of cytochrome P-450 does not result from alkylation by a radical that must diffuse back into the active site. Three mechanistic pathways can be considered (Scheme 90). Since N-ethyl-**9.70** destroys cytochrome P-450 and generates ethyl radicals as effectively as does **9.70**, pathways b and c were excluded. Also, 3,5-bis(ethoxycarbonyl)-2,6-dimethylpyridine (Structural Formula 9.74) was identified as a reaction product. The only pathway consistent with these data is pathway a.

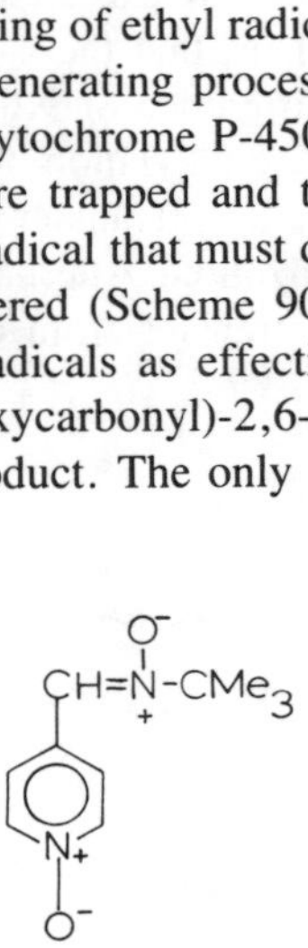

Scheme 89. Structural Formula 9.73.

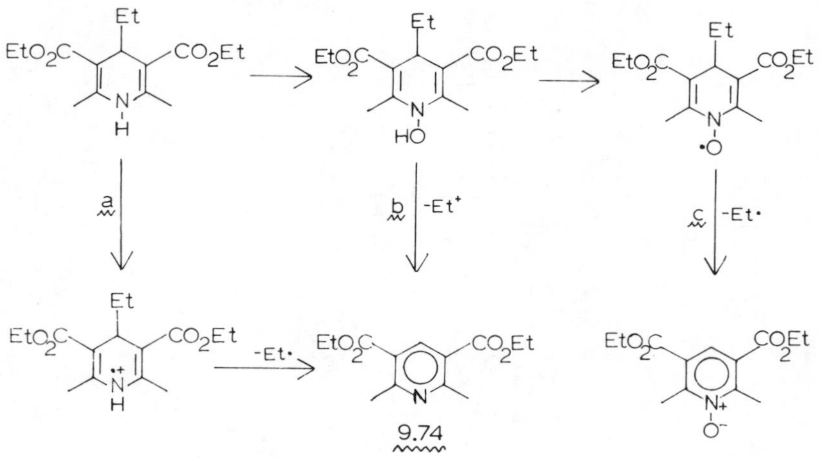

Scheme 90.   Containing Structural Formula 9.74.

Consistent with the work of Ortiz de Montellano and co-workers[5,129] and DeMatteis et al.,[128] Marks et al.[130] found that when the the 4-substituent was Me, Et, Pr, i-Pr, benzyl, cyclohexyl, and (3-cyclohexenyl), cytochrome P-450 from chick embryo liver microsomes was inactivated; only Me, Et, and Pr caused heme destruction. Phenyl- or non- (i.e., H) substituted analogues do not inactivate the enzyme, nor does the oxidized 4-methyl compound (i.e., the pyridine). All of these results are consistent with the radical transfer mechanism of Augusto et al.[129] A correlation was found between the ability of the compound to alkylate the heme of cytochrome P-450 and its ability to produce ferrochelatase inhibition in vivo. Further work by McCluskey et al.[130a] showed that the 4-butyl, 4-pentyl, 4-hexyl, 4-chloromethyl, and 4-cyclopropylmethyl analogues of 3,5-diethoxycarbonyl-2,4,6-trimethyl-1,4-dihydropyridine inactivate cytochrome P-450 in chick embryo liver cells with concomitant heme destruction and inhibit ferrochelatase activity. The 4-isobutyl analogue produces N-isobutylporphyrin in rat liver, but does not inhibit ferrochelatase.

A comparison of the circular dichroism spectra of the C-ring N-ethylprotoporphyrin IX isomers arising from the reaction of ethylhydrazine with human hemoglobin and from the reaction of cytochrome P-450 with **9.70** proved them to have identical structures.[131] Since alkylation in both cases occurs from the face of the heme to which oxygen (molecular oxygen in the case of hemoglobin or the activated oxygen species with cytochrome P-450) is bound, the absolute configuration of the N-ethylprotoporphyrin IX isomer was deduced by Ortiz de Montellano et al.[131] as **9.75** (Scheme 91). The circular dichroism spectrum for the ring A isomer of N-(2-oxopropyl)protoporphyrin IX obtained from propyne is very similar to that of the ring A isomer of the N-ethyl isomer from **9.70**. Also, the spectrum of the ring D isomer of N-(2-hydroxyethyl)protoporphyrin IX obtained from ethylene is the same as the ring D isomer of the N-ethyl isomer from **9.70**.

The four isomers of N-methylprotoporphyrin IX (dimethyl esters) were prepared by methylation of the dimethyl ester of protoporphyrin IX.[132] The isomers were separated and purified by HPLC and their structures determined by 360-MHz NMR spectroscopy. Differentiation was achieved by Kunze and Ortiz de Montellano[132] through a combined use of deuterium labeling, proton relaxation time measurements, and nuclear Overhauser effect determinations. These structures were useful in the identification of the absolute configuration of **9.75**.

4-Substituted-3,5-dicarbethoxy-2,6-dimethyl-1,4-dihydropyridine analogues can be synthesized by the route shown in Scheme 92.[129]

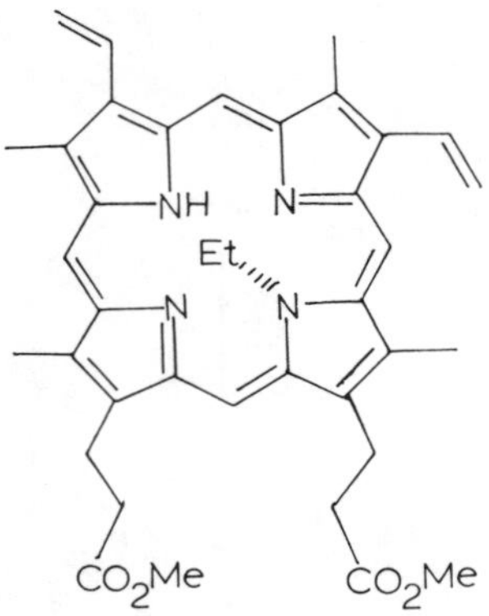

Scheme 91.   Structural Formula 9.75.

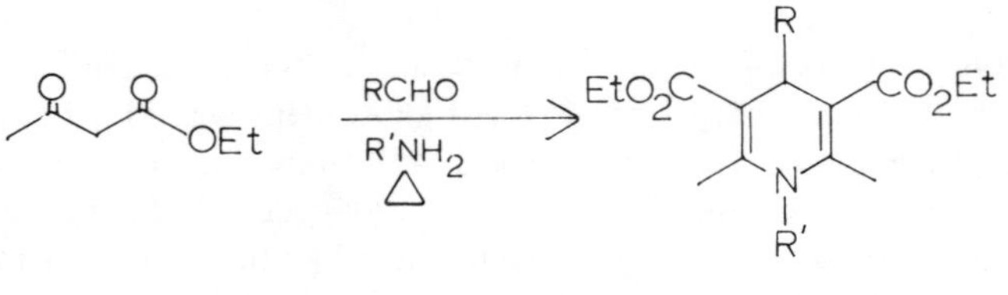

Scheme 92.

d. Phenylethylenediamine and Related

A detailed investigation of the inactivation of bovine adrenal dopamine β-hydroxylase by phenylethylenediamine (Structural Formula 9.75a (R = $CH_2CH_2NH_2$), Scheme 92 and the corresponding N-methyl and β-methyl analogues was carried out by Wimalasena and May.[132a] A partition ratio for inactivation of 1750 ws observed for phenylethylenediamine, and this is pH independent, indicating that N-dealkylation and inactivation proceed through a common step. Inactivation cannot be reversed by dialysis. [ring-³H]Phenylethylenediamine inactivates the enzyme with incorporation of 0.74 molecules of radioactivity per subunit, only in the presence of ascorbate. The N-methyl analogue has a partition ratio of 1650 and the β-methyl analogue about 60. 5-Hydroxyindole, a compound without an abstractable α-H, also inactivates the enzyme, as does aniline. β,β-Dideuteriophenylethylenediamine inactivates the enzyme with an isotope effect on the partition ratio and on V_{max} of 2.8. These results suggest that C–H bond cleavage is involved in N-dealkylation, but not inactivation. Since 20% of the tritiumin the [ring-³H]-inactivator is at the *para*-position, 72% *ortho*, and 8% *meta*, it was suggested[132a] that inactivation results from attachment at the *para*-position (Scheme 92A).

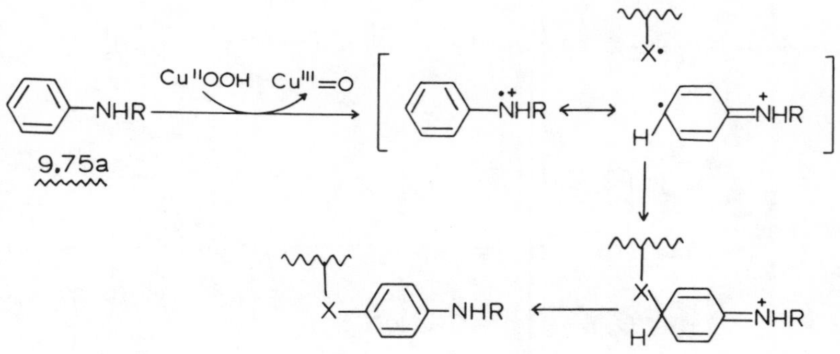

Scheme 92A.   Containing Structural Formula 9.75a.

3. With Carbon-Silicon Bond Cleavage
a. (Aminoalkyl) trimethylsilanes

(Aminoalkyl) trimethylsilanes (Structural Formulas 9.75b (n = 1 to 3), Scheme 92B) were synthesized by Silverman and Banik[132b] and shown to be time-dependent inactivators of beef liver MAO. β-Mercaptoethanol does not protect the enzyme from inactivation. Inactivations by **9.75b** (n = 1 and 2) are virtually irreversible; extended dialyses slowly ($t_{1/2}$ = 5.5 days) reactivate the enzyme. Dialysis of **9.75b** (n = 3)-inactivated MAO results in return of enzyme activity with $t_{1/2}$ = 13 hr. Proposed mechanisms for inactivation of MAO by these compounds ae shown in Schemes 92C to 92E.

$$Me_3Si(CH_2)_nNH_3^+$$

Scheme 92B.   Structural Formula 9.75b.

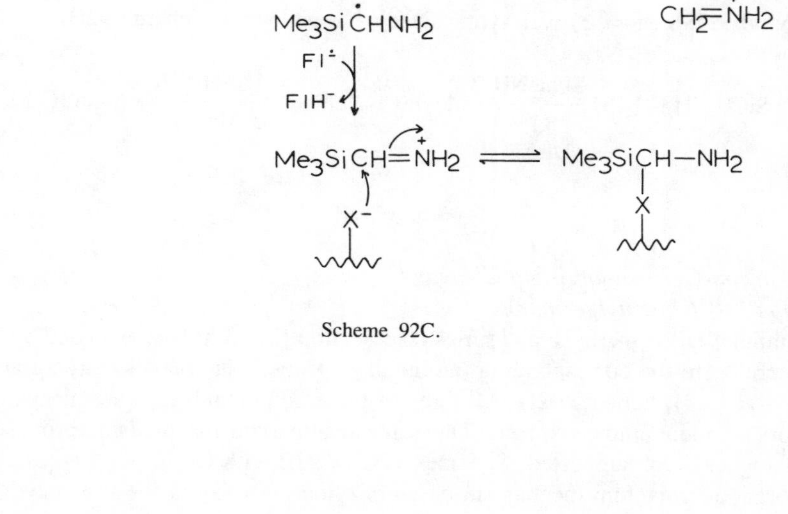

Scheme 92C.

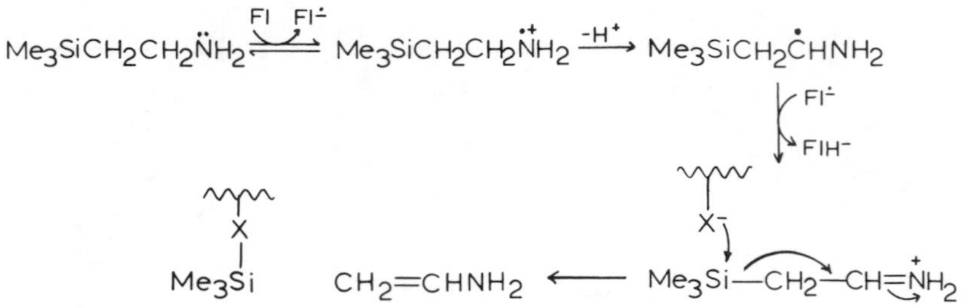

Scheme 92D.

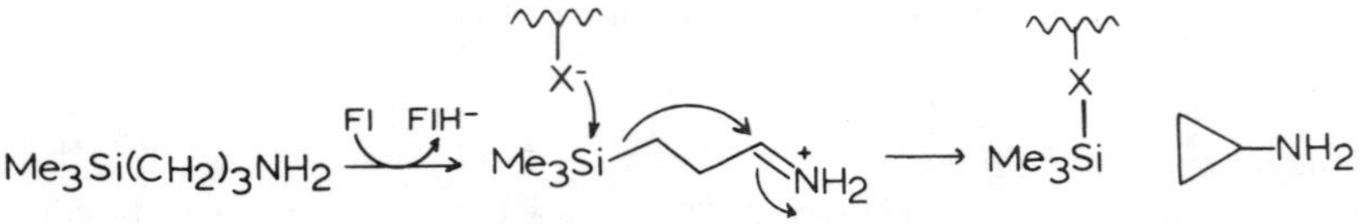

Scheme 92E.

Compound **9.75b** (n = 1) was synthesized[132b] as shown in Scheme 92F.

$$\text{Me}_3\text{SiCH}_2\text{Cl} \xrightarrow[\substack{\text{K}_2\text{CO}_3 \\ \text{DMF}}]{\text{PhthNH}} \text{Me}_3\text{SiCH}_2\text{NPhth} \xrightarrow[\text{2. H}_3\text{O}^+]{\text{1. NH}_2\text{NH}_2} \text{Me}_3\text{SiCH}_2\overset{+}{\text{N}}\text{H}_3$$

Scheme 92F.

Compound **9.75b** (n = 2) was synthesized[132b] as shown in Scheme 92G.

$$\text{Me}_3\text{SiCH}_2\text{CN} \xrightarrow[\text{2. H}_3\text{O}^+, \Delta]{\text{1. LiAlH}_4} \text{Me}_3\text{SiCH}_2\text{CH}_2\overset{+}{\text{N}}\text{H}_3$$

Scheme 92G.

Compound **9.75b** (n = 3) was synthesized[132b] as shown in Scheme 92H.

$$\text{Me}_3\text{SiCH}_2\text{CH}_2\text{CH}_2\text{OH} \xrightarrow[\substack{\text{Ph}_3\text{P} \\ \text{DEAD}}]{\text{PhthNH}} \text{Me}_3\text{Si(CH}_2)_3\text{NPhth} \xrightarrow[\text{2. H}_3\text{O}^+, \Delta]{\text{1. NH}_2\text{NH}_2} \text{Me}_3\text{Si(CH}_2)_3\overset{+}{\text{N}}\text{H}_3$$

Scheme 92H.

4. With Carbon-Germanium Bond Cleavage
a. (Aminoalkyl) trimethylgermanes

The (aminoalkyl) trimethylsilane series (see Section II.B.3.a) was extended by Silverman and Vadnere[132c] to the corresponding (aminoalkyl) trimethylgermanes (Structural Formulas 9.75c (n = 1 to 3), Scheme 92I). All three of these compounds also are mechanism-based inactivators of monoamine oxidase. The same inactivation mechanisms proposed for the silane analogues were suggested (Schemes 92C to 92E with Ge in place of Si). These are the first organogermanium mechanism-based enzyme inactivators for any enzyme.

$$Me_3Ge(CH_2)_nNH_3^+$$

Scheme 92I.   Structual Formula 9.75c.

The synthesis of **9.75c** (n = 1) is shown in Scheme 92J.

$$CH_2I_2 + ZnCu \rightarrow ICH_2ZnI \xrightarrow{GeCl_4} Cl_3GeCH_2I \xrightarrow[MeMgBr]{excess}$$

$$Me_3GeCH_2NH_3^+ \xleftarrow[\text{2. HCl}]{\text{1. } NH_2NH_2} Me_3GeCH_2NPhth \xleftarrow[\substack{K_2CO_3 \\ DMF, \Delta}]{PhthNH} Me_3GeCH_2I$$

Scheme 92J.

The synthesis of **9.75c** (n = 2) is shown in Scheme 92K.

$$Me_3GeBr + BrCH_2CN \xrightarrow{Zn} Me_3GeCH_2CN \xrightarrow[\text{2. HCl}]{\text{1. } NaBH_4/CoCl_2} Me_3GeCH_2CH_2\overset{+}{N}H_3$$

Scheme 92K.

C. Carbon-Oxygen Bond Oxidation
1. With Carbon-Hydrogen Bond Cleavage
a. Acetylene-Containing Inactivators
(1) 2-Hydroxy-3-butynoic Acid and Related

2-Hydroxy-3-butynoic acid (Structural Formula 9.76, Scheme 93) was briefly mentioned in Chapter 2 in the deprotonation reactions section because inactivation of the various flavoenzymes by this compound has been suggested to proceed by initial α-carbon deprotonation. The resulting stabilized carbanion can then add to the electrophilic oxidized flavin (Scheme 93; Fl is oxidized flavin). The alternative mechanism is initial oxidation of the alcohol to the acetylenic ketone with concomitant reduction of the flavin; reduced flavin can then add to the oxidized form of the inactivator (Scheme 94; FlH⁻ is reduced flavin). Since the mechanisms of oxidation by these enzymes are not well understood, **9.76** is discussed in detail in this section.

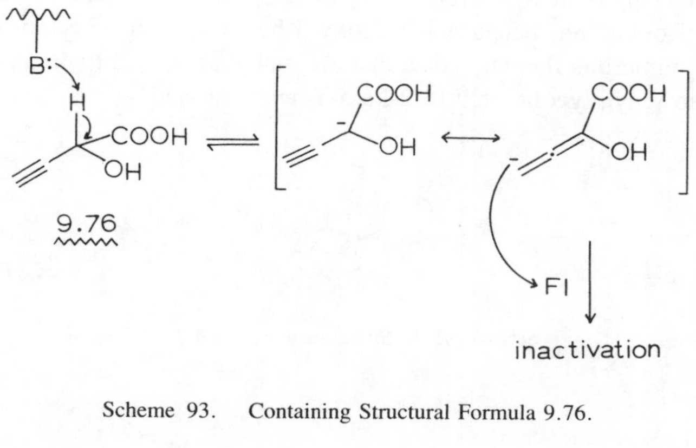

Scheme 93.   Containing Structural Formula 9.76.

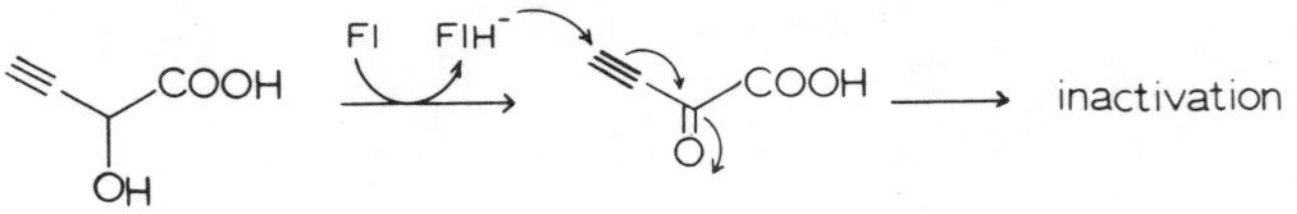

Scheme 94.

D,L-2-Hydroxy-3-butynoic acid (Structural Formula 9.76, Scheme 93) inactivates the flavoenzyme, L-lactate oxidase, from *Mycobacterium smegmatis*[28] with a partition ratio of 110.[133] Following inactivation, dissociation of the FMN gives an apoenzyme that is fully reactivatable by added FMN.[28] The released flavin, however, has been modified to a fluorescent form. Inactivation with 2-hydroxy-3-[4-^{3}H]butynoate results in the incorporation of 1 mol of [^{3}H] per mole of active site; all of the radioactivity was removed from the enzyme by Walsh et al.[28] by resolving the flavin from the apoenzyme. The acetylenic linkage is no longer intact because the tritium cannot be exchanged out of the modified flavin under conditions where it exchanges in the inactivator. Inactivation with 2-hydroxy-3-[2-^{3}H]butynoate results in no [^{3}H] incorporation into the enzyme. Ghisla et al.[133] showed that adduct formation also results from the reaction of reduced lactate oxidase (reduced photochemically under nitrogen in the presence of oxalate) with 2-keto-3-butynoate (prepared *in situ* from lactate dehydrogenase-catalyzed oxidation of 2-hydroxy-3-butynoate in the presence of NAD$^+$). The change in the absorption spectrum after inactivation appears to produce a 4a,5-dihydroflavin derivative on the basis of model compounds. Compound **9.77** (R = ribitol phosphate, R′ = H), shown in Scheme 95, was prepared as a model for the inactivated enzyme adduct, and its spectrum in apolactate oxidase was taken; the spectrum resembles that of inactivated enzyme. Treatment of inactivated enzyme with sodium borohydride results in a change in the absorption and emission spectra. Similar spectral changes were observed upon reduction of 4a,5-dihydro-4a,5-propano-3-methyllumiflavin (**9.77**, R = R′ = Me) and this further confirms the enzyme adduct structure. Denaturation of inactivated enzyme with methanol gives a complex mixture of products.[134] However, reduction with sodium borohydride prior to denaturation gives a major product whose structure was identified by isotopic labeling experiments, NMR and UV-Vis spectroscopy, chemical reactivity, and comparisons with synthesized model compounds. The structure of the reduced and released product is **9.78**, which is believed to be derived from hydrolysis of the pyrimidine ring of the reduced enzyme-bound adduct (**9.78a**; see Scheme 96). The mechanism proposed by Schonbrunn et al.[134] to account for the inactivated enzyme adduct (Structural Formula 9.79) is shown in Scheme 97. An alternative mechanism is oxidation of the inactivator to 2-keto-3-butynoic acid followed by reduced flavin addition to oxidized inactivator (Scheme 98). Although these two mechanisms cannot be distinguished by the work carried out, the authors favor the former mechanism (which, therefore, would put this inactivator in the deprotonation chapter instead of oxidation) because 2-hydroxy-3-butenoic acid (vinylglycolic acid) is not an inactivator of numerous flavoenzymes that are inactivated by 2-hydroxy-3-butynoic acid but which oxidize vinylglycolic acid to 2-keto-3-butenoic acid.

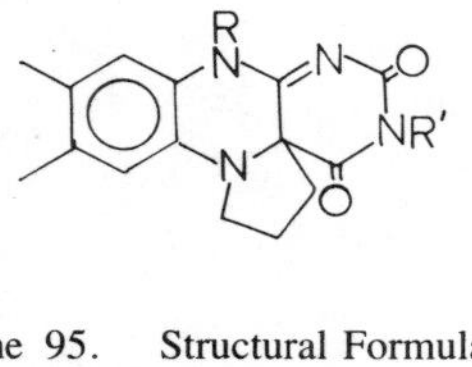

Scheme 95.   Structural Formula 9.77.

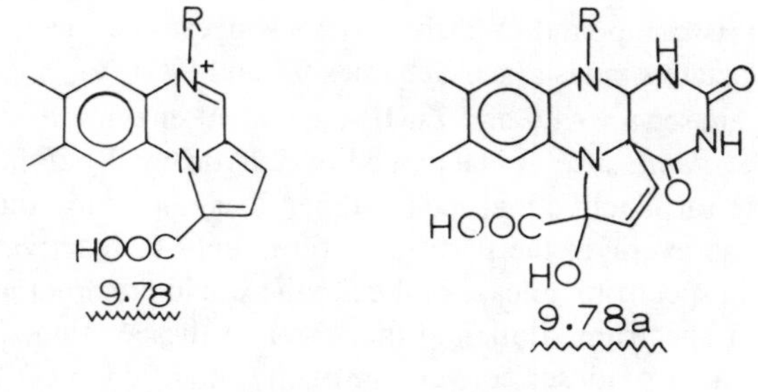

Scheme 96.   Structural Formulas 9.78 and 9.78a.

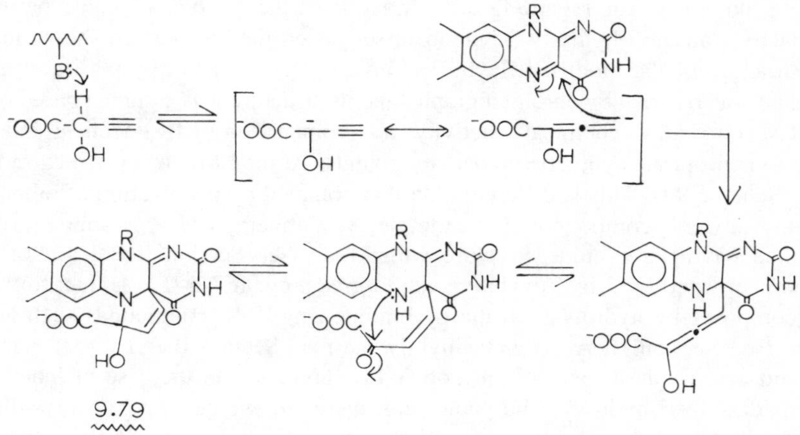

Scheme 97.   Containing Structural Formula 9.79.

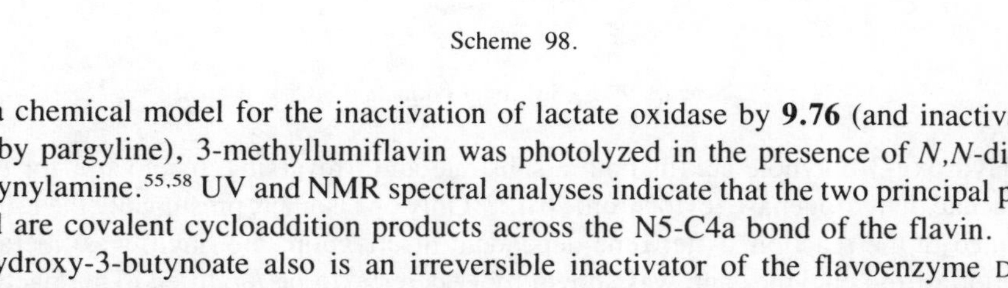

Scheme 98.

As a chemical model for the inactivation of lactate oxidase by **9.76** (and inactivation of MAO by pargyline), 3-methyllumiflavin was photolyzed in the presence of *N*,*N*-dimethyl-2-propynylamine.[55,58] UV and NMR spectral analyses indicate that the two principal products formed are covalent cycloaddition products across the N5-C4a bond of the flavin.

2-Hydroxy-3-butynoate also is an irreversible inactivator of the flavoenzyme D-lactate dehydrogenase and D-lactate-dependent lactose, proline, and valinomycin-induced rubidium transport in isolated membrane vesicles and in solubilized, partially purified preparation from *Escherichia coli*.[135] There are 15 to 30 turnovers prior to inactivation. The oxidation of succinate and NADH, and transport stimulated by these compounds, are not affected. The olefin, 2-hydroxy-3-butenoate is a substrate, but not an inactivator. Inactivation of D-lactate-dependent transport by 2-hydroxy-3-butynoate is prevented by D-lactate, but not by succinate and NADH. A NADH-dependent lactate dehydrogenase from rabbit heart muscle is not inactivated by this compound. Walsh et al.[135] suggested a mechanism similar to that for inactivation of L-lactate oxidase (see Schemes 97 and 98).

D-Lactate dehydrogenase, a FAD- and Zn(II)-dependent enzyme from the anaerobic bacterium *Megasphaera elsdenii*, also is inactivated by 2-hydroxy-3-butynoic acid.[136] Only the D-isomer is responsible for inactivation; the L-isomer does not inhibit the rate of inactivation by the D-isomer. On the average, the partition ratio is only 5. Inactivation is accompanied by a change in the flavin spectrum; release of the flavin by acid precipitation and neutralization gives a product with a spectrum similar to that prior to dissociation. The modified flavin binds to apo-D-amino acid oxidase; conversion of the modified FAD to a FMN analogue with snake venom phosphodiesterase gives a product that binds to the FMN-specific apoflavodoxin. As in the case of inactivation of *M. smegmatis* L-lactate oxidase by 2-hydroxy-3-butynoate, no conclusion regarding differentiation of the mechanism could be made, but lack of inactivation by 2-hydroxy-3-butenoate suggested the mechanism shown in Scheme 97. The structure of the modified flavin was investigated in detail by Ghisla et al.[137] The flavin adduct was purified by chromatography and its structure was assigned based on optical and NMR spectroscopy, chemical reactivity, and comparison of its chemical and physical properties to appropriate synthetic model compounds. At the FMN level, structure **9.80** was proposed (Scheme 99). This is different than that obtained by inactivation of other flavoenzymes with acetylenic compounds. For example, *N,N*-dimethylpropargylamine forms an N5 flavin adduct having a cyanine structure with MAO (see Section II.B.1.a.(1)); L-lactate oxidase is inactivated by 2-hydroxybutynoate to give a cyclic N5-C4a adduct to the flavin, which decomposes by hydrolysis of the pyrimidine ring;[133,134] compounds such as propargylglycine (see Section III.A.1.) and vinylglycine (see Section II.B.1.c.(6)) inactivate D- and L-amino acid oxidase, but do not modify the flavin. As in the case of inactivation of L-lactate oxidase by 2-hydroxy-3-butynoic acid, there are two general pathways that can be considered (Scheme 100). Oxidation to 2-ketobutynoate could occur by a covalent intermediate (pathway a-a″) or noncovalent route (pathway a′); pathway b, however, is favored.[137] A radical mechanism also cannot be excluded. Since inactivation of L-lactate oxidase to give **9.79** is different than the product of inactivation of D-lactate dehydrogenase, it suggests that L- and D-2-hydroxy-3-butynoate bind to the two enzyme active sites in a diastereotopic manner with respect to the flavin.

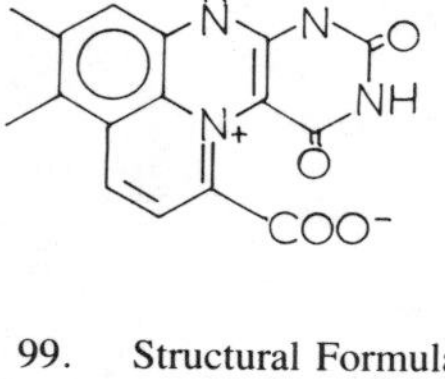

Scheme 99.   Structural Formula 9.80.

DL-2-Hydroxy-3-butynoic acid is both a substrate and irreversible inactivator for bakers' yeast L-lactate dehydrogenase (cytochrome b₂).[138] Only one isomer, presumably the L-isomer, is consumed in the reaction. After time-dependent inactivation, the flavin was liberated by precipitation of the enzyme, and was shown by Lederer[138] to be modified. The inactivation of bakers' yeast flavocytochrome b₂ by 2-hydroxy-3-butynoate was reinvestigated by Pompon and Lederer.[139] Inactivation does not take place under anaerobic conditions. In the presence of the monoelectronic electron acceptor, ferricyanide, the inactivation rate dramatically increases. The partition ratio is independent of substrate concentration and is 3200. No inactivation of oxidized enzyme occurs by 2-keto-3-butynoate, but rapid time-dependent inactivation of reduced enzyme is observed. The partition ratio for reduction to 2-hydroxy-3-butynoate and inactivation is 5. Inactivation followed by sodium borohydride reduction

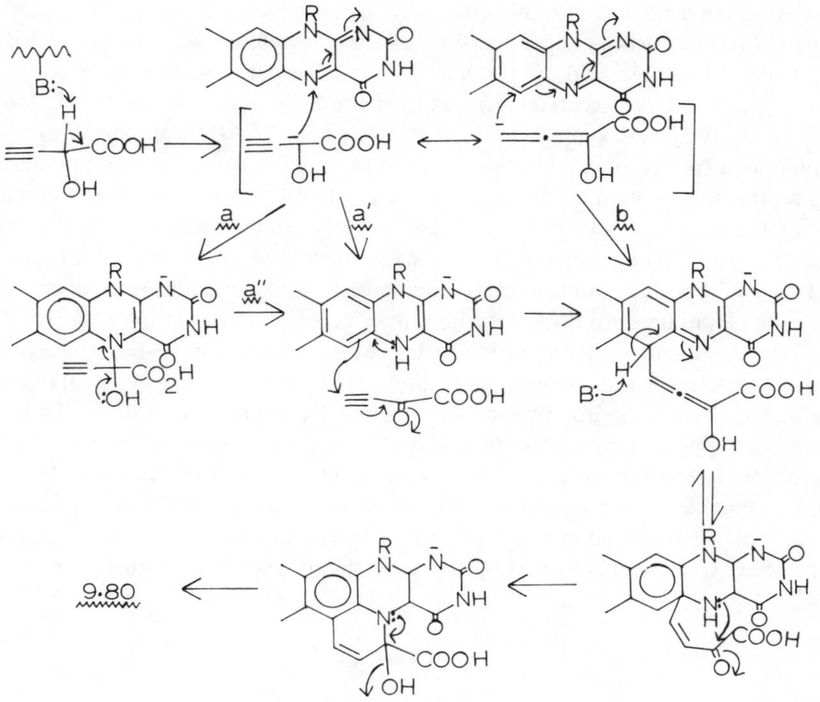

Scheme 100.

was carried out in parallel with lactate oxidase and flavocytochrome b_2 and several flavin-derived products were obtained. One of the products was identical to the only flavin product formed when lactate oxidase is inactivated by 2-hydroxy-3-butynoate and then $NaBH_4$ reduced (see **9.78**, Scheme 96). If $NaBH_4$ reduction is carried out after the flavin adduct is removed from the enzyme, lactate oxidase produces another compound, identified as the decarboxylated product (Structural Formula 9.81). Under these conditions, flavocytochrome b_2 produces two adducts, one is **9.81** and the second was identified as **9.82** (Scheme 101). This was one of the flavin adducts produced from $NaBH_4$ reduction of the enzyme-bound modified flavin, but is formed prior to the borohydride treatment. The mechanism proposed is the carbanionic mechanism in Scheme 100 based on enzyme mechanistic studies by Urban and Lederer[140] with halolactate derivatives.

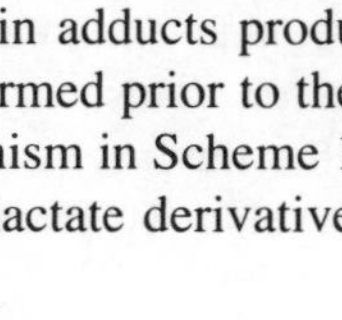

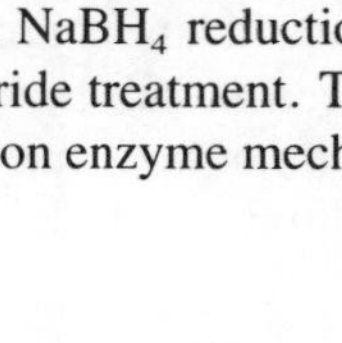

Scheme 101.   Structural Formulas 9.81 and 9.82.

2-Hydroxy-3-butynoic acid also irreversibly inactivates glycollate oxidase from pea leaves.[141] Nonpseudo, first-order, time-dependent loss of enzyme activity was observed. 3-Butynoate and 2-hydroxy-3-butenoic acid are not inactivators. The partition ratio is > 6.[142]

A series of 2-hydroxy-3-acetylenic acids (Structural Formula 9.83, R = H, Me, Et, Pr, Bu; see Scheme 102) were synthesized by Cromartie et al.[143] All 5 compounds are substrates for L-amino acid (α-hydroxy acid) oxidase from rat kidney. The parent compound (**9.83**, R = H) is an inactivator with a partition ratio of 25. [1-^{14}C]-Inactivator gives stoichiometric labeling of the flavin coenzyme;[143,144] 1 mol of inactivator becomes bound per 100,000 grams of enzyme.[144] The other compounds show much less inactivation (partition ratios > 2000). This could be the result of steric hindrance to Michael addition or decreased allene isomerization. Dithiothreitol does not affect the rate of inactivation by **9.83** (R = H), but it slows down the inactivation rate by **9.83** (R = Me).[145] 2-Hydroxy-3-butenoate is oxidized, but does not inactivate this enzyme until after standing; i.e., nonspecific alkylation from solution occurs. Dithiothreitol protect the enzyme from this inactivation. This result is, again, used in support of the carbanion mechanism shown in Scheme 93 and would explain why amino acids are not labeled. Also, substituents at the 4-position would destabilize carbanionic character at the 4-position and would result in higher partition ratios. Although the structure of the flavin adduct is unknown, it has an absorbance spectrum identical with that of L-lactate oxidase inactivated by the same compound;[144] the structure of that adduct is **9.79**.[134]

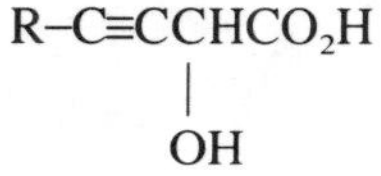

$$R-C\equiv CCHCO_2H$$
$$|$$
$$OH$$

Scheme 102.   Structural Formula 9.83.

Meyer and Cromartie[146] used [^{14}C]diethyl pyrocarbonate to show that two histidines of L-α-hydroxy acid oxidase are labeled, but only one is essential for activity. Substrates and competitive inhibitors decreased the rate of inactivation, suggesting that an active-site histidine is labeled. Treatment of the 2-hydroxy-3-butynoate-inactivated enzyme with [^{14}C]diethyl pyrocarbonate, results in incorporation of radioactivity to the same extent as enzyme not previously inactivated by 2-hydroxy-3-butynoate. To explain this, it was proposed that the histidine is involved in abstracting a proton from another amino acid near the flavin which is the actual base involved in substrate deprotonation. This way, the histidine would be essential for catalysis, but not be located close to the flavin.

Rabbit kidney D-hydroxy acid dehydrogenase also is inactivated by 2-hydroxy-3-butynoate.[145]

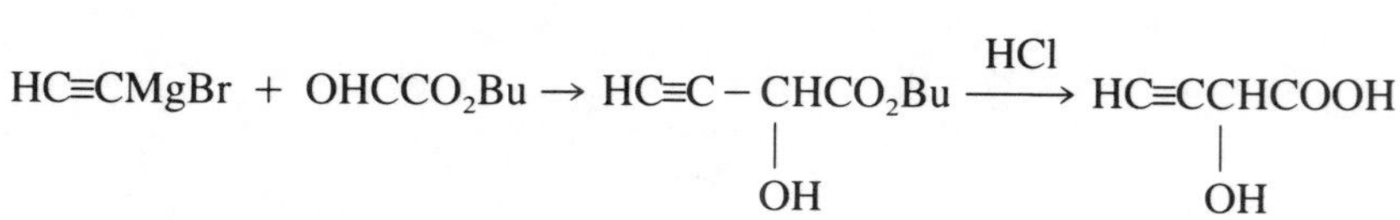

$$HC\equiv CMgBr \;+\; OHCCO_2Bu \rightarrow HC\equiv C-CHCO_2Bu \xrightarrow{HCl} HC\equiv CCHCOOH$$

with OH groups below the respective CHCO_2Bu and CHCOOH carbons

Scheme 103.

$$RC\equiv CCHO \xrightarrow{HCN} RC\equiv CCHCN \xrightarrow{H_3O^+} RC\equiv C-CHCOOH$$

with OH groups below the respective CHCN and CHCOOH carbons

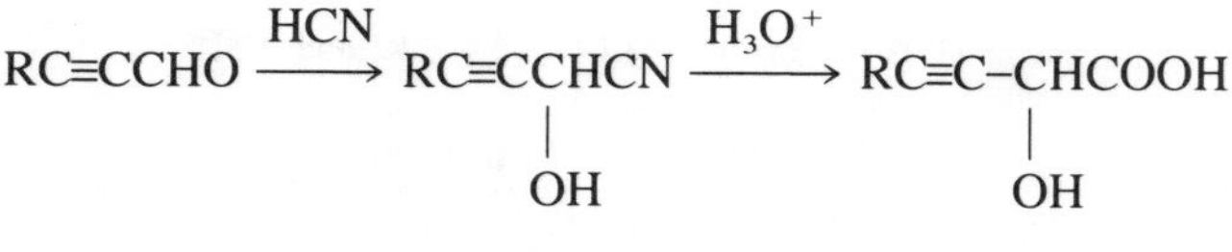

Scheme 104.

2-Hydroxy-3-butynoic acid has been synthesized by Verny and Vessière[147] (Scheme 103) and by Cromartie et al.[143] (Scheme 104).

(2) (1R)- and (1S)-17β-[1-Hydroxy-2-propynyl]androst-4-en-3-ones

17β-[1(R)-Hydroxy-2-propynyl]androst-4-en-3-one (Structural Formula 9.84, Scheme 105) is an irreversible inactivator of both the 17β-estradiol dehydrogenase and 20α-hydroxysteroid dehydrogenase activities that have been shown to be present in the human placental enzyme purified to apparent homogeneity;[148] inactivation requires NAD$^+$. An identical, time-dependent process for both enzyme activities was observed by Tobias et al.[148] The 1S-isomer is not oxidized, and also is not an inactivator. The α,β-alkynyl ketone, 17β-(1-oxo-2-propynyl)androst-4-en-3-one, is a potent affinity labeling agent. Protection of inactivation by substrates for either activity was observed. These results support the notion that the one enzyme catalyzes two different stereospecific reactions. This is similar to the bifunctional activity of 3α,20β-hydroxysteroid dehydrogenase described by Strickler et al.[149] In this case, the 1S-isomer is the inactivator, but β-mercaptoethanol completely protects the enzyme from inactivation. Therefore, this is discussed in Section V.D. No β-mercaptoethanol experiment was mentioned when the 1R-isomer inactivated the 17β-estradiol-20α-hydroxysteroid dehydrogenase,[148] so it is not clear if this is not mechanism-based as well. The R- and S-isomers were synthesized by Covey[150] (Scheme 106).

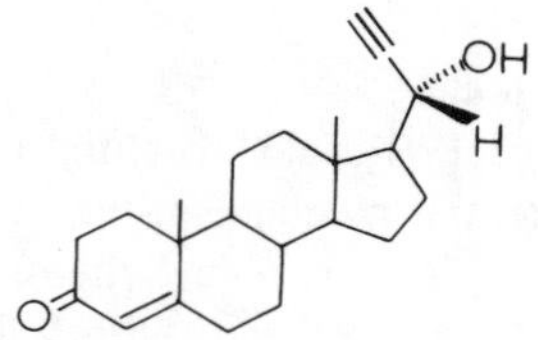

Scheme 105.   Structural Formula 9.84.

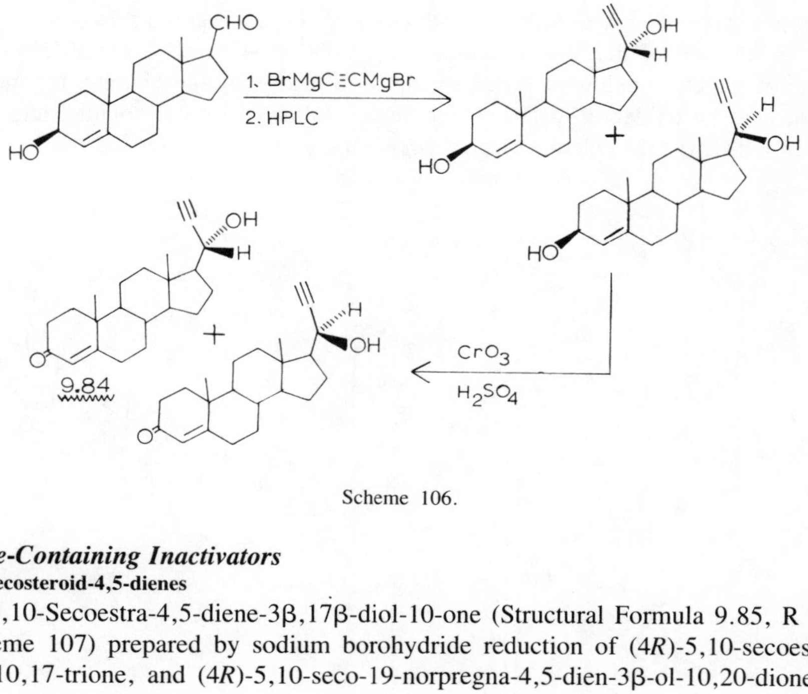

Scheme 106.

b. Allene-Containing Inactivators

(1) 5,10-Secosteroid-4,5-dienes

(4R)-5,10-Secoestra-4,5-diene-3β,17β-diol-10-one (Structural Formula 9.85, R = OH; see Scheme 107) prepared by sodium borohydride reduction of (4R)-5,10-secoestra-4,5-diene-3,10,17-trione, and (4R)-5,10-seco-19-norpregna-4,5-dien-3β-ol-10,20-dione (**9.85,**

R = Ac), prepared by sodium borohydride reduction of (4R)-5,10-secopregna-4,5-diene-3,10,20-trione, inactivate solubilized Δ^5-3β-hydroxysteroid dehydrogenase from bovine adrenals and human placenta; no inactivation of 3β-hydroxysteroid dehydrogenase from *Pseudomonas testosteroni* was observed by Balasubramanian and Robinson.[151] Oxidation of the 3β-hydroxyl group was shown by identification of the corresponding 3-oxo compound in solution; the 3-oxo compound is an inactivator of the enzyme. Since inactivation of the enzyme is subsequent to oxidation of the 3β-hydroxyl group and release of the allenic ketone, it is not clear if these compounds are mechanism-based inactivators. Gel filtration does not regenerate enzyme activity. The mechanism of inactivation proposed[151] is shown in Scheme 107.

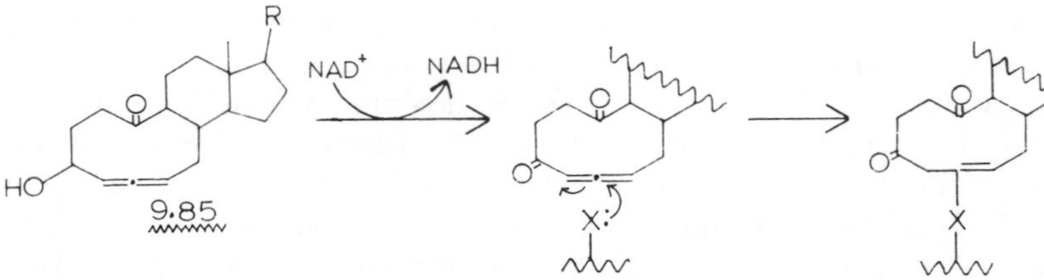

Scheme 107.   Containing Structural Formula 9.85.

(2) 4-Ethenylidene-5α-androstane-3β-17β-diol

4-Ethenylidene-5α-androstane-3β,17β-diol (Structural Formula 9.86) was synthesized by Balasubramanian et al.[152] as a potential mechanism-based inactivator of 3β-hydroxysteroid dehydrogenase. The synthetic route is shown in Scheme 108. Incubation of 3β,17β-hydroxysteroid dehydrogenase from *P. testosteroni* with **9.86** produces rapid oxidation only of the 17β-hydroxyl group and, therefore, no inactivation of the enzyme. When solubilized beef adrenal microsomal Δ^5-3β-hydroxysteroid dehydrogenase is used, a slow time-dependent inactivation occurs. Inactivation of the enzyme by the allenic ketone occurs three to four times faster than with **9.86**. In both cases NAD$^+$ is required for inactivation, even though, in the case of the allenic ketone, no oxidation is required for affinity labeling. Presumably, cofactor binding is a prerequisite for effective steroid binding. the mechanism is presumed to be oxidation to the allenic ketone followed by nucleophilic attack on the allene. It is not clear if the reactive species escapes the active site prior to enzyme inactivation.

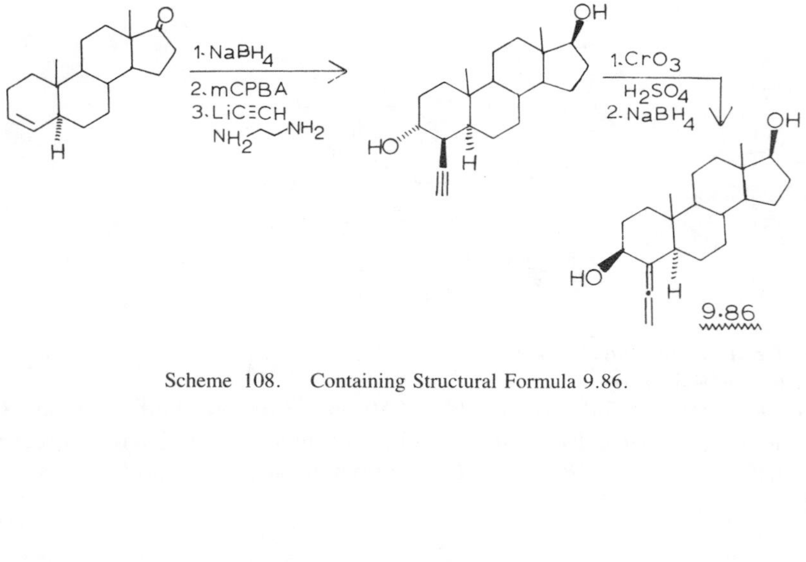

Scheme 108.   Containing Structural Formula 9.86.

c. Olefin-Containing Inactivators
(1) 16-Methylene estra-1,3,5(10)-triene-3,17β-diol

The first estrogen mechanism-based inactivator for human placental 17β-20α-hydroxy-steroid dehydrogenase was 16-methylene estradiol (16-methylene estra-1,3,5(10)-triene-3,17β-diol; Structural Formula 9.87, Scheme 109) which inactivates both the 17β- and 20α-hydroxysteroid dehydrogenase activities.[153] The time-dependent inactivation requires NAD$^+$ and is protected by estradiol-17β, estrone, and progesterone. The corresponding ketone, 16-methylene estrone is an affinity-labeling agent that alkylates the enzyme with a 1:1 stoichiometry. The implied mechanism for inactivation is oxidation of the allylic alcohol to an α,β-unsaturated ketone followed by Michael addition of an active site nucleophile (Scheme 109). However, Covey noted (personal communication) that inactivation of steroid dehydrogenases by many of the α,β-unsaturated alcohols is completely prevented by scavenger nucleophiles. If this is the case here, then **9.87** would not be a mechanism-based inactivator.

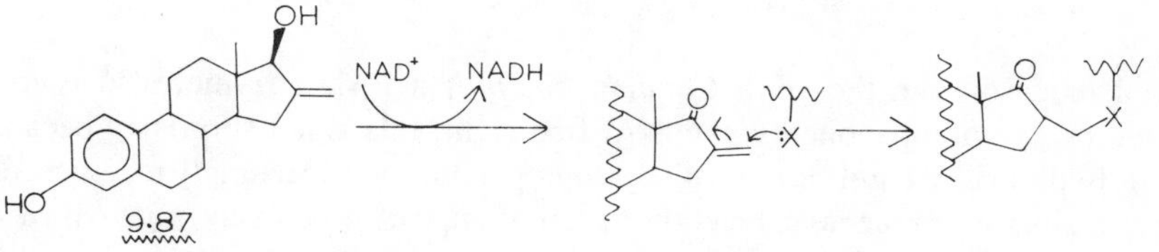

Scheme 109.   Containing Structural Formula 9.87.

The synthesis of **9.87** was reported in a patent by Ringold and Rosenkranz[154] (Scheme 110).

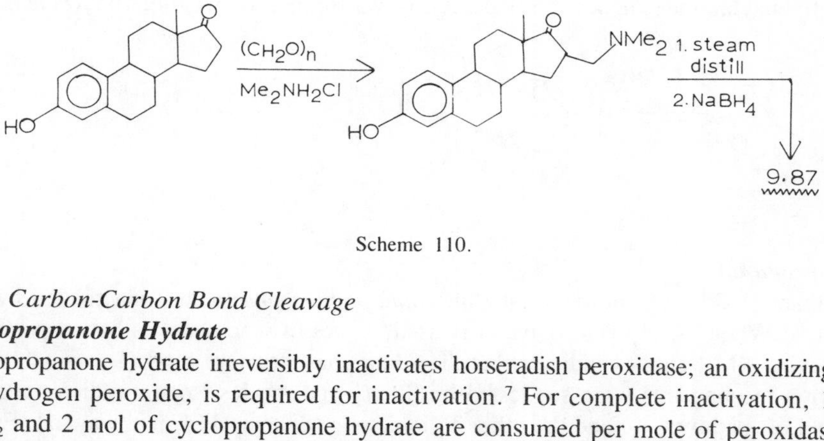

Scheme 110.

2. With Carbon-Carbon Bond Cleavage
a. Cyclopropanone Hydrate

Cyclopropanone hydrate irreversibly inactivates horseradish peroxidase; an oxidizing agent, e.g., hydrogen peroxide, is required for inactivation.[7] For complete inactivation, 1.4 mol of H_2O_2 and 2 mol of cyclopropanone hydrate are consumed per mole of peroxidase. This suggests that the inactivator is, in actuality, reacting with compound I, the two-electron oxidized form of peroxidase (Structural Formula 9.87a; see Scheme 111). No reactivation occurs after gel filtration. When the heme is dissociated from inactivated enzyme, reconstitution of the apoenzyme gives fully active enzyme. Fast atom bombardment mass spectrometry of the heme shows a total addition of $C_3H_5O_2$. The NMR spectrum reveals a propionic acid side chain on the heme. Changes in the optical spectrum initially indicate formation of a π-cation isoporphyrin that is converted to a mesosubstituted porphyrin. The mechanism proposed by Wiseman et al.[7] is shown in Scheme 111.

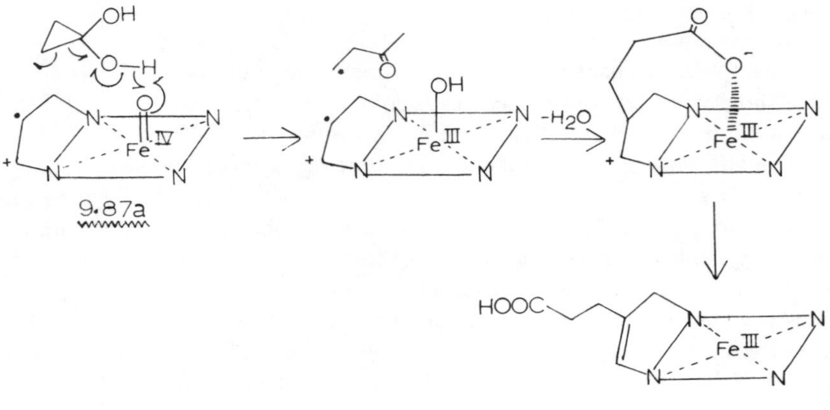

Scheme 111.   Containing Structural Formula 9.87a.

Alcohol oxidase from the yeast *Candida boidinii*, a FAD enzyme, also is irreversibly inactivated by cyclopropanone hydrate.[8] At first sight, this is not surprising because cyclopropanone hydrate is in equilibrium with cyclopropanone, an exceedingly reactive compound. However, a simple bimolecular reaction does not appear to be occurring. All of the usual criteria for mechanism-based inactivation are met in addition to the finding that during inactivation, the flavin is reduced and remains reduced following denaturation and acid treatment. This indicates a stable covalent linkage to the flavin. If simple reaction with the cyclopropanone at the highly electrophilic carbonyl had occurred, denaturation and acid treatment would be expected to cleave the bond. In order to rationalize the stability of the adduct, a mechanism similar to that proposed for inactivation of MAO by *N*-(1-methylcyclopropyl)benzylamine (see Section II.B.2.a.(5)) was proposed by Cromartie[8] (Scheme 112).

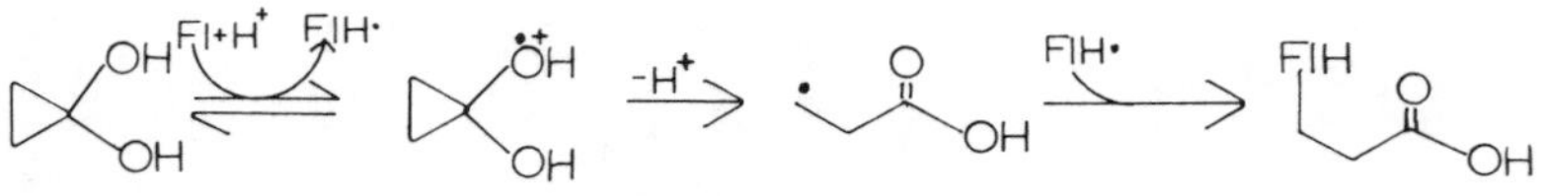

Scheme 112.

b. Cyclopropanol

A methanol oxidase from the yeast *Hansenula polymorpha* DL-1 is inactivated by cyclopropanol.[155] When 1-[³H]cyclopropanol is used, 1 mol of inactivator is bound per mole of active flavin. NMR studies indicate that the adduct has structure **9.88** (Scheme 113). Two inactivation mechanisms were suggested by Sherry and Abeles,[9] one a radical mechanism (Scheme 114, pathway a) and the other an ionic one (Scheme 114, pathway b). Evidence in favor of the radical mechanism was obtained: treatment of deazaflavin reconstituted apoenzyme with cyclopropanol does not give a covalent adduct; also, the rate of enzyme-catalyzed cyclopropyl ring opening is much faster than would be expected for an ionic mechanism.

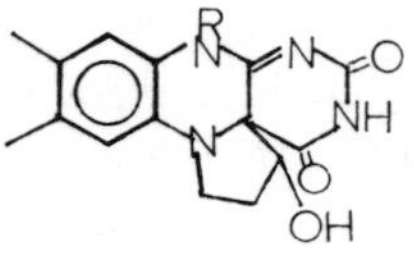

Scheme 113.   Structural Formula 9.88.

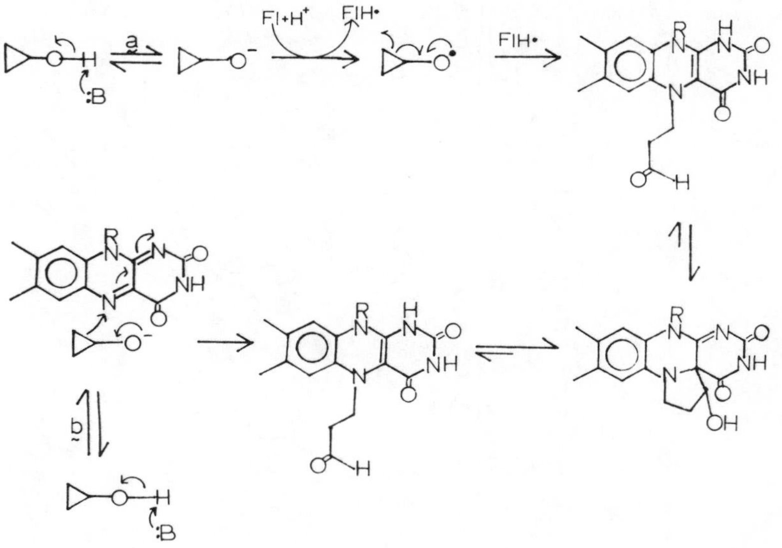

Scheme 114.

A methoxatin-(pyrroloquinoline quinone) dependent alcohol dehydrogenase from the bacterium *Methylomonas methanica* is inactivated by cyclopropanol; no enzyme activity is recovered upon dialysis.[156] [1-^{2}H]Cyclopropanol inactivates the enzyme with a kinetic isotope effect of 4. An amount of cyclopropanol that is only 14% of the total amount of enzyme is all that is required to completely inactivate the enzyme. This amount corresponds to the amount of radical present in native enzyme, and it is believed that only 14% of the enzyme is catalytically active. There is no change in the absorption spectrum during inactivation. When the enzyme is inactivated by [1-^{3}H]cyclopropanol, no [^{3}H]water is generated. Upon urea denaturation or methanol precipitation, 60 to 70% of the bound radioactivity remains bound and 30 to 40% is released as an unknown product. It is believed that the cyclopropanol is oxidized to some species that inactivates the enzyme, but Mincey et al.[156] did not suggest a mechanism. Further inactivation studies on this methoxatin-requiring enzyme were carried out by Parkes and Abeles.[157] A lag time for cyclopropanol inactivation was rationalized in terms of the inactivation occurring with the enzyme-electron acceptor dye complex and not with free enzyme. The reaction with [1-^{14}C]cyclopropanol results in incorporation of 0.13 mol of radioactivity per mole of enzyme with complete inactivation of the enzyme. The product of inactivation is not known, but it attaches to the cofactor, and the optical spectrum suggests an adduct with one of the carbonyl groups. Although acrolein was not observed as a product, it was suggested in the reaction shown in Scheme 115 to account for only 0.13 mol of radioactivity bound. The results and conclusions of Abeles and co-workers[156,157] were contradicted by Dijkstra et al.[157a] Cyclopropanol and cyclopropanone hydrate are time-dependent inactivators of the pyrroloquinoline quinone containing methanol dehydrogenase from dimeric *Hyphomicrobium* X (in its oxidized form) and from the monomeric and dimeric enzyme from *Pseudomonas* BB$_1$. Cyclopropanone ethyl hemiketal also inactivates the *Pseudomonas* enzyme, but not the *Hyphomicrobium* enzyme. Cyclopropylamine, cyclopropanemethanol, cyclobutanol, and cyclohexanol have no effect. The cofactor has to be in the oxidized form (PQQ), not the semiquinone form (PQQH·) suggested by Abeles and co-workers,[156,157] for inactivation to take place. Complete inactivation of the monomeric enzymes occurs with one equivalent of inactivator; the dimeric enzymes require two equivalents of inactivator. During inactivation, protons and electrons are neither consumed nor produced. The mechanism proposed[157a] is shown in Scheme 116.

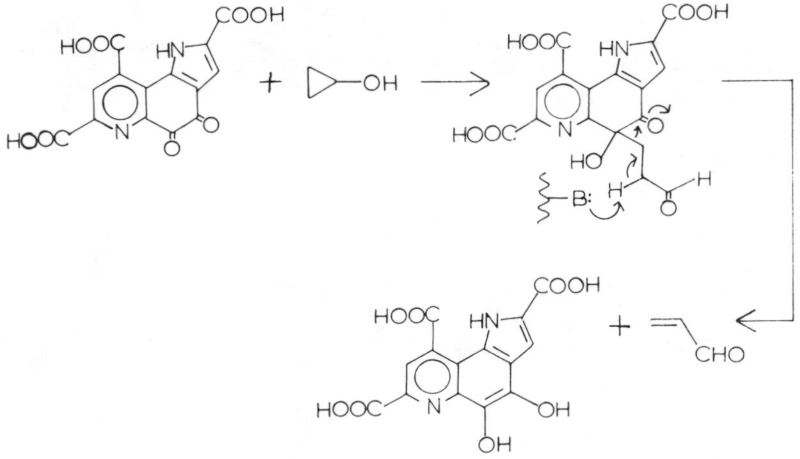

Scheme 115.

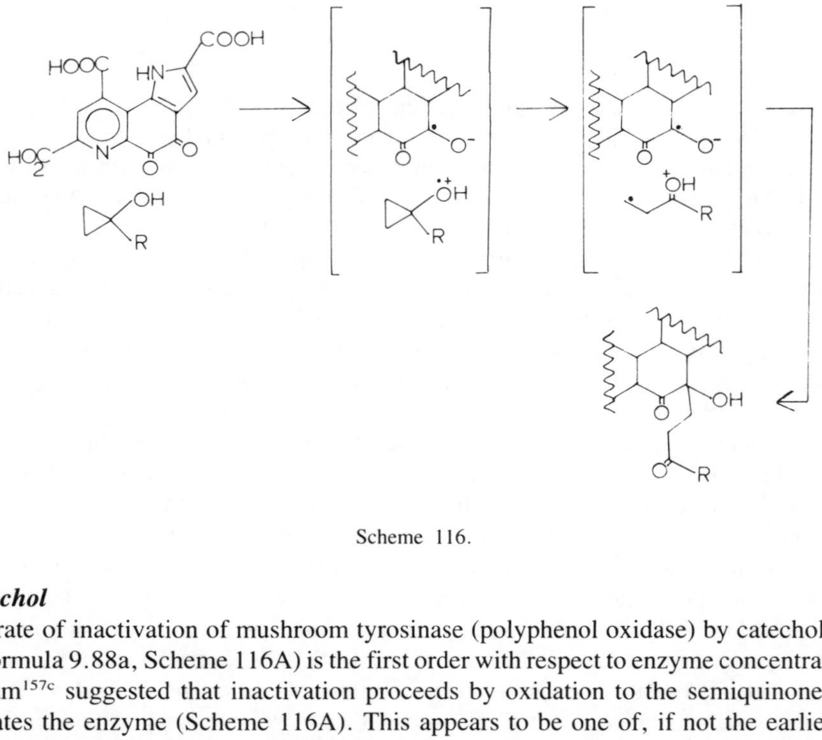

Scheme 116.

c. Catechol

The rate of inactivation of mushroom tyrosinase (polyphenol oxidase) by catechol (Structural Formula 9.88a, Scheme 116A) is the first order with respect to enzyme concentration.[157b] Ingraham[157c] suggested that inactivation proceeds by oxidation to the semiquinone, which inactivates the enzyme (Scheme 116A). This appears to be one of, if not the earliest, case where a mechanism-based enzyme inactivator is proposed to inactivate via a radical intermediate. Inactivation of polyphenol oxidase by catechol and oxidation of catechol have the same Michaelis-Menten dependence on catechol concentration and oxygen pressure.[157d] This supports the notion that the two processes are the same up to the formation of the enzyme-semiquinone complex, at which point partitioning between product release and enzyme inactivation can occur. This also appears to be one of, if not the earliest, references to a partition ratio.

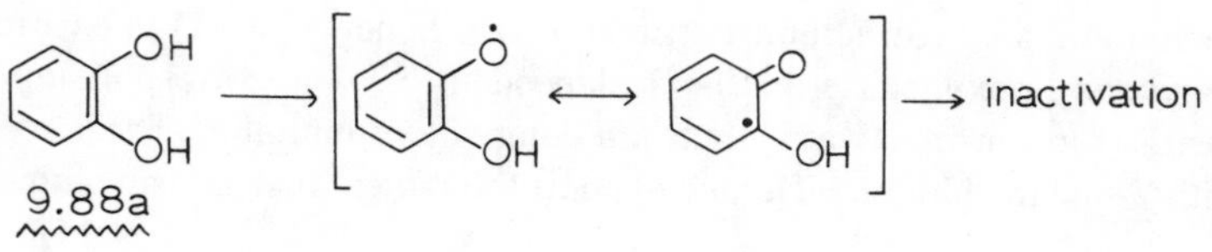

Scheme 116A.   Containing Structural Formula 9.88a.

The transient phase approach of Tudela et al.[157e] was applied to the study of the inactivation of mushroom tyrosinase by *o*-diphenols.[157f] Equations of product vs. time were developed for the multisubstrate mechanism and the kinetic parameters for inactivation were determined.

D. Carbon-Sulfur Bond Oxidation
1. With Carbon-Carbon Bond Cleavage
a. 7-Thiaarachidonic Acid and Related

7-Thiaarachidonic acid (Structural Formula 9.89) and related compounds (Structural Formulas 9.90 and 9.91), shown in Scheme 117, were synthesized by Corey et al.[11] (Scheme 118). Pseudo first-order, time-dependent inactivation of 5-lipoxygenase from RBL-1 cells occurs in the presence of O_2, but not in its absence. The inactivation mechanism proposed in shown in Scheme 119.

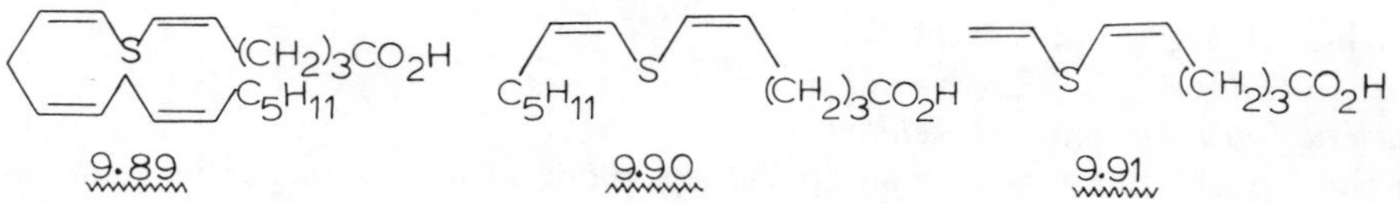

Scheme 117.   Structural Formulas 9.89 to 9.91.

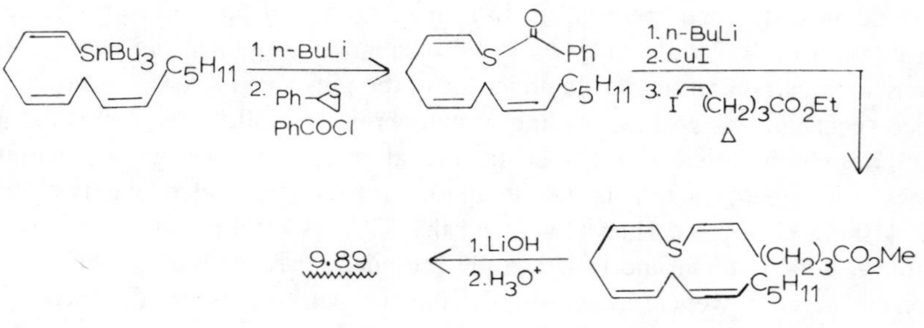

Scheme 118.

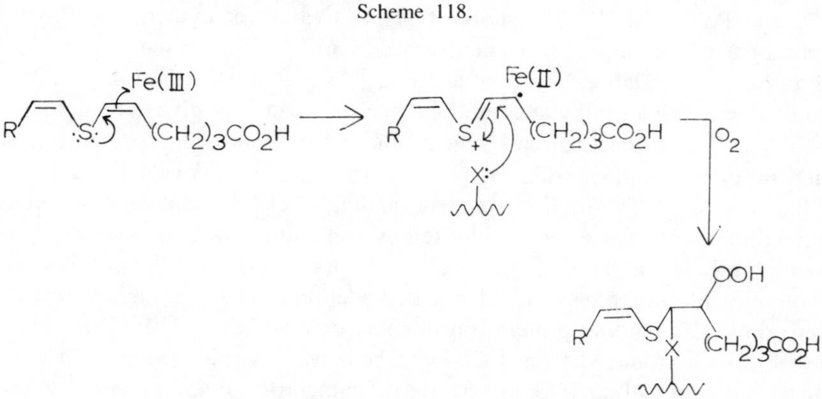

Scheme 119.

13-Thiaarachidonic acid (Structural Formula 9.91a, Scheme 119A) is an irreversible in-activator of soybean lipoxygenase that is O_2 dependent.[157g] The corresponding sulfoxide is neither a substrate nor an inactivator, but is a competitive inhibitor. These results parallel the results with 7-thiaarachidonic acid and support the radical mechanism shown in Scheme 119.

The synthesis of **9.91a** by Corey et al.[157h] is shown in Scheme 119A.

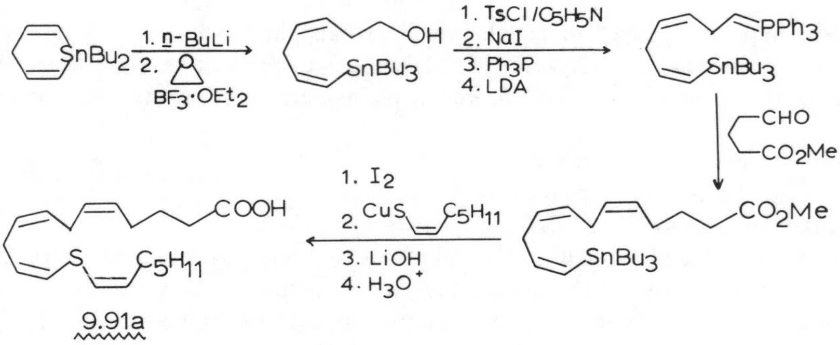

Scheme 119A.   Containing Structural Formula 9.91a.

E. Carbon-Halogen Bond Oxidation
1. Carbon-Halogen Bond Cleavage
a. Carbon Tetrachloride and Related

Carbon tetrachloride produces an optical absorption change in rabbit liver microsomes under anaerobic conditions in the presence of NADPH that was suggested by Reiner and Uehleke[158] to be the result of complex formation of CCl_4 with the divalent iron of heme in cytochrome P-450. Under anaerobic conditions in the presence of NADPH, $^{14}CCl_4$ produces a time-dependent difference spectrum in liver microsomes of phenobarbital-treated rabbits and incorporation of [^{14}C] into proteins.[159,160] Incubation of guinea pig adrenal or hepatic microsomes with carbon tetrachloride under air in the presence of NADPH produces a type I difference spectrum, suggesting binding to cytochrome P-450.[161] NADPH is required for inactivation of cytochrome P-450. Effects in adrenal microsomes was greater than in hepatic microsomes. Lipid peroxidation is not obligatory for CCl_4-mediated inactivation of cyto-chrome P-450. DeGroot and Haas[162] showed that CCl_4 is a time-dependent inactivator of rat liver microsomal cytochrome P-450 in the presence or absence of O_2 with concomitant destruction of heme; however, increasing the oxygen partial pressure decreases the inacti-vation of cytochrome P-450 by CCl_4 and increases lipid peroxidation.[163] Guzelian and Swisher[164] and Poli et al.[165] have shown that inactivation of cytochrome P-450 and the degradation of the heme in rat liver microsomes by CCl_4 in the presence of NADPH is not the result of lipid peroxidation. DeGroot and Haas[166] suggested that CCl_4 is reduced to $^{\cdot}CCl_3$ or dichlorocarbene, which alkylates the heme. No evidence is given, though, to exclude metabolically activated inactivation. Cytochrome P-450 from phenobarbital-treated rats is inactivated by carbon tetrachloride in a process that results in covalent attachment of the heme to the protein.[166a] The in vitro covalent binding of carbon tetrachloride metabolites to microsomal proteins in mouse liver is blocked by pretreatment with methoxsalen, indicating that cytochrome P-450 is the site of activation.[166b] As a model for the covalent binding of [^{14}C]carbon tetrachloride to heme and heme degradation products of an NADPH-dependent liver microsomal CO-binding protein (presumably cytochrome P-450), Fernández et al.[167] showed that $^{\cdot}CCl_3$, produced from CCl_4 with benzoyl peroxide, reacts with the heme of hemin. In an earlier model study for inactivation of cytochrome P-450 by carbon tetrachloride,

Mansuy et al.[168] showed that tetraphenylporphyrin iron(II) reacts with carbon tetrachloride in the presence of a reducing agent to give a complex having the properties of that of an iron-carbene complex, $(TPP)Fe(II)(CCl_2)$. This is believed to be the first isolated carbene complex of an iron porphyrin. Other polyhalogenated compounds react similarly. An X-ray structure of the proposed dichlorocarbene complex with $(TPP)Fe(II)$ confirmed that it was $(TPP)Fe(II)(CCl_2)(H_2O)$.[169] As a model for cytochrome P-450 inactivation by polyhalogenated compounds in general, Lange and Mansuy[170] showed that porphyrin-carbene complexes react with acids or one-electron oxidants to give N-substituted porphyrins (Scheme 120).

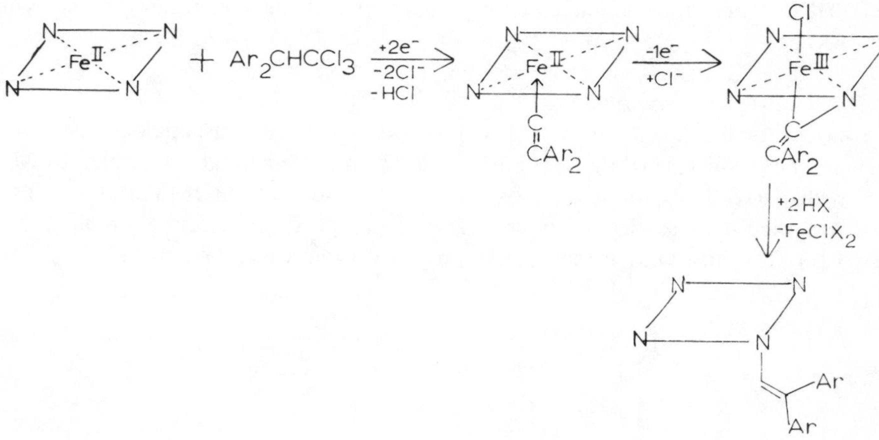

Scheme 120.

Various polyhalogenated methanes (CCl_4, CBr_4, CCl_3F, CCl_3Br, CCl_3CN, CHI_3, $CHBr_3$, and $CHCl_3$) react with liver microsomal ferrous cytochrome P-450 from phenobarbital-treated rats under reducing conditions (NADPH); cytochrome P-450 carbene complexes were suspected by Wolf et al.[171]

b. Halothane and Related

The inhalation anesthetic, halothane (1,1,1-trifluoro-2-bromo-2-chloroethane), irreversibly binds to proteins in liver microsomes of phenobarbital-treated rabbits in the presence of NADPH under anaerobic incubation.[172] An unusual difference spectrum is obtained when halothane reacts anaerobically with reduced cytochrome P-450 in liver microsomes of phenobarbital-treated rats.[173] This same spectrum can be simulated by addition of trifluoro diazoethane to dithionite-reduced microsomes. On the basis of model reactions, it was suggested by Mansuy et al.[173] that in both cases a trifluoromethyl carbene complex with the reduced hemoprotein is formed. The formation of trifluoromethyl carbene from halothane was proposed to occur by the addition of two electrons and loss of chloride- and bromide ion.

Inactivation of cytochrome P-450 is not as pronounced in 3-methylcholanthrene-pretreated or untreated rats as in phenobarbital-treated rats.[174] This indicates a different sensitivity of various isozymes of cytochrome P-450 to halothane. In contrast to CCl_4, there appears to be no evidence for carbene formation from halothane; inactivation was suggested by deGroot et al.[174] to be the result of $CF_3\dot{C}HCl$ formation. Increasing the O_2 pressure decreases the amount of cytochrome P-450 that is inactivated. This was explained as resulting from a stimulation of oxidative metabolism of halothane to CF_3COOH. An alternate explanation is that the $CF_3\dot{C}HCl$ is released into solution and is trapped by O_2, in which case halothane is not a mechanism-based inactivator. The hepatotoxicity noted in some cases of halothane treatment may be the result of reactive halothane radicals. The aerobic inactivation of cytochrome P-450 by halothane is reversed upon dialysis, but the anaerobic inhibition is only partially reversed.[175] Heme *N*-alkylation does not occur; Krieter and Van Dyke[175] suggested that a reactive metabolite binds to the protein portion.

As a model for cytochrome P-450 inactivation by halothane, the reaction of tetraphenylporphyrin iron(II) with CF_3CCl_3 and with halothane was studied by Mansuy and Battioni[176] under reducing conditions. The former compound produces $(TPP)Fe(II)(CClCF_3)$ carbene complex and the latter produces the σ-alkyl complex $(TPP)Fe(III)(CHClCF_3)$.

1,1-Dichloroethane inactivates microsomal cytochrome P-450 from rats in the presence of NADPH.[176a] This effect is enhanced by induction with phenobarbital, but not with β-naphthoflavone.

c. 2,2-Dichloro-1,3-benzodioxole

As a model for the inactivation of cytochrome P-450 by the insecticide synergist, 2,2-dichloro-1,3-benzodioxole (Structural Formula 9.92), this compound was treated by Mansuy et al.[177] with Fe(II)(TPP). Immediate oxidation occurs to give Fe(III)(TPP)(Cl). The final product was identified as iron(porphyrin)carbene complex (Structural Formula 9.93, see Scheme 121). The inactivation mechanism proposed is shown in Scheme 122.

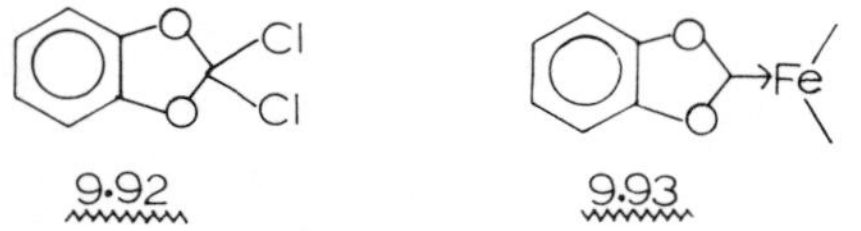

9.92   9.93

Scheme 121.   Structural Formulas 9.92 and 9.93.

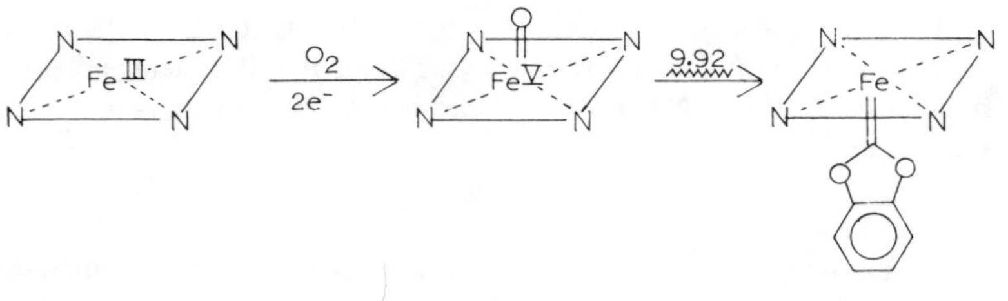

Scheme 122.

F. Nitrogen-Nitrogen Bond Oxidation

1. 1-Aminobenzotriazole and Related

Because of the ease of chemical oxidation of 1-aminobenzotriazole (Structural Formula 9.94, Scheme 123) to benzyne,[178] this compound was tried as an inactivator of cytochrome P-450 by Ortiz de Montellano and Mathews.[179] Inactivation of hepatic cytochrome P-450 by **9.94** is highly effective and results in destruction of the microsomal heme, but not loss of cytochrome b_5. NADPH and O_2 are required; the presence of glutathione does not prevent enzyme inactivation. Compound **9.94** inactivates both phenobarbital- and 3-methylcholanthrene-induced cytochrome P-450 isozymes.[180] The porphyrin generated by inactivation is radically different from *N*-monoalkylated protoporphyrin IX derivatives.[179] The electronic absorption spectrum of the alkylated heme, however, is identical with that of synthetic *N,N*-dimethylprotoporphyrin IX. It also has similar properties to the dialkylated heme. The structure proposed is **9.95** (Scheme 124). The inactivator was synthesized by the route shown in Scheme 125. Further studies with **9.94** were carried out by Ortiz de Montellano et al.[181] NMR spectra analysis of the modified heme (isolated from livers of rats treated with **9.94**) indicates that two vicinal nitrogens of the porphyrin ring are bound to vicinal carbons of a benzene ring. This is consistent with the addition of benzyne across two nitrogens of the heme; however, a mixture of isomers was obtained that could not be resolved. Another minor product obtained was *N*-phenylprotoporphyrin IX. Inactivation with modified heme formation also occurs when the 1-amino group is substituted with either methyl, acetyl, or both. Benzotriazole, 2-aminobenzotriazole, 1-hydroxy-, and 1-nitrobenzotriazole are ineffective. 1-Amino-1*H*-naphtho(2,3-d)triazole (Structural Formula 9.96, Scheme 126) also inactivates the enzyme and arylates the heme.

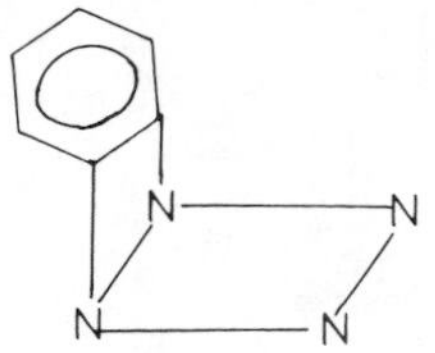

Scheme 123.   Structural Formula 9.94. 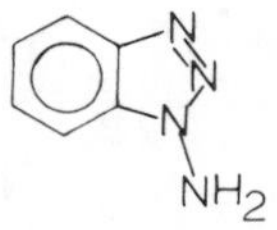   Scheme 124.   Structural Formula 9.95.

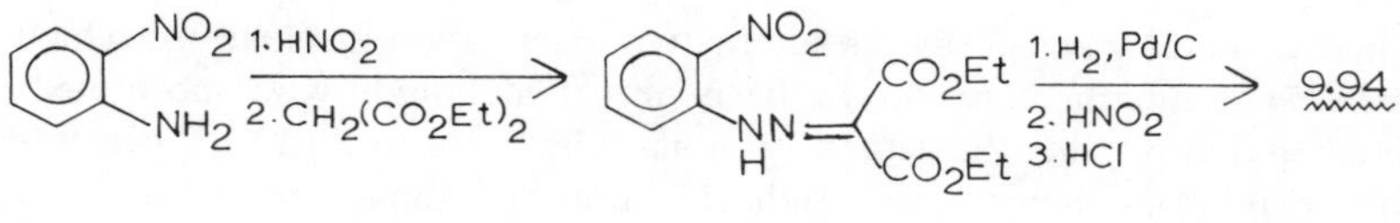

Scheme 125.

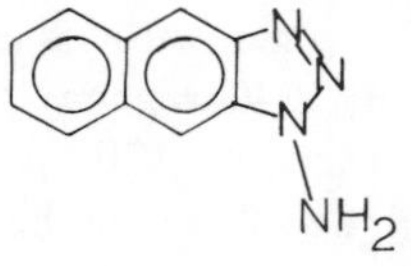

Scheme 126.   Structural Formula 9.96.

Compound **9.94** is a time-dependent inactivator of chloroperoxidase as well, and produces phenol as the major metabolite.[181] This is further support for a benzyne intermediate. The mechanism preferred is a one-electron mechanism (Scheme 127). The benzyne could add directly across two of the pyrrole nitrogens to give the product. An alternative mechanism that also would account for the small amount of *N*-phenylprotoporphyrin IX would be an initial addition across the iron and one of the pyrrole nitrogens. As was shown for phenylhydrazine inactivation of cytochrome P-450 (see Section IV.A.1.a.), an initial iron-phenyl complex is formed that is oxidized to an *N*-phenyl complex. In the case of **9.94**, the iron-pyrrole-substituted benzene could be oxidized to the *N*,*N*-bridged product (Scheme 128) or nonoxidatively hydrolyzed to *N*-phenylprotoporphyrin IX.

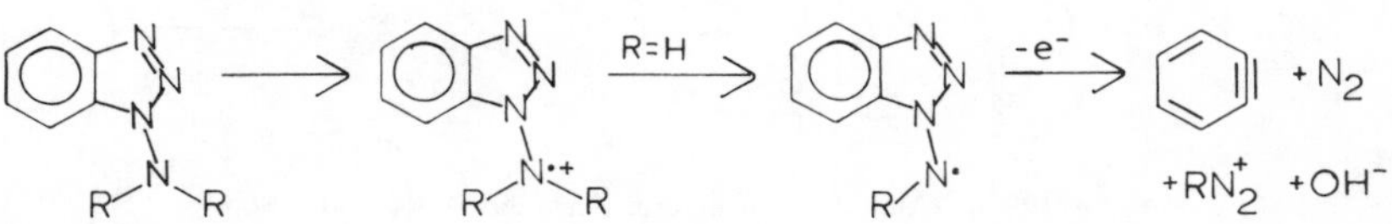

Scheme 127.

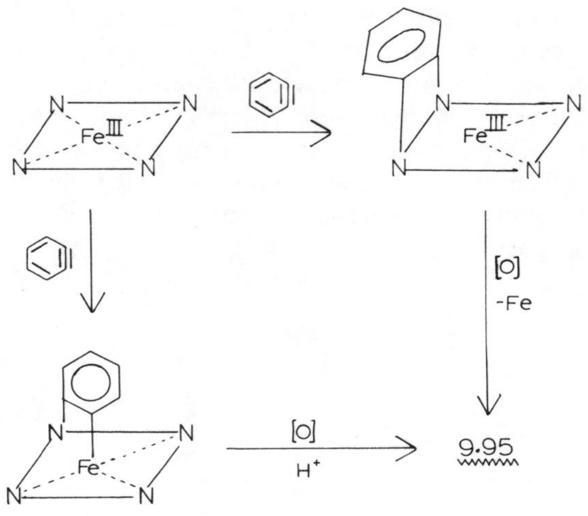

Scheme 128.

Callot and Cromer[181a] suggested an alternative inactivation mechanism, which does not involve a benzyne intermediate, on the basis of a model study with substituted vinylcobalt(III)tetraphenylporphyrins (Structural Formula 9.96a, Scheme 129). When **9.96a** is oxidized with tri(*p*-bromophenyl)amine radical cation, rearrangement to the *N*-substituted porphyrin occurs (Scheme 129). Further oxidation with the same oxidant results in cyclization to the bridged compound (Structural Formula 9.96b). In support of this mechanism, *N,N'*-phenyleneporphyrins (**9.95**) were prepared from Co(II)- and chloro Mn(II)-*N*-phenyltetraphenylporphyrin with $(p\text{-BrC}_6\text{H}_4)_3\text{N} \cdot +$ or electrolysis.[181b] Under forcing conditions (90 to 147°C, air, 1 to 3 days), chlorocobalt(III)-tetraphenylporphyrin reacts with 1-aminobenzotriazole to give low yields of **9.95** (Scheme 130). Furthermore, when this reaction was carried out in the presence of a large excess of the benzyne trap, tetraphenylcyclopentadienone, no expected benzyne trapping product, tetraphenylnaphthalene, was observed by Callot et al.[181b]

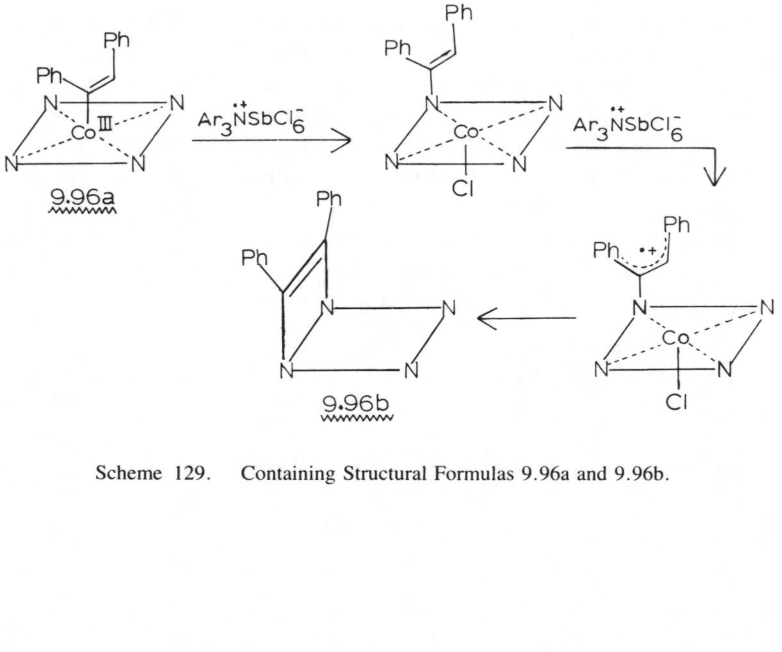

Scheme 129.   Containing Structural Formulas 9.96a and 9.96b.

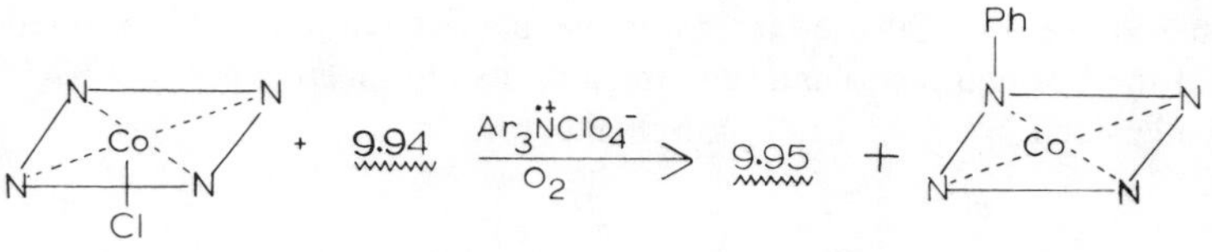

Scheme 130.

1-Aminobenzotriazole also is a time-dependent inactivator of pulmonary microsomal cytochrome P-450 from β-naphthoflavone-treated rabbits; inactivation depends upon NADPH.[182] Both isozymes 2 and 5 are inactivated, but no inactivation of flavin-dependent monooxygenase activity is evident.

Microsomal cytochrome P-450 in the housefly (*Musca domestica*) is inactivated by **9.94** in a NADPH-, O_2-, and pseudo first-order, time-dependent process.[183] The electronic absorption spectrum of the modified porphyrin isolated is identical with that of the green pigment isolated from inactivation of rat liver cytochrome P-450 by the same compound.[179] Aldrin epoxidation by housefly microsomes (a cytochrome P-450 activity) is inactivated by **9.94** in a pseudo first-order, time-dependent process. The amount of enzyme destroyed depends on the strain of housefly and on whether the insects are pretreated with phenobarbital. Cytochrome P-450 from phenobarbital-treated resistant houseflies is most sensitive to inactivation.

1-Aminobenzotriazole also is a specific time-dependent, pseudo first-order inactivator of cinnamic acid 4-hydroxylase, a cytochrome P-450 monooxygenase from *Helianthus tuberosus* L (Jerusalem artichoke) microsomes.[184] Unlike the lack of specificity of **9.94** for liver cytochrome P-450 induced in rats by phenobarbital and 3-methylcholanthrene,[180] it is quite specific in plant tissue. Cytochrome b_5 and NADPH cytochrome C reductase are not affected; lauric acid hydroxylase is only weakly inhibited.[184]

N-Alkylaminobenzotriazoles (Structural Formula 9.96c, R = PhCH$_2$, PhCHMe, Me, *n*-Bu, Scheme 130A) were prepared by Mathews and Bend[184a] as isozyme-selective, mechanism-based inactivators of different rabbit lung cytochrome P-450 isozymes. The *N*-benzyl- and *N*-α-methylbenzyl analogues are the most potent and isozyme selective. At 1 μ*M* concentration, the *N*-α-methylbenzyl analogue inhibits 80% of isozyme 2 and 20% of isozyme 6, but does not inhibit isozyme 5. The partition ratio for this analogue is 11 ± 2. The mechanism of inactivation appears to be the same as that for the parent compound (see Schemes 127 and 128).

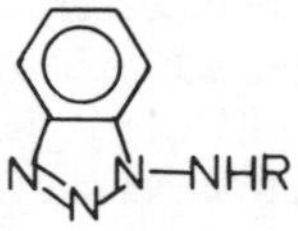

Scheme 130A.   Structural Formula 9.96c.

2. 5,6-Dichloro-1,2,3-benzothiadiazole

Incubation of liver microsomes from phenobarbital-treated rats with 5,6-dichloro-1,2,3-benzothiadiazole (Structural Formula 9.97, see Scheme 131) results in rapid loss of cytochrome P-450 only in the presence of NADPH and O_2.[185] Glutathione does not attenuate loss of enzyme. No green pigment formation, however, was observed, similar to the inactivation of cytochrome P-450 by allenes (see Chapter 10, Section II.A.2.) and vinyl chloride (Chapter 10, Section II.A.3.d.(2)) in which heme loss also is not coupled with pigment formation. Possible reactive intermediates suggested by Ortiz de Montellano and Mathews[185] are shown in Scheme 131. Purified liver microsomal cytochrome P-450 isozymes from both phenobarbital- and 3-methylcholanthrene-treated rats are inactivated by **9.97**.[180] Compound **9.97** was synthesized[185] by the route in Scheme 132.

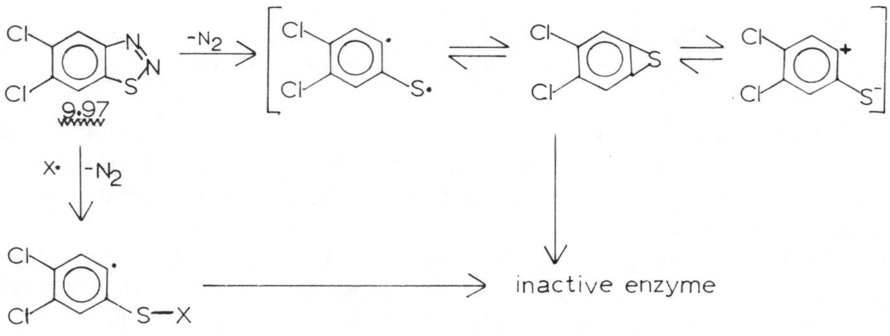

Scheme 131.   Containing Structural Formula 9.97.

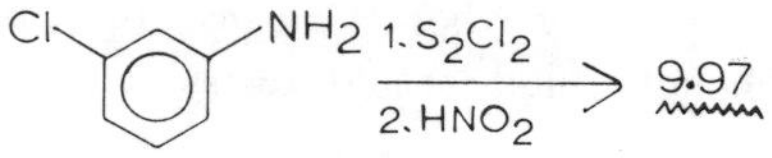

Scheme 132.

III. OXIDATION/ISOMERIZATION

A. Oxidation/Isomerization/Addition

1. Propargylglycine

D-Propargylglycine (Structural Formula 9.98, see Scheme 133) was shown by Horiike et al.[186] to be a time-dependent inactivator of D-amino acid oxidase; 600 to 800 turnovers occur prior to irreversible inactivation. L-Propargylglycine has no effect. After inactivation, the FAD was removed and shown to be unmodified. Enzyme activity could not be regenerated after inactivation even when the apoenzyme was prepared and exogenous FAD added. Modification of an active site amino acid residue, therefore, was suggested. This work was confirmed by Marcotte and Walsh[82] who showed that a product is generated during inactivation by D-propargylglycine and that L-propargylglycine is a substrate, but not an inactivator for L-amino acid oxidase. In both cases, the product formed was identified as acetopyruvate (Structural Formula 9.99, see Scheme 133). Scavenger molecules partially protect the enzyme. With D,L-propargyl[^{14}C]glycine followed by exhaustive dialysis, 1.7 mol of radioactivity are incorporated per active site, indicating that even after modification, the enzyme may be capable of carrying out an oxidation or that some labeling is derived from reaction of an activated species released into solution. The inactivation mechanism proposed[82] is shown in Scheme 133. Further characterization of D-propargylglycine-inactivated D-amino acid oxidase was described by Marcotte and Walsh.[187] The enzyme is not really inactivated, but rather is converted into a mixture of at least five species which have catalytic capabilities that are markedly different than those of the native enzyme. The modified enzyme has a preference for oxidation of hydrophobic amino acids. There is a primary kinetic isotope effect on oxidation of amino acids that is different than with native enzyme, indicating a change in the rate-determining step upon modification. An intense absorbance at 318 to 320 nm also is apparent in modified enzyme that is not present in the native enzyme. This chromophore can be reduced by 75% with hydrazine without effect on its enzyme activity. It was suggested that this chromophore results from peripheral alkylations which are not responsible for the observed modified kinetic behavior. During inactivation of D-amino acid oxidase by propargylglycine, two noncovalent inhibitors are generated.[188] Previously, it was reported[82] that **9.99** is the major product, but Marcotte and Walsh[188] modified that suggestion. The initial product is 3-amino-5-methylene-2(5*H*)-furanone (Structural Formula 9.100), produced by an intramolecular attack of the carboxylate ion on C-4 of the intermediate (Scheme 134). Two other species have been identified that form "charge-transfer" complexes with the enzyme. One is 2-amino-4-keto-2-pentenoate (Structural Formula 9.101), which is derived from the lactone by hydrolysis (Scheme 135); it was independently synthesized. The second species is believed to be 2-amino-2-penten-4-ynoate (Structural Formula 9.102, Scheme 136) the enamine derived from the initial oxidation product.

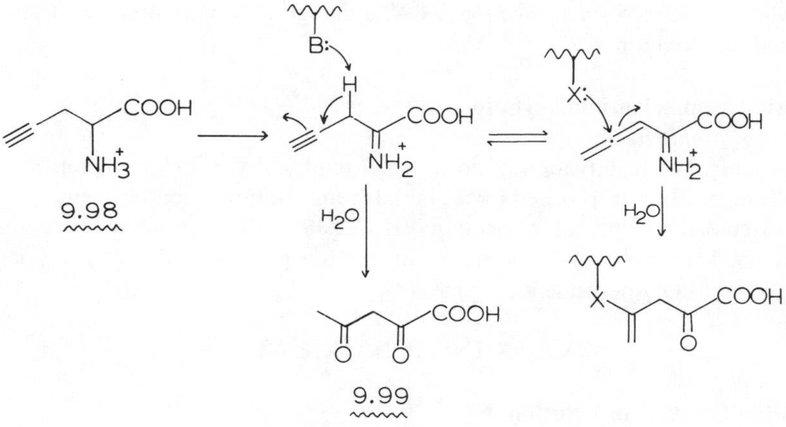

Scheme 133. Containing Structural Formulas 9.98 and 9.99.

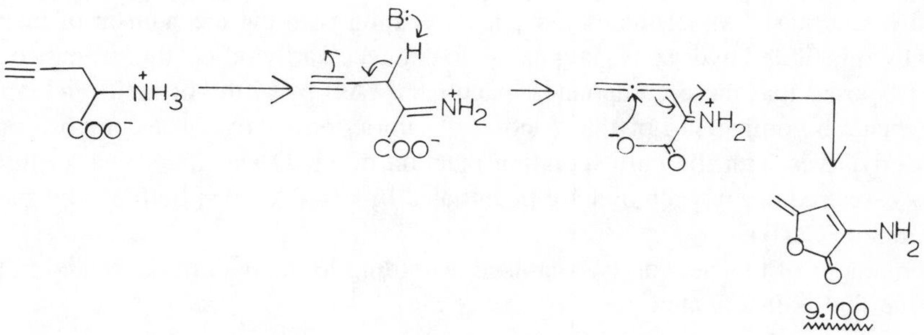

Scheme 134. Containing Structural Formula 9.100.

$$\underset{\sim\sim}{9.100} \xrightarrow{H_2O} CH_3\underset{\underset{O}{\|}}{C}CH{=}\underset{\underset{NH_2}{|}}{C}COO^-$$

9.101

Scheme 135. Containing Structural Formula 9.101.

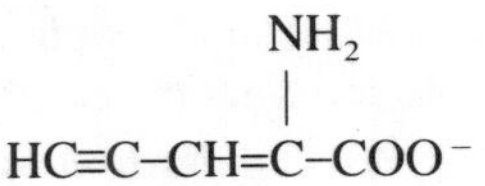

$$NH_2$$
$$|$$
$$HC \equiv C-CH = C-COO^-$$

9.102

Scheme 136.   Structural Formula 9.102.

2. *2'-Azido-2'-deoxynucleoside 5'-diphosphates*

Between the time that the manuscript for this book was submitted and the galley proofs were prepared, Stubbe and co-workers reported further studies with 2'-azido-2'-deoxynucleoside 5'-diphosphates that shed considerable light on a possible mechanism of inactivation of *E. coli* ribonucleoside diphosphate reductase.[188a] This work has been added to the previous studies, which are described in Section VI.A, although it is more appropriate that Section VI.A should be moved here.

B. Oxidation/Isomerization/Acylation

1. *β-Aminopropionitrile*

The mechanism of inactivation of aortic lysyl oxidase by β-aminopropionitrile was proposed by Tang et al.[189] to proceed by an initial isomerization reaction; therefore, this compound is discussed in Chapter 7, Section VII.C.2. Since the oxidation mechanism of the enzyme is not known, it could be placed in this chapter as well. Lysyl oxidase is now known[98a] to be a PQQ-dependent enzyme.

IV. OXIDATION/OXIDATION

A. Oxidation/Oxidation/Addition

1. *Substituted Hydrazines*

The fact that substituted hydrazines are discussed in this section reflects a bias by the author of this book regarding the general mechanism of these compounds and may not be completely accurate. Two of the earlier papers dealing with the mechanism of inactivation of MAO by substituted hydrazines in general do not necessarily reflect this prejudice. Carbon et al.[190] proposed that these compounds inactivate MAO by virtue of their nucleophilicity. Green[191] made a comparison of the reaction of cupric-ion catalyzed decomposition of hydrazine derivatives with the corresponding reaction of MAO and found many similarities. Both processes require oxygen and are potentiated by cyanide ions; both can be suppressed by chelators of Cu(II).

The remainder of this section is organized according to the hydrazine studied rather than the enzyme that is inactivated.

a. *Phenylhydrazine*

Phenylhydrazine is an irreversible inactivator of bovine kidney mitochondrial MAO[51,192] that appears to be oxidized to phenyldiazene with concomitant flavin reduction.[192] Phenyldiazene rapidly inactivates MAO. A Hammett plot of the σ values for *meta-* and *para*-substituted phenylhydrazines vs. the log of the rate of inactivation gives a ρ value of -1.9; this suggests that phenylhydrazine reduction of the flavin is rate determining. Although Patek and Hellerman[192] did not mention it, ρ values of this magnitude are consistent with a radical mechanism as Fitzpatrick et al.[6] found when they used substituted 3-phenylpropenes to inactivate dopamine β-hydroxylase (see Section II.A.1.c.). [1-^{14}C]Phenylhydrazine inactivates MAO with incorporation of 1.4 equivalents of [^{14}C] per enzyme subunit. Tryptic digestion gives [^{14}C]-labeled, flavin-containing peptides in which the flavin spectrum is reduced. This suggests that the phenylhydrazine reacts with the flavin cofactor. Sulfhydryl analysis indicates that phenylhydrazine does not become attached to a cysteine residue.[51]

Human brain MAO B is selectively inactivated by phenylhydrazine.[193] Two different inactivation processes occur, one temperature sensitive and the other temperature insensitive. These two processes may account for the greater than 1 mol of inactivator bound per mole of enzyme. Tipton[194] found that phenylhydrazine also is a time-dependent inactivator of the pig brain enzyme; benzylamine does not protect the enzyme from inactivation by phenyl-hydrazine, and O_2 is not required. The hydrazone mechanism suggested earlier[191] was proposed.[194]

Phenylhydrazine is a time-dependent, irreversible inactivator of beef liver mitochondrial MAO.[110] With [^{14}C]phenylhydrazine, 1.1 to 1.4 mol of [^{14}C] are incorporated into the enzyme after complete inactivation, and the radioactivity is stable to acid precipitation of the enzyme. Following tryptic-chymotryptic digestion, the labeled flavin peptide was characterized as was done for the product of inactivation of trimethylamine dehydrogenase by phenylhydrazine (see below). The proposed structure is **9.103** (Scheme 137).

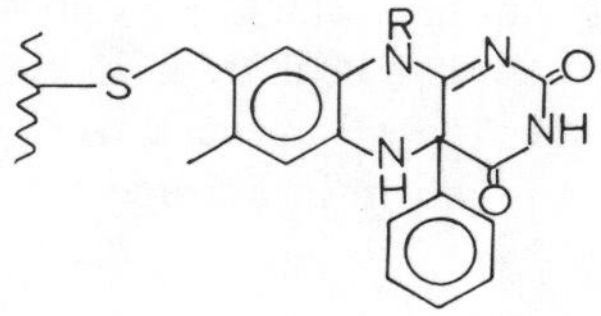

Scheme 137.   Structural Formula 9.103.

Trimethylamine dehydrogenase from bacterium W3A1 undergoes time-dependent inactivation by phenylhydrazine.[195] The rate of inactivation is greatly accelerated by dyes capable of reoxidizing the enzyme; only the oxidized form of the enzyme is inactivated. Inactivation is accompanied by bleaching of the flavin spectrum. Under anaerobic conditions, phenyl-diazene, but not phenylhydrazine, inactivates the enzyme. When [^{14}C]phenylhydrazine was used to inactivate the enzyme, a 1:1 stoichiometry of inactivator to enzyme was observed by Nagy et al.[195] after acid precipitation. Tryptic and chymotryptic digestion, followed by purification of the peptide fragments, still showed a 1:1 stoichiometry of [^{14}C] to flavin. On the basis of the absorption spectra (pH 7 and in 5 N HCl) of the radioactive flavin peptide compared with the corresponding spectra of synthetic N5- and C4a-substituted flavins, the labeled flavin peptide appears to be a C4a-substituted flavin (**9.103**). The product of the dark reaction of FMN with phenylhydrazine has all of the spectral characteristics of the product of the inactivated enzyme flavin peptide. A mass spectrum of the synthetic product indicated an incorporation of a phenyl group into the flavin. The mechanism proposed is shown in Scheme 138. Although it was not mentioned how phenyldiazene reacts with the flavin to form **9.103**, one-electron transfer would produce the phenyldiazenium radical cation which would rapidly decompose to phenyl radical and nitrogen. The enzyme from bacterium 4B6 also is inactivated by phenylhydrazine; dialysis does not regenerate enzyme activity.[196]

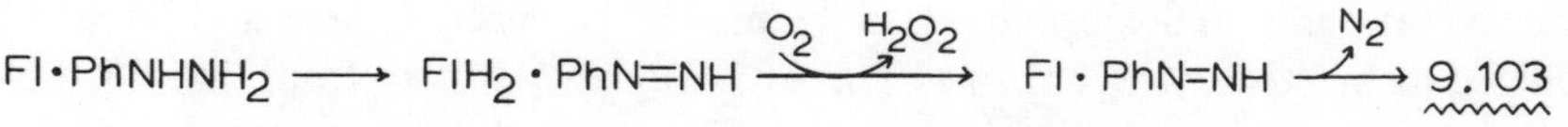

$$Fl \cdot PhNHNH_2 \longrightarrow FlH_2 \cdot PhN{=}NH \xrightarrow[]{O_2 \quad H_2O_2} Fl \cdot PhN{=}NH \xrightarrow{N_2} 9.103$$

Scheme 138.

Amine oxidase from *Aspergillus niger* is irreversibly inactivated by 2 mol of phenylhydrazine per dimer of enzyme.[197] It was suggested by Suzuki et al.[197] that the enzyme contains two enzyme-bound carbonyl groups at the active site that react to form hydrazones. Since the active site groups are not known (it is not **not** a flavoenzyme, but does contain Cu(II)), a mechanism involving oxidation of the phenylhydrazine cannot be excluded.

Phenylhydrazine was shown by Falk[198] to be an irreversible inactivator of pig plasma amine oxidase. One mole of commercially available [*ring*-¹⁴C(U)] phenylhydrazine is incorporated into the dimeric enzyme. The adduct is stable at room temperature for 72 hr. However, the enzyme mechanism is unclear, although an initial oxidation is reasonable.

Dopamine β-hydroxylase from bovine adrenal medulla undergoes time-dependent inactivation by phenylhydrazine.[198a] Inactivation with [¹⁴C]phenylhydrazine results in the incorporation of 0.94 mol of radioactivity per enzyme subunit. In the presence of the reductant, ascorbate, inactivation is slower, suggesting that inactivation requires oxidized copper. The presence of substrate increases the inactivation rate presumably by increasing the rate of reoxidation of reduced copper. A carbon-centered radical can be observed in the ESR spectrum of the enzyme inactivated by phenyhydrazine in the presence of the radical trap, α-(4-pyridyl-1-oxide)-N-*tert*-butylnitrone. Since 5,5-dimethyl-1-pyrroline-1-oxide is a reversible inhibitor that also acts as a radical trap, it was suggested by Fitzpatrick and Villafranca[198a] that the radicals are trapped at the active site by the bound spin trap. The inactivation mechanism proposed is shown in Scheme 139 (R = Ph).

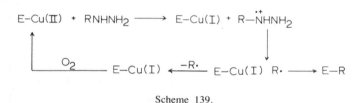

Scheme 139.

Lactoperoxidase from raw skim milk and the acyl phosphatase activity of glyceraldehyde-3-phosphate dehydrogenase are inactivated by phenylhydrazine.[199] The mechanism proposed by Allison et al.[199] is not what is expected for a mechanism-based inactivator, namely displacement of hydroxyl from an active site sulfenic acid to give a sulfenyl hydrazide. However, since phenylhydrazine is now known to inactivate several enzymes that catalyze oxidation reactions by one-electron mechanisms, a similar radical-mediated inactivation mechanism may be involved in this case as well.

Phenylhydrazine also inactivates catalase in a pseudo first-order, time dependent process that requires hydrogen peroxide.[200] For every 3 molecules of phenylhydrazine and 52 molecules of H_2O_2, 2 molecules of benzene are formed with concomitant heme alkylation. The product was identified by Ortiz de Montellano and Kerr[200] as N-phenylprotoporphyrin IX when an acidic methanol work-up was employed. Extraction of the phenylhydrazine-inactivated catalase with 2-butanone under argon in the presence of BHT gives the σ-phenyl iron complex, thus providing evidence that this is the initial product of alkylation.

Phenylhydrazine rapidly inactivates horseradish peroxidase in the presence of H_2O_2.[200a] Titration experiments indicate that 11 to 13 mol of phenylhydrazine and 25 mol of H_2O_2 are required to inactivate the enzyme. This stoichiometry differs from that for the reaction with hemoglobin and myoglobin[200,201] (6 mol of phenylhydrazine) or for the inactivation of catalase[200] (3 mol of phenylhydrazine and 52 mol of H_2O_2). The heme is converted to a δ-meso-phenyl derivative. Phenyl radical coupling products also are isolated. The inactivation mechanism proposed by Ator and Ortiz de Montellano[200a] involves oxidation of phenylhydrazine to phenyldiazene which decomposes to N_2 and phenyl radical. The phenyl radical adds to the δ-meso position of the porphyrin in Compound II, and subsequently deprotonation gives the modified heme.

Liver microsomal cytochrome P-450 of phenobarbital-treated rats is inactivated by phenylhydrazine; NADPH and O_2 are required for inactivation.[201] Phenyldiazine, generated chemically, has the same effect and gives a product with the same optical spectrum as when phenylhydrazine is used. The formation of a binary complex of phenyldiazene and the heme of oxidized cytochrome P-450 was suggested by Jonen et al.[201] as the cause for irreversible inactivation. [¹⁴C]Phenylhydrazine inactivation leads to incorporation of radioactivity into the enzyme.

Phenylhydrazine is a time-dependent inactivator of the catechol oxidases from grapes, apples, potatoes, sugarbeets, and apricots.[201a] It also inactivates mushroom tyrosinase and cucumber ascorbic acid oxidase. In all cases Lerner et al.[201a] found that dialysis does not regenerate enzyme activity and oxygen is required for inactivation. A laccase-like enzyme from peaches and cytochrome C reductase from lettuce are not inhibited by 1 and 0.5 m*M* phenylhydrazine, respectively.

A series of monosubstituted hydrazines and diazenes were studied by Battioni et al.[202] as inactivators of liver microsomal cytochrome P-450 of phenobarbital-treated rats in order to compare with results obtained with hemoglobin (see Section IV.A.1.h.). The complexes formed from $RNHNH_2$ (R = Me, Et, Ph, CH_2Ph) give optical spectra and have properties very similar to those prepared with ferric tetraporphyrin chloride (see Section IV.A.1.h.). The mechanisms proposed are shown in Scheme 140. Preference was given to radical formation rather than iron-nitrene complex formation. Phenyldiazene gives the same complex with cytochrome P-450 as does phenylhydrazine. Maloney et al.[203] suggested that a phenyldiazene-heme complex is produced which can decompose to an iron-carbon bonded complex, then to an alkylated heme.

$$\text{P-450 Fe(III)} + RNHNH_2 \rightarrow \text{P-450 Fe(II)RN=NH} \xrightarrow{O_2} \text{P-450 Fe(III)RN=NH}$$

$$\text{P-450 Fe(III)N=NR} \xleftarrow{?} \text{P-450 Fe(II)RN=N}^{\cdot}$$

$$\downarrow -N_2$$

$$\text{P-450 Fe(II) R}^{\cdot}$$

$$\downarrow$$

$$\text{P-450 Fe(III)–R}$$

Scheme 140.

b. Benzylhydrazine

Benzylhydrazine irreversibly inactivates human brain MAO A and B with some selectivity for the B form.[204] It also inactivates bovine kidney MAO,[192] liver cytochrome P-450 from phenobarbital-treated rats,[202] and trimethylamine dehydrogenase from bacterium 4B6.[196]

Dopamine β-hydroxylase from bovine adrenal medulla also is inactivated by benzylhydrazine.[198a] In contrast to inactivation by phenyl-, phenylethyl-, and methylhydrazine (see Sections IV.A.1.a., c., and f., respectively), the rate of inactivation by benzylhydrazine is accelerated in the presence of ascorbate and retarded by substrate. In the presence of the radical trap, α-(4-pyridyl-1-oxide)-*N-tert*-butylnitrone, a carbon-centered radical can be observed for all of the substituted hydrazines. Since 5,5-dimethyl-1-pyrroline-1-oxide is a reversible inhibitor of the enzyme that also acts as a radical trap, it was suggested by Fitzpatrick and Villafranca[198a] that the radicals are trapped at the active site by the bound spin trap. The partition ratio for inactivation by benzylhydrazine was determined to be 160. α,α-Dideuteriobenzylhydrazine inactivation of dopamine β-hydroxylase in the presence of ascorbate results in a 4-fold **increase** in the second-order rate constant for inactivation, an isotope effect of 13 on V_{max}, and an 11-fold decrease in the partition ratio (to 14) relative to the nondeuterated compound. The products formed during inactivation, namely, benzaldehyde and benzaldehyde hydrazone, indicate that α-hydroxylation occurs. These data suggest that product formation for benzylhydrazine results from carbon-hydrogen bond cleavage and inactivation occurs by carbon-nitrogen bond cleavage (Scheme 141).

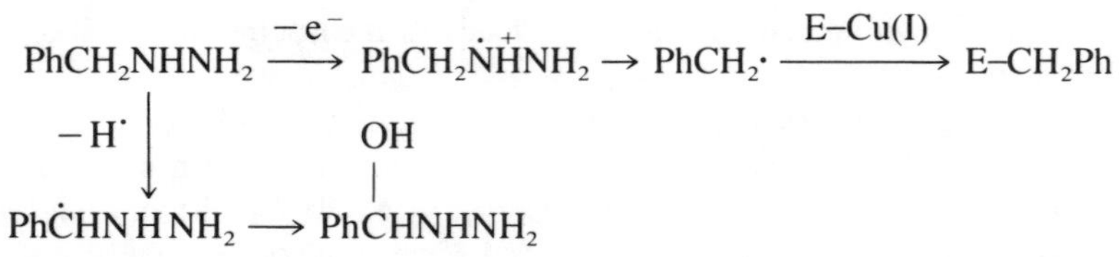

$$\text{PhCH}_2\text{NHNH}_2 \xrightarrow{-\text{e}^-} \text{PhCH}_2\overset{\cdot}{\overset{+}{\text{N}}}\text{HNH}_2 \rightarrow \text{PhCH}_2\cdot \xrightarrow{\text{E}-\text{Cu(I)}} \text{E}-\text{CH}_2\text{Ph}$$

$$-\text{H}^\cdot \downarrow \qquad\qquad\qquad \overset{\text{OH}}{\underset{|}{}}$$

$$\text{Ph}\overset{\cdot}{\text{C}}\text{HN H NH}_2 \longrightarrow \text{PhCHNHNH}_2$$

Scheme 141.

c. 2-Phenylethylhydrazine (Phenelzine) and Related

Phenelzine (2-phenylethylhydrazine) is both an irreversible inactivator of rat liver mitochondrial MAO and a substrate; [1-^{14}C]phenelzine is converted into phenylacetic acid.[205] A 1:1 molar ratio of [^{14}C] to enzyme results after time-dependent inactivation.[206] Tipton[194] found that 2-phenylethylhydrazine is a time-dependent irreversible inactivator of MAO from pig brain and rat liver mitochondria. The failure of hydralzine (see Section IV.A.1.d.) and acylhydrazides to act as irreversible inactivators of MAO is consistent with oxidation to hydrazones as a mechanism of inactivation of MAO by hydrazines. However, it was proposed that phenelzine and phenylhydrazine have two different mechanisms of inactivation. Phenelzine and 1-methyl-2-phenylethylamine inactivate bovine kidney MAO;[192] phenelzine and 1-phenylethylhydrazine inactivate trimethylamine dehydrogenase from bacterium 4B6.[196]

In the presence of NADPH there is a time-dependent inactivation of liver microsomal cytochrome P-450 from phenobarbital-treated rats by phenelzine.[207,208] Heme content is destroyed concomitantly;[207,208] N-(2-phenylethyl)protoporphyrin IX, where the substituent is either on the pyrrole ring C or D, is formed.[208] The 2-phenylethyl radical was identified as an intermediate by spin trapping with α-(4-pyridyl-1-oxide) N-t-butylnitrone.[208] If all of the radicals are released into solution prior to inactivation, then this is not a mechanism-based inactivation.

Phenelzine also inactivates dopamine β-hydroxylase from bovine adrenal medulla.[198a] In the presence of ascorbate, inactivation is slower, suggesting that inactivation requires oxidized copper. A carbon-centered radical can be detected during inactivation in the presence of a spin trap. It was suggested that the radicals are trapped at the active site by bound spin trap. The inactivation mechanism is the same as that shown in Scheme 139 (Section IV.A.1.a.; R = PhCH$_2$CH$_2$).

d. Hydralazine

The antihypertensive agent hydralazine (Structural Formula 9.103a, Scheme 142) is an irreversible inactivator of chick embryo aorta lysyl oxidase,[208a] of the semicarbazide-sensitive, clorgyline-resistant amine oxidase in rat aorta,[208b] and of rabbit-, guinea pig-, and baboon liver aldehyde oxidases.[208c] Bovine milk xanthine oxidase is unaffected.[208c]

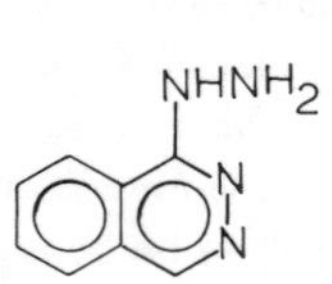

Scheme 142.   Structural Formula 9.103a.

e. Isopropylhydrazine (from Iproniazid)

Inactivation by iproniazid (1-isonicotinyl-2-isopropylhydrazine) of rat liver mitochondrial MAO requires O_2; isopropylhydrazine is a more potent time-dependent inactivator that also requires O_2.[209] Since the apparent energy of activation for isopropylhydrazine is about the same as that for iproniazid, it was concluded that formation of isopropylhydrazine was **not** the cause for inactivation by iproniazid. Also, dialysis of iproniazid-inactivated enzyme results in some return of enzyme activity, but much less reactivation results when isopropylhydrazine is used. Davison[209] concluded that inactivation by the two compounds, therefore, is different. 1-Isopropylidene-2-isonicotinoylhydrazine is not an inhibitor, suggesting that oxidation of iponiazid is not the first step in the inactivation.

Iproniazid is converted by rat brain mitochondria into isopropylhydrazine and isonicotinic acid; isopropylhydrazine, then irreversibly inactivates MAO.[210] Evidence to support this mechanism is a lag time for inactivation of MAO by iproniazid, whereas isopropylhydrazine inactivates MAO without a lag period. Seiden and Westley[210] suggested that isocarboxazid probably also inhibits MAO by a similar mechanism. Taylor et al.[27] also found iproniazid to be a time-dependent inactivator of mouse brain and liver mitochondrial MAO. McEwen et al.[26] noted that isopropylhydrazine causes time-dependent inactivation of human liver MAO that is reversed upon dilution and noncompetitive in nature.

Iproniazid inactivates guinea pig liver mitochondrial MAO in the presence of O_2; Smith et al.[211] suggested that iproniazid is not the active inhibitor, but it is converted, by a different enzyme, into the active form. Isopropylhydrazine, or a more highly oxidized derivative of isopropylhydrazine, is the active substance; isopropylhydrazine is a more potent inactivator than is iproniazid.

Iproniazid and other hydrazines react covalently and stoichiometrically with the nonheme portion of horseradish peroxidase; the rate of enzyme inactivation corresponds to the rate of incorporation of radioactivity from [14C]iproniazid.[212]

Iodide peroxidase-tyrosine iodinase from calf thyroid is irreversibly inactivated by iproniazid; when [14C]iproniazid is used, the rate of inactivation corresponds to the rate of radioactivity incorporation into the enzyme.[213] Spectral shifts concomitant with inactivation are comparable to those observed with horseradish peroxidase.[212]

Isoniazid (isonicotinic acid hydrazide) also is a time-dependent inactivator of cytochrome P-450 in rat liver microsomes in the presence of NADPH; both the hydrazine and aromatic moieties are essential for inactivation.[214] Potassium ferricyanide reverses the inhibition of cytochrome P-450 by isoniazid.[215] This suggests that isoniazid forms a metabolic intermediate complex with P-450, since $K_3Fe(CN)_6$ is known to dissociate nitrogenous P-450 complexes.

Isopropylhydrazine is an irreversible inactivator of trimethylamine dehydrogenase from bacterium 4B6.[196]

f. Ethyl-, Methyl-, and Acetylhydrazine

Ethylhydrazine and acetylhydrazine cause inactivation of hepatic microsomal cytochrome P-450.[216] In the presence of the spin trap, α-phenyl-t-butylnitrone, the ethyl radical produced during inactivation is captured and observed by ESR spectroscopy. If all of the ethyl radical is released prior to inactivation, then this is not a mechanism-based inactivation. Battioni et al.[202] studied a series of hydrazine derivatives, including ethyl- and methylhydrazine, as inactivators of liver cytochrome P-450 from phenobarbital-treated rats in order to compare with results obtained with hemoglobin (see Section IV.A.1.h.). The properties of the complexes formed are similar to those prepared with ferric tetraporphyrin chloride.

When ethylhydrazine was used to inactivate catalase, all four regioisomers of N-ethyl-protoporphyrin IX were formed.[200] The absolute configuration of the major isomer (ring C alkylated) is the same as the major product of the reaction of hemoglobin with ethylhydrazine (see Section IV.A.1.h.), and therefore the prosthetic heme has the same chiral orientation in both enzymes.

Methylhydrazine is a time-dependent inactivator of dopamine β-hydroxylase from bovine adrenal medulla.[198a] In the presence of ascorbate, inactivation is slower, suggesting that inactivation requires oxidized copper. Consistent with that hypothesis is the observation that the presence of the substrate, tyramine, increases the inactivation rate, presumably by increasing the rate of reoxidation of reduced copper. Inactivation in the presence of a spin trap produces a carbon-centered radical. The radicals generated are believed to be trapped at the active site by the bound spin trap. The inactivation mechanism is the same as that shown in Scheme 139 (Section IV.A.1.a.; R = Me).

g. 1,1- and 1,2-Disubstituted Hydrazines

A series of 1,1-disubstituted hydrazines was shown to inhibit liver microsomal cytochrome P-450 from phenobarbital-treated rats in a time-dependent manner; NADPH and O_2 are required for inhibition.[217] Only the 1,1-disubstituted hydrazines, not the 1,2-disubstituted- or monosubstituted hydrazines, give rise to a time-dependent formation of a 438-nm absorption in the optical spectrum. A nitrene-heme complex proposed by Hines and Prough[217] as the inactivated enzyme product. Support for an intermediate containing an iron-nitrene bond was obtained by Mansuy et al.[218] who isolated the first nitrene complex (Fe(IV)=NR) of a metalloporphyrin. Ferric tetraphenylporphyrin chloride reacted with 1-amino-2,2,6,6-tetramethylpiperidine aerobically to give **9.104** (Scheme 143).

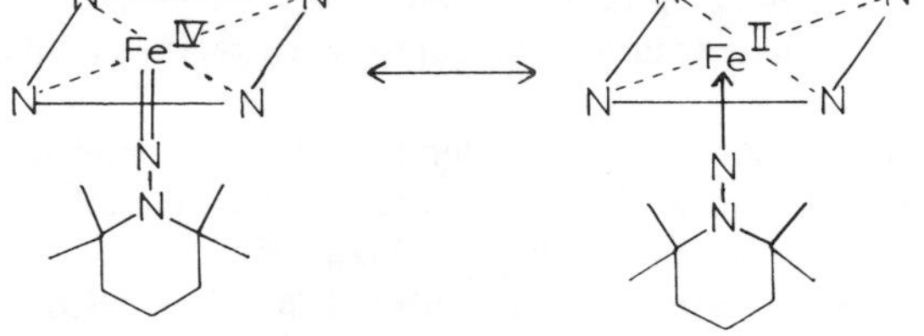

9.104

Scheme 143.   Structural Formulas 9.104.

A comparison by Maloney et al.[203] of monosubstituted, 1,1- and 1,2-disubstituted hydrazines showed that monosubstituted and 1,2-disubstituted hydrazines and hydrazides, procarbazine, and iproniazid cause loss of CO-reactive cytochrome P-450 and heme destruction. 1,1-Disubstituted hydrazines inactivate the enzyme with no heme destruction.

h. Models for Inactivation of Oxidases by Hydrazines. Reactions of Hydrazines with Proteins

By definition, a mechanism-based enzyme inactivator must be converted by the target enzyme via its normal mechanism of action into the activated species. Noncatalytic proteins, therefore, do not really qualify as targets for this class of enzyme inactivators. However, various heme-containing proteins, e.g., hemoglobin, oxyhemoglobin, methemoglobin, and myoglobin, react with hydrazines in a similar way to that which heme itself reacts. It is not clear how important binding of the hydrazine to the protein is prior to oxidation, but as models for the mechanism of hydrazine-induced inactivation of oxidases, these reactions are very informative and, therefore, are included here.

Cupric ion-catalyzed oxidation of β-phenylisopropylhydrazine was carried out under conditions resembling those in biological systems as a model for inactivation of MAO by hydrazines.[219] The data indicate a radical mechanism, initiated by transfer of one electron from the hydrazine to the Cu(II). Eberson and Persson[219] proposed that substituted hydrazines transfer an electron to the enzyme with subsequent oxidation of the hydrazine radical by molecular oxygen. The intermediate radicals produced react with the enzyme, perhaps a sulfhydryl group, to give irreversible inactivation (Scheme 144, R = $PhCH_2CH(CH_3)$).

$$R-NHNH_2 + Cu(II) \rightarrow RN\dot{H}NH_2 + Cu(I)$$

$$\xrightarrow{O_2} R-N=\dot{N}H \xrightarrow{Cu(I)} R-N=NH_2$$

$$\downarrow O_2$$

$$R^{\cdot} + N_2 + HOO^{\cdot}$$

Scheme 144.

Misra and Fridovich[220] showed that phenylhydrazine is oxidized by oxyhemoglobin; after a lag time, oxygen consumption decreases. This bimolecular reaction causes oxidative denaturation of hemoglobin.[221] In the presence of phenylhydrazine in air, methemoglobin is converted to the same complex as is produced when phenyldiazene is added to methemoglobin in the absence of air. Treatment of human hemoglobin with phenylhydrazine gives a green-brown pigment that was identified by Ortiz de Montellano and Kunze,[222] with the use of field desorption mass spectroscopy and NMR spectroscopy, as N-phenylprotoporphyrin IX. The pigment can be resolved into two fractions, each of which consisting of two similar isomeric structures. Therefore, all four N-phenyl isomers appear to be formed. The protein, apparently, does not play a role in this reaction, since hemin reacts with phenylhydrazine in the presence of O_2 to give the same four isomers. The aerobic reaction of oxyhemoglobin and oxymyoglobin with phenylhydrazine and p-tolylhydrazine produces a blue pigment and a green pigment which were identified by Saito and Itano[223] as β-meso-arylbiliverdin IXα and N-arylprotoporphyrin IX, respectively. The mechanism proposed is shown in Scheme 145. Oxidation of phenylhydrazine by oxyhemoglobin in the presence of the spin trap, 5,5-dimethyl-1-pyrroline-1-oxide, produces a stable radical; Hill and Thornalley[224] suggested that the phenyl radical is trapped.

$$Fe(II)O_2 + ArNHNH_2 \rightarrow [Fe(II)H_2O_2] + ArN=NH$$

$$\downarrow O_2$$

$$Ar^{\cdot} + N_2 \leftarrow ArN=N^{\cdot} + H^+ + O_2^{-}$$

Scheme 145.

A detailed study by Augusto et al.[225] of the reaction of phenylhydrazine with a hemoglobin showed that inactivation of the prosthetic heme groups occurs with a partition ratio of six. Each heme catalyzes the consumption of six phenylhydrazines and six O_2 molecules and produces five benzene molecules. The reactions of oxy- and methemoglobin or -myoglobin with phenylhydrazine are similar. The sixth phenyl group was found as N-phenylprotoporphyrin IX. Alkylhydrazines undergo a similar reaction, but much slower. Analogous to phenylhydrazine, *para-* and *meta*-substituted phenylhydrazines give rise to N-aryl heme derivatives, but *ortho*-substituted phenylhydrazines do not, suggesting a strong steric effect. Nonetheless, the *ortho*-substituted compounds still inactivate the heme without formation of an N-aryl heme derivative. In some cases the inactivated heme is slowly regenerated with time. Evidence is presented that supports a hypothesis in which the initial complex formed is an iron-aryl adduct that is converted, in most of the cases studied, to an N-phenyl heme (Scheme 146). Chemical model studies were carried out by Ortiz de Montellano et al.[226] for the reaction of phenylhydrazine with hemoglobin that support the hypothesis of an initial formation of an iron-phenyl complex, of regeneration of the iron porphyrin from the complex, and of a migration of the phenyl from iron to nitrogen aerobically. Treatment of iron tetraphenylporphyrin with phenylmagnesium bromide in the presence of 2,6-di-*t*-butyl-4-methylphenol gives the corresponding iron-phenyl complex which is stable in air but rapidly decomposes in solution (in the absence of BHT) to iron tetraphenylporphyrin and benzene. Treatment of the iron-phenyl complex with 5% H_2SO_4 in methanol (conditions resembling the work-up conditions following phenylhydrazine treatment of hemoglobin) gives *N*-phenyl tetraphenylporphyrin in a 68% yield. As in the case with hemoglobin, this reaction is oxygen dependent. Although these model studies suggest an initial σ-bonded aryliron complex, a phenyldiazene-iron complex[227,228] or a reversible *N*-phenylheme complex also are possible. In order to differentiate these, Kunze and Ortiz de Montellano[229] showed that the 240-MHz NMR spectra and the electronic absorption spectra of *p*-tolylhydrazine-inactivated myoglobin or phenylhydrazine-inactivated hemoglobin are identical to those of the reaction products of *p*-tolylmagnesium bromide and phenylmagnesium bromide, respectively, with hemin. This confirms that the initial products with hemoglobin and myoglobin are σ-bonded aryliron complexes. An X-ray crystallographic structure of myoglobin that has been inactivated by phenylhydrazine showed that the phenyl group is bound to the iron atom.[230] This attachment forces aside several surrounding amino acid side chains, resulting in an open channel to the surface, and may give a clue as to how substrates enter myoglobin.

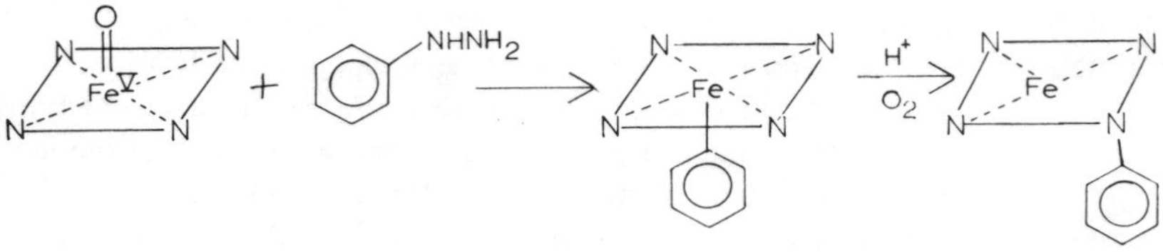

Scheme 146.

Ortho-substituted arylhydrazines inactivate myoglobin to form aryl-iron adducts.[231] Unlike phenylhydrazine and *m*- and *p*-substituted phenylhydrazines that undergo a migration from iron to nitrogen of the porphyrin upon denaturation,[225,226,229,230,232] *o*-substituted phenylhydrazines do not migrate. An NMR study was carried out by Ortiz de Montellano and Kerr[231] to differentiate between a different complex (e.g., aryldiazenyl- rather than aryl-iron complexes) and a stable aryl-iron complex. The results indicate a normal aryl-iron complex, where the aryl group is conformationally locked so that migration to the porphyrin nitrogen is prohibited.

Methylhydrazine reacts with hemoglobin and myoglobin in the presence of limited amounts of oxygen with quantitative formation of methyldiazene-Fe(II) complexes.[227] These complexes are oxidized by oxygen to stable complexes similar to those described for the aerobic reaction of phenylhydrazine with hemoglobin or myoglobin.[225,226,229,230] As a model for the migration of σ-alkyl iron(III) complexes to *N*-alkyl complexes, Mansuy et al.[232] showed that (TPP)Fe(III)R (prepared from (TPP)Fe(III)Cl + RMgX, where R = CH=CPh$_2$, CH$_3$, Ph), under anaerobic conditions in the presence of FeCl$_3$ at −20°C gives the *N*-alkyl complex. In the presence of sodium dithionite, this *N*-alkyl complex reverts back to the σ-alkyliron(III) complex (Scheme 147).

The reactions of iron porphyrins (ferric tetraporphyrin chloride) with hydrazines and diazenes were studied by Battioni et al.[233] as models for the corresponding reactions of heme proteins. The diazene complex formed with phenyldiazene is much less stable than the one with methyldiazene, and this may explain why myoglobin-Fe(II)–PhN=NH complexes have not been observed, but the corresponding methyl diazene complex has.[227]

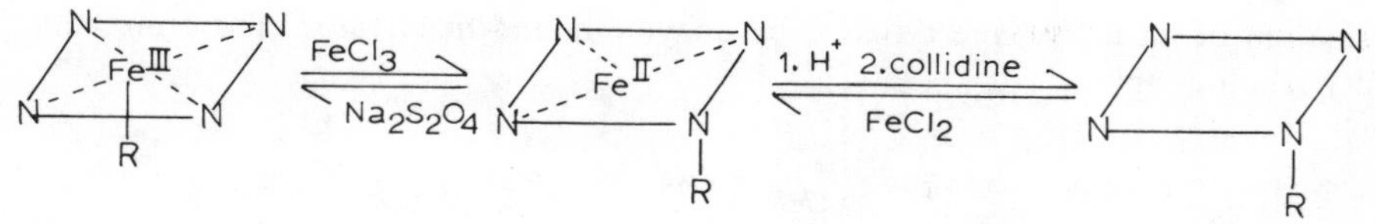

Scheme 147.

A comparison of the circular dichroism spectra of the C ring *N*-ethylprotoporphyrin IX isomers arising from the reaction of ethylhydrazine with human hemoglobin and from the reaction of cytochrome P-450 with 3,5-bis(carbethoxy)-2,6-dimethyl-4-ethyl-1,4-dihydropyridine proved them to have identical structures[131] (see Section II.B.2.c.).

V. NOT MECHANISM-BASED INACTIVATION

A. (1-Phenylcyclopropyl)methylamine

Since doubly-activated cyclopropanes are known[234] to be susceptible to nucleophilic attack, (1-phenylcyclopropyl)methylamine (Structural Formula 9.105) was designed by Silverman and Zieske[126] as an inactivator of monoamine oxidase (Scheme 148). Although oxidation to the imine occurs rapidly, no inactivation results.

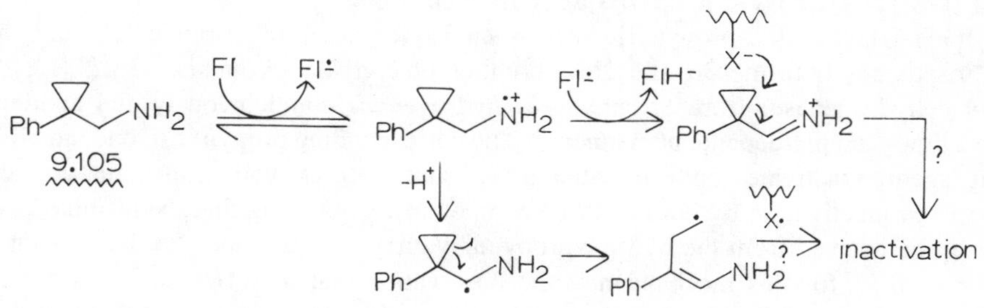

Scheme 148.   Containing Structural Formula 9.105.

B. *N*-Chloro-D-leucine

N-Chloro-D-leucine is an irreversible inactivator of hog kidney D-amino acid oxidase; neither dialysis nor gel filtration regenerates enzyme activity.[235] Inactivation by *N*-[^{36}Cl]chloro-D-leucine showed that two equivalents of chlorine per flavin are bound, but none of the remainder of the molecule. The partition ratio is close to 0 since very little O_2 consumption results from inactivation. After resolution of the labeled enzyme into FAD and apoenzyme, 94% of the [^{36}Cl] was found in the protein. Reconstitution with FAD gives a haloenzyme with catalytic and spectral properties the same as modified enzyme. Titration of the enzyme with *N*-chloro-D-leucine gives an end point of two equivalents per flavin. Porter and Bright[235] suggested that inactivation does not result from simple transchlorination on the basis of the following experiments: (1) the *N*-chloro-L-amino acids are competitive reversible inhibitors; (2) neither NH_2Cl nor *N*-chloro-D-alanine or -valine inactivate the enzyme; and (3) no nonenzymatic transchlorination between *N*-chloro-D-leucine and other α-amino acids was observed. The mechanism of inactivation suggested was oxidation to the *N*-chloroimmonium derivative (Scheme 149). Perhaps a tyrosine residue is chlorinated. Further studies by Rudie et al.[236] and Ronchi et al.[237] revealed that the two equivalents of chlorine are incorporated into an active site tyrosine as 3,5-dichlorotyrosine. However, several lines of evidence indicate that the mechanism of inactivation does **not** involve a prior oxidation to the *N*-chloroimmonium compound (see Scheme 149). The mechanism supported involves direct chlorination of an active site tyrosine by enzyme-bound inactivator. Therefore, this is an affinity labeling agent.

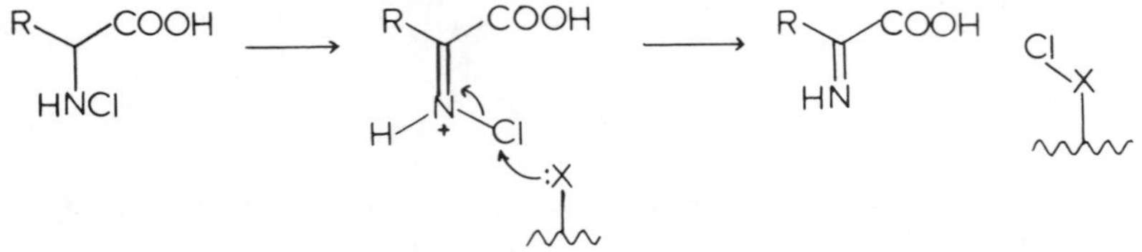

Scheme 149.

C. 2-Hydroxy-3-butynoate

2-Hydroxy-3-butynoate produces a time-dependent and covalent inactivation of *Escherichia coli* phosphoenolpyruvate-dependent hexose uptake, but this was shown by Kaczorowski et al.[238] to be the result of D-lactate dehydrogenase-catalyzed oxidation to 2-keto-3-butynoate and diffusion in the membrane. Inactivation of Enzyme I of the phosphoenol-pyruvate-phosphate transferase system by this compound is the result of lactate dehydrogenase oxidation and diffusion of the ketobutynoate to Enzyme I.[238,239]

D. 17β-[(1S)-1-Hydroxy-2-propynyl]androst-4-en-3-one

17β-[(1S)-1-Hydroxy-2-propynyl]androst-4-en-3-one (Structural Formula 9.105a, Scheme 149A) inactivates both the 3α and 20β activities of cortisone reductase (3α,20β-hydroxysteroid dehydrogenase) from *Streptomyces hydrogenans* simultaneously and at identical rates in a time- and pH-dependent manner.[240] The corresponding propynyl ketone, an affinity-labeling agent, inactivates both activities at the same rate as well. Substrates protect the enzyme from inactivation by both of these compounds, suggesting that the affinity-labeling agent may be derived from the hydroxypropyne by enzyme activation. Further evidence by Strickler et al.[240] for this mechanism is the observation that the (1R)-isomer is not enzymatically oxidized and does not inactivate the enzyme. These results suggest that the enzyme has a single active site for oxidation at both the 3α and 20β positions of substrates. [4-¹⁴C]-**9.105a** inactivates the enzyme with the incorporation of one equivalent of radioactivity per tetrameric enzyme.[240a] The low incorporation was explained by Covey et al.[240a] as the result of low specific activity enzyme. However, β-mercaptoethanol in large excess completely protects the enzyme from inactivation, indicating release of the reactive species from the active site prior to inactivation. Therefore, this is not a mechanism-based inactivation.

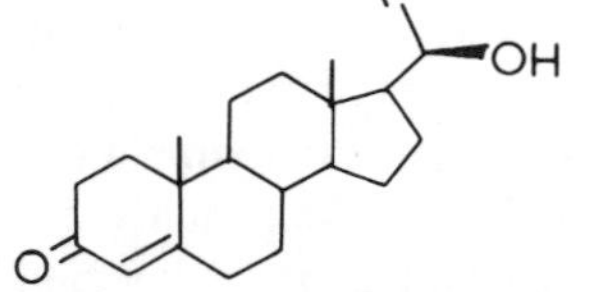

Scheme 149A.   Structural Formula 9.105a.

E. Allyl Alcohol

Allyl alcohol is oxidized to acrolein by both yeast and horse liver alcohol dehydrogenase, which leads to their inactivation.[241] However, the enzyme is protected from inactivation when dithiothreitol is in the incubation buffer.[242] Furthermore, most of the acrolein that can be formed in the presence of limiting NAD^+ is formed prior to inactivation; therefore, this is not a mechanism-based inactivator. Rando[242] reported that horse liver alcohol dehydrogenase is not inactivated by allyl alcohol.

F. 2-Hydroxy-3-butenoic Acid (Vinylglycolic Acid)

2-Hydroxy-3-butenoate (vinylglycolic acid) inactivates Enzyme I of the phosphoenolpyruvate-phosphate transferase system in *E. coli;* the alkene is 50 to 100 times more potent than the corresponding alkyne.[239] However, lactate dehydrogenase is responsible for oxidation of the vinylglycolate to the α,β-unsaturated ketone, which is the inactivating species for Enzyme I. This reaction occurs in a wide variety of bacteria.[243]

G. 2'-Chloro-2'-deoxyuridine 5'-diphosphate and Related

2'-Chloro-2'-deoxyribonucleoside diphosphates (Structural Formula 9.106, Scheme 150) irreversibly inactivate the B1 subunit of *E. coli* ribonucleoside diphosphate reductase without affecting the B2 subunit.[244] Inactivation requires the presence of active B2, is controlled by allosteric effectors, and the corresponding monophosphates have no effect. Two moles of inactivator per mole of B1 are required for complete inactivation. Modification of the active dithiols was suggested by Thelander et al.[244] as the cause for inactivation. The products of the enzyme reaction are the free base, chloride ion, and 2'-deoxyribose 5'-diphosphate. The nature of the inactivation was not known. A reinvestigation of the reaction of 2'-chloro-2'-deoxyuridine 5'-diphosphate with *E. coli* ribonucleoside diphosphate reductase by Stubbe and Kozarich[245] showed that chloride ion, the free base, and **inorganic pyrophosphate** are released, not a phosphosugar. They suggested that these products are derived from a 3'-ketonucleoside 5'-diphosphate (Structural Formula 9.107), as shown in Scheme 151. How **9.107** is formed was not known, but they suggested a radical hydrogen abstraction. Inactivation is believed to result from sulfhydryl attack on the unsaturated ketoribose moiety (Structural Formula 9.108, see Scheme 151). [^{14}C]-2'-Chloro-2'-deoxyuridine 5'-diphosphate inactivates the enzyme with incorporation of radioactivity, consistent with this mechanism. A mechanism by Stubbe and Ackles[246] to explain the formation of **9.107** during inactivation of ribonucleoside diphosphate reductase by **9.106** is shown in Scheme 152. This mechanism involves cleavage of the C–H bond at the 3'-position. Evidence to support this hypothesis is that [3'-^{3}H]uridine 5'-diphosphate shows a selection against [^{3}H] of approximately 3.3 upon reduction. A detailed study by Stubbe et al.[247] of the reaction of [3'-^{3}H]uridine 5'-diphosphate with ribonucleoside diphosphate reductase supports cleavage of the C–H bond at the 3'-position. Cleavage of this bond results in release of only about 1% of the tritium (presumably as ^{3}H$_2$O). When the enzyme is first inactivated with either 2'-chloro- or 2'-azido-2'-deoxyuridine 5'-diphosphate, then no tritium is released from [3'-^{3}H]uridine 5'-diphosphate. Therefore, the B1 and B2 subunits are required for 3'-C–H bond cleavage. Inactivation of *E. coli* ribonucleoside diphosphate reductase (B1 subunit) by [U-^{14}C]-2'-chloro-2'-deoxyuridine diphosphate results in the incorporation of [^{14}C] that is stable to gel filtration and NaDodSO$_4$-polyacrylamide gel electrophoresis.[248] Inactivation in the presence of dithiothreitol occurs at a reduced rate, suggesting that the reactive species, at least partially, is released into solution prior to inactivation. 2'-Chloro-2'-deoxy[3'-^{3}H]uridine 5'-diphosphate inactivates *E. coli* ribonucleoside diphosphate reductase with the release of 6 equivalents of pyrophosphate, 5 equivalents of ^{3}H$_2$O, and the incorporation of 0.96 equivalent of [^{3}H] per protomer of B1 subunit.[249] These results suggest that the hydrogen abstracted from the 3'-position is exchanged with solvent five times for every time it is returned to the 2'-position. Inactivation with the corresponding [5'-^{3}H] compound leads to incorporation of 4.6 equivalents of [^{3}H] per protomer of B1 subunit. Inactivation, however, is prevented by inclusion of 10 mM dithiothreitol in the preincubation buffer, indicating that the reactive species is released prior to inactivation. A product isolated in the presence of dithiothreitol has the characteristics of the product of 2-methylene-3(2*H*)-furanone (**9.108**) with ethane-thiol. The mechanism proposed by Stubbe and co-workers[248,249] to account for these observations is shown in Scheme 153. Further studies to test this mechanistic hypothesis for *E. coli* ribonucleotide reductase and to gain information regarding the mechanism of inactivation of the enzyme by 2′-chloro-2′-deoxyuridine 5′-diphosphate were carried out by Ator and Stubbe.[250] When the inactivation was performed in the presence of sodium borohydride, the rate of inactivation was greatly diminished. The products isolated, after conversion to nucleosides, are 2′-deoxyuridine and 2′-deoxy-3′-epiuridine, the products expected for NaBH₄ reduction of the hypothesized intermediate, 2′-deoxy-3′-ketonucleotide (**9.107** in Scheme 151). When the inactivation was carried out with [3′-³H] inactivator, [2′-³H]-2′-deoxyuridine was isolated and the tritium was located at the β-face of the 2′-position of **9.107**. As suggested previously[249] one out of six turnovers results in incorporation of the 3′-hydrogen into the 2′-position; the enzyme, therefore, catalyzes a [1,2]hydrogen shift. Inactivation of the enzyme by 2′-chloro-2′-deoxyuridine 5′-diphosphate is much slower in the presence of etha-nethiol. It was previously suggested[249] that the intermediate **9.107** nonenzymatically decomposes to 2-methylene-3(2*H*)-furanone (**9.108**), which inactivates the enzyme. The products isolated in the presence of ethanethiol are 2-[(ethylthio)methyl]-3(2*H*)-furanone (Structural Formula 9.109) and 5-(ethylthio)-2-[(ethylthiomethyl]-4,5-dihydro-3(2*H*)-fura-none (Structural Formula 9.109a; see Scheme 154). When [5′-³H]-inactivator is used, both subunits (B1 and B2) are labeled in the ratio of 3:2. These results suggest that **9.108** is generated in solution and nonspecific alkylation occurs. The same results are obtained when **9.106** inactivates prereduced enzyme in the absence of a reductant or with oxidized enzyme, suggesting that the redox-active thiols of the B1 subunit are not involved in the inactivation mechanism.

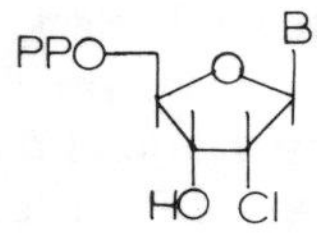

Scheme 150.   Structural Formula 9.106.

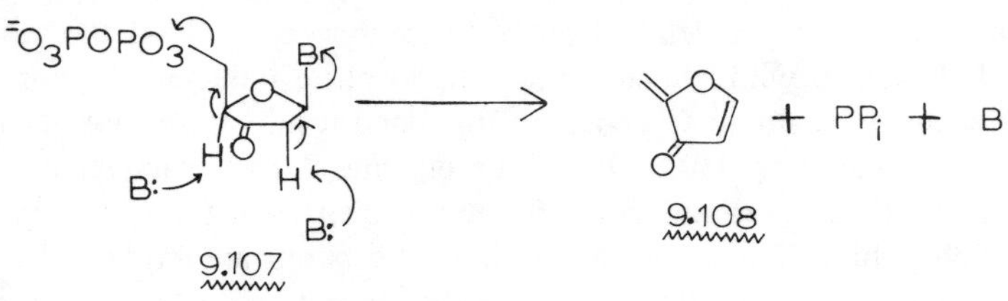

Scheme 151.   Containing Structural Formulas 9.107 and 9.108.

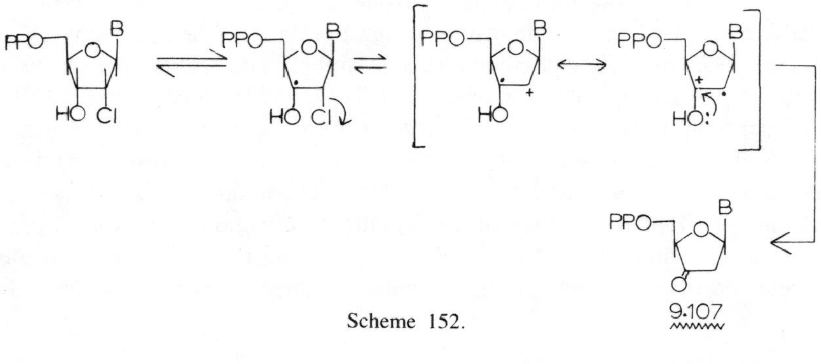

Scheme 152.

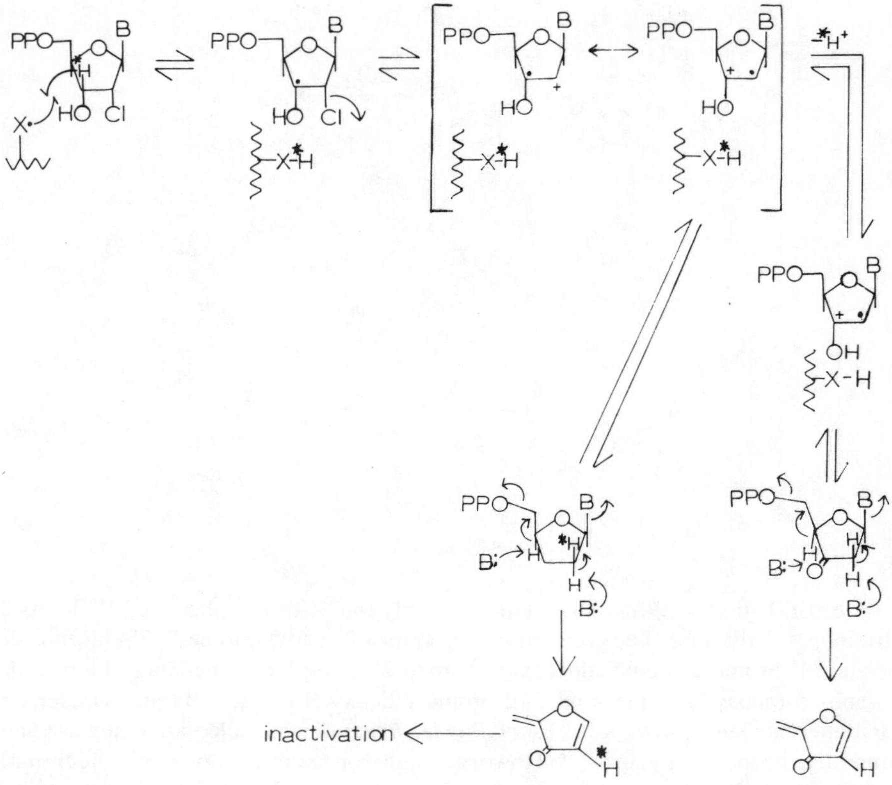

Scheme 153.

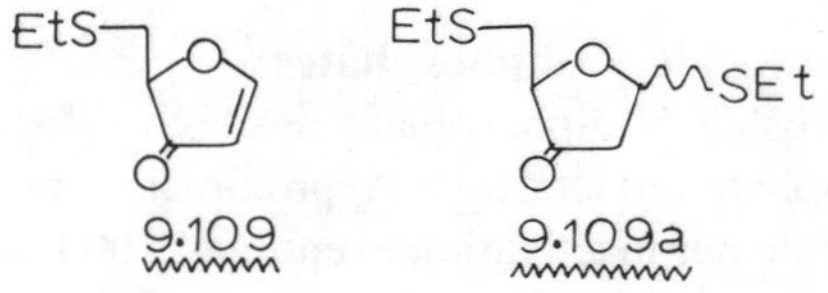

9.109   9.109a

Scheme 154.   Structural Formulas 9.109 and 9.109a.

The synthesis[244] of **9.106** is shown in Scheme 155.

2′-Chloro-2′-deoxyuridine 5′-diphosphate is a time-dependent inactivator of the herpes simplex virus type I ribonucleotide reductase.[250a] When [3′-³H]2′-chloro-2′-deoxyuridine 5′-diphosphate was used, quantitative and time-dependent release of tritium as ³H₂O was observed by Ator et al.[250a] The reaction of the enzyme with [β-³²P]-**9.106** leads to release of [³²P]pyrophosphate. All of these results are analogous to those obtained with the corresponding *E. coli* enzyme and suggest similar mechanisms (see Scheme 153).

Similar results were obtained by Stubbe and co-workers[249,251,252] for the inactivation of the coenzyme B₁₂-dependent ribonucleotide reductase from *Lactobacillus leichmannii* by 2′-chloro-2′-deoxyuridine 5′-triphosphate. Again, the presence of thiols protects the enzyme from inactivation. With [3-³H] inactivator, the 3′-C–H bond is cleaved[251] and ³H₂O, uracil, tripolyphosphate, 4′-deoxyadenosine, and cob(II)alamin are produced.[252]

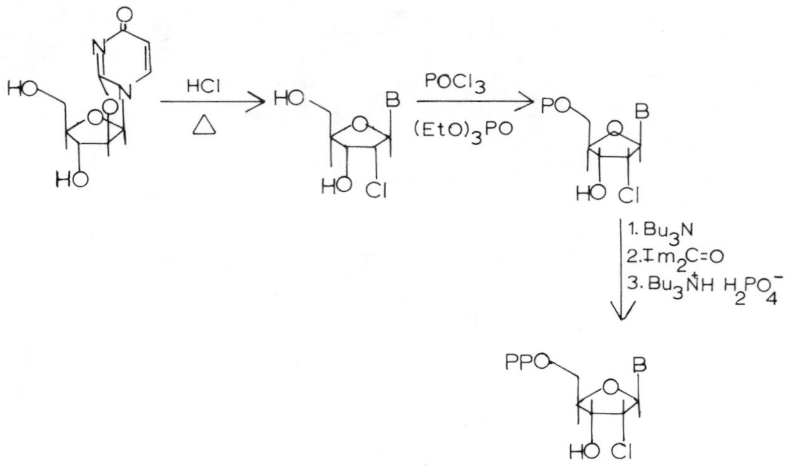

Scheme 155.

A series of 2'-deoxy-2'-halonucleotides, namely the 5'-triphosphates of 2'-fluoro-2'-de-oxyuridine, 2'-fluoro-2'-deoxycytidine, 2'-chloro-2'-deoxyuridine, 2'-chloro-2'deoxy-adenosine, 2'-bromo-2'-deoxyadenosine, 2'-iodo-2'-deoxyadenosine, 9-(2-chloro-2-deoxy-β-D-arabinofuranosyl)adenine, and 9-(2-bromo-2-deoxy-β-D-arabinofuranosyl)adenine, are the first alternate substrates for *Lactobacillus leichmannii* ribonucleotide reductase and also are time-dependent inactivators.[252a] However, high concentrations of thiols block inactiva-tion, suggesting that the reactive species, shown by Harris et al.[252a] to be 2-methylene-3(2H)-furanone (**9.108**), may be released from the active site prior to inactivation. The radical cation mechanism (Scheme 153) is further supported by these studies.

H. 2'-Fluoro-2'-deoxynucleoside 5'-diphosphates

2'-Fluoro-2'-deoxynucleoside 5'-diphosphates are both substrates and inactivators of *E. coli* ribonucleoside-diphosphate reductase.[253] Approximately two fluoride ions are released from the adenine dinucleotide per inactivation event, and 100 fluoride ions are released from the cytosine analogue per inactivation. Concomitant with fluoride ion release is an equivalent amount of adenine or cytosine and pyrophosphate. A radical mechanism was suggested by Stubbe and Kozarich,[253] but no specific details were offered. Inactivation is hypothesized to be the result of the generation of a Michael acceptor (see **9.108** and Scheme 151). The inactivator was prepared by phosphorylation of 2'-deoxy-2'-fluorocytidine[254] via the same route as that shown in Scheme 155 or by the route in Scheme 156. Similar results were obtained by Ator et al.[250a] for inactivation of the herpes simplex virus type I ribonucleotide reductase by 2'-fluoro-2'-deoxyuridine 5'-diphosphate.

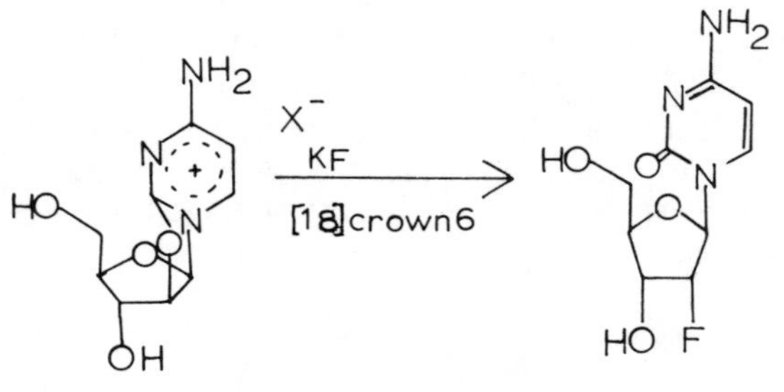

Scheme 156.

2'-Fluoro-2'-deoxyuridine 5'-triphosphate and 2'-fluoro-deoxycytidine 5'-triphosphate are time-dependent inactivators of *Lactobacillus leichmannii* ribonucleotide reductase; high concentrations of thiols block inactivation, suggesting that the reactive species, shown to be 2-methylene-3(2*H*)-furanone (**9.108**), may be released from the active site prior to inactivation.[252a]

I. 3-Butyn-1-ol

The inactivation of horse liver and yeast alcohol dehydrogenase by 3-butyn-1-ol (Structural Formula 9.110, Scheme 157) is protected by thiols.[255]

$$HC{\equiv}CCH_2CH_2OH$$

Scheme 157.   Structural Formula 9.110.

J. 14,15-Secoestra-1,3,5(10)-trien-15-yne-3,17β-diol

14,15-Secoestra-1,3,5(10)-trien-15-yne-3,17β-diol (Structural Formula 9.110a, Scheme 157A) is a time-dependent inactivator of human placental 17β,20α-hydroxysteroid dehydrogenase that requires NAD^+-dependent oxidation; however, glutathione completely protects the enzyme from inactivation.[255e] [³H]-Labeled inactivator inactivates the enzyme with incorporation of 1.4 moles of tritium per subunit. In order to determine what active site residue becomes attached to **9.110a** during inactivation, Auchus and Covey[255b] incorporated ^{13}C into the ethynyl group and the ^{13}C NMR spectrum of the inactivated enzyme was recorded. Model compounds having the Structural Formula 9.110b (Scheme 157B), where R = $Me_2NCH_2CH_2$ and X = O, PhO,S,NH, Im, NHC(=NH)NH or R = Et and X = COO, were prepared and the ^{13}C NMR spectra were compared with that of the inactivated and pronase-digested enzyme. The only model compound whose spectrum corresponded to that of the labeled enzyme was **9.110b**, where R = $Me_2NCH_2CH_2$ and X = NH. This suggests that a Michael addition mechanism is involved and the nucleophile is a lysine residue. The reason the lysine residue can compete effectively with other active site nucleopohiles may be because the pH of the inactivation reaction was 9.2 and the pK_a of $\epsilon\text{-}NH_3^+$ group is 9 to 10.

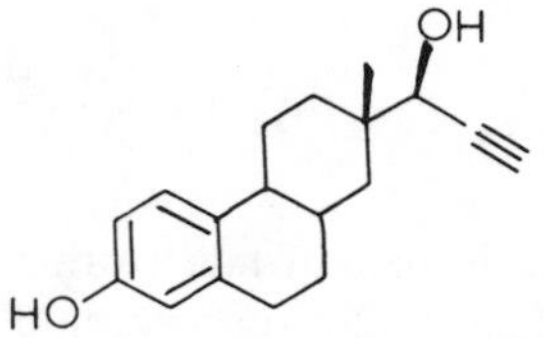

Scheme 157A.   Structural Formula 9.110a. 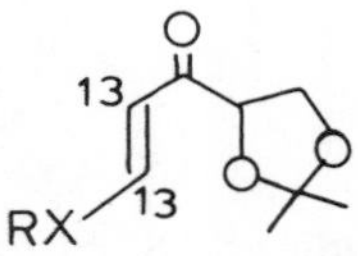   Scheme 157B.   Structural Formula 9.110b.

VI. MECHANISM UNKNOWN

A. 2'-Azido-2'-deoxynucleoside 5'-diphosphates

2'-Azido-2'-deoxyribonucleoside diphosphates (Structural Formula 9.111, see Scheme 158) irreversibly inactivate subunit B2 of *E. coli* ribonucleoside diphosphate reductase without affecting B1.[244] This is the converse of the effect of the 2'-chloro analogue (see Section V.G). Inactivation leads to destruction of the free radical of B2. Since the free radical content is low, only about 0.2 mol of inactivator is needed for inactivation. This azido compound also inactivates the coenzyme B_{12}-dependent ribonucleoside triphosphate reductase from *Lactobacillus leichmannii*,[244] a ribonucleotide reductase from calf thymus,[244] and the ribonucleotide reductase from herpes simplex virus type I.[250a] The nature of the irreversible inactivation is not known. Further studies by Sjöberg et al.[256] revealed that inactivation of *E. coli* ribonucleoside diphosphate reductase by 2′-azido-2′-deoxynucleoside diphosphates produces a new-transient radical in the enzyme concomitant with decay of the tyrosine radical signal in the native enzyme. On the basis of the ESR characteristics, a new radical species with the localization of the unpaired electron at the sugar moiety of the nucleotide was suggested. The radical shows hyperfine couplings to a hydrogen and a nitrogen nucleus of the azido group. The mechanism proposed for formation of the transient new radical is shown in Scheme 158. An alternative explanation for the new species is attachment of the active site tyrosine radical to the inactivator molecule that has lost a molecule of N_2 from its azido group (Structural Formula 9.112, Scheme 159). The same radical is observed in the ESR spectrum during reaction of this inactivator with bacteriophage T4 ribonucleotide reductase. [2′-^{2}H]- and [2′-^{15}N] 2′-Azido-2′-deoxyuridine diphosphate were prepared by Ator et al.[257] in order to elucidate the structure of this transient radical. The new radical is shown to reside on a nitrogen originally at the 2′-position and the observed coupling of this species to hydrogen is **not** caused by the hydrogen on the 2′-carbon, as was suggested by Sjöberg et al.[256]

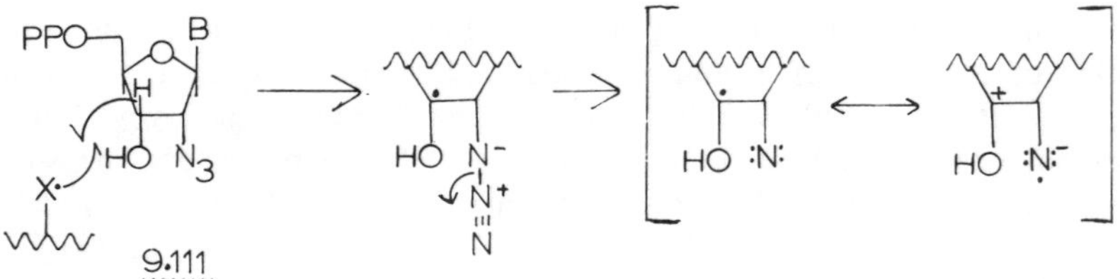

Scheme 158.   Containing Structural Formula 9.111.

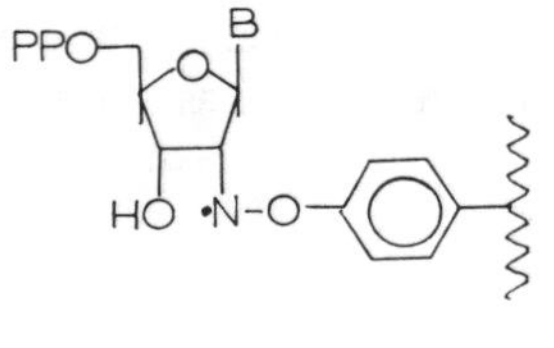

Scheme 159.   Structural Formula 9.112.

Further studies by Stubbe and co-workers[188a] on the mechanism of inactivation of *E. coli* ribonucleoside diphosphate reductase by **9.111** were published after submission of the manuscript for this book which shed considerable light on a possible inactivation mechanism. Incubation of the enzyme with 1 equivalent of **9.111** results in inactivation with subunit B2 tyrosyl radical loss. When [3′-^{3}H] **9.111** is used, 0.2 mol of ^{3}H is released to the medium per mole of enzyme inactivated, indicating that cleavage of the 3′ carbon-hydrogen bond occurs during inactivation. In order to determine the decomposition products of the nucleotide, the enzyme was incubated with [β-^{32}P] **9.111**. One equivalent each of inorganic pyrophosphate and uracil are produced, as is the case with 2′-chloro-2′-deoxyuridine 5′-diphosphate. Inactivation with [^{15}N$_3$] **9.111** produces 0.9 mol of ^{15}N$_2$. Unlike the conclusion of Thelander et al.[244] that B1 is not affected during inactivation, Salowe et al.[188a] show that the specific activity of the B1 subunit is reduced by half. Incubation of the enzyme with [5′-^{3}H] **9.111** results in stoichiometric covalent labeling of the enzyme. Separation of the subunits showed that the B1 subunit was modified. Chemical models suggest that the product is an adduct similar to that produced with **9.108** (Scheme 151). Unlike the inactivation of ribonucleotide reductase by **9.106** which is protected by thiols, inactivation by **9.111** is not protected by thiols and, therefore, is a true mechanism-based inactivator. Consequently, **9.111** belongs in Section III.A.2., not here. The inactivation mechanism proposed by Salowe et al.[188a] is shown in Scheme 159A.

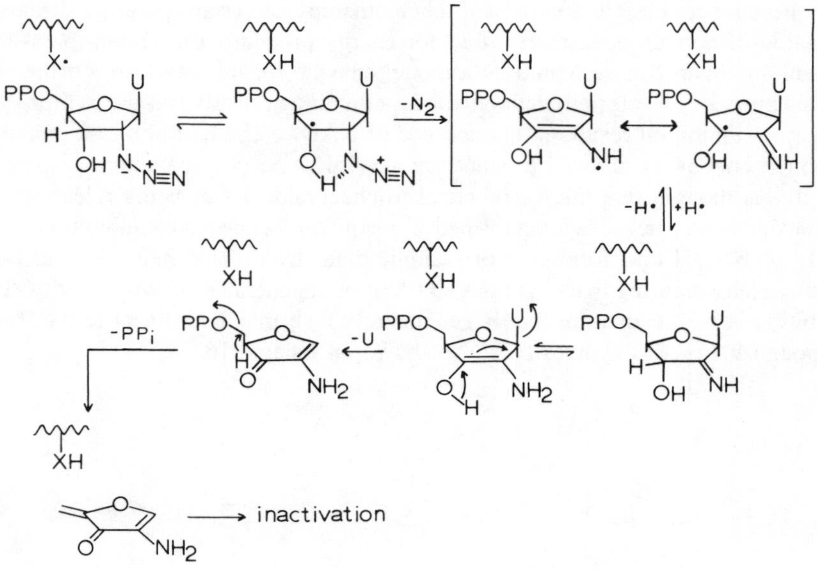

Scheme 159A.

B. 4-Methylbenzaldehyde

4-Methylbenzaldehyde, in the presence of NADPH, is a time-dependent inactivator of rabbit pulmonary cytochrome P-450.[258] An equimolar amount of heme is destroyed in the process, but only 50% of the enzyme is destroyed. The 50% not destroyed appears to be protected by some product of metabolism of 4-methylbenzaldehyde. Lipid peroxidation is not responsible for loss of enzyme activity.

C. Piperonyl Butoxide

Piperonyl butoxide (Structural Formula 9.113, Scheme 160) is a time-dependent inactivator of liver microsomal cytochrome P-450 from phenobarbital-treated rats[259] and from mice;[260] NADPH is required for inactivation.

$$CH_2O(CH_2CH_2O)_2C_4H_9$$

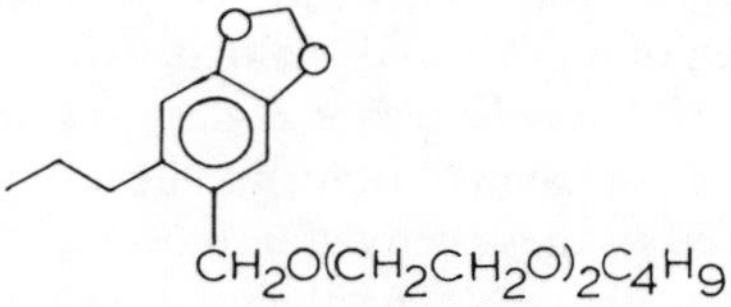

Scheme 160.   Structural Formula 9.113.

VII. NONCOVALENT INACTIVATION

A. Oxidation/Addition/Elimination

1. 3-Ethylthioprop-2-en-1-ol

3-Ethylthioprop-2-en-1-ol (Structural Formula 9.114, see Scheme 161) is a time-dependent inactivator of liver alcohol dehydrogenase; NAD$^+$ is required for inactivation.[261] With the

1,1-dideuterio analogue, inactivation occurs with a primary kinetic isotope effect (k_H/k_D) of 1.4. The corresponding aldehyde also is an inactivator. Upon dialysis or gel filtration, enzyme activity is recovered. The mechanism proposed by Schorstein and Suckling[261] is shown in Scheme 161. Further studies on this inactivation mechanism were carried out by MacInnes et al.,[262] because a possible approach to chemotherapy of certain parasitic diseases would be inhibition of dehydrogenases required for energy production in certain parasites; horse liver alcohol dehydrogenase would be a model enzyme for this reaction. Of the several 3-substituted prop-2-en-1-ols prepared, only the 3-ethylthio derivative produces time-dependent inhibition; no inhibition results in the absence of NAD$^+$. The partition ratio is only 0.1. If the inhibited enzyme is allowed to stand for several days, enzyme activity is regenerated. At first it was thought that the hypothetical Michael adduct was being released; however, it was shown that no Michael adduct formed. Concomitant with enzyme inhibition is reduction of NAD$^+$ to NADH **and** formation of malonic dialdehyde. This indicates that the carbon skeleton is released during inactivation. Gel filtration regenerates enzyme activity. The cause for inhibition is the ethanethiol that is generated, which may coordinate to the Zn(II).

Compound **9.114** was synthesized[258] as shown in Scheme 162.

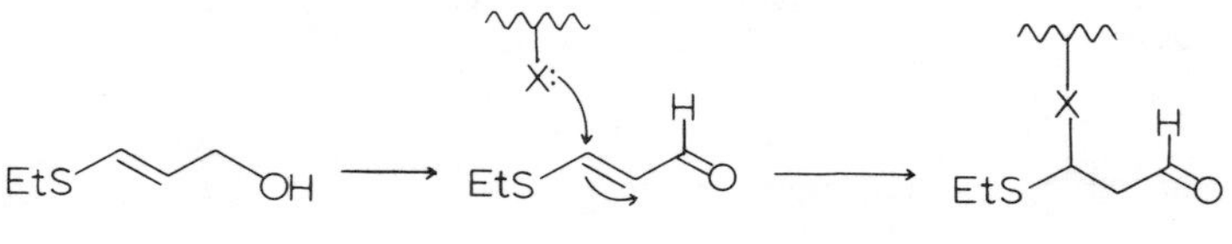

Scheme 161.   Containing Structural Formula 9.114.

Scheme 162.

B. Oxidation/Elimination

1. 2'-Deoxyadenosine and Related

S-Adenosylhomocysteine hydrolase from human splenic lymphoblasts is inactivated in a time-dependent, irreversible process by 2'-deoxyadenosine and adenine arabinoside.[263] When 2'-deoxy-[^{3}H]adenosine is used, the tritium remains tightly bound to the enzyme after gel filtration. Inactivation was suggested by Hershfield[263] to result from irreversible inactivation of the cofactor or from reaction of an oxidized analogue with the enzyme. Other analogues that inactivate the enzyme are 5'-deoxydenosine, 5'-deoxy-4'-thiomethyladenosine, and 3'-deoxyadenosine. Since the latter compound inactivates the enzyme, it suggests that oxidation at the 3-position is not involved in the inactivation process. Abeles et al.[264] found that *S*-adenosylhomocysteinase from calf liver is inactivated by [2-^{3}H] 2'-deoxyadenosine (label is in the adenine group) with concomitant and stoichiometric incorporation of radioactivity into the enzyme. After acid or ethanol denaturation of the labeled enzyme, 77% of the bound radioactivity was isolated as adenine. Also during inactivation there is a change in the cofactor absorption spectrum from that of NAD$^+$ to NADH. The mechanism proposed is shown in Scheme 163. Inactivation could result either because the cofactor cannot be reoxidized or an enzyme nucleophile adds to the α,β-unsaturated ketone sugar moiety (Structural Formula 9.115). The reason 2'-deoxyadenosine and adenine arabinoside are inactivators, but not adenosine, may be related to the fact that the former two compounds, after oxidation, can undergo *trans*-elimination of adenine, but adenosine cannot. Further support for the mechanism in Scheme 163 was presented by Abeles et al.[265] The formation of adenine was confirmed. Inactivation of *S*-adenosylhomocysteinase with deoxy[2'(*R*)-³H]adenosine results in release of ³H₂O, consistent with *trans*-elimination of [³H] and adenine. When the enzyme is reduced to the NADH form, no ³H₂O is released. The stoichiometry of ³H₂O released suggests that two of the four subunits are involved. Furthermore, only two of the four enzyme-bound NAD⁺ molecules are reduced. When adenine-labeled 2'-deoxyadenosine is used, 0.5 to 1.0 mol of radioactivity is bound per mole of subunit; the radioactivity is not released upon dialysis, but is released by unlabeled 2'-deoxyadenosine or adenine or by denaturation. The radioactivity released is adenine. [¹⁴C]Adenine binds tightly to enzyme in the NADH form, but not in the NAD⁺ form. The observation of Guranowski et al.[266] that carbocyclic adenosine inactivates *S*-adenosylhomocysteinase was confirmed, and adenine is bound to the enzyme. If the mechanism of this inactivation is the same as for deoxyadenosine, then a *cis*-elimination would be involved. The refined mechanism for 2'-deoxyadenosine inactivation is shown in Scheme 164.[265] An additional mechanism was suggested on the basis of the observation that adenine is produced, even if the enzyme is in the NADH form, in which case no [³H] is released from deoxy [2'(*R*)-³H]adenosine. The reaction is an enzyme-catalyzed hydrolysis of the glycosidic bond.[265]

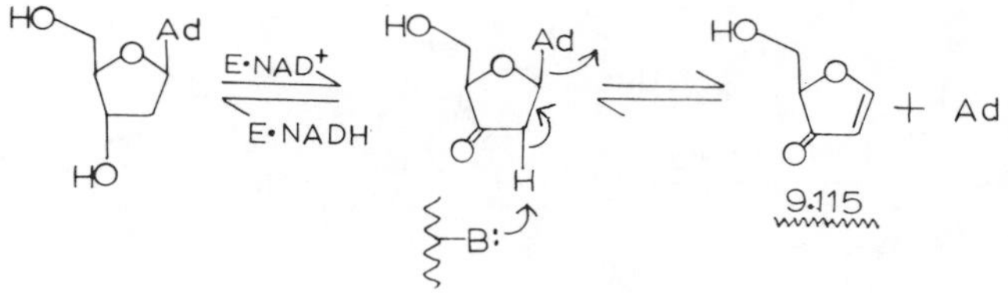

Scheme 163.   Containing Structural Formula 9.115.

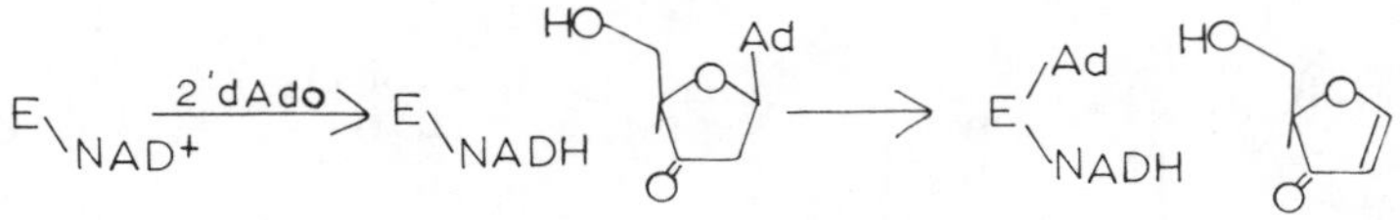

Scheme 164.

Other analogues are known to inactivate this enzyme. A large number of nucleoside analogues irreversibly inactivate beef liver *S*-adenosylhomocysteine hydrolase in a time-dependent manner.[267] There are two classes of analogues that inactivate the enzyme. One class is devoid of any side chain at the 5'-OH group and inactivate in a pseudo first-order process; the other class generally has a side chain at the 5'-position and inactivates in a nonpseudo first-order process. The most potent inactivators are 2-chloroadenosine, 9-β-D-arabinofuranosyladenine (Ara-A), and (±) aristeromycin. The adenine base is released from adenosine and Ara-A during inactivation. The results support the mechanism in Scheme 164. Rat liver *S*-adenosyl-L-homocysteine hydrolase is irreversibly inactivated by adenosine.[268] *erythro*-9-(2-Hydroxynon-3-yl)adenine and adenosine also are time-dependent inactivators of rat liver *S*-adenosylhomocysteine hydrolase.[269] Ara-A, 9-β-D-arabinofuranosyladenine 5'-monophosphate, and 9-β-D-arabinofuranosyladenine 5'-triphosphate are time-dependent irreversible inactivators of mouse liver *S*-adenosylhomocysteinase.[270] 9-β-D-Arabinofuranosyl-2-fluoroadenine, a derivative of Ara-A that is resistant to enzymatic deamination and selectively inhibits DNA synthesis, is a time-dependent inactivator of *S*-adenosylhomocysteine hydrolase from mouse L1210 tumor cells.[271] Unlike inactivation by Ara-A, which is biphasic, inactivation by 2-fluoro-Ara-A is pseudo first-order. The enzyme from hamster and bovine liver is irreversibly inactivated by various adenosine analogues substituted in the 5'- and 2-positions.[272] 5'-Cyano-5'-deoxyadenosine is as potent as Ara-A. Unlike that observed previously,[264,265,267] enzyme inactivation is **not** accompanied by [³H]adenine release from [³H]Ara-A. From analogue studies, it appears that there is a strict requirement for an adenine moiety in order for inactivation to result.

Neplanocin A (Structural Formula 9.116, to Scheme 165) is a time-dependent irreversible inactivator of bovine liver *S*-adenosylhomocysteine hydrolase.[273] Less than 10% of the enzyme activity is recovered after extensive dialysis. The enzyme is completely inactivated by the addition of 0.5 equivalent of **9.116**; 0.5 equivalent each of NADH and adenine are produced.[274] A mechanism similar to that shown in Scheme 164 was proposed by Wolfson et al.[274] Inactivation of bovine liver *S*-adenosylhomocysteine hydrolase by neplanocin A can be reversed by the addition of exogenous NAD⁺.[275] With the use of a fluorescence technique, NAD⁺/NADH content of native enzyme and neplanocin A-inactivated enzyme was determined by Matuszewska and Borchardt.[275] Upon neplanocin A inactivation, complete reduction of NAD⁺ to NADH occurs. Neplanocin A is a natural product isolated from *Actinoplanacea ampulariella* cultures[276] and has been synthesized by Lim and Marquez[277] (Scheme 165) and by Arita et al.[278]

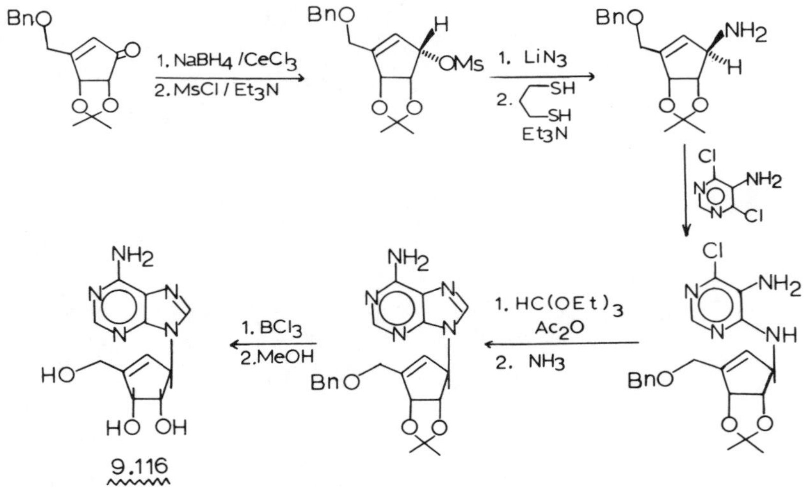

Scheme 165.   Containing Structural Formula 9.11b.

REFERENCES

1. **Silverman, R. B., Hoffman, S. J., and Catus, W. B., III,** A mechanism for mitochondrial monoamine oxidase catalyzed amine oxidation, *J. Am. Chem. Soc.,* 102, 7126, 1980.
2. **Simpson, J. T., Krantz, A., Lewis, F. D., and Kokel, B.,** Photochemical and photophysical studies of amines with excited flavins. Relevance to the mechanism of action of the flavin-dependent monoamine oxidase, *J. Am. Chem. Soc.,* 104, 7155, 1982.
3. **Hanzlik, R. P. and Tullman, R. H.,** Suicidal inactivation of cytochrome P-450 by cyclopropylamines. Evidence for cation-radical intermediates, *J. Am. Chem. Soc.,* 104, 2048, 1982.
4. **Macdonald, T. L., Zirvi, K., Burka, L. T., Peyman, P., and Guengerich, F. P.,** Mechanism of cytochrome P-450 inhibition by cyclopropylamines, *J. Am. Chem. Soc.,* 104, 2050, 1982.

5. **Ortiz de Montellano, P. R., Beilan, H. S., and Kunze, K. L.,** *N-* Alkylprotoporphyrin IX formation in 3,5-dicarbethoxy-1,4-dihydrocollidine-treated rats. Transfer of the alkyl group from the substrate to the porphyrin, *J. Biol. Chem.,* 256, 6708, 1981.

6. **Fitzpatrick, P. F., Flory, D. R., Jr., and Villafranca, J. J.,** 3-Phenylpropenes as mechanism-based inhibitors of dopamine β-hydroxylase: evidence for a radical mechanism, *Biochemistry,* 24, 2108, 1985.

7. **Wiseman, J. S., Nichols, J. S., and Kolpak, M. X.,** Mechanism of inhibition of horseradish peroxidase by cyclopropanone hydrate, *J. Biol. Chem.,* 257, 6328, 1982.

8. **Cromartie, T. H.,** Irreversible inactivation of the flavoenzyme alcohol oxidase by cyclopropanone, *Biochem. Biophys. Res. Commun.,* 105, 785, 1982.

9. **Sherry, B. and Abeles, R. H.,** Mechanism of action of methanol oxidase, reconstitution of methanol oxidase with 5-deazaflavin, and inactivation of methanol oxidase by cyclopropanol, *Biochemistry,* 24, 2594, 1985.

10. **Harris, G., Ator, M., and Stubbe, J.,** Mechanism of inactivation of *Escherichia coli* and *Lactobacillus leichmannii* ribonucleotide reductases by 2'-chloro-2'-deoxynucleotides: Evidence for generation of 2-methylene-3(2H)-furanone, *Biochemistry,* 23, 5214, 1984.

11. **Corey, E. J., Cashman, J. R., Eckrich, T. M., and Corey, D. R.,** A new class of irreversible inhibitors of leukotriene biosynthesis, *J. Am. Chem. Soc.,* 107, 713, 1985.

12. **Alston, T. A., Mela, L., and Bright, H. J.,** 3-Nitropropionate, the toxic substance of *Indigofera,* is a suicide inactivator of succinate dehydrogenase, *Proc. Natl. Acad. Sci. U.S.A.,* 74, 3767, 1977.

13. **Coles, C. J., Edmondson, D. E., and Singer, T. P.,** Inactivation of succinate dehydrogenase by 3-nitropropionate, *J. Biol. Chem.,* 254, 5161, 1979.

14. **Buntain, I. G., Suckling, C. J., and Wood, H. C. S.,** Irreversible inhibition of dihydro-orotate dehydrogenase by hydantoins derived from amino acids, *J. Chem. Soc. Chem. Commun.,* p. 242, 1985.

15. **Rajashekhar, B., Fitzpatrick, P. F., Colombo, G., and Villafranca, J. J.,** Synthesis of several 2-substituted 3-(p-hydroxyphenyl)-1-propenes and their characterization as mechanism-based inhibitors of dopamine β-hydroxylase, *J. Biol. Chem.,* 259, 6925, 1984.

16. **Colombo, G., Rajashekhar, B., Giedroc, D. P., and Villafranca, J. J.,** Mechanism-based inhibitors of dopamine β-hydroxylase: inhibition by 2-bromo-3-(p-hydroxyphenyl)-1-propene, *Biochemistry,* 23, 3590, 1984.

17. **Ash, D. E., Papadopoulos, N. J., Colombo, G., and Villafranca, J. J.,** Kinetic and spectroscopic studies of the interaction of copper with dopamine β-hydroxylase, *J. Biol. Chem.,* 259, 3395, 1984.

18. **Fitzpatrick, P. F. and Villafranca, J. J.,** Mechanism-based inhibitors of dopamine β-hydroxylase containing acetylenic or cyclopropyl groups, *J. Am. Chem. Soc.,* 107, 5022, 1985.

19. **Fong, J. C. and Schulz, H.,** On the rate-determining step of fatty acid oxidation in heart. Inhibition of fatty acid oxidation by 4-pentenoic acid, *J. Biol. Bhem.,* 253, 6917, 1978.

20. **Holland, P. C., Senior, A. E., and Sherratt, H. S. A.,** Biochemical effects of the hypoglycaemic compound pent-4-enoic acid and related non-hypoglycaemic fatty acids. Effects of their coenzyme A esters on enzymes of fatty acid oxidation, *Biochem. J.,* 136, 173, 1973.

20a. **Shaw, L. and Engel, P. C.,** The suicide inactivation of ox liver short-chain acyl-CoA dehydrogenase by propionyl-CoA, *Biochem. J.,* 230, 723, 1985.

21. **MacInnes, I., Nonhebel, D. C., Orszulik, S. T., Suckling, C. J., and Wrigglesworth, R.,** *exo -* Bicyclo[4.1.0]heptane-7-methanol: a novel latent inhibitor of liver alcohol dehydrogenase, *J. Chem. Soc. Chem. Commun.,* p. 1068, 1980.

22. **MacInnes, I., Nonhebel, D. C., Orszulik, S. T., and Suckling, C. J.,** Latent inhibitors. III. The inhibition of lactate dehydrogenase and alcohol dehydrogenase by cyclopropane-containing compounds, *J. Chem. Soc. Perkin Trans. 1,* p. 2771, 1983.

22a. **Breckenridge, R. J. and Suckling, C. J.,** Enzyme inhibition by electrophilic cyclopropane derivatives, *Tetrahedron,* 42, 5665, 1986.

23. **Stearns, R. A. and Ortiz de Montellano, P. R.,** Inactivation of cytochrome P-450 by a catalytically generated cyclobutadiene species, *J. Am. Chem. Soc.,* 107, 234, 1985.

24. **Nagahisa, A., Orme-Johnson, W. H., and Wilson, S. R.,** Silicon-mediated suicide inhibition: an efficient mechanism-based inhibitor of cytochrome P-450$_{scc}$ oxidation of cholesterol, *J. Am. Chem. Soc.,* 106, 1166, 1984.

25. **Wilson, S. R. and Shedrinsky, A.,** [β-(Trimethylsilyl)ethyl]lithium: a new reagent for carbonyl reductive vinylation, *J. Org. Chem.,* 47, 1983, 1982.

25a. **Corey, E. J. and d'Alarcao, M.,** 12-Methylidene-10(Z),13(Z)-nonadecadienoic acid, a new irreversible inhibitor of soybean lipoxygenase, *Tetrahedron Lett.,* 27, 3589, 1986.

25b. **Ortiz de Montellano, P. R. and Stearns, R. A.,** Timing of the radical recombination step in cytochrome P-450 catalysis with ring-strained probes, *J. Am. Chem. Soc.,* 109, 3415, 1987.

26. **McEwen, C. M., Jr., Sasaki, G., and Jones, D. C.,** Human liver mitochondrial monoamine oxidase. III. Kinetic studies concerning time-dependent inhibitions, *Biochemistry,* 8, 3963, 1969.

27. **Taylor, J. D., Wykes, A. A., Gladish, Y. C., and Martin, W. B.,** New inhibitor of monoamine oxidase, *Nature (London),* 187, 941, 1960.
28. **Walsh, C. T., Schonbrunn, A., Lockridge, O., Massey, V., and Abeles, R. H.,** Inactivation of a flavoprotein, lactate oxidase, by an acetylenic substrate, *J. Biol. Chem.,* 247, 6004, 1972.
29. **Kraus, J.-L., Yaouanc, J.-J., and Sturtz, G.,** Rôle des substituants *N*-allyl et *N*-propargyl dans le mécanisme d'inhibition d'enzymes flavoprotéiques: *N*-déméthylase, monoamineoxydase, *Eur. Med. Chem. Chim. Ther.,* 10, 507, 1975.
30. **Kraus, J. L. and Yaouanc, J. J.,** Inactivation of a flavin-linked oxidase, *N,N*-dimethylglycine oxidase, in vivo and in vitro, *Mol. Pharmacol.,* 13, 378, 1977.
31. **Swett, L. R., Martin, W. B., Taylor, J. D., Everett, G. M., Wykes, A. A., and Gladish, Y. C.,** Structure-activity relations in the pargyline series, *Ann. N.Y. Acad. Sci.,* 107, 891, 1963.
32. **Coulson, C. J.,** The inactivation of monoamine oxidase by clorgyline (M & B 9302), *Biochem. J.,* 121, 38p, 1971.
33. **Williams, C. H. and Lawson, J.,** Monoamine oxidase-II. Time-dependent inhibition by propargylamines, *Biochem. Pharmacol.,* 23, 629, 1974.
34. **Parkinson, D. and Callingham, B. A.,** The binding of [³H]pargyline to rat liver mitochondrial monoamine oxidase, *J. Pharm. Pharmacol.,* 32, 49, 1980.
35. **McCauley, R.,** 7[¹⁴C]Pargyline binding to mitochondrial outer membranes, *Biochem. Pharmacol.,* 25, 2214, 1976.
36. **Johnston, J. P.,** Some observations upon a new inhibitor of monoamine oxidase in brain tissue, *Biochem. Pharmacol.,* 17, 1285, 1968.
37. **Maître, L.,** Monoamine oxidase inhibiting properties of SU-11,739 in the rat. Comparison with pargyline, tranylcypromine and iproniazid, *J. Pharmacol. Exp. Ther.,* 157, 81, 1967.
38. **Youdim, M. B. H., Collins, G. G. S., and Sandler, M.,** Monoamine oxidase: multiple forms and selective inhibitors, *Biochem. J.,* 121, 34p, 1971.
39. **Oreland, L., Kinemuchi, H., and Stigbrand, T.,** Pig liver monoamine oxidase: studies on the subunit structure, *Arch. Biochem. Biophys.,* 159, 854, 1973.
40. **Fowler, C. J., Oreland, L., Magnusson, T., and Winblast, B.,** Titration of human brain monoamine oxidase-A and -B by clorgyline and L-deprenil, *Naunyn Schmiedeberg's Arch. Pharmacol.,* 311, 263, 1980.
41. **Fowler, C. J., Mantle, T. J., and Tipton, K. F.,** The nature of the inhibition of rat liver monoamine oxidase types A and B by the acetylenic inhibitors clorgyline, *l*-deprenyl and pargyline, *Biochem. Pharmacol.,* 31, 3555, 1982.
41a. **Dupont, H., Davies, D. S., and Strolin-Benedetti, M.,** Inhibition of cytochrome P-450-dependent oxidation reactions by MAO inhibitors in rat liver microsomes, *Biochem. Pharmacol.,* 36, 1651, 1987.
42. **Castro Costa, M. R. and Breakefield, X. O.,** Electrophoretic characterization of monoamine oxidase by [³H]pargyline binding in rat hepatoma cells with A and B activity, *Mol. Pharmacol.,* 16, 242, 1979.
43. **Edwards, D. J. and Pak, K. Y.,** Selective radiochemical labeling of types A and B active sites of rat liver monoamine oxidase, *Biochem. Biophys. Res. Commun.,* 86, 350, 1979.
44. **Williams, C. H.,** A New specific inhibitor of monoamine oxidase A, *Biochem. Pharmacol.,* 31, 2305, 1982.
45. **Fowler, C. J., Oreland, L., Wiberg, Å., Carlsson, A., and Magnusson, T.,** The inhibition of rat brain monoamine oxidase B by J-508 (*N*-methyl-*N*-propargyl-(1-indanyl)ammonium hydrochloride), and its use for the titration of this enzyme form, *Med. Biol.,* 57, 406, 1979.
46. **Fowler, C. J., Wiberg, Å., Oreland, L., and Winblad, B.,** Titration of human brain type-B monoamine oxidase, *Neurochem. Res.,* 5, 697, 1980.
47. **Kalir, A., Sabbagh, A., and Youdim, M. B. H.,** Selective acetylenic "suicide" and reversible inhibitors of monoamine oxidase types A and B, *Br. J. Pharmacol.,* 73, 55, 1981.
48. **Tipton, K. F., McCrodden, J. M., Kalir, A. S., and Youdim, M. B. H.,** Inhibition of rat liver monoamine oxidase by α-methyl- and *N*-propargyl-amine derivatives, *Biochem. Pharmacol.,* 31, 1251, 1982.
49. **Williams, C. H.,** Selective inhibitors of monoamine oxidases A and B, *Biochem. Pharmacol.,* 33, 334, 1984.
50. **Rando, R. R.,** The fluorescent labeling of mitochondrial monoamine oxidase, *Mol. Pharmacol.,* 13, 726, 1977.
51. **Hellerman, L. and Erwin, V. G.,** Mitochondrial monoamine oxidase II. Action of various inhibitors for the bovine kidney enzyme. Catalytic mechanism, *J. Biol. Chem.,* 243, 5234, 1968.
52. **Chuang, H. Y. K., Patek, D. R., and Hellerman, L.,** Mitochondrial monoamine oxidase. Inactivation by pargyline. Adduct formation, *J. Biol. Chem.,* 249, 2381, 1974.
53. **Oreland, L., Kinemuchi, H., and Yoo, B. Y.,** The mechanism of action of the monoamine oxidase inhibitor pargyline, *Life Sci.,* 13, 1533, 1973.
54. **Yu, P. H.,** Studies on the pargyline-binding of different types of monoamine oxidase, *Can. J. Biochem.,* 59, 30, 1981.

55. **Zeller, E. A., Gärtner, B., and Hemmerich, P.,** 4a, 5-Cycloaddition reactions of acetylenic compounds at the flavoquinone nucleus as mechanisms of flavoprotein inhibitions, *Z. Naturforsch.,* 27b, 1050, 1972.
56. **Maycock, A. L.,** Structure of a flavoprotein-inactivator model compound, *J. Am. Chem. Soc.,* 97, 2270, 1975.
57. **Gärtner, B. and Hemmerich, P.,** Inhibition of monoamine oxidase by propargylamine: structure of the inhibitor complex, *Angew. Chem. Int. Ed. Engl.,* 14, 110, 1975.
58. **Gärtner, B., Hemmerich, P., and Zeller, E. A.,** Structure of flavin adducts with acetylenic substrates. Chemistry of a monoamine oxidase and lactate oxidase inhibition, *Eur. J. Biochem.,* 63, 211, 1976.
59. **Maycock, A. L., Abeles, R. H., Salach, J. I., and Singer, T. P.,** The structure of the covalent adduct formed by the interaction of 3-dimethylamino-1-propyne and the flavine of mitochondrial amine oxidase, *Biochemistry,* 15, 114, 1976.
60. **White, R. L., Smith, R. A., and Krantz, A.,** Differential inactivation of mitochondrial monoamine oxidase by stereoisomers of allenic amines, *Biochem. Pharmacol.,* 32, 3661, 1983.
61. **Kraus, J.-L. and Belleau, B.,** The bioorganic chemistry of N-allyl and N-propargyl substituents in drug interactions with flavin-linked oxidases, *Can. J. Chem.,* 53, 3141, 1975.
62. **Yaouanc, J. J., Dugenet, P., and Kraus, J. L.,** Observations on the "in vitro" inhibition of dimethyl-glycine oxidase by "suicide-substrates," *Pharmacol. Res. Commun.,* 11, 115, 1979.
63. **Kraus, J. L. and Yaouanc, J. J.,** Inhibition of the cytochrome P-450 multifunctional oxidase by N-propargyl analogs of SKF-525A and acetylmethadol, *Eur. J. Drug Metab. Pharmacokinet.,* 4, 219, 1979.
64. **Kraus, J. L., Yaouanc, J. J., and Sturtz, G.,** Synthèse, étude structurale en RMN ^{13}C et réactivité de nouveaux modèles isoalloxazines (flavines), *Bull. Soc. Chim. Fr.,* II-230, 1979.
65. **Bey, P. Bolkenius, F. N., Seiler, N., and Casara, P.,** N -2,3-Butadienyl-1,4-butanediamine derivatives: potent irreversible inactivators of mammalian polyamine oxidase, *J. Med. Chem.,* 28, 1, 1985.
66. **Claesson, A. and Sahlberg, C.,** Allenes and acetylenes. XXIII. Synthesis of α-allenic amines via allenic imines obtained from organocopper reactions, *Tetrahedron,* 38, 363, 1982.
67. **Sahlberg, C., Ross, S. B., Fagerwall, I., Ask, A-L., and Claesson, A.,** Synthesis and monoamine oxidase inhibitory activities of α-allenic amines in vivo and in vitro. Different activities of two enantiomeric allenes, *J. Med. Chem.,* 26, 1036, 1983.
68. **Sahlberg, C. and Claesson, A.,** Allenes and acetylenes. XXIV. Synthesis of α-allenic amines by organ-ocuprate reactions of acetylenic aminoethers, *Acta Chem. Scand. Ser. B,* 36, 179, 1982.
69. **Halliday, R. P., Davis, C. S., Heotis, J. P., Pals, D. T., Watson, J., and Bickerton, R. K.,** Allenic amines: A new class of nonhydrazine MAO inhibitors, *J. Pharm. Sci.,* 57, 430, 1968.
69a. **McCarthy, J. R., Barney, C. L., Matthews, D. P., and Bargar, T. M.,** A facile synthesis of β-substituted-α-allenyl primary amines, *Tetrahedron Lett.,* 28, 2207, 1987.
70. **Krantz, A. and Lipkowitz, G. S.,** Studies of mitochondrial monoamine oxidase. Inactivation of the enzyme by isomeric acetylenic and allenic amines yielding mutually exclusive products, *J. Am. Chem. Soc.,* 99, 4156, 1977.
71. **Krantz, A., Kokel, B., Sachdeva, Y. P., Salach, J., Detmer, K., Claesson, A., and Sahlberg, C.,** Inactivation of mitochondrial monoamine oxidase by α,γ,δ-allenic amines, in *Monoamine Oxidase: Structure, Function, and Altered Functions,* Singer, T. P., Von Korff, R. W., and Murphy, D. L., Eds., Academic Press, New York, 1979, 51.
72. **Rando, R. R. and Eigner, A.,** The pseudoirreversible inhibition of monoamine oxidase by allylamine, *Mol. Pharmacol.,* 13, 1005, 1977.
73. **Silverman, R. B., Hiebert, C. K., and Vazquez, M. L.,** Inactivation of monoamine oxidase by allylamine does not result in flavin attachment, *J. Biol. Chem.,* 260, 14648, 1985.
74. **Silverman, R. B. and Yamasaki, R. B.,** Mechanism-based inactivation of mitochondrial monoamine oxidase by N-(1-methylcyclopropyl)benzylamine, *Biochemistry,* 23, 1322, 1984.
75. **Langston, J. W., Ballard, P., Tetrud, J. W., and Irwin, I.,** Chronic parkinsonism in humans due to a product of meperidine-analog synthesis, *Science,* 219, 979, 1983.
76. **Lanston, J. W., Irwin, I., Langston, E. B., and Forno, L. S.,** Pargyline prevents MPTP-induced parkinsonism in primates, *Science,* 225, 1480, 1984.
77. **Salach, J. I., Singer, T. P., Castagnoli, N., Jr., and Trevor, A.,** Oxidation of the neurotoxic amine 1-methyl-4-phenyl-1,2,3,6-tetrahydropyridine (MPTP) by monoamine oxidases A and B and suicide inactivation of the enzymes by MPTP, *Biochem. Biophys. Res. Commun.,* 125, 831, 1984.
78. **Singer, T. P., Salach, J. I., and Crabtree, D.,** Reversible inhibition and mechanism-based irreversible inactivation of monoamine oxidases by 1-methyl-4-phenyl-1,2,3,6-tetrahydropyridine (MPTP), *Biochem. Biophys. Res. Commun.,* 127, 707, 1985.
78a. **Singer, T. P., Salach, J. I., Castagnoli, Jr., N., and Trevor, A.,** Interactions of the neurotoxic amine 1-methyl-4-phenyl-1,2,3,6-tetrahydropyridine with monoamine oxidases, *Biochem. J.,* 235, 785, 1986.
79. **Fuller, R. W. and Hemrick-Leucke, S. K.,** Inhibition of types A and B monoamine oxidase by 1-methyl-4-phenyl-1,2,3,6-tetrahydropyridine, *J. Pharmacol. Exp. Ther.,* 232, 696, 1985.

80. **Buckman, T. D. and Eiduson, S.,** Photoinactivation of B-type monoamine oxidase by a 1-methyl-4-phenyl-1,2,3,6-tetrahydropyridine metabolite, *J. Biol. Chem.,* 260, 11899, 1985.

80a. **Kinemuchi, H., Arai, Y., and Toyoshima, Y.,** Participation of brain monoamine oxidase B form in the neurotoxicity of 1-methyl-4-phenyl-1,2,3,6-tetrahydropyridine: relationship between the enzyme inhibition and the neurotoxicity, *Neurosci. Lett.,* 58, 195, 1985.

80b. **Tipton, K. F., McCrodden, J. M., and Youdim, M. B. H.,** Oxidation and enzyme-activated irreversible inhibition of rat liver monoamine oxidase-B by 1-methyl-4-phenyl-1,2,3,6-tetrahydropyridine (MPTP), *Biochem. J.,* 240, 379, 1986.

81. **Fritz, R. R., Abell, C. W., Patel, N. T., Gessner, W., and Brossi, A.,** Metabolism of the neurotoxin in MPTP by human liver monoamine oxidase B, *FEBS Lett.,* 186, 224, 1985.

82. **Marcotte, P. and Walsh, C.,** Vinylglycine and propargylglycine: complementary suicide substrates for L-amino acid oxidase and D-amino acid oxidase, *Biochemistry,* 15, 3070, 1976.

83. **Baldwin, J. E., Haber, S. B., Hoskins, C., and Kruse, L. I.,** Synthesis of β,γ-unsaturated amino acids, *J. Org. Chem.,* 42, 1239, 1977.

84. **Hanessian, S. and Sahoo, S. P.,** A novel and efficient synthesis of L-vinylglycine, *Tetrahedron Lett.,* 25, 1425, 1984.

85. **Afzali-Ardakani, A. and Rapoport, H.,** L-Vinylglycine, *J. Org. Chem.,* 45, 4817, 1980.

86. **Bey, P., Fozard, J., Lacoste, J. M., McDonald, I. A., Zreika, M., and Palfreyman, M. G.,** (E)-2-(3,4-Dimethoxyphenyl)-3-fluoroallylamine: a selective, enzyme activated inhibitor of type B monoamine oxidase, *J. Med. Chem.,* 27, 9, 1984.

87. **Zreika, M., McDonald, I. A., Bey, P., and Palfreyman, M. G.,** MDL 72145, an enzyme-activated irreversible inhibitor with selectivity for monoamine oxidase type B, *J. Neurochem.,* 43, 448, 1984.

88. **McDonald, I. A., Lacoste, J. M., Bey, P., Palfreyman, M. G., and Zreika, M.,** Enzyme-activated irreversible inhibitors of monoamine oxidase: phenylallylamine structure-activity relationships, *J. Med. Chem.,* 28, 186, 1985.

89. **McDonald, I. A. and Bey, P.,** A general preparation of fluoroallylamine enzyme inhibitors incorporating a β-substituted heteroatom, *Tetrahedron Lett.,* 26, 3807, 1985.

90. **Knoll, J., Ecsery, Z., Magyar, K., and Satory, E.,** Novel (−) deprenyl-derived selective inhibitors of B-type monoamine oxidase. The relation of structure to their action, *Biochem. Pharmacol.,* 27, 1739, 1978.

91. **McDonald, I. A., Palfreyman, M. G., Zreika, M., and Bey, P.,** (Z)-2-(2,4-Dichlorophenoxy)methyl-3-fluoroallylamine (MDL 72638): a clorgyline analogue with surprising selectivity for monoamine oxidase type B, *Biochem. Pharmacol.,* 35, 349, 1986.

92. **McDonald, I. A., Lacoste, J. M., Bey, P., Wagner, J., Zreika, M., and Palfreyman, M. G.,** (E)-β-(Fluoromethylene)-*m*-tyrosine: A substrate for aromatic L-amino acid decarboxylase liberating an enzyme-activated irreversible inhibitor of monoamine oxidase, *J. Am. Chem. Soc.,* 106, 3354, 1984.

92a. **Palfreyman, M. G., McDonald, I. A., Fozard, J. R., Mely, Y., Sleight, A. J., Zreika, M., Wagner, J., Bey, P., and Lewis, P. J.,** Inhibition of monoamine oxidase selectively in brain monoamine nerves using the bioprecursor (E)-β-fluoromethylene-*m*-tyrosine (MDL 72394), a substrate for aromatic L-amino acid decarboxylase, *J. Neurochem.,* 45, 1850, 1985.

92b. **McDonald, I. A., Lacoste, J. M., Bey, P., Wagner, J., Zreika, M., and Palfreyman, M. G.,** Dual enzyme-activated irreversible inhibition of monoamine oxidase, *Bioorg. Chem.,* 14, 103, 1986.

93. **Clavier, A.,** Préparation de l'allylamine à partir de bromure et de chlorure d'allyle: méthode dérivée de la méthode de Delépine, *Bull. Soc. Chim. Fr.,* 21, 646, 1954.

94. **Rando, R. R.,** 3-Bromoallylamine induced irreversible inhibition of monoamine oxidase, *J. Am. Chem. Soc.,* 95, 4438, 1973.

95. **Hamilton, G. A.,** The proton in biological redox reactions, *Prog. Bioorg. Chem.,* 1, 83, 1971.

96. **Neumann, R., Hevey, R., and Abeles, R. H.,** The action of plasma amine oxidase on β-haloamines. Evidence for proton abstraction in the oxidative reaction, *J. Biol. Chem.,* 250, 6362, 1975.

97. **Suva, R. H. and Abeles, R. H.,** Studies on the mechanism of action of plasma amine oxidase, *Biochemistry,* 17, 3538, 1978.

98. **Tang, S-S., Simpson, D. E., and Kagan, H. M.,** β-Substituted ethylamine derivatives as suicide inhibitors of lysyl oxidase, *J. Biol. Chem.,* 259, 975, 1984.

98a. **Williamson, P. R. and Kagan, H. M.,** α-Proton abstraction and carbanion formation in the mechanism of action of lysyl oxidase, *J. Biol. Chem.,* 262, 8196, 1987.

98b. **Weyler, W.,** 2-Chloro-2-phenylethylamine as a mechanistic probe and active site-directed inhibitor of monoamine oxidase from bovine liver mitochondria, *Arch. Biochem. Biophys.,* 255, 400, 1987.

99. **Tipton, K. F., Fowler, C. J., McCrodden, J. M., and Strolin Benedetti, M.,** The enzyme-activated irreversible inhibition of type B monoamine oxidase by 3-{4-[(3-chlorophenyl)methoxy]phenyl}-5-[(methylamino)methyl]-2-oxazolidinone methanesulphonate (compound MD 780236) and the enzyme-catalysed oxidation of this compound as competing reactions, *Biochem. J.,* 209, 235, 1983.

100. **Dostert, P., Strolin Benedetti, M., and Guffroy, C.,** Different stereoselective inhibition of monoamine oxidase-B by the R- and S-enantiomers of MD 780236, *J. Pharm. Pharmacol.,* 35, 161, 1983.

101. **Strolin Benedetti, M., Dow, J., Boucher, T., and Dostert, P.,** Metabolism of the monoamine oxidase-B inhibitor, MD 780236 and its enantiomers by the A and B forms of the enzyme in the rat, *J. Pharm. Pharmacol.*, 35, 837, 1983.

102. **Silverman, R. B. and Hoffman, S. J.,** Mechanism of inactivation of mitochondrial monoamine oxidase by *N*-cyclopropyl-*N*-arylalkyl amines, *J. Am. Chem. Soc.*, 102, 884, 1980.

103. **Paech, C., Salach, J. I., and Singer, T. P.,** Suicide inactivation of monoamine oxidase by *trans*-2-phenylcyclopropylamine, *J. Biol. Chem.*, 255, 2700, 1980.

104. **Hanzlik, R. P., Kishore, V., and Tullman, R.,** Cyclopropylamines as suicide substrates for cytochrome P-450, *J. Med. Chem.*, 22, 759, 1979.

105. **Zeller, E. A., Sarkar, S., and Reinen, R. M.,** Amine oxidases. XIX. inhibition of monoamine oxidase by phenylcyclopropylamines and iproniazid, *J. Biol. Chem.*, 237, 2333, 1962.

106. **Barbato, L. M. and Abood, L. G.,** Purification and properties of monoamine oxidase, *Biochim. Biophys. Acta*, 67, 531, 1963.

107. **Guha, S. R.,** Inhibition of purified rat liver mitochondrial monoamine oxidase by *trans*-2-phenylcyclopropylamine, *Biochem. Pharmacol.*, 15, 161, 1966.

108. **Belleau, B. and Moran, J.,** The mechanism of action of 2-phenylcyclopropylamine type of monoamine oxidase inhibitors, *J. Med. Pharm. Chem.*, 5, 215, 1962.

109. **Burger, A. and Nara, S.,** *In vitro* inhibition studies with homogeneous monoamine oxidases, *J. Med. Chem.*, 8, 859, 1965.

110. **Kenney, W. C., Nagy, J., Salach, J. I., and Singer, T. P.,** Structure of the covalent phenylhydrazine adduct of monoamine oxidase, in *Monoamine Oxidase: Structure, Function, and Altered Functions*, Singer, T. P., Von Korff, R. W., and Murphy, D. L., Eds., Academic Press, New York, 1979, 25.

111. **Silverman, R. B.,** Mechanism of inactivation of monoamine oxidase by *trans*-2-phenylcyclopropylamine and the structure of the enzyme-inactivator adduct, *J. Biol. Chem.*, 258, 14766, 1983.

112. **Fuller, R. W., Marsh, M. M., and Mills, J.,** Inhibition of monoamine oxidase by *N*-(phenoxyethyl)cyclopropylamines. Correlation of inhibition with Hammett constants and partition coefficients, *J. Med. Chem.*, 11, 397, 1968.

113. **Fuller, R. W.,** Kinetic studies and effects *in vivo* of a new monoamine oxidase inhibitor, *N*-[2-(*o*-chlorophenoxy)-ethyl]cyclopropylamine, *Biochem. Pharmacol.*, 17, 2097, 1968.

114. **Mills, J., Kattau, R., Slater, I. H., and Fuller, R. W.,** *N*-Substituted cyclopropylamines as monoamine oxidase inhibitors. Structure-activity relationships. Dopa potentiation in mice and *in vitro* inhibition of kynuramine oxidation, *J. Med. Chem.*, 11, 95, 1968.

115. **Fuller, R. W., Hemrick-Luecke, S. K., and Molloy, B. B.,** *N*-[2-(*o*-Iodophenoxy)-ethyl]cyclopropylamine hydrochloride (LY121768), a potent and selective irreversible inhibitor of type A monoamine oxidase, *Biochem. Pharmacol.*, 32, 1243, 1983.

116. **Mantle, T. J., Wilson, K., and Long, R. F.,** Studies on the selective inhibition of membrane-bound rat liver monoamine oxidase, *Biochem. Pharmacol.*, 24, 2031, 1975.

117. **Long, R. F., Mantle, T. J., and Wilson, K.,** Substrate-selective inhibition of monoamine oxidase by some cyclopropylamino substituted oxadiazoles, *Biochem. Pharmacol.*, 25, 247, 1976.

118. **Yamasaki, R. B. and Silverman, R. B.,** Mechanism for the reactivation of *N*-cyclopropylbenzylamine-inactivated monoamine oxidase by amines, *Biochemistry*, 24, 6543, 1985.

119. **Vazquez, M. L. and Silverman, R. B.,** Revised mechanism for inactivation of mitochondrial monoamine oxidase by *N*-cyclopropylbenzylamine, *Biochemistry*, 24, 6538, 1985.

120. **Silverman, R. B.,** Effect of α-methylation on inactivation of monoamine oxidase by *N*-cyclopropylbenzylamine, *Biochemistry*, 23, 5206, 1984.

121. **Silverman, R. B. and Hoffman, S. J.,** Syntheses of *N*-[1-^{2}H]- and *N*-[1-^{3}H]-cyclopropylbenzylamine and [phenyl-^{14}C]-*N*-cyclopropylbenzylamine, *J. Labelled Compds. Radiopharm.*, 18, 781, 1981.

122. **Guengerich, F. P., Willard, R. J., Shea, J. P., Richards, L. E., and Macdonald, T. L.,** Mechanism-based inactivation of cytochrome P-450 by heteroatom-substituted cyclopropanes and formation of ring-opened products, *J. Am. Chem. Soc.*, 106, 6446, 1984.

123. **Silverman, R. B. and Hoffman, S. J.,** *N*-(1-Methyl)cyclopropylbenzylamine: a novel inactivator of mitochondrial monoamine oxidase, *Biochem. Biophys. Res. Commun.*, 101, 1396, 1981.

124. **Silverman, R. B. and Zieske, P. A.,** Mechanism of inactivation of monoamine oxidase by 1-phenylcyclopropylamine, *Biochemistry*, 24, 2128, 1985.

125. **Silverman, R. B. and Zieske, P. A.,** Identification of the amino acid bound to the labile adduct formed during inactivation of monoamine oxidase by 1-phenylcyclopropylamine, *Biochem. Biophys. Res. Commun.*, 135, 154, 1986.

126. **Silverman, R. B. and Zieske, P. A.,** 1-Benzylcyclopropylamine and (1-phenylcyclopropyl)methylamine, an inactivator and a substrate of monoamine oxidase, *J. Med. Chem.*, 28, 1953, 1985.

127. **Silverman, R. B. and Zieske, P. A.,** 1-Phenylcyclobutylamine: the first in a new class of monoamine oxidase inactivators. Further evidence for a radical intermediate, *Biochemistry*, 25, 341, 1986.

128. **DeMatteis, F., Hollands, C., Gibbs, A. H., deSa, N., and Rizzardini, M.,** Inactivation of cytochrome P-450 and production of *N*-alkylated porphyrins caused in isolated hepatocytes by substituted dihydropyridines. Structural requirements for loss of haem and alkylation of the pyrrole nitrogen atom, *FEBS Lett.*, 145, 87, 1982.

129. **Augusto, O., Beilan, H. S., and Ortiz de Montellano, P. R.,** The catalytic mechanism of cytochrome P-450. Spin-trapping evidence for one-electron substrate oxidation, *J. Biol. Chem.*, 257, 11288, 1982.

130. **Marks, G. S., Allen, D. T., Johnson, C. T., Sutherland, E. P., Nakatsu, K., and Whitney, R. A.,** Suicidal destruction of cytochrome P-450 and reduction of ferrochelatase activity by 3,5-diethoxycarbonyl-1,4-dihydro-2,4,6-trimethylpyridine and its analogues in chick embryo liver cells, *Mol. Pharmacol.*, 27, 459, 1985.

130a. **McCluskey, S. A., Marks, G. S., Sutherland, E. P., Jacobsen, N., and Ortiz de Montellano, P. R.,** Ferrochelatase-inhibitory activity and *N*-alkylprotoporphyrin formation with analogues of 3,5-diethoxycarbonyl-1,4-dihydro-2,4,6-trimethylpyridine (DDC) containing extended 4-alkyl groups: implications for the active site of ferrochelatase, *Mol. Pharmacol.*, 30, 352, 1986.

131. **Ortiz de Montellano, P. R., Kunze, K. L., and Beilan, H. S.,** Chiral orientation of prosthetic heme in the cytochrome P-450 active site, *J. Biol. Chem.*, 258, 45, 1983.

132. **Kunze, K. L. and Ortiz de Montellano, P. R.,** *N*- Methylprotoporphyrin. IX. Identification by NMR of the nitrogen alkylated in each of the four isomers, *J. Am. Chem. Soc.*, 103, 4225, 1981.

132a. **Wimalasena, K. and May, S. W.,** Mechanistic studies on dopamine β-monooxygenase catalysis: *N*-dealkylation and mechanism-based inhibition by benzylic-nitrogen-containing compounds. Evidence for a single-electron-transfer mechanism, *J. Am. Chem. Soc.*, 109, 4036, 1987.

132b. **Silverman, R. B. and Banik, G. M.,** (Aminoalkyl)trimethylsilanes. A new class of monoamine oxidase inactivators, *J. Am. Chem. Soc.*, 109, 2219, 1987.

132c. **Silverman, R. B. and Vadnere, M. K.,** Aminoalkyltrimethylgermanes: the first organogermanium mechanism-based enzyme inactivators. A new class of monoamine oxidase inactivators, *Bioorg. Chem.*, 15, 328, 1987.

133. **Ghisla, S., Ogata, H., Massey, V., Schonbrunn, A., Abeles, R. H., and Walsh, C. T.,** Kinetic studies on the inactivation of L-lactate oxidase by [the acetylenic suicide substrate] 2-hydroxy-3-butynoate, *Biochemistry*, 15, 1791, 1976.

134. **Schonbrunn, A., Abeles, R. H., Walsh, C. T., Ghisla, S., Ogata, H., and Massey, V.,** The structure of the covalent flavin adduct formed between lactate oxidase and the suicide substrate 2-hydroxy-3-butynoate, *Biochemistry*, 15, 1798, 1976.

135. **Walsh, C. T., Abeles, R. H., and Kaback, H. R.,** Mechanisms of active transport in isolated bacterial membrane vesicles. X. Inactivation of D-lactate dehydrogenase and D-lactate dehydrogenase-coupled transport in *Escherichia coli* membrane vesicles by an acetylenic substrate, *J. Biol. Chem.*, 247, 7858, 1972.

136. **Olson, S. T., Massey, V., Ghisla, S., and Whitfield, C. D.,** Suicide inactivation of the flavoenzyme D-lactate dehydrogenase by α-hydroxybutynoate, *Biochemistry*, 18, 4724, 1979.

137. **Ghisla, S., Olson, S. T., Massey, V., and Lhoste, J-M.,** Structure of the flavin adduct formed in the suicide reaction of α-hydroxybutynoate with D-lactate dehydrogenase, *Biochemistry*, 18, 4733, 1979.

138. **Lederer, F.,** On the first steps of lactate oxidation of bakers' yeast L-(+)-lactate dehydrogenase (cytochrome b₂), *Eur. J. Biochem.*, 46, 393, 1974.

139. **Pompon, D. and Lederer, F.,** On the mechanism of flavin modification during inactivation of flavocytochrome b₂ from bakers' yeast by acetylenic substrates, *Eur. J. Biochem.*, 148, 145, 1985.

140. **Urban, P. and Lederer, F.,** Baker's yeast flavocytochrome b₂. A mechanistic study of the dehydrohalogenation reaction, *Eur. J. Biochem.*, 144, 345, 1984.

141. **Jewess, P. J., Kerr, M. W., and Whitaker, D. P.,** Inhibition of glycollate oxidase from pea leaves, *FEBS Lett.*, 53, 292, 1975.

142. **Fendrich, G. and Ghisla, S.,** Studies on glycollate oxidase from pea leaves. Determination of stereospecificity and mode of inhibition by α-hydroxybutynoate, *Biochim. Biophys. Acta*, 702, 242, 1982.

143. **Cromartie, T., Fisher, J., Kaczorowski, G., Laura, R., Marcotte, P., and Walsh, C.,** Synthesis of α-hydroxy-β-acetylenic acids and their oxidation by and inactivation of flavoprotein oxidases, *J. Chem. Soc. Chem. Commun.*, p. 597, 1974.

144. **Cromartie, T. H. and Walsh, C. T.,** Rat kidney L-α-hydroxy acid oxidase: isolation of enzyme with one flavine coenzyme per two subunits, *Biochemistry*, 14, 2588, 1975.

145. **Cromartie, T. H. and Walsh, C.,** Mechanistic studies on the rat kidney flavoenzyme L-α-hydroxy acid oxidase, *Biochemistry*, 14, 3482, 1975.

146. **Meyer, S. E. and Cromartie, T. H.,** Role of essential histidine residues in L-α-hydroxy acid oxidase from rat kidney, *Biochemistry*, 19, 1874, 1980.

147. **Verny, M. and Vessière, R.,** Transposition propargylique sur les halogéno-2-butyne-3-oates d'alcoyle, *Bull. Soc. Chim. Fr.*, p. 2210, 1967.

148. **Tobias, B., Covey, D. F., and Strickler, R. C.,** Inactivation of human placental 17β-estradiol dehydrogenase and 20α-hydroxysteroid dehydrogenase with active-site directed 17β-propynyl-substituted progestin analogs, *J. Biol. Chem.*, 257, 2783, 1982.

149. **Strickler, R. C., Covey, D. F., and Tobias, B.,** Study of 3α, 20β-hydroxysteroid dehydrogenase with an enzyme-generated affinity alkylator: dual enzyme activity at a single active site, *Biochemistry*, 19, 4950, 1980.

150. **Covey, D. F.,** Synthesis of 17β-[1S)-1-hydroxy-2-propynyl]- and 17β-[(1R)-1-hydroxy-2-propynyl]androst-4-en-3-one. Potential suicide substrates of 20α- and 20β-hydroxysteroid dehydrogenases, *Steroids*, 34, 199, 1979.

151. **Balasubramanian, V. and Robinson, C. H.,** Irreversible inactivation of mammalian Δ⁵-3β-hydroxysteroid dehydrogenases by 5,10-secosteroids. Enzymatic oxidation of allenic alcohols to the corresponding allenic ketones, *Biochem. Biophys. Res. Commun.*, 101, 495, 1981.

152. **Balasubramanian, V., McDermott, I. R., and Robinson, C. H.,** 4-Ethenylidene steroids as mechanism-based inactivators of 3β-hydroxysteroid dehydrogenases, *Steroids*, 40, 109, 1982.

153. **Thomas, J. L., LaRochelle, M. C., Covey, D. F., and Strickler, R. C.,** Inactivation of human placental 17β, 20α-hydroxysteroid dehydrogenase by 16-methylene estrone, an affinity alkylator enzymatically generated from 16-methylene estradiol-17β, *J. Biol. Chem.*, 258, 11500, 1983.

154. **Ringold, H. J. and Rosenkranz, G.,** 16-Methyl derivatives of estrone and estradiol, *Chem. Abst.*, 61, 5720h, 1964.

155. **Mincey, T., Tayrien, G., Mildvan, A. S., and Abeles, R. H.,** Presence of a flavin semiquinone in methanol oxidase, *Proc. Natl. Acad. Sci. U.S.A.*, 77, 7099, 1980.

156. **Mincey, T., Bell, J. A., Mildvan, A. S., and Abeles, R. H.,** Mechanism of action of methoxatin-dependent alcohol dehydrogenase, *Biochemistry*, 20, 7502, 1981.

157. **Parkes, C. and Abeles, R. H.,** Studies on the mechanism of action of methoxatin-requiring methanol dehydrogenase: reaction of enzyme with electron-acceptor dye, *Biochemistry*, 23, 6355, 1984.

157a. **Dijkstra, M., Frank, J., Jongejan, J. A., and Duine, J. A.,** Inactivation of quinoprotein alcohol dehydrogenases with cyclopropane-derived suicide substrates, *Eur. J. Biochem.*, 140, 369, 1984.

157b. **Ingraham, L. L., Corse, J., and Makower, B.,** Enzymatic browning of fruits. III. Kinetics of the reaction inactivation of polyphenoloxidase, *J. Am. Chem. Soc.*, 74, 2623, 1952.

157c. **Ingraham, L. L.,** Reaction-inactivation of polyphenoloxidase: temperature dependence, *J. Am. Chem. Soc.*, 76, 3777, 1954.

157d. **Ingraham, L. L.,** Reaction-inactivation of polyphenol oxidase: catechol and oxygen dependence, *J. Am. Chem. Soc.*, 77, 2875, 1955.

157e. **Tudela, J., García Cánovas, F., Varón, R., García Carmona, F., Gálvez, J., and Lozano, J. A.,** Transient-phase kinetics of enzyme inactivation induced by suicide substrates, *Biochim. Biophys. Acta*, 912, 408, 1987.

157f. **García Cánovas, F., Tudela, J., Martinez Madrid, C., Varón, R., García Carmona, F., and Lozano, J. A.,** Kinetic study on the suicide inactivation of tyrosinase induced by catechol, *Biochim. Biophys. Acta*, 912, 417, 1987.

157g. **Corey, E. J., d'Alarcao, M., and Matsuda, S. P. T.,** A new irreversible inhibitor of soybean lipoxygenase: relevance to mechanism, *Tetrahedron Lett.*, 27, 3585, 1986.

157h. **Corey, E. J., d'Alarcao, M., and Kyler, K. S.,** Synthesis of new lipoxygenase inhibitors 13-thia- and 10-thiaarachidonic acids, *Tetrahedron Lett.*, 26, 3919, 1985.

158. **Reiner, O. and Uehleke, H.,** Bindung von Tetrachlorkohlenstoff an reduziertes mikrosomales Cytochrom P-450 und an Häm, *Hoppe-Seyler's Z. Physiol. Chem.*, 352, 1048, 1971.

159. **Uehleke, H., Hellmer, K. H., and Tabarelli, S.,** Binding of ¹⁴C-carbon tetrachloride to microsomal proteins *in vitro* and formation of CHCl₃ by reduced liver microsomes, *Xenobiotica*, 3, 1, 1973.

160. **Uehleke, H. and Werner, T.,** A comparative study on the irreversible binding of labeled halothane trichlorofluoromethane, chloroform, and carbon tetrachloride to hepatic protein and lipids *in vitro* and *in vivo*, *Arch. Toxicol.*, 34, 289, 1975.

161. **Colby, H. D., Brogen, W. C., III, and Miles, P. R.,** Carbon tetrachloride-induced changes in adrenal microsomal mixed-function oxidases and lipid peroxidation, *Toxicol. Appl. Pharmacol.*, 60, 492, 1981.

162. **deGroot, H. and Haas, W.,** O₂-Independent damage of cytochrome P-450 by CCl₄-metabolites in hepatic microsomes, *FEBS Lett.*, 115, 253, 1980.

163. **Noll, T. and deGroot, H.,** The critical steady-state hypoxic conditions in carbon tetrachloride-induced lipid peroxidation in rat liver microsomes, *Biochim. Biophys. Acta*, 795, 356, 1984.

164. **Guzelian, P. S. and Swisher, R. W.,** Degradation of cytochrome P-450 haem by carbon tetrachloride and 2-allyl-2-isopropylacetamide in rat liver *in vivo* and *in vitro*. Involvement of non-carbon monoxide-forming mechanisms, *Biochem. J.*, 184, 481, 1979.

165. **Poli, G., Cheeseman, K., Slater, T. F., and Dianzani, M. U.,** The role of lipid peroxidation in CCl₄-induced damage to liver microsomal enzymes: comparative studies in vitro using microsomes and isolated liver cells, *Chem. Biol. Interact.*, 37, 13, 1981.

166. **deGroot, H. and Haas, W.,** Self-catalysed, O_2-independent inactivation of NADPH- or dithionite-reduced microsomal cytochrome P-450 by carbon tetrachloride, *Biochem. Pharmacol.*, 30, 2343, 1981.

166a. **Davies, H. W., Britt, S. G., and Pohl, L. R.,** Carbon tetrachloride and 2-isopropyl-4-pentenamide-induced inactivation of cytochrome P-450 leads to hemederived protein adducts, *Arch. Biochem. Biophys.*, 244, 387, 1986.

166b. **Labbe, G., Descatoire, V., Letteron, P., Degott, C., Tinel, M., Larrey, D., Carrion-Pavlov, Y., Geneve, J., Amouyal, G., and Pessayre, D.,** The drug methoxsalen, a suicide substrate for cytochrome P-450, decreases the metabolic activation, and prevents the hepatotoxicity, of carbon tetrachloride in mice, *Biochem. Pharmacol.*, 36, 907, 1987.

167. **Fernández, G., Villarruel, M. C., deToranzo, E. G. D., and Castro, J. A.,** Covalent binding of carbon tetrachloride metabolites to the heme moiety of cytochrome P-450 and its degradation products, *Res. Chem. Commun. Pathol. Pharmacol.*, 35, 283, 1982.

168. **Mansuy, D., Lange, M., Chottard, J.-C., Guerin, P., Morliere, P., Brault, D., and Rougee, M.,** Reaction of carbon tetrachloride with 5,10,15,20-tetraphenylporphinatoiron(II) [(TPP)Fe^{II}]: evidence for the formation of the carbene complex [(TPP)Fe^{II}(CCl$_2$)], *J. Chem. Soc. Chem. Commun.*, p. 648, 1977.

169. **Mansuy, D., Lange, M., Chottard, J. C., Bartoli, J. F., Chevrier, B., and Weiss, R.,** Dichlorocarbene complexes of iron(II)-porphyrins-crystal and molecular structure of Fe(TPP)(CCl$_2$)(H$_2$O), *Angew. Chem. Int. Ed. Engl.*, 17, 781, 1978.

170. **Lange, M. and Mansuy, D.,** N- Substituted porphyrins formation from carbene iron-porphyrin complexes: a possible pathway for cytochrome P-450 heme destruction, *Tetrahedron Lett.*, 22, 2561, 1981.

171. **Wolf, C. R., Mansuy, D., Nastainczyk, W., Deutschmann, G., and Ullrich, V.,** The reduction of polyhalogenated methanes by liver microsomal cytochrome P-450, *Mol. Pharmacol.*, 13, 698, 1977.

172. **Uehleke, H., Hellmer, K. H., and Tabarelli-Poplawski, S.,** Metabolic activation of halothane and its covalent binding to liver endoplasmic proteins in vitro, *Naunyn Schmiedeberg's Arch. Pharmacol.*, 279, 39, 1973.

173. **Mansuy, D., Nastainczyk, W., and Ullrich, V.,** The mechanism of halothane binding to microsomal cytochrome P-450, *Naunyn Schmiedeberg's Arch. Pharmacol.*, 285, 315, 1974.

174. **deGroot, H., Harnisch, U., and Noll, T.,** Suicidal inactivation of microsomal cytochrome P-450 by halothane under hypoxic conditions, *Biochem. Biophys. Res. Commun.*, 107, 885, 1982.

175. **Krieter, P. A. and Van Dyke, R. A.,** Cytochrome P-450 and halothane metabolism. Decrease in rat liver microsomal P-450 in vitro, *Chem. Biol. Interact.*, 44, 219, 1983.

176. **Mansuy, D. and Battioni, J.-P.,** Isolation of σ-alkyl-iron(III) or carbene-iron(II) complexes from reduction of polyhalogenated compounds by iron(II)-porphyrins: the particular case of halothane CF$_3$CHClBr, *J. Chem. Soc. Chem. Commun.*, p. 638, 1982.

176a. **McCall, S. N., Jurgens, P., and Ivanetich, K. M.,** Hepatic microsomal metabolism of the dichloroethanes, *Biochem. Pharmacol.*, 32, 207, 1983.

177. **Mansuy, D., Battioni, J.-P., Chottard, J.-C., and Ullrich, V.,** Preparation of a porphyrin-iron-carbene model for the cytochrome P-450 complexes obtained upon metabolic oxidation of the insecticide synergists of the 1,3-benzodioxole series, *J. Am. Chem. Soc.*, 101, 3971, 1979.

178. **Campbell, C. D. and Rees, C. W.,** Reactive intermediates. I. Synthesis and oxidation of 1- and 2-aminobenzotriazole, *J. Chem. Soc. C*, p. 742, 1969.

179. **Ortiz de Montellano, P. R. and Mathews, J. M.,** Autocatalytic alkylation of the cytochrome P-450 prosthetic haem group by 1-aminobenzotriazole. Isolation of an N,N-bridged benzyne-protoporphyrin IX adduct, *Biochem. J.*, 195, 761, 1981.

180. **Ortiz de Montellano, P. R., Mico, B. A., Mathews, J. M., Kunze, K. L., Miwa, G. T., and Lu, A. Y. H.,** Selective inactivation of cytochrome P-450 isozymes by suicide substrates, *Arch. Biochem. Biophys.*, 210, 717, 1981.

181. **Ortiz de Montellano, P. R., Mathews, J. M., and Langry, K. C.,** Autocatalytic inactivation of cytochrome P-450 and chloroperoxidase by 1-aminobenzotriazole and other aryne precursors, *Tetrahedron*, 40, 511, 1984.

181a. **Callot, H. J. and Cromer, R.,** Oxidative cyclization of substituted N-vinylporphyrin cobalt complexes. Synthesis of N,N'-(1,2-vinylidene) and N,N'-(1,2-phenylene)porphyrins, *Tetrahedron Lett.*, 26, 3357, 1985.

181b. **Callot, H. J., Cromer, R., Louati, A., and Gross, M.,** Synthesis of N,N'-phenyleneporphyrins. Models for cytochrome P-450-1-aminobenzotriazole inactivation products, *J. Chem. Soc. Chem. Commun.*, p. 767, 1986.

182. **Mathews, J. M., Dostal, L. A., and Bend, J. R.,** Inactivation of rabbit pulmonary cytochrome P-450 in microsomes and isolated perfused lungs by the suicide substrate 1-aminobenzotriazole, *J. Pharmacol. Exp. Ther.*, 235, 186, 1985.

183. **Feyereisen, R., Langry, K. C., and Ortiz de Montellano, P. R.,** Self-catalyzed destruction of insect cytochrome P-450, *Insect Biochem.*, 14, 19, 1984.

184. **Reichhart, D., Simon, A., Durst, F., Mathews, J. M., and Ortiz de Montellano, P. R.**, Autocatalytic inactivation of plant cytochrome P-450 enzymes: selective inactivation of cinnamic acid 4-hydroxylase from *Helianthus tuberosus* by 1-aminobenzotriazole, *Arch. Biochem. Biophys.*, 216, 522, 1982.

184a. **Mathews, J. M. and Bend, J. R.**, *N*-Alkylaminobenzotriazoles as isozyme-selective suicide inhibitors of rabbit pulmonary microsomal cytochrome P-450, *Mol. Pharmacol.*, 30, 25, 1986.

185. **Ortiz de Montellano, P. R. and Mathews, J. M.**, Inactivation of hepatic cytochrome P-450 by a 1,2,3-benzothiadiazole insecticide synergist, *Biochem. Pharmacol.*, 30, 1138, 1981.

186. **Horiike, K., Nishina, Y., Miyake, Y., and Yamano, T.**, Affinity labeling of D-amino acid oxidase with an acetylenic substrate, *J. Biochem.*, 78, 57, 1975.

187. **Marcotte, P. and Walsh, C.**, Properties of D-amino acid oxidase covalently modified upon its oxidation of D-propargylglycine, *Biochemistry*, 17, 2864, 1978.

188. **Marcotte, P. and Walsh, C.**, Sequence of reactions which follows enzymatic oxidation of propargylglycine, *Biochemistry*, 17, 5613, 1978.

188a. **Salowe, S. P., Ator, M. A., and Stubbe, J.**, Products of the inactivation of ribonucleoside diphosphate reductase from *Escherichia coli* with 2'-azido-2'-deoxyuridine 5'-diphosphate, *Biochemistry*, 26, 3408, 1987.

189. **Tang, S-S., Trackman, P. C., and Kagan, H. M.**, Reaction of aortic lysyl oxidase with β-aminopropionitrile, *J. Biol. Chem.*, 258, 4331, 1983.

190. **Carbon, J. A., Burkard, W. P., and Zeller, E. A.**, Über die Wirkung von symmetrischen 1,2-dialkyl-hydrazinen auf Carbonylverbindungen und Amin-oxydasen, *Helv. Chim. Acta*, 41, 1883, 1958.

191. **Green, A. L.**, Studies on the mechanism of inhibition of monoamine oxidase by hydrazine derivatives, *Biochem. Pharmacol.*, 13, 249, 1964.

192. **Patek, D. R. and Hellerman, L.**, Mitochondrial monoamine oxidase. Mechanism of inhibition by phenylhydrazine and by aralkylhydrazines. Role of enzymatic oxidation, *J. Biol. Chem.*, 249, 2373, 1974.

193. **Roth, J. A., Eddy, B. J., Pearce, L. B., and Mulder, K. B.**, Phenylhydrazine: selective inhibition of human brain type B monoamine oxidase, *Biochem. Pharmacol.*, 30, 945, 1981.

194. **Tipton, K. F.**, Inhibition of monoamine oxidase by substituted hydrazines, *Biochem. J.*, 128, 913, 1972.

195. **Nagy, J., Kenney, W. C., and Singer, T. P.**, The reaction of phenylhydrazine with trimethylamine dehydrogenase and with free flavins, *J. Biol. Chem.*, 254, 2684, 1979.

196. **Colby, J. and Zatman, L. J.**, Purification and properties of the trimethylamine dehydrogenase of bacterium 4B6, *Biochem. J.*, 143, 555, 1974.

197. **Suzuki, H., Ogura, Y., and Yamada, H.**, Stoichiometry of the reaction by amine oxidase from *Aspergillus niger*, *J. Biochem.*, 69, 1065, 1971.

198. **Falk, M. C.**, Stoichiometry of phenylhydrazine inactivation of pig plasma amine oxidase, *Biochemistry*, 22, 3740, 1983.

198a. **Fitzpatrick, P. F. and Villafranca, J. J.**, The mechanism of inactivation of dopamine β-hydroxylase by hydrazines, *J. Biol. Chem.*, 261, 4510, 1986.

199. **Allison, W. S., Swain, L. C., Tracy, S. M., and Benitez, L. V.**, The inactivation of lactoperoxidase and the acyl phosphatase activity of oxidized glyceraldehyde-3-phosphate dehydrogenase by phenylhydrazine and phenyldiimide, *Arch. Biochem. Biophys.*, 155, 400, 1973.

200. **Ortiz de Montellano, P. R. and Kerr, D. E.**, Inactivation of catalase by phenylhydrazine. Formation of a stable aryl-iron heme complex, *J. Biol. Chem.*, 258, 10558, 1983.

200a. **Ator, M. A. and Ortiz de Montellano, P. R.**, Protein control of prosthetic heme reactivity. Reaction of substrates with the heme edge of horseradish peroxidase, *J. Biol. Chem.*, 262, 1542, 1987.

201. **Jonen, H. G., Werringloer, J., Prough, R. A., and Estabrook, R. W.**, The reaction of phenylhydrazine with microsomal cytochrome P-450. Catalysis of heme modification, *J. Biol. Chem.*, 257, 4404, 1982.

201a. **Lerner, H. R., Harel, E., Lehman, E., and Mayer, A. M.**, Phenylhydrazine, a specific irreversible inhibitor of catechol oxidase, *Phytochemistry*, 10, 2637, 1971.

202. **Battioni, P., Mahy, J. P., Deleforge, M., and Mansuy, D.**, Reaction of monosubstituted hydrazines and diazenes with rat liver cytochrome P-450. Formation of ferrous-diazene and ferric σ-alkyl complexes, *Eur. J. Biochem.*, 134, 241, 1983.

203. **Moloney, S. J., Snider, B. J., and Prough, R. A.**, The interactions of hydrazine derivatives with rat-hepatic cytochrome P-450, *Xenobiotica*, 14, 803, 1984.

204. **Roth, J. A.**, Benzylhydrazine — a selective inhibitor of human and rat brain monoamine oxidase, *Biochem. Pharmacol.*, 28, 729, 1979.

205. **Clineschmidt, B. V. and Horita, A.**, The monoamine oxidase catalyzed degradation of phenelzine-1-^{14}C, an irreversible inhibitor of monoamine oxidase-I. Studies *in vitro*, *Biochem. Pharmacol.*, 18, 1011, 1969.

206. **Collins, G. G. S. and Youdim, M. B. H.**, The binding of ^{14}C]phenethylhydrazine to rat liver monoamine oxidase, *Biochem. Pharmacol.*, 24, 703, 1975.

207. **Muakkassah, S. F. and Yang, W. C. T.**, Mechanism of the inhibitory action of phenelzine on microsomal drug metabolism, *J. Pharmacol. Exp. Ther.*, 219, 147, 1981.

208. **Ortiz de Montellano, P. R., Augusto, O., Viola, F., and Kunze, K. L.,** Carbon radicals in the metabolism of alkyl hydrazines, *J. Biol. Chem.,* 258, 8623, 1983.

208a. **Numata, Y., Takei, T., and Hayakawa, T.,** Hydralazine as an inhibitor of lysyl oxidase activity, *Biochem. Pharmacol.,* 30, 3125, 1981.

208b. **Lyles, G. A. and Callingham, B. A.,** Hydralazine is an irreversible inhibitor of the semicarbazide-sensitive, clorgyline-resistant amine oxidase in rat aorta homogenates, *J. Pharm. Pharmacol.,* 34, 139, 1982.

208c. **Johnson, C., Stubley-Beedham, C., and Stell, G. P.,** Hydralazine: a potent inhibitor of aldehyde oxidase activity *in vitro* and *in vivo, Biochem. Pharmacol.,* 34, 4251, 1985.

209. **Davison, A. N.,** The mechanism of the irreversible inhibition of rat-liver monoamine oxidase by iproniazid (Marsilid), *Biochem. J.,* 67, 316, 1957.

210. **Seiden, L. S. and Westley, J.,** Mechanism of iproniazid inhibition of brain monoamine oxidase, *Arch. Intern. Pharmacodyn.,* 146, 145, 1963.

211. **Smith, T. E., Weissbach, H., and Udenfriend, S.,** Studies on monoamine oxidase: the mechanism of inhibition of monoamine oxidase by iproniazid, *Biochemistry,* 2, 746, 1963.

212. **Hidaka, H. and Udenfriend, S.,** Evidence of a hydrazine-reactive group of the active site of the nonheme portion of horseradish peroxidase, *Arch. Biochem. Biophys.,* 140, 174, 1970.

213. **Hidaka, H., Udenfriend, S., Nagasaka, A., and DeGroot, L. J.,** Inhibition of thyroid iodide peroxidase *in vivo* and *in vitro* by iproniazid, *Biochem. Biophys. Res. Commun.,* 40, 103, 1970.

214. **Muakkassah, S. F., Bidlack, W. R., and Yang, W. C. T.,** Mechanism of the inhibitory action of isoniazid on microsomal drug metabolism, *Biochem. Pharmacol.,* 30, 1651, 1981.

215. **Muakkassah, S. F., Bidlack, W. R., and Yang, W. C. T.,** Reversal of the effects of isoniazid on hepatic cytochrome P-450 by potassium ferricyanide, *Biochem. Pharmacol.,* 31, 249, 1982.

216. **Augusto, O. and Ortiz de Montellano, P. R.,** Spin-trapping of free radicals formed during microsomal metabolism of ethylhydrazine and acetylhydrazine, *Biochem. Biophys. Res. Commun.,* 101, 1324, 1981.

217. **Hines, R. N. and Prough, R. A.,** The characterization of an inhibitory complex formed with cytochrome P-450 and a metabolite of 1,1-disubstituted hydrazines, *J. Pharmacol. Exp. Ther.,* 214, 80, 1980.

218. **Mansuy, D., Battioni, P., and Mahy, J. P.,** Isolation of an iron-nitrene complex from the dioxygen and iron porphyrin dependent oxidation of a hydrazine, *J. Am. Chem. Soc.,* 104, 4487, 1982.

219. **Eberson, L. E. and Persson, K.,** Studies on monoamine oxidase inhibitors. I. The autoxidation of β-phenylisopropylhydrazine as a model reaction for irreversible monoamine oxidase inhibition, *J. Med. Pharm. Chem.,* 5, 738, 1962.

220. **Misra, H. P. and Fridovich, I.,** The oxidation of phenylhydrazine: superoxide and mechanism, *Biochemistry,* 15, 681, 1976.

221. **Itano, H. A. and Matteson, J. L.,** Mechanism of initial reaction of phenylhydrazine with oxyhemoglobin and effect of ring substitutions on the bimolecular rate constant of this reaction, *Biochemistry,* 21, 2421, 1982.

222. **Ortiz de Montellano, P. R. and Kunze, K. L.,** Formation of *N*-phenylheme in the hemolytic reaction of phenylhydrazine with hemoglobin, *J. Am. Chem. Soc.,* 103, 6534, 1981.

223. **Saito, S. and Itano, H. A.,** β-*meso*-Phenylbiliverdin IXα and *N*-phenylprotoporphyrin IX, products of the reaction of phenylhydrazine with oxyhemoproteins, *Proc. Natl. Acad. Sci. U.S.A.,* 78, 5508, 1981.

224. **Hill, H. A. O. and Thornalley, P. J.,** Phenyl radical production during the oxidation of phenylhydrazine and in phenylhydrazine-induced haemolysis, *FEBS Lett.,* 125, 235, 1981.

225. **Augusto, O., Kunze, K. L., and Ortiz de Montellano, P. R.,** *N*-Phenylprotoporphyrin IX formation in the hemoglobin-phenylhydrazine reaction. Evidence for a protein-stabilized iron-phenyl intermediate, *J. Biol. Chem.,* 257, 6231, 1982.

226. **Ortiz de Montellano, P. R., Kunze, K. L., and Augusto, O.,** Hemoprotein destruction. Iron-nitrogen shift of a phenyl group in a porphyrin complex, *J. Am. Chem. Soc.,* 104, 3545, 1982.

227. **Mansuy, D., Battioni, P., Mahy, J.-P., and Gillet, G.,** Comparison of the hemoglobin reactions with methyl- and phenyl-hydrazine: intermediate formation of a hemoglobin Fe(II)-methyldiazine complex, *Biochem. Biophys. Res. Commun.,* 106, 30, 1982.

228. **Itano, H. A., Hirota, K., and Vedvick, T. S.,** Ligands and oxidants in ferrihemochrome formation and oxidative hemolysis *Proc. Natl. Acad. Sci. U.S.A.,* 74, 2556, 1977.

229. **Kunze, K. L. and Ortiz de Montellano, P. R.,** Formation of a σ-bonded aryliron complex in the reaction of arylhydrazines with hemoglobin and myoglobin, *J. Am. Chem. Soc.,* 105, 1380, 1983.

230. **Ringe, D., Petsko, G. A., Kerr, D. E., and Ortiz de Montellano, P. R.,** Reaction of myoglobin with phenylhydrazine: a molecular doorstop, *Biochemistry,* 23, 2, 1984.

231. **Ortiz de Montellano, P. R. and Kerr, D. E.,** Inactivation of myoglobin by ortho-substituted arylhydrazines. Formation of prosthetic heme aryl-iron but not *N*-aryl adducts, *Biochemistry,* 24, 1147, 1985.

232. **Mansuy, D., Battioni, J.-P., Dupré, D., and Sartori, E.,** Reversible iron-nitrogen migration of allyl, aryl, or vinyl groups in iron porphyrins: a possible passage between σ Fe^{III}(porphyrin)(R) and Fe^{II}(N-R)(porphyrin) complexes, *J. Am. Chem. Soc.,* 104, 6159, 1982.

233. **Battioni, P., Mahy, J. P., Gillet, G., and Mansuy, D.,** Iron porphyrin dependent oxidation of methyl- and phenylhydrazine: isolation of iron(II)-diazene and σ-alkyliron(III) (or aryliron (III)) complexes. Relevance to the reactions of hemoproteins with hydrazines, *J. Am. Chem. Soc.,* 105, 1399, 1983.

234. **Danishefsky, S.,** Electrophilic cyclopropanes in organic synthesis, *Acc. Chem. Res.,* 12, 66, 1979.

235. **Porter, D. J. T. and Bright, H. J.,** Active site chlorination of D-amino acid oxidase by *N*-chloro-D-leucine, *J. Biol. Chem.,* 251, 6150, 1976.

236. **Rudie, N. G., Porter, D. J. T., and Bright, H. J.,** Chlorination of an active site tyrosyl residue in D-amino acid oxidase by *N*-chloro-D-leucine, *J. Biol. Chem.,* 255, 498, 1980.

237. **Ronchi, S., Galliano, M., Minchiotti, L., Curti, B., Rudie, N. R., Porter, D. J. T., and Bright, H. J.,** An active site tyrosine-containing heptapeptide from D-amino acid oxidase, *J. Biol. Chem.,* 255, 6044, 1980.

238. **Kaczorowski, G., Kaback, H. R. and Walsh, C.,** Reversible inactivation of vectorial phosphorylation of hydroxybutynoate in *Escherichia coli* membrane vesicles, *Biochemistry,* 14, 3903, 1975.

239. **Walsh, C. T. and Kaback, H. R.,** Vinylglycolic acid. An inactivator of the phosphoenolpyruvate-phosphate transferase system in *Escherichia coli, J. Biol. Chem.,* 248, 5456, 1973.

240. **Strickler, R. C., Covey, D.F. and Tobias, B.,** Study of 3α, 20β-hydroxysteroid dehydrogenase with an enzyme-generated affinity alkylator: dual enzyme activity at a single active site, *Biochemistry,* 19, 4950, 1980.

240a. **Covey, D. F., McMullan, P. C., Weaver, A. J., and Chien, W. W.,** Inactivation of *Streptomyces hydrogenans* 20β-hydroxysteroid dehydrogenase by an enzyme-generated ethoxyacetylenic ketone in the presence of a thiol scavenger, *Biochemistry* 25, 7288, 1986.

241. **Rando, R. R.,** *In situ* generation of irreversible enzyme inhibitors, *Nature (London) New Biol.,* 237, 53, 1972.

242. **Rando, R. R.,** Allyl alcohol-induced irreversible inhibition of yeast alcohol dehydrogenase, *Biochem. Pharmacol.,* 23, 2328, 1974.

243. **Synder, M. A., Kaczorowski, G. J., Barnes, E. M., Jr., and Walsh, C.,** Inactivation of the phosphoenolpyruvate-dependent phosphotransferase system in various species of bacteria by vinylglycolic acid, *J. Bacteriol.,* 127, 671, 1976.

244. **Thelander, L., Larsson, B., Hobbs, J., and Eckstein, F.,** Active site of ribonucleoside diphosphate reductase from *Escherichia coli.* Inactivation of the enzyme by 2'-substituted ribonucleoside diphosphates, *J. Biol. Chem.,* 251, 1398, 1976.

245. **Stubbe, J. and Kozarich, J. W.,** Inorganic pyrophosphate is released from 2'-chloro-2'-deoxyuridine 5'-diphosphate by ribonucleoside diphosphate reductase, *J. Am. Chem. Soc.,* 102, 2505, 1980.

246. **Stubbe, J. and Ackles, D.,** On the mechanism of ribonucleoside diphosphate reductase from *Escherichia coli.* Evidence for 3'-C-H bond cleavage, *J. Biol. Chem.,* 255, 8027, 1980.

247. **Stubbe, J., Ator, M., and Krenitsky, T.,** Mechanism of ribonucleoside diphosphate reductase from *Escherichia coli.* Evidence for 3'-C-H bond cleavage, *J. Biol. Chem.,* 258, 1625, 1983.

248. **Stubbe, J.,** Mechanism of B_{12}-dependent ribonucleotide reductase, *Mol. Cell. Biochem.,* 50, 25, 1983.

249. **Harris, G., Ator, M., and Stubbe, J.,** Mechanism of inactivation of *Escherichia coli* and *Lactobacillus leichmannii* ribonucleotide reductases by 2'-chloro-2'-deoxynucleotides: evidence for generation of 2-methylene-3(2H)-furanone, *Biochemistry,* 23, 5214, 1984.

250. **Ator, M. A. and Stubbe, J.,** Mechanism of inactivation of *Escherichia coli* ribonucleotide reductase by 2'-chloro-2'-deoxyuridine 5'-diphosphate: evidence for generation of a 2'-deoxy-3'ketonucleotide via a net 1,2 hydrogen shift, *Biochemistry,* 24, 7214, 1985.

250a. **Ator, M. A., Stubbe, J., and Spector, T.,** Mechanism of ribonucleotide reductase from herpes simplex virus type I. Evidence for 3' carbon-hydrogen bond cleavage and inactivation by nucleotide analogs, *J. Biol. Chem.,* 261, 3595, 1986.

251. **Stubbe, J., Ackles, D., Segal, R., and Blakley, R. L.,** On the mechanism of ribonucleoside triphosphate reductase from *Lactobacillus leichmannii.* Evidence for 3'C-H bond cleavage, *J. Biol. Chem.,* 256, 4843, 1981.

252. **Stubbe, J., Smith, G., and Blakley, R. L.,** Interaction of 3'-[³H]2'-chloro-2'-deoxyuridine 5'-triphosphate with ribonucleotide reductase from *Lactobacillus leichmannii, J. Biol. Chem.,* 258, 1619, 1983.

252a. **Harris, G., Ashley, G. W., Robins, M. J., Tolman, R. L., and Stubbe, J.,** 2'-Deoxy-2'-halonucleotides as alternate substrates and mechanism-based inactivators of *Lactobacillus leichmannii* ribonucleotide reductase, *Biochemistry,* 26, 1895, 1987.

253. **Stubbe, J. and Kozarich, J. W.,** Fluoride, pyrophosphate, and base release from 2'-deoxy-2'-fluoronucleoside 5'-diphosphates by ribonucleoside-diphosphate reductase, *J. Biol. Chem.,* 255, 5511, 1980.

254. **Mengel, R. and Guschlbauer, W.,** A simple synthesis of 2'deoxy-2'-fluorocytidine by nucleophilic substitution of 2,2'-anhydrocytidine with potassium fluoride/crown ether, *Angew. Chem. Int. Ed. Engl.,* 17, 525, 1978.

255. **Alston, T. A., Mela, L., and Bright, H. J.,** Inactivation of alcohol dehydrogenase by 3-butyn-1-ol, *Arch. Biochem. Biophys.,* 197, 516, 1979.

255a. **Auchus, R. J. and Covey, D. F.,** Mechanism-based inactivation of 17β, 20α-hydroxysteroid dehydrogenase by an acetylenic secoestradiol, *Biochemistry,* 25, 7295, 1986.

255b. **Auchus, R. J. and Covey, D. F.,** Dehydrogenase inactivation by an enzyme-generated acetylenic ketone: identification of a lysyl enaminone by ^{13}C NMR, *J. Am. Chem. Soc.,* 109, 280, 1987.

256. **Sjöberg, B.-M., Gräslund, A., Eckstein, F.,** A substrate radical intermediate in the reaction between ribonucleotide reductase from *Escherichia coli* and 2'-azido-2'-deoxynucleoside diphosphates, *J. Biol. Chem.,* 258, 8060, 1983.

257. **Ator, M., Salowe, S. P., Stubbe, J., Emptage, M. H., and Robins, M. J.,** 2'-Azido-2'-deoxynucleotide interaction with *E. coli* ribonucleotide reductase: generation of a new radical species, *J. Am. Chem. Soc.,* 106, 1886, 1984.

258. **Patel, J. M., Wolf, C. R., and Philpot, R. M.,** Interaction of 4-methylbenzaldehyde with rabbit pulmonary cytochrome P-450 in the intact animal, microsomes, and purified systems. Destructive and protection reactions, *Biochem. Pharmacol.,* 28, 2031, 1979.

259. **Franklin, M. R.,** Inhibition of hepatic oxidative xenobiotic metabolism by piperonyl butoxide, *Biochem. Pharmacol.,* 21, 3287, 1972.

260. **Philpot, R. M. and Hodgson, E.,** The effect of piperonyl butoxide concentration on the formation of cytochrome P-450 difference spectra in hepatic microsomes from mice, *Mol. Pharmacol.,* 8, 204, 1972.

261. **Schorstein, D. and Suckling, C. J.,** Latent inhibition of liver alcohol dehydrogenase by a substituted allyl alcohol, *J. Chem. Soc. Chem. Commun.,* p. 795, 1978.

262. **MacInnes, I., Schorstein, D. E., Suckling, C. J., and Wrigglesworth, R.,** Latent inhibitors. II. Allylic inhibitors of alcohol dehydrogenase, *J. Chem. Soc. Perkin Trans. I,* p. 1103, 1981.

263. **Hershfield, M. S.,** Apparent suicide inactivation of human lymphoblast *S*-adenosylhomocysteine hydrolase by 2'-deoxyadenosine and adenine arabinoside. A basis for direct tonic effects of analogs of adenosine, *J. Biol. Chem.,* 254, 22, 1979.

264. **Abeles, R. H., Tashjian, A. H., Jr., and Fish, S.,** The mechanism of inactivation of *S*-adenosylhomocysteinase by 2'-deoxyadenosine, *Biochem. Biophys. Res. Commun.,* 95, 612, 1980.

265. **Abeles, R. H., Fish, S., and Lapinskas, B.,** *S*- Adenosylhemocysteinase: mechanism of inactivation by 2'-deoxyadenosine and interaction with other nucleosides, *Biochemistry,* 21, 5557, 1982.

266. **Guranowski, A., Montgomery, J. A., Cantoni, G. L., and Chiang, P. K.,** Adenosine analogues as substrates and inhibitors of *S*-adenosylhomocysteine hydrolase, *Biochemistry,* 20, 110, 1981.

267. **Chiang, P. K., Guranowski, A., and Segall, J. E.,** Irreversible inhibition of *S*-adenosylhomocysteine hydrolase by nucleoside analogs, *Arch. Biochem. Biophys.,* 207, 175, 1981.

268. **Kajander, E. O. and Raina, A. M.,** Affinity-chromatographic purification of *S*-adenosyl-L-homocysteine hydrolase. Some properties of the enzyme from rat liver, *Biochem. J.,* 193, 503, 1981.

269. **Kajander, E. O.,** Inactivation of liver *S*-adenosylhomocysteine hydrolase *in vitro* of rats treated with *erythro*-9-(2-hydroxynon-3-yl)adenine, *Biochem. J.,* 205, 585, 1982.

270. **Helland, S. and Ueland, P. M.,** Interaction of 9-β-D-arabionofuranosyladenine, 9-β-D-arabionofurano-syladenine 5'-monophosphate, and 9-β-D-arabinofuranosyladenine 5'-triphosphate with *S*-adenosylhomo-cysteinase, *Cancer Res.,* 41, 673, 1981.

271. **White, E. L., Shaddix, S. C., Brockman, R. W., and Bennett, L. L., Jr.,** Comparison of the actions of 9-β-D-arbinofuranosyl-2-fluoroadenine and 9-β-D-arabinofuranosyladenine on target enzymes from mouse tumor cells, *Cancer Res.,* 42, 2260, 1982.

272. **Kim, I.-Y., Zhang, C.-Y., Cantoni, G. L., Montgomery, J. A., and Chiang, P. K.,** Inactivation of *S*-adenosylhomocysteine hydrolase by nucleosides, *Biochim. Biophys. Acta,* 829, 150, 1985.

273. **Borchardt, R. T., Keller, B. T., and Patel-Thombre, U.,** Neplanocin A, a potent inhibitor of *S*-adenosylhomocysteine hydrolase and of vaccina virus multiplication in mouse L929 cells, *J. Biol. Chem.,* 259, 4353, 1984.

274. **Wolfson, G., Chisholm, J., Tashjian, A. H., Jr., Fish, S., and Abeles, R. H.,** Neplanocin A, actions on *S*-adenosylhomocysteine hydrolase and on hormone synthesis by GH_4C_1 cells, *J. Biol. Chem.,* 261, 4492, 1986.

275. **Matuszewska, B. and Borchardt, R. T.,** The role of nicotinamide adenine dinucleotide in the inhibition of bovine liver *S*-adenosylhomocysteine hydrolase by neplanocin A, *J. Biol. Chem.,* 262, 265, 1987.

276. **Yaginuma, S., Muto, N., Tsujino, M., Sudate, Y., Hayashi, M., and Otani, M.,** Studies on neplanocin A, new antitumor antibiotic. I. Producing organism, isolation, and characterization, *J. Antibiot.,* 34, 359, 1981.

277. **Lim, M.-I. and Marquez, V. E.,** Total synthesis of (−)-neplanocin A, *Tetrahedron Lett.,* 24, 5559, 1983.

278. **Arita, M., Adachi, K., Ito, Y., Sawai, H., and Ohno, M.,** Enantioselective synthesis of the carbocyclic nucleosides (−)-aristeromycin and (−)-neplanocin A by a chemicoenzymatic approach, *J. Am. Chem. Soc.,* 105, 4049, 1983.

Chapter 10

OXYGENATION REACTIONS*

I. INTRODUCTION

Oxygenation inactivation reactions comprise a subclass of oxidation reactions in which at least one oxygen atom is incorporated into the inactivator. The largest section in this chapter, Oxygenation/Addition, is divided according to acetylene-containing compounds, then allene- and alkene-containing compounds. The compounds are arranged in increasing complexity within each section.

II. OXYGENATION/ADDITION

A. Oxygenation/Addition

Ivanetich et al.[1] studied the effects of more than 50 organic compounds on the levels of hepatic microsomal cytochrome P-450, cytochrome b_5, NADPH-cytochrome C reductase, and heme from phenobarbital-treated rats. Many alkyl compounds containing allyl, alkynyl, halo, nitrilo, or nitro functional groups degrade the heme moiety of cytochrome P-450 without affecting the other enzymes; these reactions require NADPH. The compounds appear to require metabolic activation by the cytochrome P-450, but it is not clear if the reactive species diffuse from the enzyme that generates them.

1. Acetylene-Containing Inactivators
a. Acetylene

Ammonia mono-oxygenase from the nitrifying bacterium *Nitrosomonas europaea* is inactivated in a time-dependent reaction by acetylene; inactivation requires O_2.[2] When cells are treated with [^{14}C]acetylene, a single membrane protein with an apparent M_r of 28,000 is labeled. No mechanism was given, but, since this enzyme is known to catalyze a variety of organic oxidations similar to that of cytochrome P-450, the mechanisms may be the same (see Scheme 1). Acetylene is a time-dependent inactivator of liver cytochrome P-450 from phenobarbital-treated rats; NADPH and O_2 are required for inactivation.[3,4] Isolation of the alkylated heme by Kunze et al.[5] established that attachment occurs to at least two different pyrrole nitrogens. The structure of the acetylene-heme adduct was determined by Ortiz de Montellano et al.[6] to be *N*-(2-oxoethyl)protoporphyrin IX. Three mechanistic possibilities can be considered (Scheme 1).

---

* A list of abbreviations and shorthand notations can be found prior to Chapter 7.

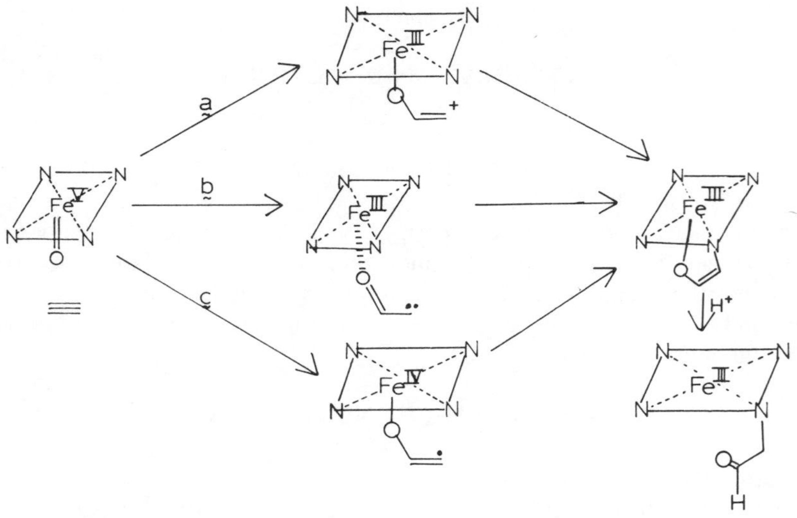

Scheme 1.

b. Monosubstituted Arylacetylenes

The generality of the reaction of mono- and disubstituted acetylenes with microsomal cytochrome P-450 from liver of phenobarbital-treated rats was established by Ortiz de Montellano and Kunze.[4] Since the carbon adjacent to the acetylene can be disubstituted, the mechanism cannot involve propargylic hydrogen atom abstraction. The adjacent carbon need not be substituted with a hydroxyl, so oxidation to ethynyl ketones is not involved in inactivation. Concomitant with inactivation by the monosubstituted acetylenes is destruction of the heme and formation of a green pigment. The pigment is not produced by disubstituted acetylenes, even though cytochrome P-450 is inactivated. Isolation and mass spectrometric analysis of the pigments produced from terminal acetylenes indicate that a 1:1 covalent alkylation of the protoporphyrin IX skeleton of the heme is the pigment. In all cases, the molecular ion equals the sum of the dimethyl ester of protoporphyrin IX (the ester is formed during work-up) plus the acetylene plus 16; presumably an oxygen atom is incorporated from O_2 as is the case with other substrates. The mechanism preferred to account for these results is shown in Scheme 2, although oxene formation is a possibility (Scheme 3). An experiment was carried out by Ortiz de Montellano and Kunze[7] to differentiate two possible mechanisms for terminal acetylene oxygenation by liver microsomes of phenobarbital-treated rats (Scheme 4). [1-^{2}H]Biphenylacetylene was used as the substrate and 2-biphenylacetic acid was isolated as the product (as its methyl ester); mass and NMR spectral analyses showed the retention of the deuterium atom in the methylene group. The same results were obtained by chemical oxidation (*m*-chloroperoxybenzoic acid) of the deuterated acetylene. Pathway a (Scheme 4), therefore, is supported, and this provides evidence for the competitive pathway of oxygenation and enzyme inactivation catalyzed by cytochrome P-450.

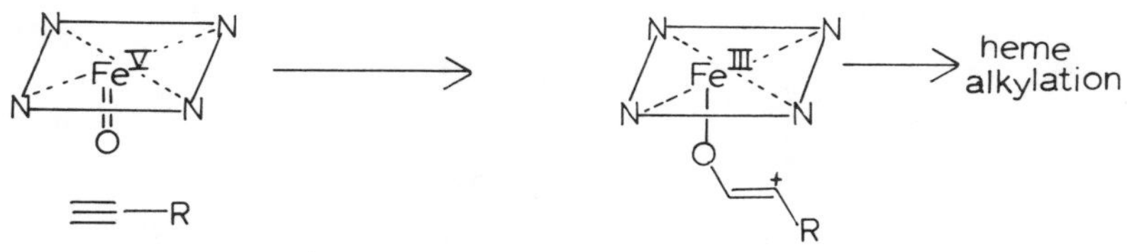

Scheme 2.

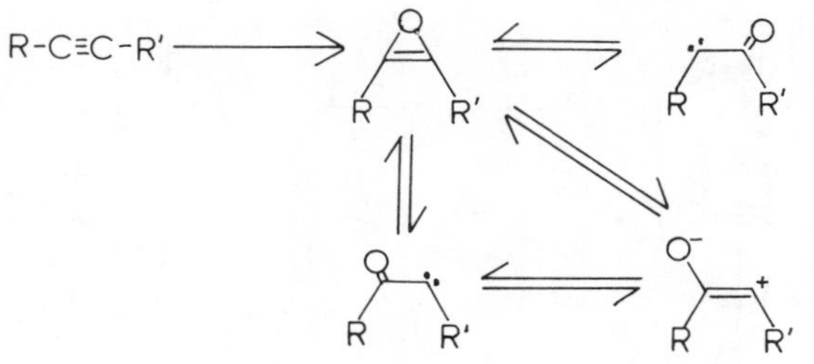

Scheme 3.

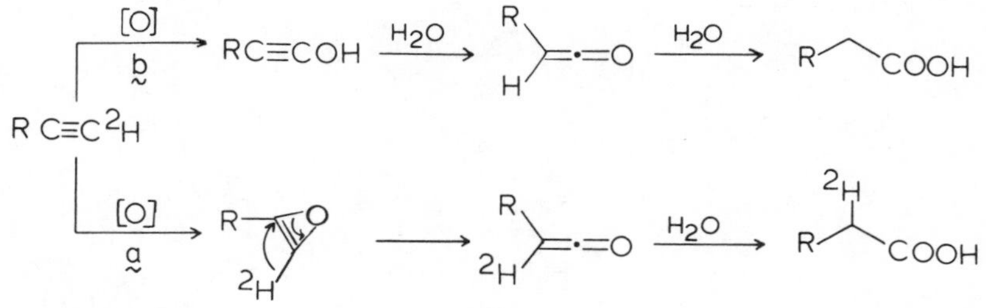

Scheme 4.

Further oxidation studies by Ortiz de Montellano and Kunze[8] with 1-[^{13}C]biphenylacetylene yielded methyl 2-biphenylacetate (after esterification) in which the biphenyl group is still attached to the labeled carbon. Therefore, it is the terminal hydrogen, and not the biphenyl group, that undergoes a 1,2-shift during cytochrome P-450-catalyzed oxidation (see Scheme 4). In order to investigate a concerted vs. nonconcerted mechanism for cytochrome P-450-catalyzed oxygenation of acetylenes, the reaction and inactivation of the enzyme were determined by Ortiz de Montellano and Komives[9] with deuterium-labeled phenylacetylene and biphenylacetylene, and by following the incorporation of [^{18}O] from ^{18}O$_2$ into products and inactivated enzyme. Oxidation of [1-^{2}H]phenylacetylene by liver microsomal cytochrome P-450 from phenobarbital-treated rats and by *m*-chloroperoxybenzoic acid gives phenylacetic acid with the acetylenic deuterium shifted to the benzylic carbon. The oxygen atom from ^{18}O$_2$ is bound to the terminal carbon in the metabolites (pathway a, Scheme 5), but to the internal carbon in the heme adducts (pathway b, Scheme 5); no hydrogen migration occurs with heme alkylation. Whereas there is a deuterium isotope effect of 1.8 for conversion of phenylacetylene to phenylacetic acid, there is no isotope effect on enzyme inactivation. Therefore, the mechanisms for these two processes must diverge prior to the point where the hydrogen shifts. This rules out a ketene as the reactive intermediate. An α-ketocarbene was ruled out, since the corresponding carbene from phenylacetylene oxidation was prepared and shown to be stable enough to be detected under the experimental conditions, yet it was not observed during inactivation. Two possible mechanisms that could account for these data are shown in Scheme 6.

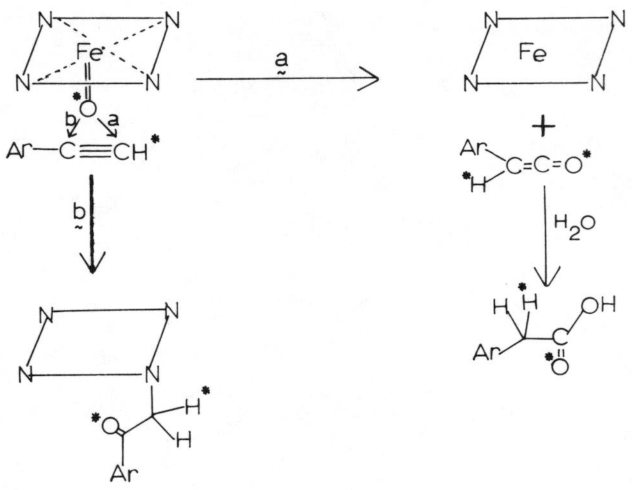

Scheme 5.

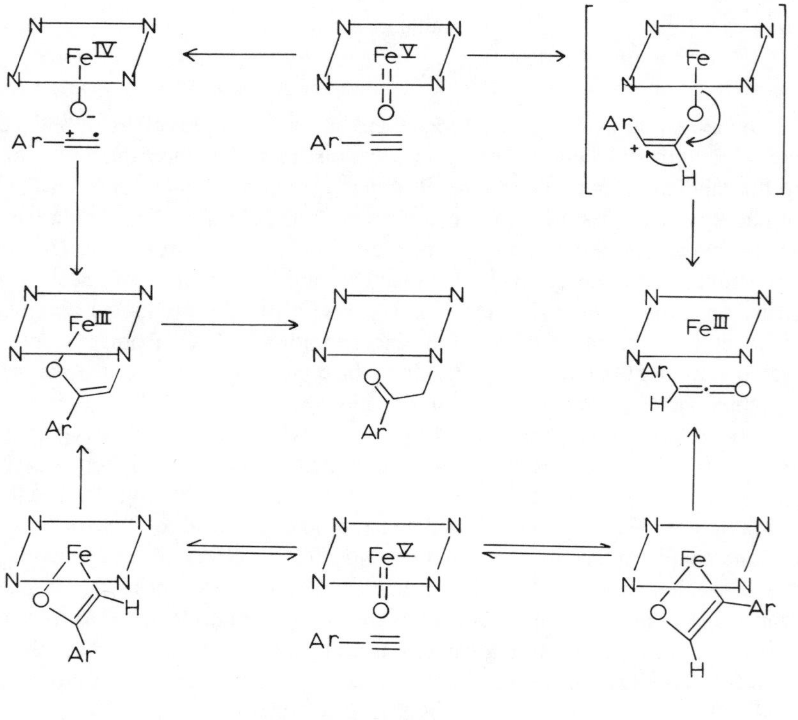

Scheme 6.

1-Ethynylpyrene (Structural Formula 10.1) was prepared by Gan et al.[10] in two steps from pyrene (Scheme 7). The design of **10.1** was based on two concepts. Ortiz de Montellano and co-workers[4-9] had shown that alkynes are potent inactivators of cytochrome P-450-dependent monooxygenases. Jerina had proposed a model for the substrate binding site of cytochrome P-450$_c$-dependent monoxygenase, the major cytochrome P-450 which metabolizes carcinogenic aryl hydrocarbons. The pyrene backbone of **10.1** should impart specificity for the cytochrome P-450$_c$-dependent enzyme and the ethynyl group should produce the desired mechanism-based inactivation. There is, indeed, specific in vitro inactivation of benzo[a]pyrene hydroxylase activity from liver microsomes of phenobarbital- or 5,6-ben-zoflavone-treated rats by **10.1** in the presence of NADPH.[10,11] However, unlike observations by Ortiz de Montellano and co-workers[4-9] with other alkynes, very little loss of intact cytochrome P-450 content was seen with **10.1**. It was concluded[10,11] that reactions of mechanism-based alkyne inactivators with nucleophiles other than heme are possible.

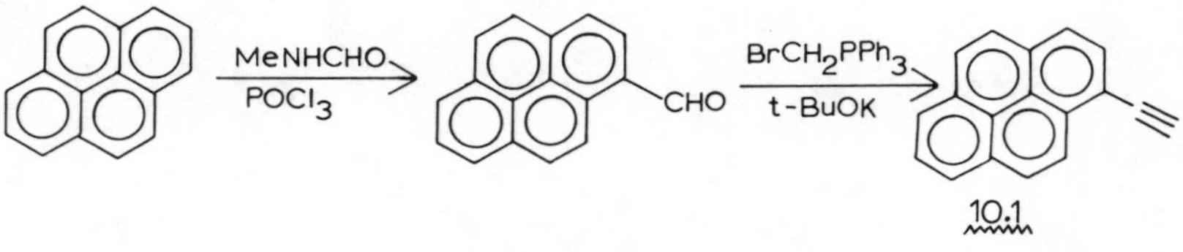

Scheme 7.   Containing Structural Formula 10.1.

Phenylacetylene also inactivates ammonia monooxygenase from the nitrifying bacterium *Nitrosomonas europaea* in the presence of oxygen.[2]

c. Monosubstituted Alkyl- or Cycloalkylacetylenes

The structure of the green pigment produced upon inactivation of liver microsomal cytochrome P-450 of phenobarbital-treated rats by propyne[5,12] and octyne[5] was determined by Ortiz de Montellano and coworkers. The electronic absorption properties, molecular weight, and NMR spectrum indicated that it is a single isomer of N-(2-oxopropyl)protoporphyrin IX in which the terminal carbon of the acetylene is attached to the nitrogen of the ring A pyrrole (Structural Formula 10.2, Scheme 8). Two general mechanistic pathways involve either oxirene formation or, preferably, an iron-coordinated enol cation or radical (Scheme 9). Theoretically, addition could have occurred at either of the acetylenic carbons; the other regioisomer would have been N-(1-methyl-2-oxoethyl)protoporphyrin IX. The isolation of only one isomer indicates that steric or electronic effects govern the reaction, and this may explain why disubstituted acetylenes do not give prosthetic heme adducts.[4]

Scheme 8.   Structural Formula 10.2.

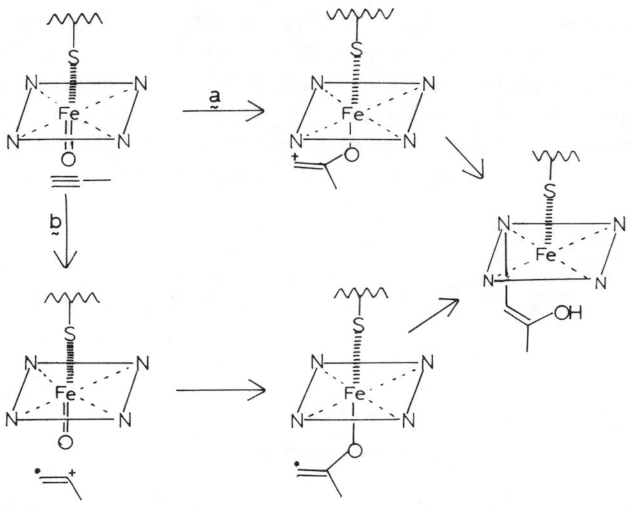

Scheme 9.

1-Ethynylcylopentanol inactivates purified phenobarbital-induced, but not 3-methylchol-anthrene-induced, rat liver cytochrome P-450.[13] 1-Ethynylcyclohexanol and substituted phenyl propynyl ethers inactivate microsomal cytochrome P-450 from the housefly (*Musca domestica*).[14] The amount of enzyme destroyed depends on the strain of housefly and on whether the insects are pretreated with phenobarbital; cytochrome P-450 from phenobarbital-treated resistant houseflies is most sensitive to inactivation.

The sedative-hypnotic agent, ethchlorvynol (1-chloro-3-ethyl-1-penten-4-yn-3-ol) (Structural Formula 10.3, Scheme 10) destroys liver microsomal cytochrome P-450 from phenobarbital-treated rats in a NADPH-, O_2-, and time-dependent process.[15] The green pigments isolated from rats given ethchlorvynol were identified as four regioisomers of N-(5-chloro-3-ethyl-3-hydroxy-2-oxo-4-pentenyl)protoporphyrin IX (Structural Formula 10.4, Scheme 11). This product confirms the greater reactivity of a terminal acetylene to that of a chlorovinyl group.

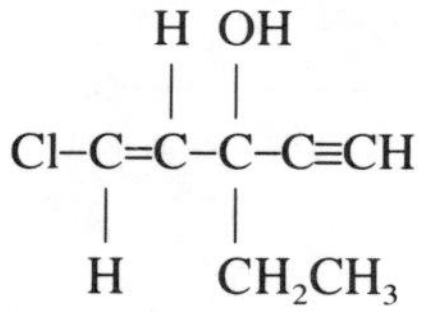

Scheme 10.   Structural Formula 10.3.

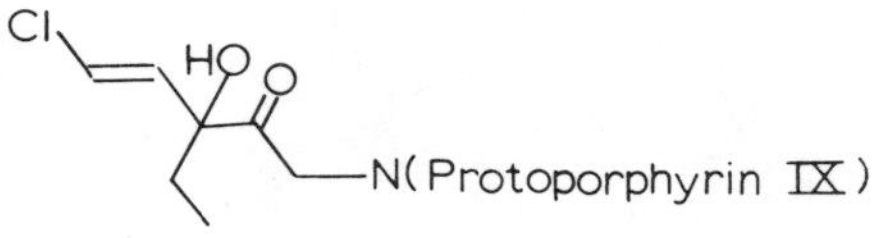

Scheme 11.   Structural Formula 10.4.

$$HC{\equiv}C(CH_2)_nCO_2H$$

Scheme 12.   Structural Formula 10.5.

The acetylenic fatty acids (see Scheme 12), 11-dodecynoic acid (Structural Formula 10.5, n = 9) and 10-undecynoic acid (**10.5**, n = 8), are specific inactivators of rat liver microsomal ω- and ω-1 lauric acid hydroxylases (cytochrome P-450 enzymes from uninduced rats).[16] Inactivation requires NADPH and O_2, and is time-dependent and pseudo first order. The specificity of the reagents for lauric acid hydroxylase is indicated by a failure to alter the spectroscopically measured cytochrome P-450 concentration or the microsomal N-demethylation of benzphetamine or N-methyl-p-chloroaniline. What this indicates is that these acetylenes inactivate a subset of cytochrome P-450 isozymes other than N-demethylases and is so small that its destruction has a negligible perturbation on the total cytochrome P-450 spectrum. Ortiz de Montellano and Reich[16] found that **10.5** (n = 9) inactivates 100% of the ω-hydroxylase activity, but only 50% of the ω-1 hydroxylase activity. This suggests that there is more than one ω-1 hydroxylase. Compound **10.5** (n = 9) also is a time-dependent inactivator of microsomal lauric acid in-chain hydroxylase from *Helianthus tuberosus* L.[17] Leukotriene B_4 ω-hydroxylase in human polymorphonuclear leukocytes is inactivated by the terminal acetylenic acids, 11-dodecynoic (**10.5**, n = 9), 15-hexadecynoic (**10.5**, n = 13), and 17-octadecynoic acid (**10.5**, n = 15).[17a] Inactivation requires O_2 and NADPH. Time-dependent inactivation only occurs when the acetylenic moiety is at the terminus. Compounds **10.5** (n = 13 and 15) inactivate the ω-hydroxylase at much lower concentrations than does **10.5** (n = 9); **10.5** (n = 8) does not inactivate it even at high concentrations. Other cytochrome P-450-dependent hydroxylases are not inactivated. For example, Shak et al.[17a] found that prostaglandin A_1 ω-hydroxylase from pig kidney is not inactivated by **10.5** (n = 13 or 15); lauric acid hydroxylase from pregnant rabbit lung is not affected by low concentrations of **10.5** (n = 15), although it is inhibited by **10.5** (n = 8). 10-Undecynoic acid inactivates lauric acid hydroxylases in vitro, but not in vivo, presumably because of rapid metabolic degradation.[17b] In order to study in vivo inactivation of these fatty acid hydroxylases, analogues of 10-undecynoic acid that were resistant to metabolic oxidation were prepared by CaJacob and Ortiz de Montellano.[17b] Thus, 2,2-dimethyl-11-dodecynoic acid and sodium 10-undecynyl sulfate inactivate lauric acid hydroxylases both in vitro and in vivo. In vitro inactivation is time- and NADPH-dependent and is selective only for the cytochrome P-450 isozymes that hydroxylate lauric acid. No apparent destruction of the heme group occurs.

$$CH_2{=}CH(CH_2)_nCO_2H \xrightarrow[\substack{2.\ NaNH_2 \\ 3.\ HCl}]{1.\ Br_2} 10.5$$

Scheme 13.

The synthesis of **10.5** from the corresponding alkene is shown in Scheme 13. A general synthesis of terminal acetylenic fatty acid is shown in Scheme 14.[17a]

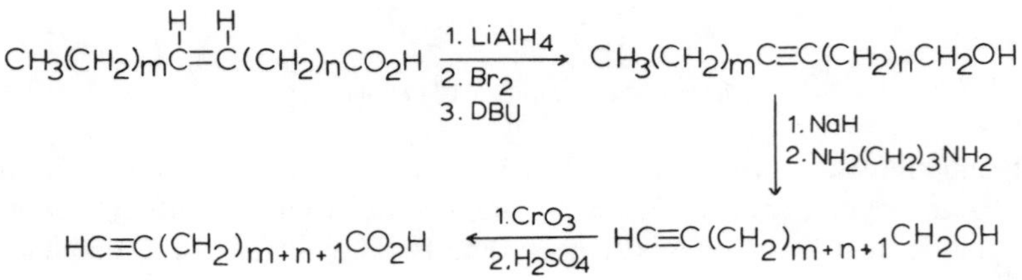

$$CH_3(CH_2)_m\overset{H}{\underset{|}{C}}=\overset{H}{\underset{|}{C}}(CH_2)_nCO_2H \xrightarrow[\substack{2.\ Br_2 \\ 3.\ DBU}]{1.\ LiAlH_4} CH_3(CH_2)_mC\equiv C(CH_2)_nCH_2OH$$

$$\downarrow \substack{1.\ NaH \\ 2.\ NH_2(CH_2)_3NH_2}$$

$$HC\equiv C(CH_2)_{m+n+1}CO_2H \xleftarrow[2.\ H_2SO_4]{1.\ CrO_3} HC\equiv C(CH_2)_{m+n+1}CH_2OH$$

Scheme 14.

$$(HC\equiv CCH_2)_2CHCONH_2$$

Scheme 15.   Structural Formula 10.6.

Dipropargylacetamide (Structural Formula 10.6, Scheme 15) and Danazol (Structural Formula 10.7; 17α-pregna-2,4-dien-20-yno[2,3-*d*]isoxazol-17-ol; Scheme 16) also inactivate liver cytochrome P-450 from phenobarbital-treated rats. NADPH and oxygen are required for inactivation.[3]

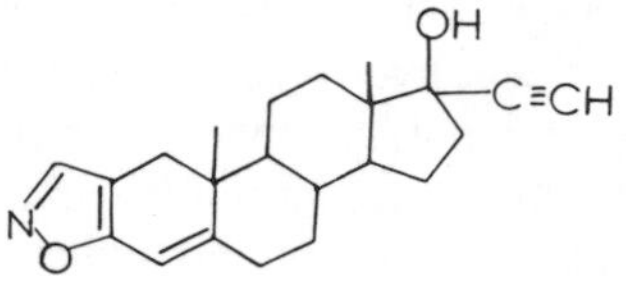

Scheme 16.   Structural Formula 10.7.

The oral contraceptive agent norethindrone (also called norethisterone; 17-hydroxy-19-norpregn-4-en-20-yn-3-one; Structural Formula 10.8, R = CH$_3$; Scheme 17) and ethynylestradiol (19-nor-17α-pregna-1,3,5(10)-trien-20-yne-3,17-diol; Structural Formula 10.9, Scheme 17) are time-dependent inactivators of liver microsomal cytochrome P-450 from phenobarbital-treated rats; NADPH and O$_2$ are required for inactivation.[18] Treatment of liver microsomes from phenobarbital-treated rats with norethindrone (**10.8**, R = Me), norgestrel (**10.8**, R = Et), or 1-ethynylcyclohexanol (Structural Formula 10.10, Scheme 17) results in destruction of cytochrome P-450 and formation of a green-brown pigment.[19] When [9,11-³H]norethisterone was used, the radioactivity was shown by Ortiz de Montellano et al.[19] to be covalently bound to the heme in a 1:1 ratio. Cytochrome P-450 also is destroyed by **10.8** (R = Et) and **10.10**, but not by **10.8** (R = Me) in which the ethynyl group is reduced to a vinyl group. This indicates the importance of the acetylenic substituent to inactivation. Because of the requirement of O$_2$ for inactivation, the mechanism shown in Scheme 18 was proposed.[19] In order to eliminate the possibility that activation of **10.8** (R = Me) by cytochrome P-450 leads to release of an activated species that alkylates nonbound heme, rats were treated with [³H]diisopropylheme in place of [³H] heme.[20] Diisopropylheme was shown not to bind to cytochrome P-450, but does bind to the microsomal membrane; therefore, any alkylated diisopropylheme produced during inactivation of cytochrome P-450 by norethindrone would be the result of a released activated species. No alkylation of diisopropylheme was detected, indicating that the activated species reacts with bound heme at the active site of cytochrome P-450.

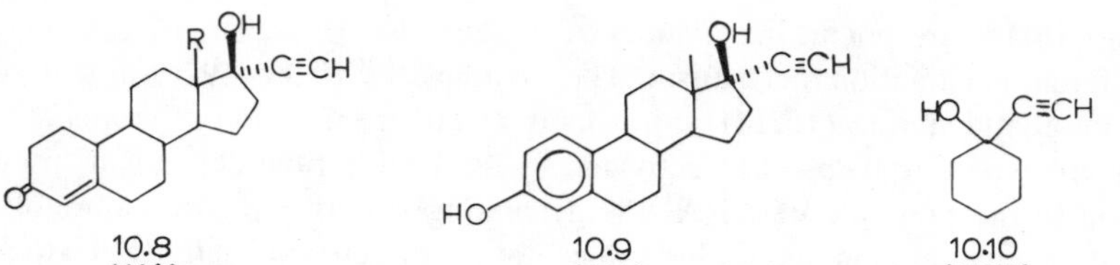

Scheme 17.   Structural Formulas 10.8 to 10.10.

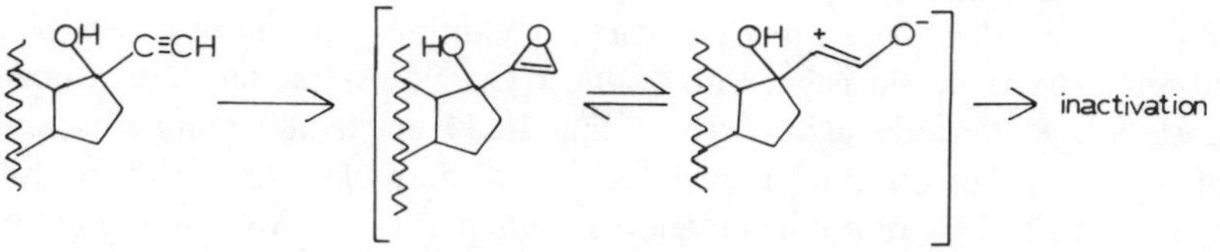

Scheme 18.

Norethisterone (**10.8**, R = Me) also irreversibly inactivates human placental aromatase with a K_1 of 1.7 μ*M*.[21] Inactivation requires NADPH and is both time- and concentration-dependent. Ethisterone (17α-ethynyltestosterone) does not inactivate the enzyme, even at a concentration of 100 μ*M*. Androgen aromatase is a molecular complex of a cytochrome P-450 and an NADPH cytochrome P-450 reductase, which probably accounts for the observed inactivation. It is not clear if inactivation of aromatase is responsible for the contraceptive properties of the drug. Norethisterone (**10.8**, R = Me) and norethynodrel (Structural Formula 10.10a, Scheme 18A) also were studied by Covey et al.[21a] as inactivators of human placental aromatase. In contrast to the findings of Osawa et al.[21] norethisterone was found *not* to inactivate aromatase. Norethynodrel also does not inactivate aromatase.[21a]

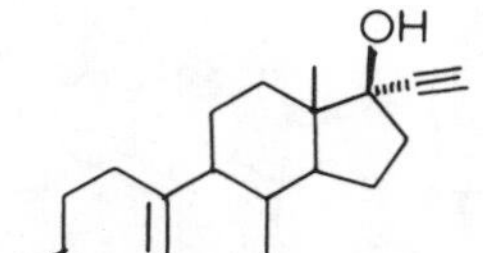

Scheme 18A.   Structural Formula 10.10a.

Norethisterone acetate is a time-dependent inactivator of aromatase that requires NADPH.[22] The K_1 is 17.7 μ*M*, suggesting that ester hydrolysis prior to inactivation does not occur since the K_1 for norethisterone is 1.7 μ*M*. 5α-Dihydronorethisterone, a nonaromatizable analogue of norethisterone, also is a time-dependent inactivator. In order to show covalent binding, [6,7-³H]norethisterone was incubated with human placental microsomes, then, after solubilization and partial purification, the tritium was observed by Osawa et al.[22] to coincide with aromatase activity and protein.

Human placental aromatase, which catalyzes the conversion of androgens to the estrogens, is irreversibly inactivated by several 10β-propynyl-substituted steroids (Structural Formulas 10.11 to 10.13, Scheme 19) in the presence of NADPH.[23] Compound **10.11** has an apparent K_1 of 23 n*M*. The mechanism proposed by Covey et al.[23] is oxygenation to the hydroxy propynyl compound (**10.12** or **10.13**) followed by the usual oxidation to an α,β-acetylenic ketone and Michael addition (Scheme 20). Evidence for this mechanism is their observation that both isomers of the proposed hydroxy propynyl intermediate and the oxo propynyl intermediate (**10.14**) are potent inactivators of the enzyme. However, Metcalf et al.[24] suggested a different inactivation mechanism. They synthesized **10.11**, 19,19-dideuterio-**10.11**, and the oxopropynyl steroid (**10.14**) and investigated their inhibition of aromatase. Human placental microsomal aromatase is inactivated by **10.11** in a time-dependent, pseudo first-order fashion in the presence of a NADPH-generating system.[24,25] Dithiothreitol does not affect the rate of inactivation and dialysis does not restore enzyme activity. Cytochrome P-450 dependent dealkylation is not inhibited, even though acetylenes have been shown to do so; also it does not inhibit steroid 11β-hydroxylase and other steroid metabolizing enzymes.[24,25] Compound **10.14** does not inactivate aromatase at a concentration of 1 μM. Also, the 19,19-dideuterio compound inactivates aromatase at the same rate as the protio compound. These results suggest that the Michael addition pathway of Covey et al.[23] is not correct and suggest a route consistent with results of Ortiz de Montellano and Kunze[4] (Scheme 21). Marcotte and Robinson[26] also observed that **10.11** inactivates human placental microsomal aromatase; various other analogues (Structural Formula 10.15, R = CH=CH$_2$, CN, CHF$_2$, CH$_2$F, and CH$_2$Cl) are not inactivators (see Scheme 22). Aromatase in human trophoblast choriocarsinoma cells in tissue culture also is inactivated by **10.11**.[27] Inactivation is time-dependent and pseudo first order; the apparent K_I is 0.6 nM compared with a K_m for androstenedione of 35 nM.

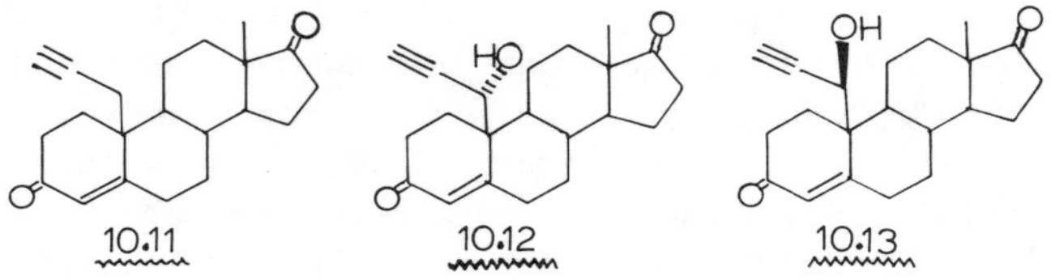

Scheme 19.   Structural Formulas 10.11 to 10.13.

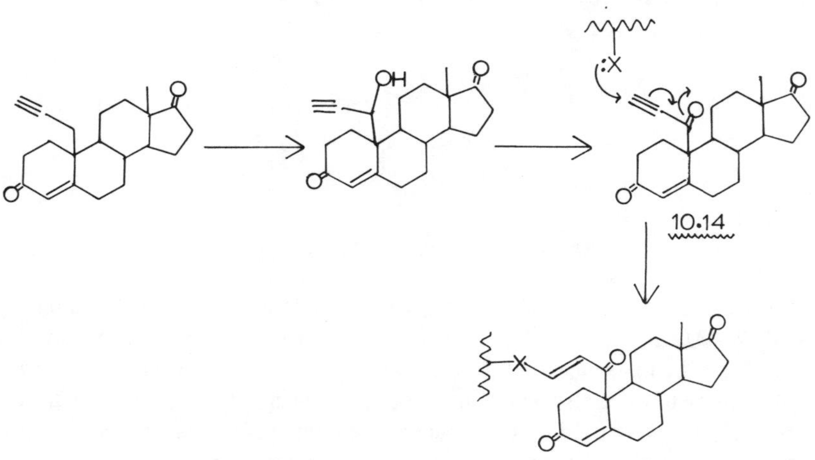

Scheme 20.   Containing Structural Formula 10.14.

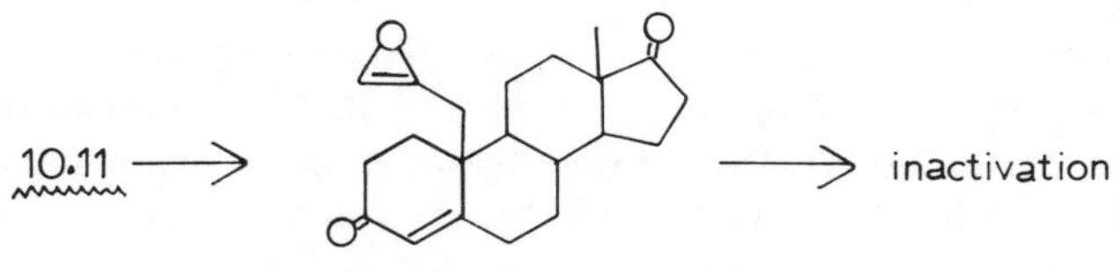

Scheme 21.

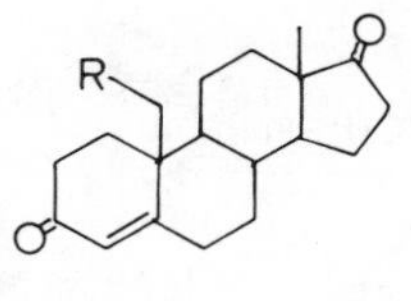

Scheme 22.   Structural Formula 10.15.

Covey et al.[23] synthesized **10.11** by the route shown in Scheme 23. Metcalf et al.[24] used the route in Scheme 24. Another route, by Marcotte and Robinson,[26] is shown in Scheme 25. Compounds **10.12** to **10.14** were synthesized by Covey et al.[28] (Scheme 26).

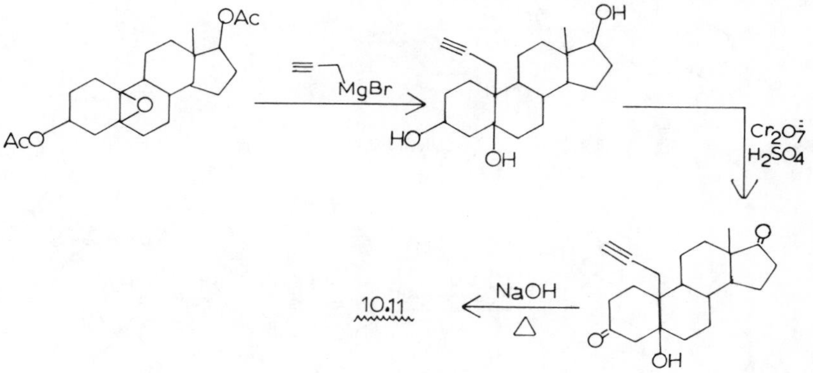

Scheme 23.

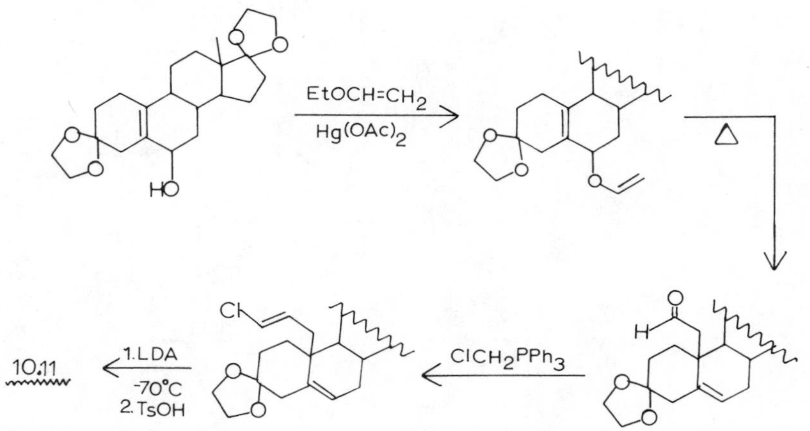

Scheme 24.

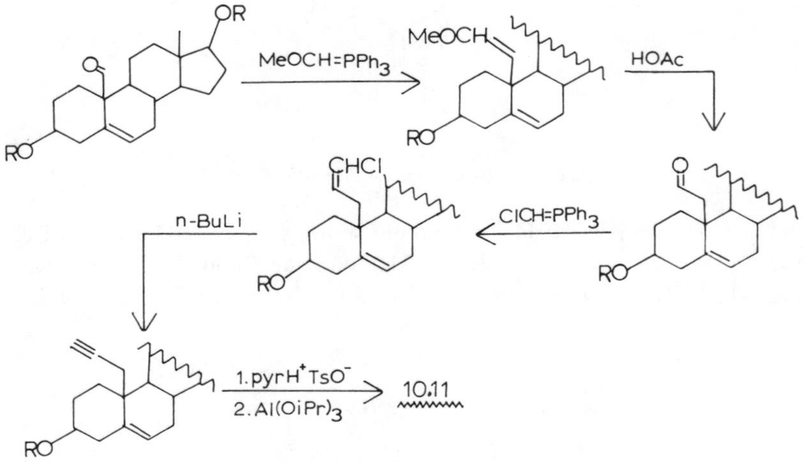

Scheme 25.

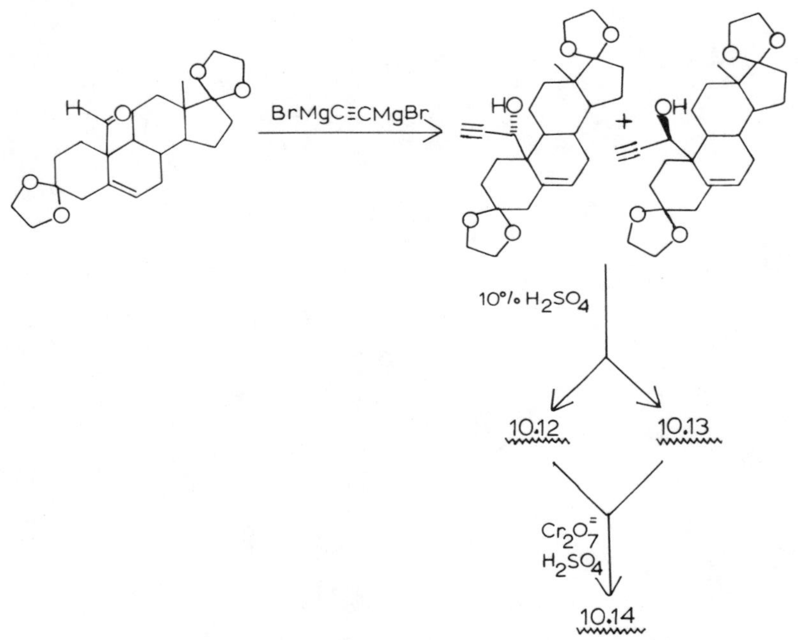

Scheme 26.

Aldosterone biosynthesis in zona glomerulosa cells of beef adrenals is inhibited by 11β-hydroxy-18-ethynylprogesterone (Structural Formula 10.16, Scheme 27); it is believed that a cytochrome P-450 is inactivated.[29]

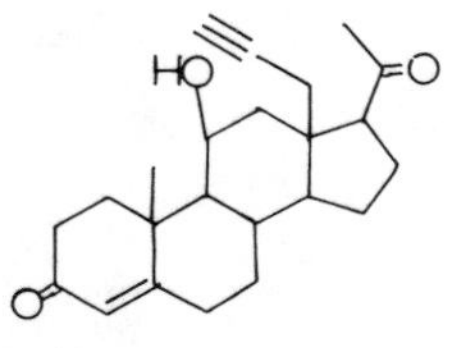

Scheme 27.   Structural Formula 10.16.

d. Disubstituted Acetylenes

20-(1-Propynyl)-5-pregnen-3β,20α-diol (**10.17**, R = CH$_3$), 20-(1-hexynyl)-5-pregnen-3β,20α-diol (**10.17**, R = (CH$_2$)$_3$CH$_3$), and 20-(1,5-hexadiynyl)-5-pregnen-3β,20α-diol (**10.17**, R = CH$_2$CH$_2$C≡CH) (Scheme 28) were shown to be mechanism-based inactivators of cytochrome P-450$_{scc}$.[30] The partition ratio for the diyne is 6. When R = H, it is neither a substrate nor inhibitor. Nagahisa et al.[30] proposed that a reactive oxirene-like species (Structural Formula 10.18, Scheme 29) is generated, but since no green pigment typical of heme alkylation was apparent, they proposed that alkylation of an amino acid residue results.

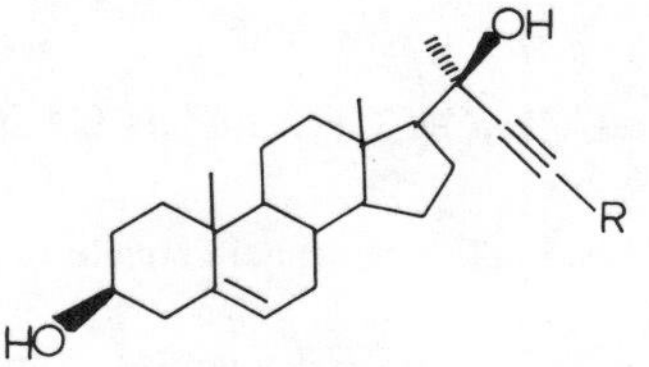

Scheme 28.   Structural Formula 10.17.

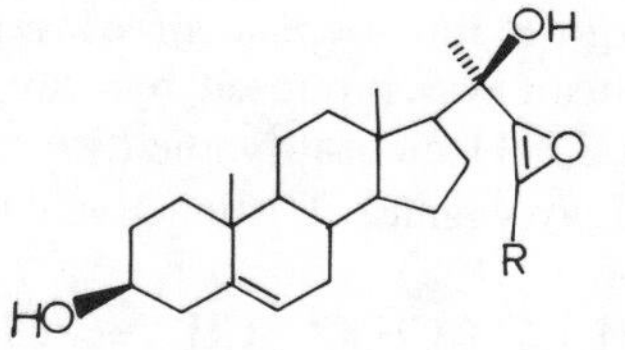

Scheme 29.   Structural Formula 10.18.

1-Phenyl-1-propyne inactivates dopamine β-hydroxylase in a pseudo first-order, time-dependent fashion with a partition ratio equal to or near 1.[31] Inactivation requires O$_2$ and ascorbate and the rate of inactivation increases with increasing O$_2$ concentration. The mechanism proposed by Colombo and Villafranca[31] is shown in Scheme 30; however, NaB^{3}H$_4$ does not result in tritium incorporation into the inactivated enzyme and, therefore, this mechanism is questionable.

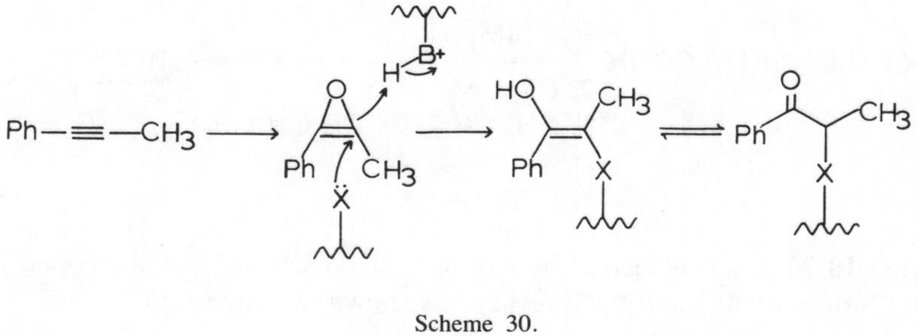

Scheme 30.

Methyl 1-pyrenylacetylene (Structural Formula 10.19, R = Me; Scheme 31) is a mechanism-based inactivator for the benzo[a]pyrene hydroxylase activity in liver microsomes of phenobarbital- or 5,6-benzoflavone treated rats.[11] Compounds **10.19** (R = Me or Ph) do not cause a loss of cytochrome P-450 content. Therefore, inactivators of cytochrome P-450 may be designed with specificity for benzo[a]pyrene hydroxylase activity without heme alkylation. Loss of cytochrome P-450 content is an insensitive measure of inactivation of benzo[a]pyrene hydroxylase.

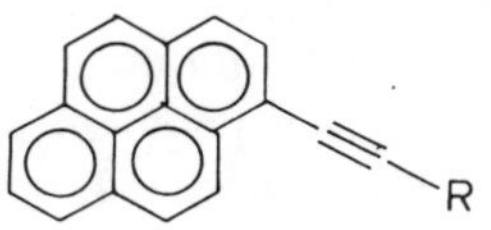

Scheme 31.   Structural Formula 10.19.

$$CH_3(CH_2)_4C{\equiv}CCH_2C{\equiv}CCH_2C{\equiv}CCH_2C{\equiv}C(CH_2)_3COOH$$

Scheme 32.   Structural Formula 10.20.

Prostaglandin biosynthesis in sheep seminal vesicles is irreversibly inactivated by eicosa-5,8,11,14-tetraynoic acid (Structural Formula 10.20, Scheme 32).[32] This compound also prevents conversion of linoleic and linolenic acids to the hydroxy acids, a reaction that also is catalyzed in the seminal vesicle. Soybean lipoxygenase and prostaglandin synthetase from sheep seminal vesicles undergo a time-dependent irreversible inactivation by **10.20**; an isomerization/addition mechanism was proposed by Downing et al.[33] Rabbit reticulocyte lipoxygenase is inactivated by 5,8,11-eicosatriynoic (Structural Formula 10.21, Scheme 33) and **10.20**, whereas soybean lipoxygenase-1 is inactivated only by the latter.[34]

$$CH_3(CH_2)_4C{\equiv}CCH_2C{\equiv}CCH_2C{\equiv}C(CH_2)_8COOH$$

Scheme 33.   Structural Formula 10.21.

In accord with the report of Downing et al.,[33] soybean lipoxygenase was inactivated by **10.20** in a time-dependent process.[35] Oxygen appears to be required for inactivation. Inactivation by [*methyl*-^{14}C]methyl 5,8,11,14-eicosatetraynoate is not accompanied by radioactive incorporation into the enzyme. Anaerobic base hydrolysis and amino acid analysis showed the formation of 1 methionine sulfoxide residue per enzyme. Kühn et al.[35] suggested that the inactivator is converted to an allene hydroperoxide and that an oxygen atom is transferred to a methionine residue in the active site.

$$HC{\equiv}CCH_2C{\equiv}C(CH_2)_3COOH \xrightarrow[\substack{2.\ Cu_2(CN)_2 \\ 3.\ CH_3(CH_2)_4C{\equiv}CCH_2C{\equiv}CCH_2I}]{1.\ 2\ EtMgBr} 10.20$$

Scheme 34.

Compound **10.20** was synthesized by Kunau et al.[36] (Scheme 34); compound **10.21** was prepared by Struijk et al.[37] and by Sprecher[38] as shown in Scheme 35.

$$CH_3(CH_2)_4C{\equiv}CCH_2C{\equiv}CCH_2Br\ +\ HC{\equiv}C-(CH_2)_8COOH \xrightarrow[\substack{2.\ Cu_2(CN)_2 \\ 3.\ H_3O^+}]{1.\ EtMgBr} 10.21$$

Scheme 35.

5,6-Dehydroarachidonic acid (Structural Formula **10.22**) is a time-dependent inactivator of 5-lipoxygenase from rat basophilic leukemia (RBL-1) cells. The mechanism proposed by Sok et al.[39] is shown in Scheme 36.

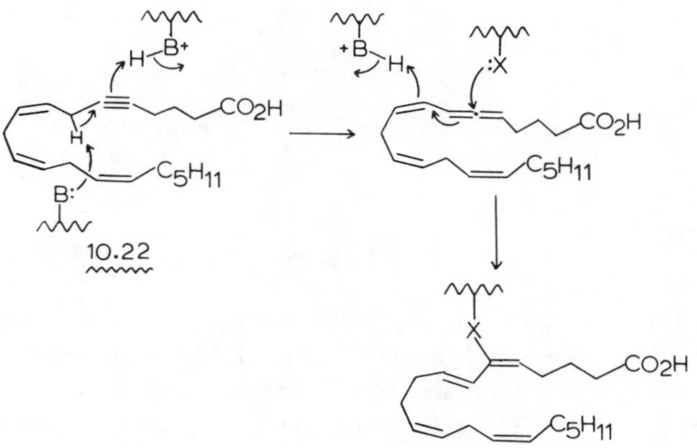

Scheme 36.   Containing Structural Formula 10.22

5,6-, 8,9-, 11,12-, and 14,15-Dehydroarachidonic acids (Structural Formulas 10.22 to 10.25, respectively; see Scheme 37), arachidonic acids in which the noted double bonds are replaced by triple bonds, were tested as inactivators of Types I and V soybean lipoxygenase.[40] Only 14,15-dehydroarachidonic acid (**10.25**) is a potent irreversible inactivator. Time-dependent inactivation of both enzymes occurs only in the presence of oxygen with a partition ratio of about 260. The presence of $NaBH_4$ does not protect the enzyme from inactivation, suggesting that the reactive intermediate that inactivates the enzyme is not released prior to inactivation. [2-^{3}H]14,15-Dehydroarachidonic acid inactivates the enzyme with 0.87 mol of [^{3}H] incorporated per mole of enzyme. The mechanism suggested by Corey and Park[40] is that shown in Scheme 38. Preliminary findings indicate that 5-lipoxygenase is irreversibly inactivated by **10.22**.

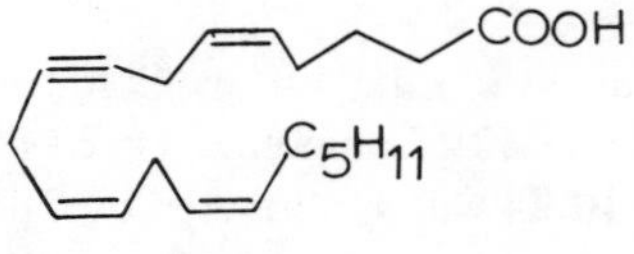

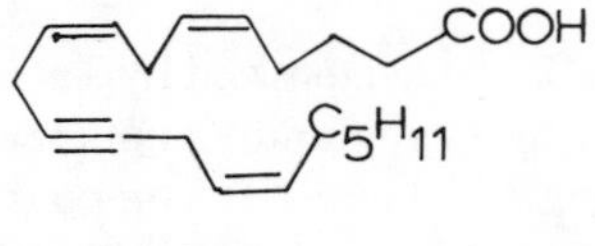

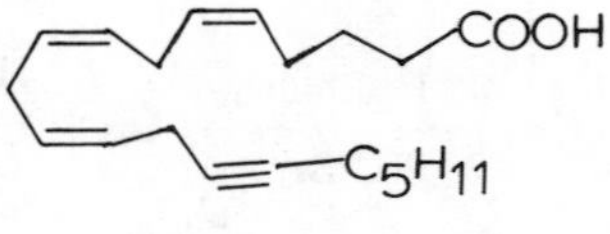

Scheme 37.   Structural Formulas 10.23 to 10.25.

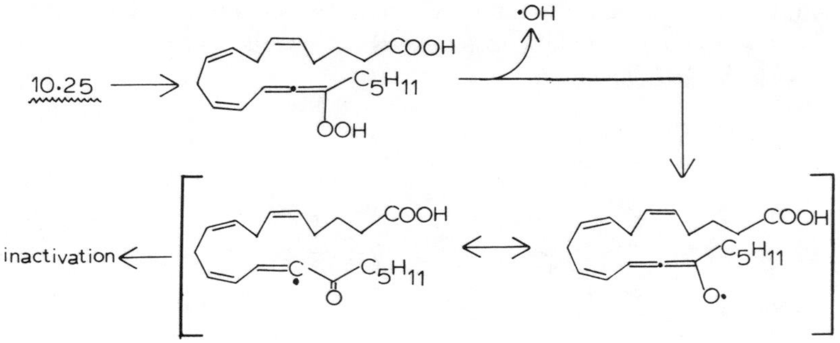

Scheme 38.

Time-dependent inactivation of prostaglandin biosynthesis in ram seminal vesicle microsomes was observed by Corey and Munroe[41] for the 11,12-dehydro compound (**10.24**). The 14,15-dehydro compound (**10.25**) is about one-fourth as potent, but the 5,6- and 8,9-isomers (**10.22** and **10.23**) are inactive. Oxygen is required for inactivation. Leukotriene biosynthesis of RBL-1 cells is blocked in a time- and oxygen-dependent manner by **10.22**. The same mechanism as shown in Scheme 38 was suggested.

Compound **10.22** irreversibly inactivates 5-lipoxygenase from RBL-1 cells in the presence of both O_2 and Ca(II) which are required for enzyme activity.[42] The rate of inactivation is shown to be subject to a primary kinetic isotope effect of about 6 when [7*R*-^{2}H] **10.22** is used. This further supports the mechanism previously proposed (Scheme 38). A series of 5,6-dehydro analogues (Structural Formula 10.26, R = PO_3H^-, $P(OMe)O_2^-$, SO_3^-, SO_2^-, $CONH_2$, $P(O)(OMe)_2$, S(O)Me, SO_2Me) were prepared and also shown to inactivate 5-lipoxygenase (see Scheme 39).

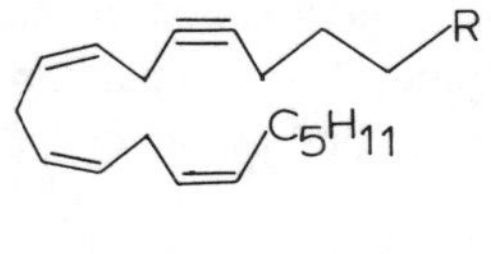

Scheme 39. Structural Formula 10.26.

Compound **10.22** was synthesized by Tai et al.[43] (Scheme 40), by Corey and co-workers[41,44] (Scheme 41) and by Corey and Kang[45] (Scheme 42); compound **10.23** also was prepared by Corey and Kang[45] (Scheme 43); compound **10.24** was synthesized by Corey and Munroe[41] (Scheme 44) and by Corey and Kang[45] (Scheme 45); and compound **10.25** was made by Corey and Park[40] (Scheme 46).

$$HOCH_2C{\equiv}CCH_2C{\equiv}CCH_2C{\equiv}C(CH_2)_4CH_3 \xrightarrow[2.\ Ph_3PBr_2]{1.\ H_2/Lindlar} BrCH_2\overset{H}{C}{=}\overset{H}{C}CH_2\overset{H}{C}{=}\overset{H}{C}CH_2\overset{H}{C}{=}\overset{H}{C}(CH_2)_4CH_3$$

$$\xrightarrow{HC{\equiv}C(CH_2)_3COOH,\ 2\ EtMgBr} \underset{\sim\sim\sim}{10.22}$$

Scheme 40.

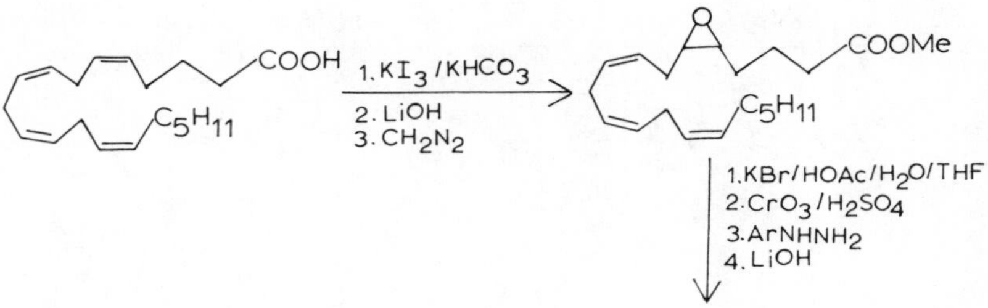

Scheme 41.

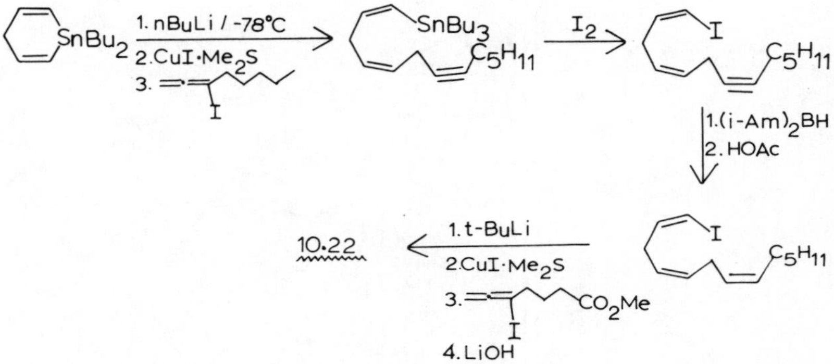

Scheme 42.

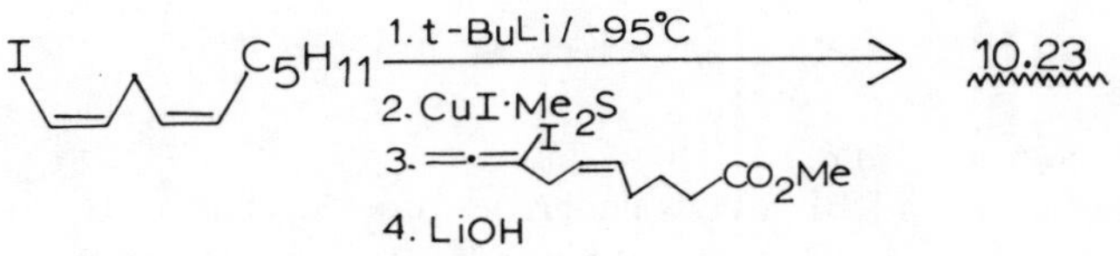

Scheme 43.

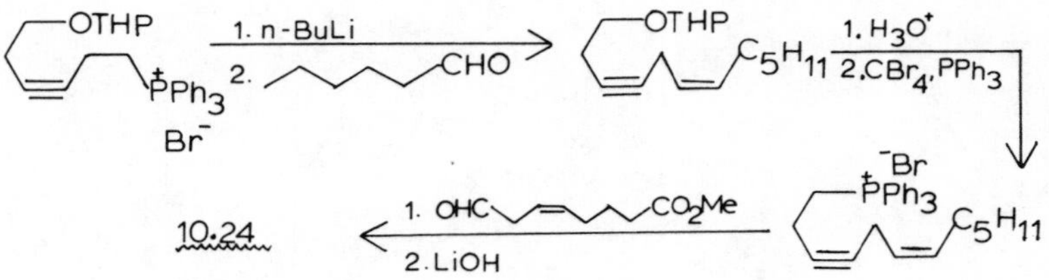

Scheme 44.

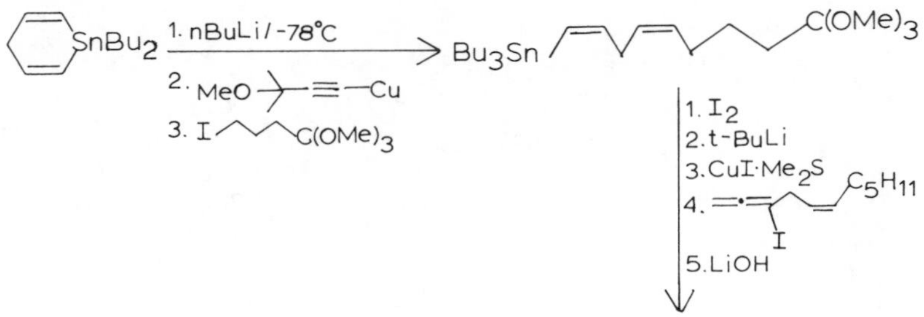

Scheme 45.

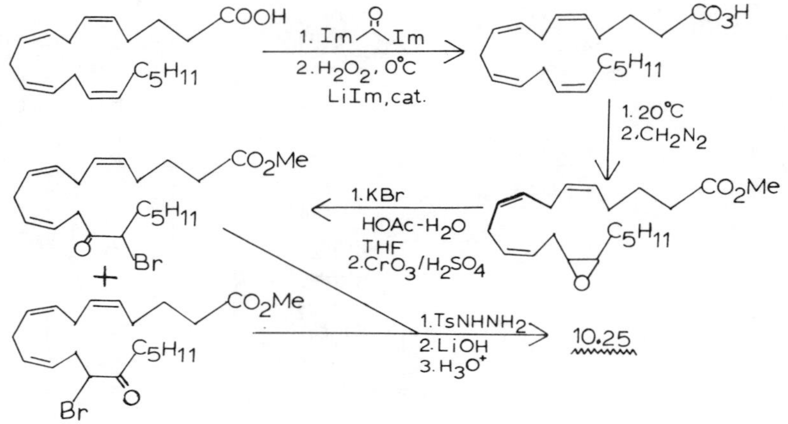

Scheme 46.

2. Allene-Containing Inactivators

The 5-lipoxygenase from RBL-1 cells undergoes time-dependent irreversible inactivation by 4,5-dehydroarachidonic acid (Structural Formula 10.27) in the presence of oxygen.[46] The synthesis by Corey et al.[46] is shown in Scheme 47.

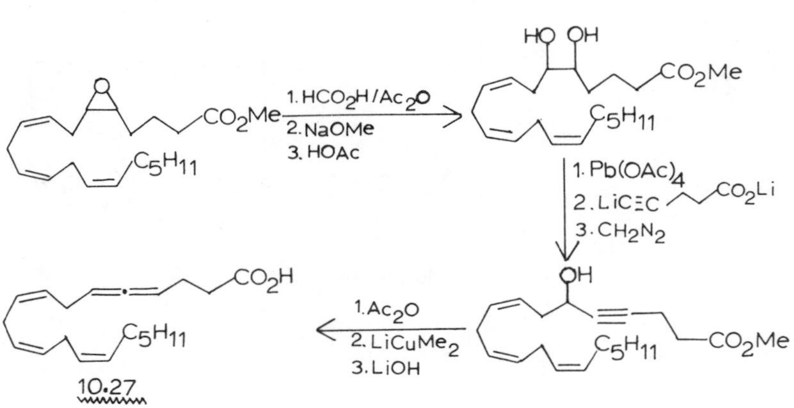

Scheme 47.   Containing Structural Formula 10.27.

Human placental microsomal aromatase is inactivated in a time-dependent, pseudo first-order fashion by allenic steroid **10.28** (Scheme 48).[24] A NADPH-generating system is required for inactivation. Dithiothreitol does not affect the rate of inactivation and dialysis does not restore enzyme activity. Cytochrome P-450-dependent dealkylation and steroid 11β-hydroxylase, however, are not inhibited. Compound **10.28** was synthesized by Metcalf et al.[24] by the route in Scheme 49.

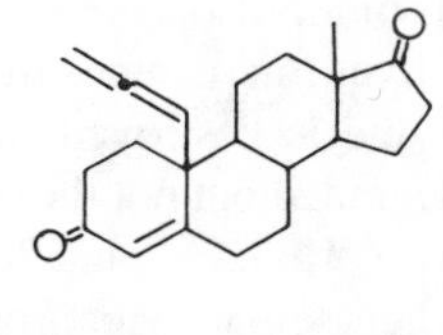

Scheme 48.   Structural Formula 10.28.

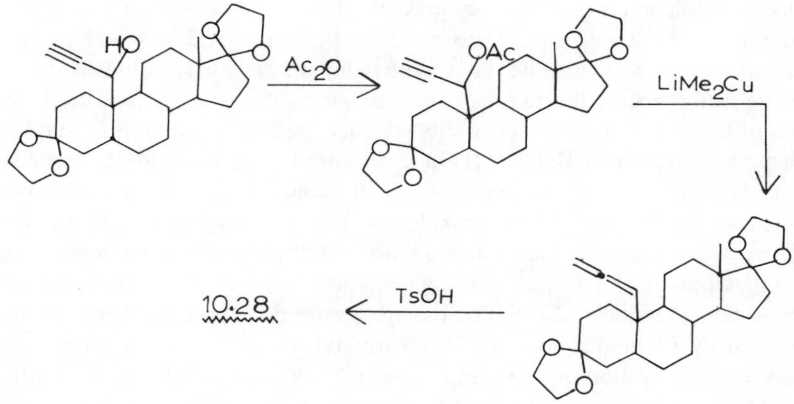

Scheme 49.

17α-Propadienyl-19-nortestosterone (Structural Formula 10.29),[13,47] 1-propadienylcyclo-hexanol (Structural Formula 10.30, see Scheme 50),[47] and 1,1-dimethylallene[47] inactivate liver microsomal cytochrome P-450 from phenobarbital-treated rats in the presence of NADPH. Glutathione does not prevent inactivation. Loss of heme is concomitant with inactivation. Unlike inactivation by terminal olefins and acetylenes, the allenes do not cause formation of abnormal hepatic pigments. This is similar to inactivation by internal acetylenes[4] (see Section II.A.1.d). An *exo*-methylene epoxide (Structural Formula 10.31, Scheme 51) was suggested by Ortiz de Montellano and Kunze[47] as the reactive intermediate. Compound **10.29** does not inactivate the 3-methylcholanthrene-induced isozyme.[13]

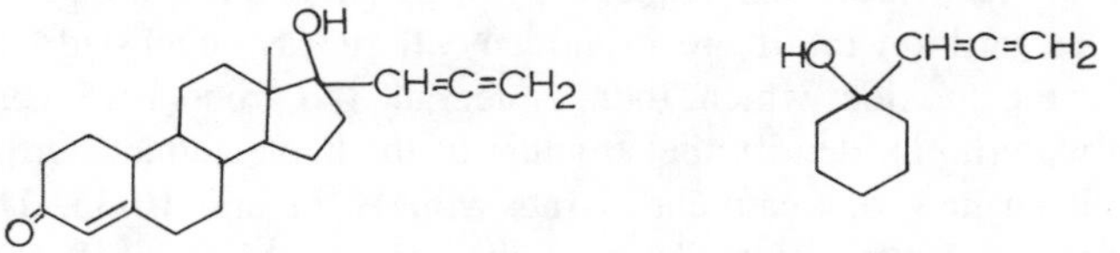

Scheme 50.   Structural Formulas 10.29 and 10.30.

Scheme 51.   Structural Formula 10.31.

3. Alkene-Containing Inactivators
a. 2-Isopropyl-4-pentenamide (Allylisopropylacetamide) and Related

Several in vivo studies of the effects of allylisopropylacetamide (Structural Formula 10.32, Scheme 52) on cytochrome P-450 in liver microsomes of phenobarbital-treated rats[48-50a] and rabbits[51] were carried out. This enzyme is destroyed, but not cytochrome b_5 or NADPH cytochrome C reductase.[48] Heme is degraded but not the apoprotein;[49,50,50a] radioactive **10.32** becomes attached to the heme.[50] In vitro, **10.32**[52,53] and 2,2-diethyl-4-pentenamide (novonal, **10.33**, Scheme 52)[54] destroy liver microsomal cytochrome P-450 of phenobarbital-treated rats only in the presence of NADPH.[52,53] This inactivation process is not the result of lipid peroxidation.[55] The partition ratio for **10.32** is 230 to 320.[13] It is not a substrate or inactivator of the 3-methylcholanthrene-induced cytochrome P-450.[13] Other inactivators of the phenobarbital-inducible isozyme, e.g., fluroxene, 5,6-dichloro-1,2,3-benzothiadiazole, and 1-aminobenzotriazole, also inactivate the 3-methylcholanthrene-inducible isozyme. This suggests that the difference in the two isozymes is not a difference in oxidation mechanism, but rather a difference in binding specificity, similar, perhaps, to that of MAO A and MAO B (see Chapter 9, Section II.B.1.a.(1)). Inactivation of the phenobarbital-induced isozyme by **10.32** leads to a 1:1 stoichiometry between **10.32** and the heme, as determined by field desorption mass spectrometry.[56] An epoxide was ruled out by Ortiz de Montellano et al.[56] as the reactive species responsible for inactivation. Methyl 4,5-epoxy-2-isopropylpentanoate and known metabolites of 2-isopropyl-4-pentenamide do not destroy cytochrome P-450, but 2-isopropyl-4-pentenamide and the corresponding ester do. The same heme adduct appears to be produced by the amide or ester. The carbonyl group is believed to be important to inactivation; the inactivation mechanism proposed[56] is shown in Scheme 53. Further studies by Ortiz de Montellano and Mico[57] support a mechanism-based inactivation by **10.32**. Davies et al.[57a] found that cytochrome P-450 from phenobarbital-treated rats is inactivated by **10.32** in a process that results in covalent attachment of the heme to the protein. Inactivation depends on O_2 and is inhibited by $NADP^+$. It is time-dependent and pseudo first order, but only 40% of enzyme activity is lost. This is presumably because only 40% of the cytochrome P-450 enzymes present are vulnerable to destruction. Methyl 4,5-epoxy-2-isopropylpentanoate, the epoxide of methyl 2-isopropyl-4-pentenoate, protects the enzyme from destruction by the inactivator. This indicates that the epoxide binds at the active site, but does not inactivate the enzyme, and again excludes an epoxide as the reactive species. The substrate and inactivator behavior of 2-[^{14}C]allylisopropylacetamide was studied by Loosemore et al.[58] in order to determine the partition ratio. In phenobarbital-induced microsomes, the partition ratio is 201, but with a pure P-450 isozyme active for oxidative N-demethylation in a reconstituted system, it is 92. Because of the previously observed O_2-dependent autoinactivation with the pure isozyme,[59] this number is low by about a factor of 2. Therefore, using that correction, the partition ratios are in agreement. It was suggested[58] that the initially released product is the epoxide, which, then, is degraded to products (Scheme 54). In order to obtain enough material to identify the structure of the heme adduct, Ortiz de Montellano et al.[60] carried out an in vivo treatment of rats with **10.32** and **10.33**. The characteristic green pigments produced were isolated and studied by 360-MHz NMR spectroscopy. The principal heme product from each compound is N-alkylated on either pyrrole ring C or D; the structures of the adducts are **10.34** (R = H; R' = i-Pr and R = R' = Et), as shown in Scheme 55.

$$CH_2=CHCH_2CHCONH_2 \qquad CH_2=CHCH_2C(C_2H_5)_2CONH_2$$
$$\underset{\displaystyle \underset{\sim\sim\sim}{10.32}}{\overset{\displaystyle |}{CH(CH_3)_2}} \qquad\qquad\qquad \underset{\sim\sim\sim}{10.33}$$

Scheme 52.   Structural Formulas 10.32 and 10.33.

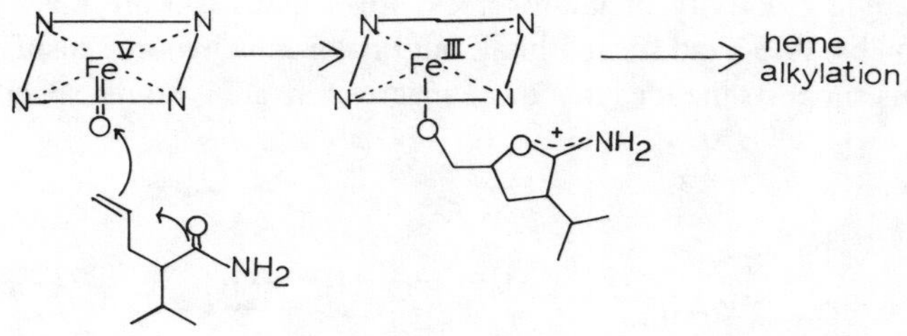

Scheme 53.

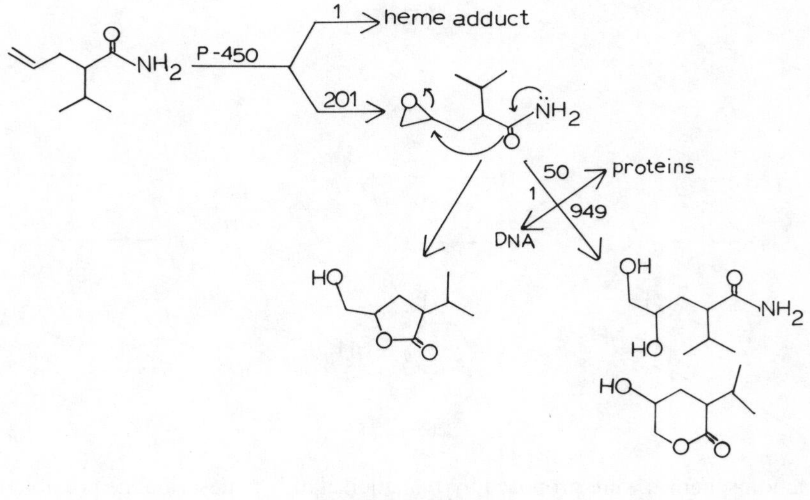

Scheme 54.

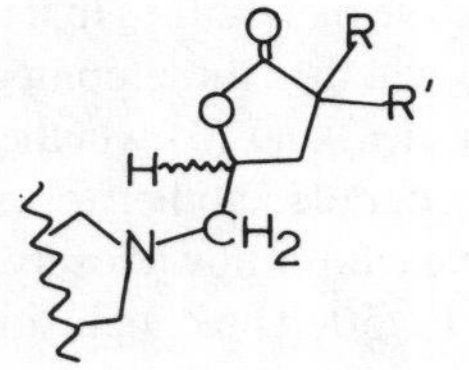

Scheme 55.   Structural Formula 10.34.

When microsomal cytochrome P-450 was inactivated with 2-isopropyl-4-pentenamide under an $^{18}O_2$ atmosphere, the modified heme showed the incorporation of one [^{18}O] into the adduct. Therefore, the mechanism previously suggested[56] was modified (Scheme 56).[60] The $^{18}O_2$ experiment excludes a mechanism in which the amide carbonyl traps a reactive intermediate, e.g., Scheme 57 (see also Section II.A.3.b.). An alternative explanation for the formation of more than one regioisomer could be the presence of more than one isozyme.[61] Two isozymes, PB-4 and PB-5, of phenobarbital-induced cytochrome P-450 in rat liver were separated and purified by Waxman and Walsh,[61] and account for the majority of dimethylaniline *N*-demethylase activity of microsomes. PB-4 and PB-5 are inactivated in a time-dependent fashion by **10.32** and secobarbital, but PB-4 is much more sensitive to inactivation than is PB-5. This supports the idea that these inactivators are isozyme specific, as suggested previously.[57,59]

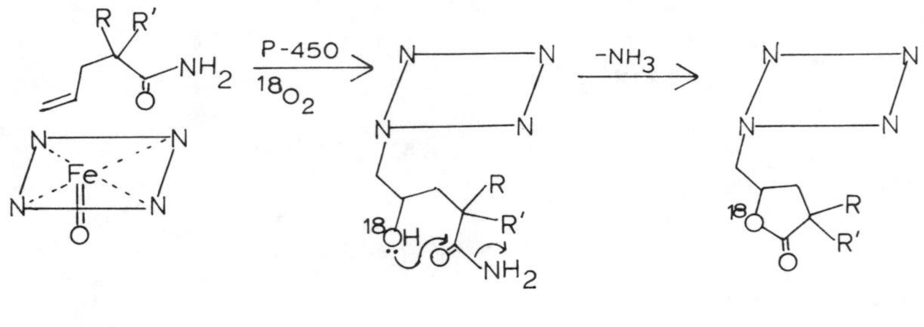

Scheme 56.

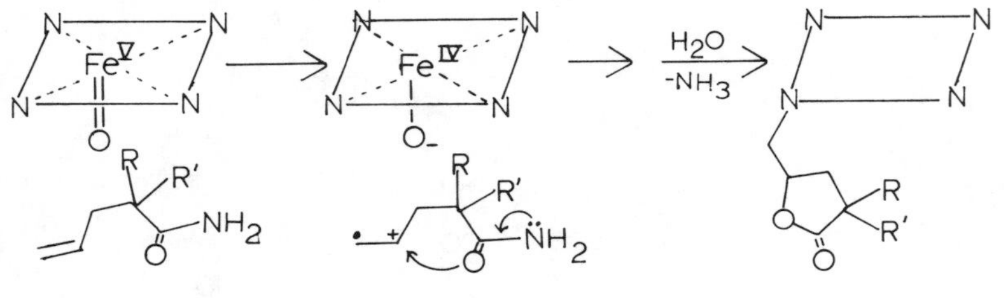

Scheme 57.

Six reaction schemes were proposed by Ivanetich et al.[62] to describe the possible reactions for destruction of various cytochrome P-450 isozymes by fluoroxene and allylisopropylacetamide. The schemes were selected to determine (1) which forms of cytochrome P-450 convert the two compounds into reactive metabolites that are capable of chemically modifying the heme of cytochrome P-450; (2) whether these compounds produce reactive metabolites that are transient enzyme-bound species; and (3) whether reactive metabolites produced by one form of cytochrome P-450 can degrade another form of the enzyme. Computer analysis of the schemes was used to evaluate each for how closely it mimicks the experimental results for the degradation of cytochromes P-450. The conclusions are (1) that the transient reactive species of the two compounds are formed by at least two forms of cytochrome P-450; (2) that the reactive species degrades the same form of the enzyme and possibly the same enzyme molecule (prior to diffusion) that produces it; and (3) that the enzyme-substrate complex giving rise to the production of the transient reactive species may be distinct from the typical cytochrome P-450-substrate complex giving rise to a Type I difference spectrum.

2-Isopropyl-4-pentenamide also inactivates cytochrome P-450 in the housefly (*Musca domestica*).[14] The amount of enzyme destroyed depends on the strain of housefly and on whether the insects are pretreated with phenobarbital. Cytochrome P-450 from phenobarbital-treated resistant houseflies is most sensitive to inactivation.

b. Ethylene and Alkyl-Substituted Ethylenes

In an in vivo study by De Matteis et al.[63] phenobarbital-treated rats were exposed to ethylene, and it was shown that cytochrome P-450 in liver was inactivated with production of a green pigment. A field desorption mass spectrum gave a molecular ion compatible with N-hydroxyethyl protoporphyrin IX dimethyl ester. An epoxide intermediate was suggested.

Parallel to the results obtained with acetylene and substituted acetylenes (see Section II.A.1.), it was found by Ortiz de Montellano and Mico[64] that ethylene and monosubstituted terminal olefins inactivate phenobarbital-inducible rat liver cytochrome P-450 in vitro. NADPH and O_2 are required for inactivation and heme alkylation results. The earlier hypothesis that a carbonyl group was required for inactivation (see Section II.A.3.a.) is proven incorrect since alkenes also inactivate the enzyme. The lactonization resulting from inactivation by 2-isopropyl-4-pentenamide, therefore, probably results subsequent to, rather than consequent with, alkylation. Alkanes do not inactivate the enzyme, indicating the importance of unsaturation in the mechanism of inactivation. Hydroxylation is not a prerequisite to inactivation as evidenced by the effectiveness of ethylene as an inactivator. The corresponding epoxides of the active olefins are inactive, although all of them bind to the active site, thus excluding epoxidation as the mechanism for inactivation. Ortiz de Montellano et al.[65] also carried out an in vivo study of the effects of ethylene (and of 4-ethyl-1-hexene) on phenobarbital-inducible rat liver cytochrome P-450. The green pigment formed by inactivation was isolated and the field desorption mass spectrum showed a molecular ion equal to the sum of the molecular weights of the dimethyl ester of protoporphyrin IX, the olefin, and an oxygen atom. The electronic absorption spectra of the adducts are superimposable with the corresponding spectra of the dimethyl ester of synthetic N-methylprotoporphyrin IX. These results suggest that heme N-alkylation is the result of enzyme inactivation, and the proposed structure for the inactivation with ethylene is N-(2-hydroxyethyl)protoporphyrin IX. The 360-MHz NMR spectrum of this adduct[66] supports this proposal and indicates that alkylation of one of the propionic acid-substituted pyrrole rings (ring C or D) has occurred. Ortiz de Montellano et al.[66] suggested the mechanism shown in Scheme 58.

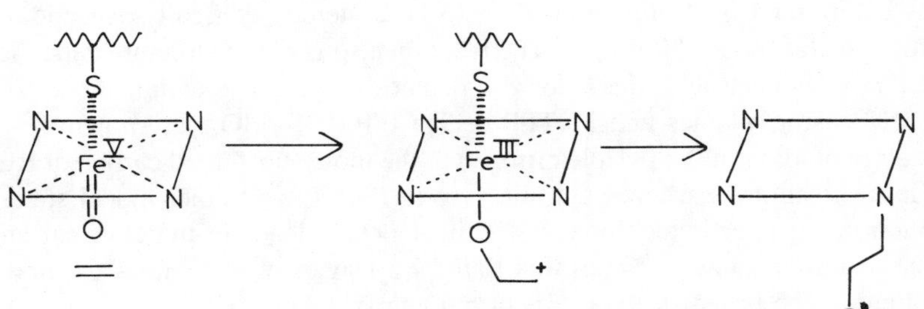

Scheme 58.

In order to probe the topology of cytochrome P-450 of phenobarbital-treated rats, a series of olefins (ethylene, propene, and octene) were used by Kunze et al.[5] to inactivate the enzyme, and the alkylated hemes were characterized. The oxygen atom introduced into the N-alkylated heme was shown to be derived from molecular oxygen by carrying out the ethylene inactivation experiment in $^{18}O_2$. This indicates that the oxygen is catalytically introduced by the enzyme. Isolation of the alkylated hemes indicates that the three olefins exclusively alkylate the nitrogen of pyrrole ring D. This is in contrast with substituted acetylenes, which alkylate pyrrole ring A (see Section II.A.1.c.). In all cases, the terminal carbon attaches to the pyrrole nitrogen. Circular dichroism spectra of the *N*-(2-hydroxy-ethyl)protoporphyrin IX from ethylene inactivated cytochrome P-450 and the D ring *N*-ethylprotoporphyrin IX isomer from 3,5-dicarbethoxy-2,6-dimethyl-4-ethyl-1,4-dihydropyr-idine (DDEP)-inactivated enzyme (see Chapter 9, Section II.B.2.c.) indicate that the ethylene reacts with the same face of the heme as does DDEP (i.e., the side having the activated oxygen species). An active-site geometry was proposed in which there is a steric encumbrance in the region over pyrrole ring B and a lipophilic binding site to accommodate chains of at least six carbon atoms over pyrrole ring C.

The stereochemistry of terminal olefin epoxidation and of prosthetic heme alkylation was determined by Ortiz de Montellano et al.[67] in order to resolve the dichotomy between retention of stereochemistry found during epoxidation of olefins and alkylation of cytochrome P-450 heme. Oxidation of 1-octene by cytochrome P-450 results in concurrent formation of 1,2-epoxyoctane and in reaction with the heme to give *N*-(2-hydroxyoctyl)protoporphyrin IX. Both processes occur stereospecifically. The stereochemistry of *trans*-1-[1-^{2}H]octene is re-tained during epoxidation and heme alkylation. This requires that addition of the pyrrole nitrogen and the activated oxygen occur on the same side of the double bond. This would not be possible if the heme were alkylated by the epoxide. Both the *R*- and *S*-epoxides are formed (*S* is slightly favored) by oxidation of the *si* and *re* faces, respectively, of the octene. However, alkylation of the heme only occurs during *re* face oxidation. The stereospecificity of the two processes requires that either they proceed by independent mechanisms or they diverge from a common acyclic intermediate. The partition ratio for inactivation, therefore, is determined by the orientation of the double bond relative to the face of the heme group.

As a model for the inactivation of cytochrome P-450 by alkenes, iron tetraarylporphyrin chloride was treated by Mansuy et al.[67a] with various alkenes in the presence of iodosyl-benzene. In addition to epoxide formation there was an oxidative destruction of the iron porphyrin. Acidic work-up produced green pigments indicative of *N*-substituted porphyrins as is observed in the enzyme system.[67] The reaction of [5,10,15,20-tetrakis(2,6-dichloro-phenyl)porphinato] iron (III) chloride and pentafluoroiodosylbenzene with 4,4-dimethyl-1-pentene was used by Mashiko et al.[67b] as a model for the inactivation of cytochrome P-450 by olefins. An *N*-alkylporphyrin is obtained, as was found by Ortiz de Montellano et al.[67] in the enzyme system. The stereochemistry of the corresponding reaction with 3-methyl-1-butene was determined by Collman et al.[67c] to be a stereospecifically *syn* addition, as is observed for cytochrome P-450 itself.[67] The reaction also is olefin structure dependent. Only one product was obtained with the *E*- or *Z*-1-deuterio olefin. The earlier model systems of Mansuy et al[67a] using alkenes and Fe(TPP or TpClPP)(Cl)-PhIO was shown by Artaud et al.[67d] to result in binding of the pyrrole nitrogen to the more substituted carbon of RCH=CH$_2$, not to the less substituted carbon, as in the case of P-450[67] and other model studies.[67b,67c]

Inactivation of rat liver cytochrome P-450 by 1-octene leads to the covalent attachment of the heme to the apoenzyme,[67e] as was found by Davies et al.[57a] for CCl$_4$ and allyliso-propylacetamide. The basis for the covalent binding is unknown.

During hemin-catalyzed epoxidation of norbornene in the presence of pentafluoroiodo-sylbenzene, the hemin was converted to a reversibly-formed *N*-alkylhemin (Structural Formula 10.34b, Scheme 58A) which is a catalyst for epoxidation.[67f] Traylor et al.[67f] suggest that enzyme inactivation may be a dead-end product from the cation intermediate (Structural Formula 10.34a, Scheme 58A) in epoxide formation.

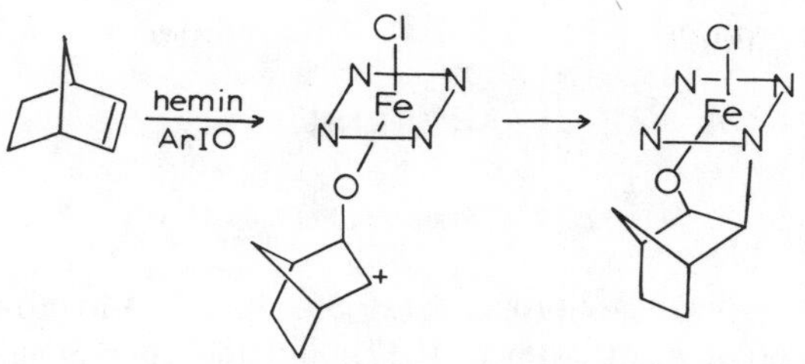

10.34a   10.34b

Scheme 58A.   Containing Structural Formulas 10.34a and 10.34b.

The time-dependent inactivation of cytochrome P-450 from liver of phenobarbital-treated mice by aliphatic olefins was studied by Testai et al.[68] Of the nine olefins tested only the vinyl alicyclic ones destroyed cytochrome P-450 in the presence of NADPH. Epoxides do not destroy the cytochrome P-450; therefore, they are not responsible for olefin inactivation. The partition ratio for epoxidation of vinylcyclooctane versus cytochrome P-450 inactivation in phenobarbital-treated mouse liver is 132.[69]

Secobarbital and other allyl-containing barbiturates produce time-dependent inactivation of liver microsomal cytochrome P-450 from phenobarbital-treated rats in the presence of NADPH.[52,70] An epoxide intermediate was suggested by Levin et al.[70]

9-Decenoic- and 11-dodecenoic acids are time-dependent inactivators of microsomal lauric acid in-chain hydroxylase from *Helianthus tuberosus* L.; the most effective is 11-dodecenoic acid.[17] Neither 9-decenoic nor 11-dodecenoic acid decreases the microsomal concentration of cytochrome b_5, NADH-cytochrome b_5 reductase, or NADPH cytochrome P-450 reductase. 9-Decenoic acid does not inhibit cinnamic acid 4-hydroxylation; 11-dodecenoic acid is less specific. These results indicate that inactivators of plant monooxygenases can be rationally derived from structure-activity relationships with hepatic enzymes. 10-Dodecenoic acid is a weak inactivator of rat liver microsomal ω- and ω-1 lauric acid hydroxylases (cytochrome P-450 enzymes from uninduced rats).[16]

c. *Aryl-Substituted Ethylenes*
(1) Safrole (4-Allyl-1,2-methylenedioxybenzene)
Formation of a safrole (4-allyl-1,2-methylenedioxybenzene)-cytochrome P-450 complex is time dependent, NADPH dependent, and stable to dialysis.[71]

(2) 1-Amino-2-aryl-2-propenes
1-Amino-2-phenyl-2-propene (Structural Formula **10.35**, Ar = Ph; see Scheme 59) is a time-dependent inactivator of dopamine β-hydroxylase from bovine adrenals.[72] Both a reductant (ascorbate or ferrocyanide) and oxygen are required for inactivation. Dialysis does not regenerate enzyme activity. The production of 1-amino-2-phenyl-2-propene oxide was determined by May et al.[72] by acid hydrolysis to 2,3-dihydroxy-2-phenyl-1-propylamine. Further studies were carried out by Padgette et al.[73] When [^{3}H] inactivator was used, 5 mol of tritium were incorporated per mol of tetrameric enzyme. The inactivator also is a substrate for the enzyme, turning over 913 times prior to inactivation. The product was identified as the corresponding epoxide. The rate of inactivation by **10.35** (Ar = Ph) is not changed by the presence of the epoxide, indicating that the epoxide is not responsible for inactivation.

$$\begin{array}{c} CH_2 \\ \| \\ ArCCH_2NH_2 \end{array}$$

Scheme 59. Structural Formula 10.35.

1-Amino-2-(4-hydroxyphenyl)-2-propene (**10.35**, Ar = 4-hydroxyphenyl) has an increased k_{inact} and apparent k_{cat} (turnover 1037), but the corresponding 3-hydroxyphenyl compound (**10.35**, Ar = 3-hydroxyphenyl) has very weak inactivation properties. 4-Hydroxy-α-methylstyrene is an inactivator, but a weak substrate (partition ratio 20).

3-Phenylpropargylamine (Structural Formula 10.36, Scheme 60) is a potent inactivator, but shows no detectable substrate activity. α-Methylstyrene, α-(cyanomethyl)styrene, and 3-hydroxy-α-methylstyrene are weak inactivators. Padgette et al.[73] suggested that the mechanisms of substrate oxygenation and enzyme inactivation are related to the corresponding cytochrome P-450 reactions (Scheme 61). *N*-Phenylethylenediamine and *N*-methyl-*N*-phenylethylenediamine also are time-dependent inactivators and substrates, but 2-phenoxyethylamine is a competitive inhibitor and not a substrate nor inactivator.

$$PhC{\equiv}CCH_2NH_3^+$$

Scheme 60. Structural Formula 10.36.

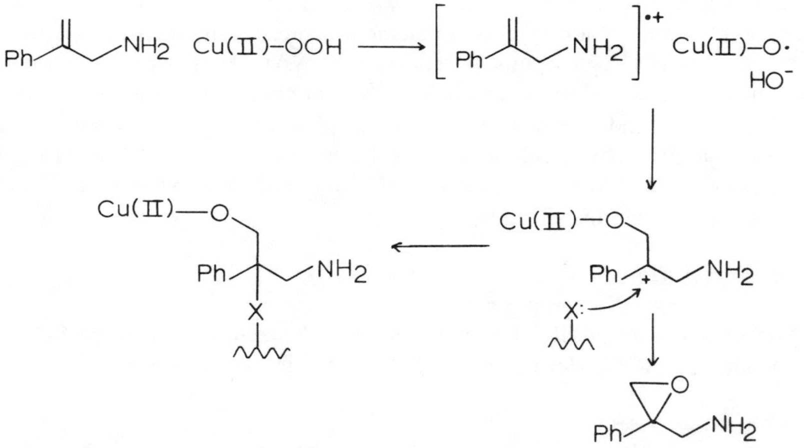

Scheme 61.

Compound **10.35** (Ar = Ph) was synthesized[73] from α-methylstyrene (Scheme 62). 3-Phenylpropargylamine was synthesized by Klemm et al.[74] (Scheme 63).

$$\begin{array}{c} CH_2 \\ \| \\ PhCCH_3 \end{array} \xrightarrow{NBS} \begin{array}{c} CH_2 \\ \| \\ PhCCH_2Br \end{array} \xrightarrow{NH_3} \begin{array}{c} CH_2 \\ \| \\ PhCCH_2NH_2 \end{array}$$

Scheme 62.

$$PhC{\equiv}CCH_2OH \xrightarrow[\text{pyr}]{SOCl_2} PhC{\equiv}CCH_2Cl \xrightarrow[\substack{\text{2. } NH_2NH_2 \\ \text{3. } HCl, \Delta}]{\text{1. } KNPhth} \text{10.36}$$

Scheme 63.

Analogues related to **10.35**, namely, **10.37** (R = H and Me, X = S,O) and **10.38** (X = S,O) (see Scheme 64) as well as compound **10.36** were prepared by Barger et al.[75] and shown to be time-dependent inactivators of beef adrenals dopamine β-hydroxylase. Inactivation is ascorbate- and oxygen-dependent and is not reversed by dialysis. Compound **10.37** (X = S, R = H) and **10.36** are equipotent and about 10^3 times more active than phenyl-allylamine. The inactivation mechanism proposed is the same as that proposed by Padgette et al.[73] The synthesis of **10.37** (X = S,O; R = H) is shown in Scheme 65.[75] Compound **10.38** (X = S) was prepared by this same route starting with 3-acetylthiophene. Compound **10.38** (X = O) was synthesized by conversion of ethyl 3-furoate with 2 equivalents of methylmagnesium bromide to the tertiary alcohol. This was dehydrated with $KHSO_4$, then the 3-(1-methylvinyl)furan was carried on to product by the same route as in Scheme 65.

Scheme 64.   Structural Formulas 10.37 and 10.38.

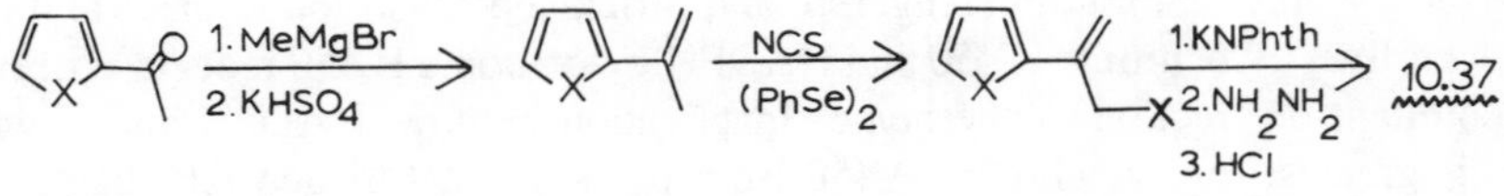

Scheme 65.

Methoxsalen (8-methoxypsoralen, (Structural Formula **10.38a** (R = OMe, R′ = H), Scheme 65A) is a time-dependent, irreversible inactivator of rat liver microsomal cytochrome P-450 in the presence of NADPH and oxygen.[75a] There was no evidence for lipid peroxidation or heme destruction. Glutathione does not inhibit the loss of cytochrome P-450. The inactivation mechanism may resemble that of other alkene-containing inactivators. Rat liver microsomal cytochrome P-450 also is irreversibly inactivated by bergapten (5-methoxyp-soralen, **10.38a**; R = H, R′ = OMe) and psoralen (**10.38a**; R = R′ = H) in the presence of NADPL and oxygen; trioxsalen (**10.38b**) is not an inactivator.[75b] Methoxsalen bergapten, and psoralen also were found to be irreversible inactivators of human liver microsomal cytochrome P-450; trioxsalen (trimethylpsoralen) does not inactivate the human liver enzyme either.[75c]

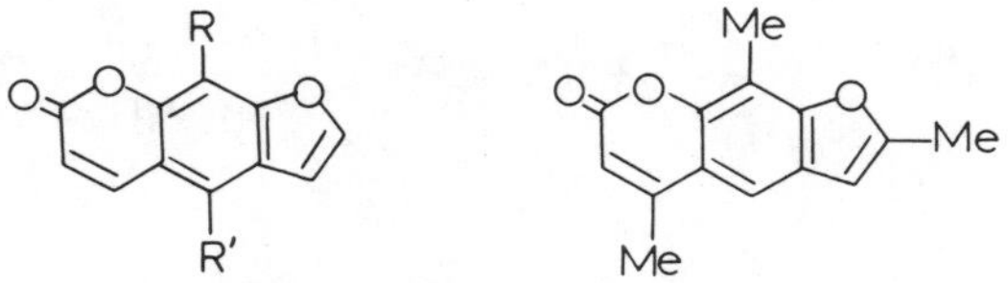

Scheme 65A.   Structural Formulas 10.38a and 10.38b.

d. Halogen- and Oxygen-Substituted Ethylenes

(1) 1-(2-Bromovinyl)pyrenes

trans-1-(2-Bromovinyl)pyrene (Structural Formula 10.39, Scheme 66 is a mechanism-based inactivator of the benzo[a]pyrene hydroxylase activity in liver microsomes of phenobarbital- or 5,6-benzoflavone-treated rats.[11] *cis*-1-(2-Bromovinyl)pyrene inactivates the hydroxylase activity only from 5,6-benzoflavone-treated rats. In the presence of NADPH, *trans*- or *cis*-1-(2-bromovinyl)pyrene causes destruction of P-450 content in microsomes from 5,6-benzoflavone- or phenobarbital-treated rats. The *cis*- and *trans*-isomers of **10.39** were prepared by a Wittig reaction of 1-formylpyrene with (bromomethyl)triphenyl-phosphonium bromide.

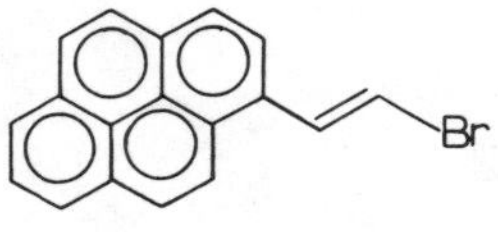

Scheme 66.   Structural Formula 10.39.

(2) Vinyl Monohalides and Ethers

Vinyl chloride decreases the levels of liver cytochrome P-450, particularly when rats are pretreated with phenobarbital.[76] Cytochrome b_5 and NADPH-cytochrome C reductase are not destroyed by vinyl chloride. Guengerich and Strickland[77] also found that vinyl chloride inactivates rat liver cytochrome P-450 and NADPH-cytochrome P-450 reductase irreversibly with concomitant destruction of the heme. Inactivation requires NADPH and oxygen. Inactivation is greatest for cytochrome P-450 from phenobarbital-treated rats than 3-methylcholanthrene-treated or untreated rats. Although a metabolite of vinyl chloride was proposed[77] to be responsible for inactivation, no metabolic products prepared were effective. Neither vinyl chloride epoxide nor chloroacetaldehyde, the potential oxygenated products of vinyl chloride, is effective in destroying cytochrome P-450. Inactivation is not prevented by the presence of radical traps. Furthermore, highly purified reconstituted enzyme systems containing NADPH-cytochrome P-450 reductase and cytochrome P-450 also are inactivated. This suggests that a metabolite may not be responsible, but, rather, inactivation occurs during oxygenation of vinyl chloride.

Ortiz de Montellano et al.[6] employed halogen- and oxygen-substituted olefins to determine the electronic effects on inactivation of cytochrome P-450. Vinyl fluoride produces an in vitro NADPH-dependent loss of enzyme activity and results in formation of a green pigment (the modified heme). Vinyl bromide causes no detectable in vitro loss of enzyme activity and gives only a very low yield of green pigment in vivo. Guengerich,[67e] however, found that vinyl bromide and fluoroxene inactivate rat liver cytochrome P-450 and result in covalent attachment of the heme to the apoenzyme, as was found by Davies et al.[57a] for CCl_4 and allylisopropylacetamide. Vinyl chloride is intermediate. Among the vinyl ethers (ethyl vinyl-, phenyl vinyl-, 2-chloromethyl vinyl- and 2,2,2-trifluoroethyl vinyl ether, vinyl acetate, and vinyl trifluoroacetate), only 2,2,2-trifluoroethyl vinyl ether (fluoroxene) is effective in destroying cytochrome P-450 and producing a green pigment.[6] The green pigments isolated with vinyl fluoride, vinyl bromide, fluroxene, and acetylene are identical after esterification and have the structure *N*-(2-oxoethyl)protoporphyrin IX (dimethyl ester). This requires the addition of an oxygen at the trifluoroethoxy or halide-substituted terminus and attachment of the unsubstituted terminus to the heme. The vinyl halides and vinyl ethers provide a more sensitive probe for the mechanism. Three alternative mechanisms considered[6] are shown in Scheme 67. For electrophilic substitution, the cation intermediate expected would have, almost exclusively, the opposite orientation to that which is shown. The same trend is true for radicals, but the preference for vinyl fluoride is only in the range of 3 to 50 in favor of the orientation opposite that shown. Since the partition ratio for fluroxene is 296, the radical or radical cation pathway is plausible, but the cation pathway is highly unlikely.

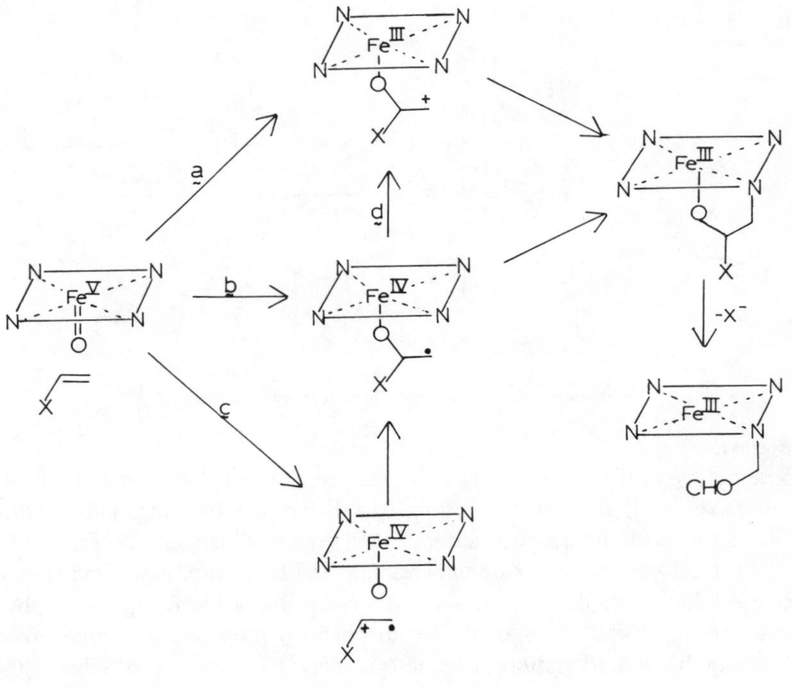

Scheme 67.

Hepatic microsomal cytochrome P-450 from rats undergoes time-dependent inactivation with destruction of the heme by fluroxene (an anesthetic) in the presence of a NADPH generating system.[78] Inactivation is enhanced by phenobarbital-induction of the cytochrome P-450,[78] but both phenobarbital- and 3-methylcholanthrene-induced cytochrome P-450 are inactivated.[13] Since 2,2,2-trifluoroethyl ethyl ether does not inactivate the enzyme, the double bond must be important.[78] Other components of the electron transport system, e.g., cytochrome b_5 and NADPH-cytochrome C reductase, are not affected, suggesting that the reactive species is not released into solution. Further studies by Ivanetich et al.[79] indicate that cytochrome P-450 enhances the rate of metabolism of the anaesthetic. In contrast, cytochrome P-448 does not metabolize fluroxene. A variety of inhibitors and inducing agents of cytochromes P-450 were used to determine if the destruction of rat liver microsomal cytochrome P-450 by fluroxene requires metabolism of the drug, i.e., cytochrome P-450 inactivation requires active enzyme.[80] The comparative studies were carried out with liver microsomes from phenobarbital- and 3-methylcholanthrene-treated and untreated rats; the partition ratios are 296, 107, and 226, respectively. Lipid peroxidation is not involved in cytochromes P-450 destruction. The total heme content is not affected by the enzyme inactivation. The results indicate that fluroxene destruction of cytochromes P-450 involves both cytochrome P-450 and cytochrome P-448, but production of trifluoroethanol from fluroxene is catalyzed only by cytochrome P-450.

(3) Vinylidene Chloride

Vinylidene chloride (Structural Formula 10.40, see Scheme 68) is oxidized by rat liver cytochrome P-450 to chloro- and dichloroacetic acid with concomitant inactivation of three of the eight cytochrome P-450 isozymes examined by Liebler and Guengerich.[81] Vinylidene chloride epoxide is not an obligate intermediate in the formation of the metabolites. On the basis of model reactions, heme does not appear to act simply as a Lewis acid in these reactions and, therefore, the mechanism shown in Scheme 68 was proposed to account for the results.

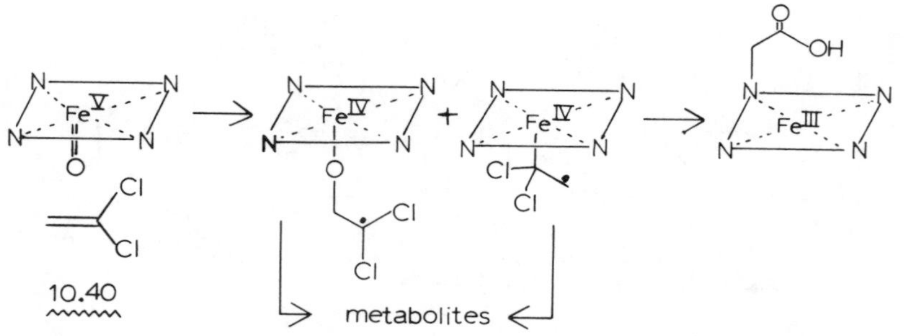

Scheme 68.   Containing Structural Formula 10.40.

(4) Trichloroethylene

Trichloroethylene (Structural Formula 10.41, see Scheme 69) is metabolized by cytochrome P-450 to chloral, glyoxylic acid, formic acid, carbon monoxide, and trichloroethylene oxide concomitant with heme destruction.[82] Miller and Guengerich[82] carried out model reactions with trichloroethylene oxide under acidic and basic conditions and showed thatchloral is not a product of its decomposition, even when iron salts, ferriprotoporphyrin IX, or purified cytochrome P-450 are added. The mechanism proposed to explain formation of chloral (pathway b) and alkylation of heme (pathway a) is shown in Scheme 69.

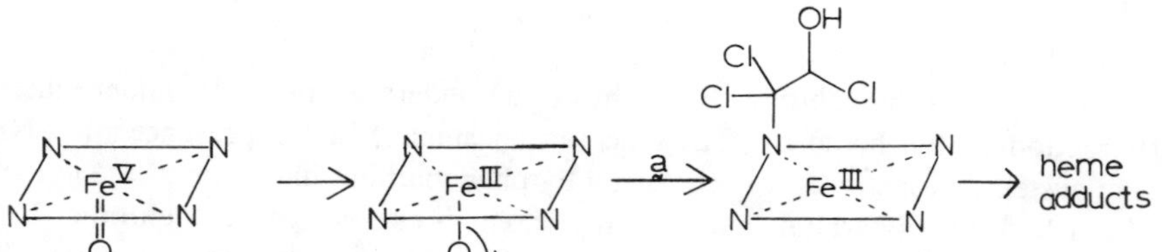

Scheme 69.   Containing Structural Formula 10.41.

Guengerich[67e] found that inactivation of cytochrome P-450 by **10.41** leads to covalent attachment of the heme to the apoenzyme, as was found by Davies et al.[57a] for CCl_4 and allylisopropylacetamide.

(5) Saturated Aliphatic Aldehydes

Various saturated aliphatic aldehydes also are time-dependent inactivators of liver microsomal cytochrome P-450 from phenobarbital-treated rats.[83] NADPH and O_2 are required for inactivation. Formaldehyde and acetaldehyde are inactive, but butanal, hexanal, octanal decanal, and dodecanal inactivate the enzyme. Inactivation involves covalent attachment to the heme. The visible spectrum is similar to that of the heme adduct formed from octyne (see Section II.A.1.c.). It is not clear if the aldehydes are metabolized to the reactive species or are mechanism-based, but there is no lag time in the inactivation of cytochrome P-450 by the aldehydes, thus supporting a mechanism-based inactivation. The corresponding alcohols are not inactivators. No evidence for ketene generation was obtained by White.[83] Although the mechanism of inactivation is not known, oxygenation of the enol forms of the compounds may be involved.

e. 2-Halo-3-arylpropenes

2-Y-3-(*p*-Hydroxyphenyl)-1-propenes (Y = H, Br, Cl) (Structural Formula 10.41a, see Scheme 70) are mechanism-based inactivators of the copper-containing enzyme, dopamine β-hydroxylase, from bovine adrenal medulla.[83a-c] Because the inactivation mechanism is not well defined, these compounds also are discussed in Chapter 9 (Section II.A.1.c.). An inactivation mechanism proposed by Colombo et al.[83b] that involves initial hydroxylation is shown in Scheme 70. Treatment of inactivated enzyme with sodium cyanoboro[³H]hydride or sodium boro[³H]hydride, however, leads to no incorporation of tritium, even under denaturing conditions. This leaves doubt regarding the proposed ketone-containing adduct.

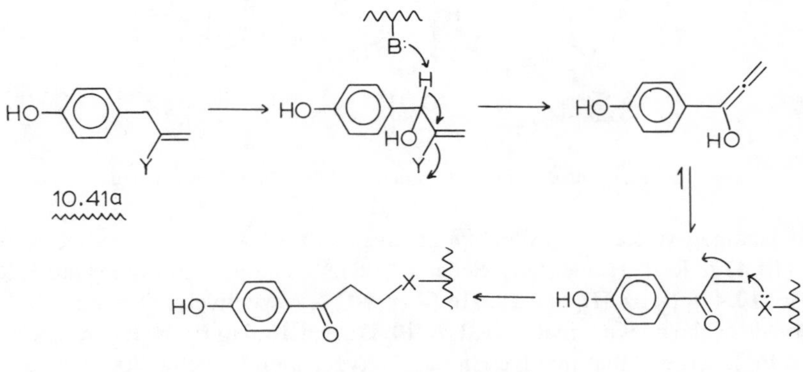

Scheme 70.   Containing Structural Formula 10.41a.

4. Thiourea Analogues (Goitrogens)

Thiourea,[83d,83e] thiouracil (Structural Formula 10.41b, R = H; Scheme 70A),[83d,83f] 6-*n*-propylthiouracil (**10.41b**, R = *n*-Pr),[83d,83f-h] and methylmercaptoimidazole (**10.41c**)[83f,83i] are time-dependent inactivators of procine thyroid peroxidase in the presence of hydrogen peroxide. Gel filtration does not regenerate enzyme activity.[83f] Methylmercaptoimidazole accelerates the conversion of Compound I to Compound II and reacts with Compound II to form the inactivated enzyme.[83i] Concomitant with a shift in the Soret spectrum and enzyme inactivation, 1 mol of [¹⁴C] **10.41c** is irreversibly incorporated per mole of enzyme.[83h] When [³⁵S]6-*n*-propylthiouracil was used by Engler et al.,[83h] only 0.5 mol of label was incorporated after 87% enzyme inactivation. Since glutathione does not inactivate the enzyme, inactivation is not simply the result of a thiol addition. 3-Amino-1,2,4-triazole and *p*-aminobenzoate also inactivate porcine thyroid peroxidase.[83d]

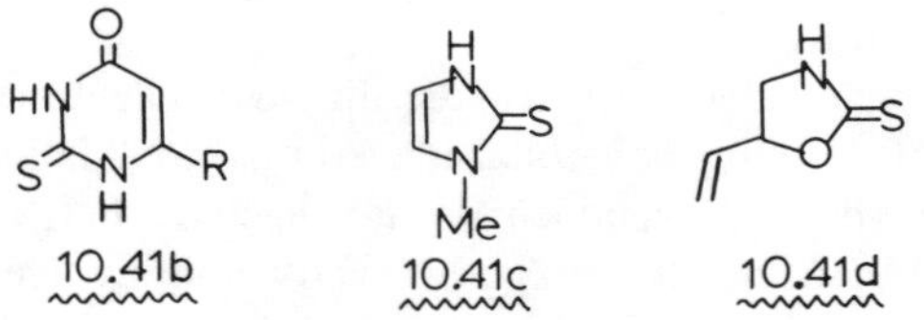

Scheme 70A.   Structural Formulas 10.41b, 10.41c, and 10.41d.

Methylmercaptoimidazole also is a time-dependent inhibitor of human thyroid peroxidase; dialysis does not restore enzyme activity.[83j] Propylthiouracil, however, does not inactivate the enzyme.

Chloroperoxidase undergoes time-dependent inactivation by various thioureylene antithyroid drugs, namely, thiourea,[83k] thiouracil,[83k] 5-vinyl-2-oxazolidinethione[83k] (Structural Formula 10.41d, Scheme 70A), and **10.41c**.[83l]

The antithyroid goitrogens, **10.41c**,[83f,83i,83m-p] **10.41b** (R = H),[83f] methylthiouracil (**10.41b**, R = Me),[83m,83n] **10.41b** (R = *n*-Pr),[83f] 2-mercaptopyrimidine[83m] (**10.41e**), and thiobarbituric acid[83m] (**10.41f**) (the latter two are shown in Scheme 70B) are time-dependent inactivators of lactoperoxidase. The potencies of **10.41c**, **10.41e**, **10.41b** (R = Me), and **10.41f** are in decreasing order of that listed, respectively.[83m] The reaction of **10.41c** with bovine lactoperoxidase resembles that with porcine thyroid peroxidase, but occurs at a slower rate.[83i]

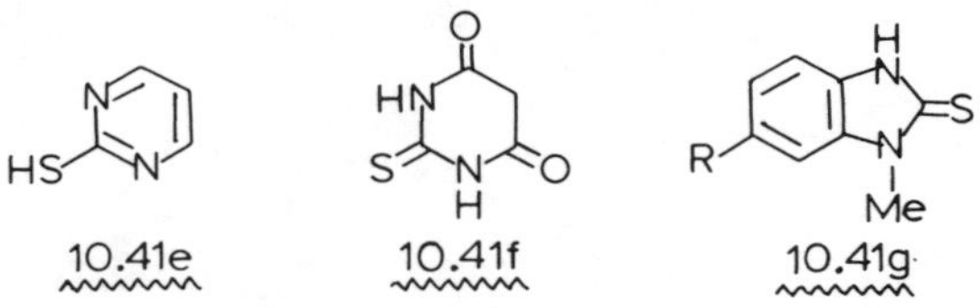

Scheme 70B.   Structural Formulas 10.41e, 10.41f, and 10.41g.

Bovine lactoperoxidase is irreversibly inactivated by **10.41c** and methylmercaptobenzimidazoles (**10.41g**, R = H and NO$_2$; Scheme 70B).[83p] The partition ratios are 9, 6, and 89 for **10.41c**, **10.41g** (R = H), and **10.41g** (R = NO$_2$), respectively. Different visible spectra are observed for the enzyme inactivated by **10.41c** and **10.41g** (R = H). A mechanism was suggested by Doerge[83p] that involves initial *S*-oxygenation (Scheme 70C).

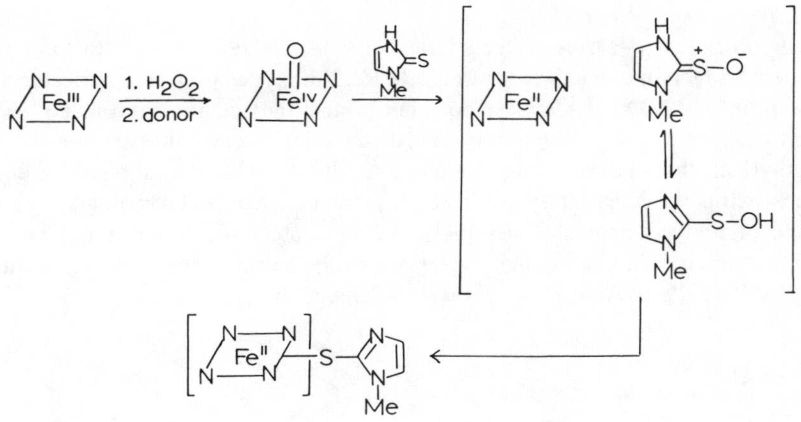

Scheme 70C.

The only product obtained by Doerge et al.[83q] from the oxidation of 2-mercaptobenzothiazole (Structural Formula 10.41h, Scheme 70D) by bovine lactoperoxidase and H_2O_2 was 2,2′-bis-dithiobenzothiazole (**10.41i**), suggested to be obtained via the sulfenic acid. This is in agreement with the inactivation mechanism shown in Scheme 70C.

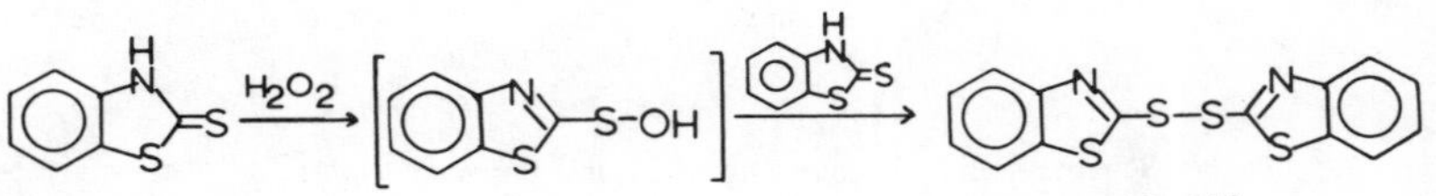

10.41h 10.41i

Scheme 70D. Containing Structural Formulas 10.41h and 10.41i.

In support of an electron transfer mechanism, phenacylphenyl sulfide (Structural Formula 10.41j) was shown by Doerge et al.[83q] to produce the corresponding sulfoxide and phenyldisulfide (Scheme 70E) upon incubation with lactoperoxidase and H_2O_2.

$$\underset{10.41j}{\mathrm{PhSCH_2C(=O)Ph}} \rightarrow \mathrm{Ph\dot{S}^{+}CH_2C(=O)Ph} \xrightarrow{-H^+} \mathrm{PhS\dot{C}HC(=O)Ph} \rightarrow \mathrm{PhSSPh}$$

$$\downarrow$$

$$\mathrm{PhS(=O)CH_2C(=O)Ph} + 1/2\ \mathrm{PhSSPh}$$

Scheme 70E. Containing Structural Formula 10.41j.

A linear free energy relationship was observed between the electrochemical oxidation potential of the inactivators and their K_I values as inactivators of lactoperoxidase.[83q] These results are further support for an one-electron S-oxidized intermediate and the formation of reactive S-oxygenated products during lactoperoxide inactivation.

5. 3-[(Methylthio)methyl]catechol

3-[(Methylthio)methyl]catechol (Structural Formula 10.41k, Scheme 70F) is a time- and O_2-dependent irreversible inactivator of catechol 2,3-dioxygenase (metapyrocatechase) with a partition ratio of 22,000.[83r] Catalase does not protect the enzyme from inactivation nor is inactivated enzyme reactivated by iron salts or reducing agents. Inactivation by 3-[(methylthio)methyl]catechol under aerobic conditions results in the incorporation of 1.6 equivalents of tritium which is not released by gel filtration, anion exchange chromatography, treatment with thiols, or dialysis against NaDodSO$_4$. 3-Propylcatechol is a substrate, but does not inactivate the enzyme, indicating the importance of the sulfur to inactivation. A mechanism suggested by Pascal and Huang[83r] is shown in Scheme 70F.

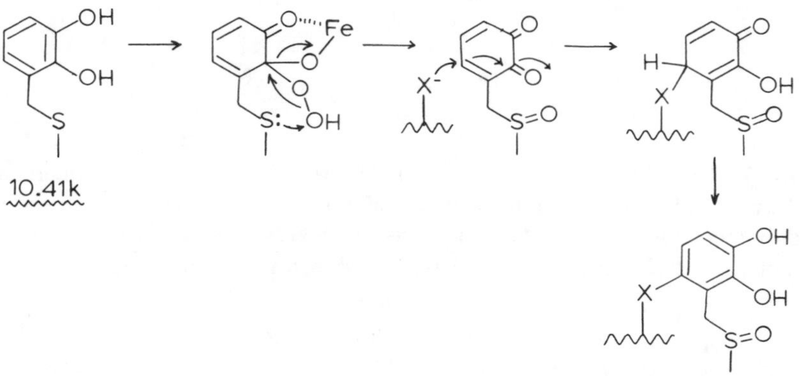

Scheme 70F.   Containing Structural Formula 10.41k.

6. p-Cresol

Numerous compounds containing various functional groups inactivate dopamine β-hydroxylase. *p*-Cresol (Structural Formula 10.41l, Scheme 70G) also is a pseudo first order time-dependent inactivator of this enzyme in the presence of O_2 and ascorbate; the partition ratio is about 1500.[83s] Since *p*-cresol lacks a latent electrophile, it was suggested by Goodhart et al.[83s] that radical stabilization may be the important factor involved in inactivation, not necessarily the presence of a latent electrophile. The presence of an amino group, apparently, is unimportant to binding because *p*-cresol has an affinity for the enzyme comparable to tyramine. The mechanism proposed is shown in Scheme 70G.

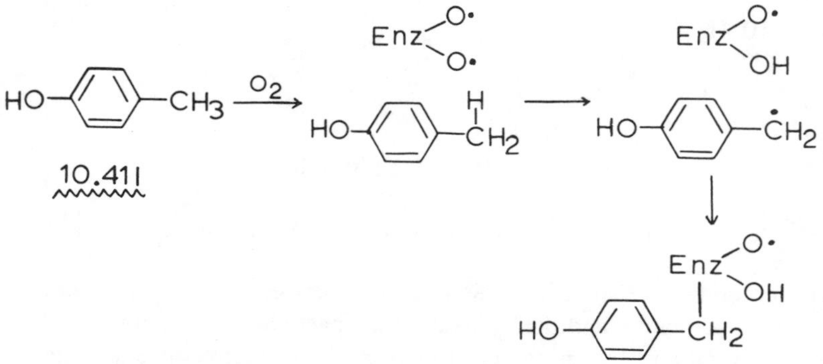

Scheme 70G.   Containing Structural Formula 10.41l.

7. 7α-(4'-Amino)phenylthio-1,4-androstadiene-3,17-dione

7α-(4'-amino)phenylthio-1,4-androstadiene-3,17-dione (Structural Formula 10.41m, Scheme 70H) is a time-dependent inactivator of human placental aromatase, only in the presence of NADPH.[83t] Cysteine does not protect the enzyme. The inactivation mechanism is unknown, but Snider and Bruggemeier[83t] suggest that inactivation may occur after hydroxylation of the C_{19} carbon.

Compound **10.41m** can be synthesized as shown in Scheme 70H.

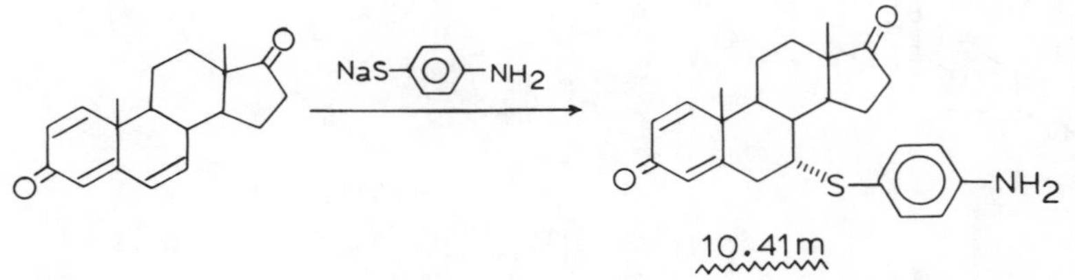

Scheme 70H.   Containing Structural Formula 10.41m.

III. OXYGENATION/REARRANGEMENT/ADDITION

A. 17β-Hydroxy-1-methylthioestra-1,4-dien-3-one

On the basis of the ease of sulfur oxygenation to sulfoxides and the [2,3] sigmatropic rearrangement of sulfoxides to sulfenyl esters, 17β-hydroxy-10-methylthioestra-1,4-dien-3-one (Structural Formula 10.42) was synthesized as a potential mechanism-based inactivator of aromatase (Scheme 71).[84] Time-dependent inactivation of human placental aromatase was observed in the presence of a NADPH-generating system. However, in the absence of a NADPH-generating system, time-dependent inactivation still occurred, albeit at one-half the rate. Both a mechanism-based pathway (pathway a) and an affinity-labeling pathway (pathway b) are possible (Scheme 72).

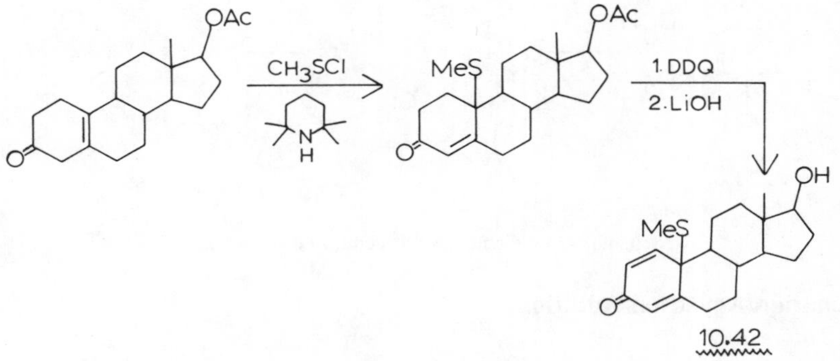

Scheme 71.   Containing Structural Formula 10.42.

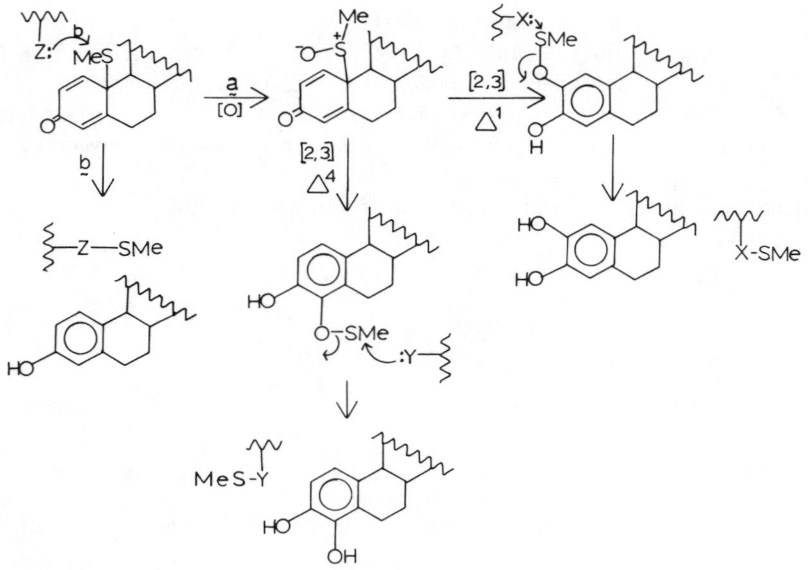

Scheme 72.

IV. OXYGENATION/ACYLATION

A. Oxygenation/Acylation

1. 4-Chloro-3-hydroxyanthranilic Acid

4-Chloro-3-hydroxyanthranilic acid (Structural Formula 10.43) is a time-dependent in-activator of 3-hydroxyanthranilic acid oxidase from rat liver.[85] Dialysis does not restore enzyme activity. The mechanism proposed is shown in Scheme 73.

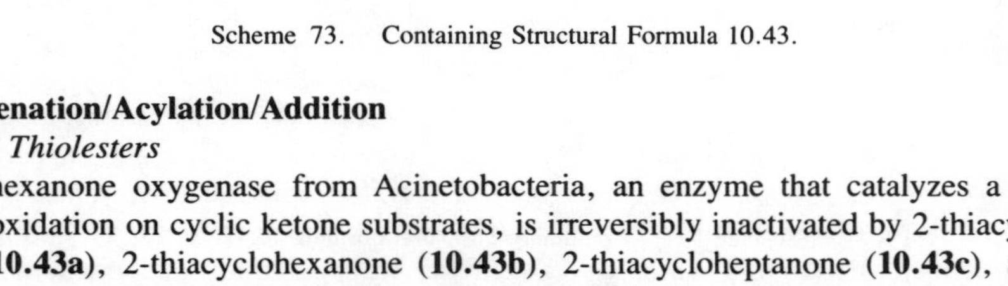

10.43

Scheme 73.   Containing Structural Formula 10.43.

B. Oxygenation/Acylation/Addition

1. Cyclic Thiolesters

Cyclohexanone oxygenase from Acinetobacteria, an enzyme that catalyzes a Baeyer-Villiger oxidation on cyclic ketone substrates, is irreversibly inactivated by 2-thiacyclopen-tanone (**10.43a**), 2-thiacyclohexanone (**10.43b**), 2-thiacycloheptanone (**10.43c**), and eth-ylene monothiocarbonate (**10.43d**) (see Scheme 74).[85a] The partition ratios are 33, 14, 17, and 110, respectively. Inactivation requires NADPH and O_2. The native enzyme reacts with one equivalent of inactivator in the absence of NADPH and O_2 without loss of enzyme activity, but with concomitant loss of one sulfhydryl group (out of a total of five in denatured enzyme). When active enzyme is treated as above and then is incubated with [^{35}S]-2-thiacyclopentanone, inactivation occurs with incorporation of 0.9 equivalent of radioactivity and loss of two more sulfhydryl groups. The inactivation mechanism proposed by Latham and Walsh[85a] for ethylene monothiocarbonate is shown in Scheme 75. The activation step is a Baeyer-Villiger oxygenation. An alternative intermediate would be initial acyl sulfoxide formation, but enzyme nucleophilic attack would then produce the same enzyme-bound thiocarbonate.[85b] When the enzyme was treated with S-phenylthioacetate, oxygenation took place, but no inactivation. Therefore, the cross-linking step appears to be important to inactivation.

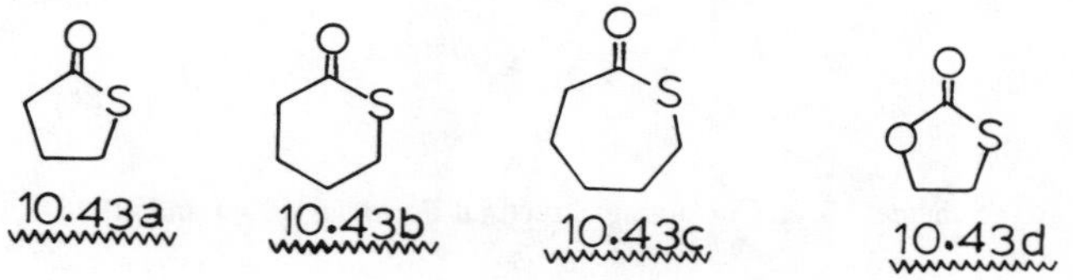

Scheme 74.   Structural Formulas 10.43a to d.

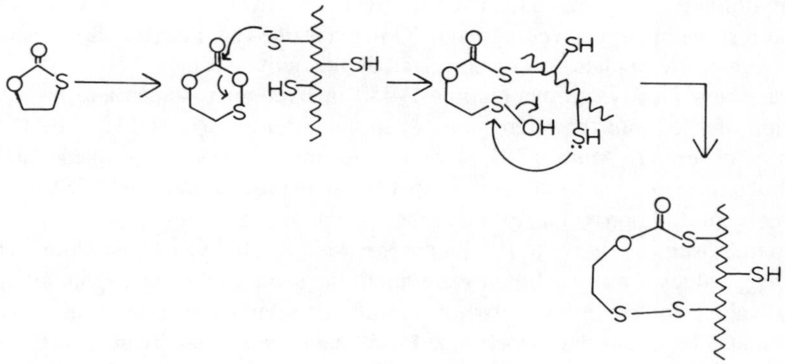

Scheme 75.

V. OXYGENATION/ELIMINATION

A. Oxygenation/Elimination/Addition

1. Parathion (Dyfonate®)

The conversion of the insecticide parathion (Structural Formula 10.44) to paraoxon (Structural Formula 10.45), catalyzed by rabbit liver microsomes, results in the incorporation of [^{18}O] from $^{18}O_2$, but not from $H_2^{18}O$.[86] NADPH is required for inactivation. A mechanism proposed by Ptashne et al.[86] is shown in Scheme 76. Concurrent with and independent of Ptashne et al.[86] the same $^{18}O_2$ and $H_2^{18}O$ experiments were carried out on parathion by McBain et al.[87] The same results were obtained and the same mechanism was suggested. A model study for the microsomal oxidation of parathion was carried out by McBain et al.[88] *m*-Chloroperoxybenzoic acid was allowed to react with parathion and the same products were obtained as in the microsomal system. A similar model study for the cytochrome P-450-catalyzed oxidation of parathion was carried out by Ptashne and Neal.[89] Parathion was treated with peroxytrifluoroacetic acid and the products obtained were the same as those from the enzyme-catalyzed reaction, namely, paraoxon and diethylphophorothoic acid.

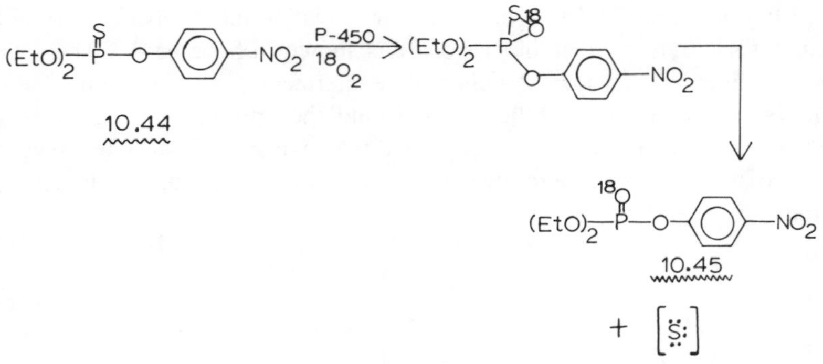

Scheme 76.   Containing Structural Formulas 10.44 and 10.45.

The reactivity of thiono sulfur-containing compounds towards cytochrome P-450 from untreated, phenobarbital-treated, and 3-methylcholanthrene-treated rats was shown by Hunter and Neal[90] to be general. The inactivation requires NADPH. De Matteis[91] found that parathion and phenylthiourea inactivate phenobarbital-treated rat liver microsomal cytochrome P-450, but the corresponding oxygen compounds (O in place of S) are inactive. Some sort of reactive sulfur generated by oxidation was suggested as the active species.

[^{35}S]Parathion inactivates cytochrome P-450 in liver microsomes and results in the incorporation of [^{35}S] into the microsomes.[92] In the absence of NADPH, little [^{35}S] incorporation was observed. More [^{35}S] is bound to microsomes from phenobarbital- or 3-methylcholanthrene-treated rats than from untreated rats. Binding of [^{35}S] is inhibited by carbon monoxide. Approximately four times as much [^{35}S] from [^{35}S]parathion is incorporated into microsomes relative to [^{14}C] incorporation from [ethyl-^{14}C]parathion. The apparent K_m and V_{max} values for sulfur binding are about the same as the corresponding apparent K_m and V_{max} values for the metabolism of parathion to paraoxon, suggesting that the sulfur bound is that released in the cytochrome P-450-catalyzed metabolism, and that it is highly reactive. Paraoxon does not inactivate cytochrome P-450 and, therefore, Norman et al.[92] suggested that the released sulfur is responsible for the inactivation. Incubation of purified liver microsomal cytochrome P-450 from phenobarbital-treated rats with [^{35}S]parathion results in the incorporation of 1 nmol of [^{35}S] per nanomole of enzyme when NADPH and O_2 are present.[93] No incorporation of [^{35}S] results in the absence of NADPH. Inactivation of the enzyme with [ethyl-^{14}C]-parathion, even in the presence of NADPH and O_2, gives no radioactive labeling of the enzyme. NADPH-cytochrome C reductase is not labeled by either radioactive compound. When microsomes are treated with [^{35}S]parathion, then treated with antibody to rat liver cytochrome P-450, 70% of the radioactivity in the microsomes is removed. This indicates that most of the radioactivity is bound to cytochrome P-450. Kamataki and Neal[93] suggested that the reactive sulfur atom becomes attached to the enzyme in a hydrodisulfide linkage to an active-site cysteine. This may lead to cross-linkage of proteins, since NaDodSO$_4$-polyacrylamide gel electrophoresis shows high-molecular-weight complexes.

The metabolism of parathion by a reconstituted mixed-function oxidase enzyme system and a cumene hydroperoxide-cytochrome P-450 system was compared by Yoshihara and Neal.[94] Earlier studies[93] indicated that most of the sulfur in a reconstituted system was bound to cytochrome P-450. This was confirmed by the observation[94] that the same percentage of sulfur released in the metabolism of parathion that becomes bound to cytochrome P-450 of the cumene-hydroperoxide-cytochrome P-450 system, also is bound to proteins of the reconstituted system. Since cumene hydroperoxide can replace NADPH and O_2 for the activation of cytochrome P-450, it was shown by Yoshihara and Neal[95] that parathion also is a time-dependent inactivator of liver microsomal cytochrome P-450 from phenobarbital-treated rats in the presence of cumene hydroperoxide. Inactivation by [^{35}S]-parathion occurs concomitant with incorporation of [^{35}S] into protein.

Rat liver microsomes were incubated with [^{35}S]parathion and the [^{35}S]-labeled protein(s) was degraded.[96] The sulfur was shown by Davis and Mende[96] to be attached to a cysteine either as a polysulfide or hydrodisulfide. Small peptides or amino acids with [^{35}S] attached are not produced by proteolytic enzymes. With the use of purified rabbit liver cytochrome P-450, it was shown that 4-5 mole-atoms of [^{35}S] are bound per mole of enzyme. At least three different [^{35}S]-containing amino acids were isolated after acid hydrolysis of [^{35}S]parathion-inactivated, purified rat liver cytochrome P-450.[97] Insertion of atomic sulfur into C–H bonds of valine, leucine, or isoleucine and into the aromatic ring of tyrosine was proposed by Halpert et al.[97] Upon inactivation, 80% of the heme is lost. Dithiothreitol releases 75% of the [^{35}S] bound to the enzyme. Inactivation of rat liver cytochrome P-450 by parathion leads to the covalent attachment of the heme to the aopenzyme,[67e] as was found by Davies et al.[57a] for CCl_4 and allylisopropylacetamide.

2. Carbon Disulfide

Cytochrome P-450 in liver microsomes is inactivated by treatment with carbon disulfide only in the presence of NADPH.[98,99] [^{35}S]Carbon disulfide inactivation leads to incorporation of [^{35}S] into the microsomes; carbon monoxide inhibits the incorporation of [^{35}S].[99] Five times as much [^{35}S] is incorporated as [^{14}C] from [^{14}C]carbon disulfide. An equivalent amount of COS is formed in the reaction as [^{35}S] bound. A mechanism proposed by Dalvi et al.[99] is shown in Scheme 77. A similar study by De Matteis[91] using [^{14}C]- and [^{35}S]-labeled carbon disulfide revealed that the [^{35}S] becomes bound only in the presence of NADPH and O_2. Some sort of reactive sulfur generated by oxidation of the compounds is suggested as the active species. Dalvi et al.[100] found that the K_m values for [^{35}S] binding to microsomes and COS formation are identical as are the V_{max} values. This indicates that the rate-limiting step in [^{35}S] binding is probably one of the reactions that leads to release of the sulfur from CS_2 and this supports a highly reactive sulfur species. The two likely candidates responsible for inactivation of cytochrome P-450 are the reactive sulfur and COS. COS is also shown to be an inactivator of the enzyme, but only in the presence of NADPH. It is suggested that a similar oxidation reaction with COS occurs as was seen with CS_2, releasing reactive S. There is not a decrease in the heme content after inactivation of the cytochrome P-450; therefore, sulfur binding must be to the protein. There are 9 nmol of sulfur attached to microsomes for every nanomole of cytochrome P-450 inactivated. Catignani and Neal[101] found that incubation of liver microsomes from phenobarbital-treated rats with [^{35}S]carbon disulfide and cyanide results in the release of 45% of the [^{35}S] as $^{35}SCN^-$. Similar results were obtained when microsomes treated with CS_2 were incubated with $^{14}CN^-$. These results support the formation of a protein-bound hydrodisulfide as the inactivation product. Since singlet sulfur is capable of C–H bond insertion, reactions analogous to those of carbenes and nitrenes, this may be responsible for the 55% of the [^{35}S] that is not released by cyanide. Obrebska et al.[102] suggested that since EDTA prevents inactivation of cytochrome P-450 in liver microsomes from untreated rats that the mechanism of destruction involves lipid peroxidation.

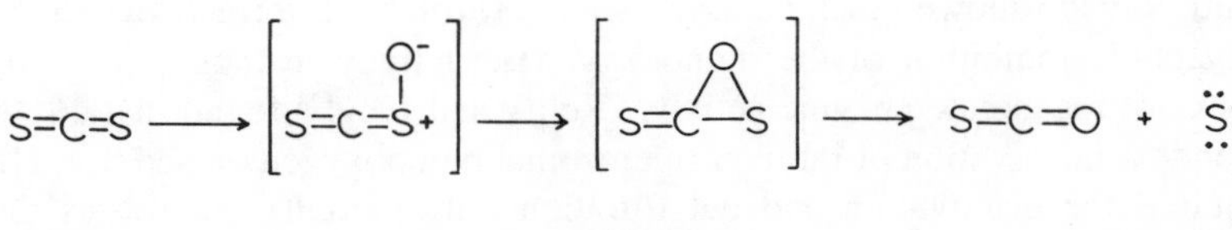

Scheme 77.

3. 3[(Arylthio)ethyl]sydnone

3-{[2-(2,4,6-Trimethylphenyl)thio]ethyl}-4-methylsydnone (Structural Formula 10.45a, Ar = 2,4,6-trimethylphenyl; Scheme 77A) causes the time- and NADPH-dependent loss of about 20% of liver microsomal cytochrome P-450 from phenobarbital-treated rats.[102a] This suggests a high degree of isozyme specificity. Pyruvic acid (16 molecules per inactivation event) is produced concomitant with inactivation, and the modified heme produced in vivo was identified by Ortiz de Montellano and Grab[102a] as a mixture of the four isomers of *N*-vinylprotoporphyrin IX (**10.45c**). On the basis of earlier model studies of White and Egger[102b] with sydnones that suggest oxidation occurs to pyruvate and the diazo intermediate (**10.45b**), the inactivation mechanism was proposed as shown in Scheme 77A.

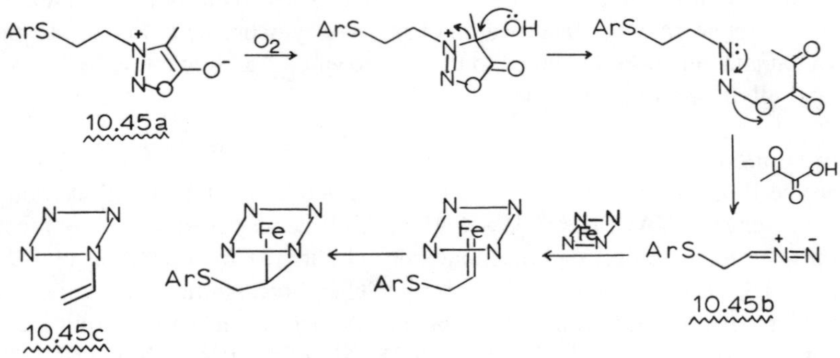

Scheme 77A. Containing Structural Formulas 10.45a, 10.45b, and 10.45c.

Compound **10.45a** was synthesized as shown in Scheme 77B.

$$\text{ArSH} \xrightarrow[\text{2. BrCH}_2\text{CH}_2\text{NH}_2,\ \Delta]{\text{1. Na}} \text{ArSCH}_2\text{CH}_2\text{NH}_2 \xrightarrow[\text{Me}_3\text{COK},\ \Delta]{\text{CH}_3\text{CH(Br)CO}_2\text{H}} \text{ArSCH}_2\text{CH}_2\text{NHCH(CH}_3\text{)CO}_2\text{H}$$

$$\xrightarrow[\text{H}_2\text{/HCl}]{\text{NaNO}_2} \text{ArSCH}_2\text{CH}_2\text{N(NO)CH(CH}_3\text{)CO}_2\text{H} \xrightarrow{\text{Ac}_2\text{O}} \textbf{10.45a}$$

Scheme 77B.

B. Oxygenation/Elimination/Acylation

1. Chloramphenicol and Related

Dixon and Fouts[103] showed that chloramphenicol (Structural Formula 10.46, Scheme 78) causes irreversible inhibition of the monooxygenase activity in mouse liver microsomes; dialysis does not regenerate enzyme activity. Reilly and Ivey[104] found that **10.46** produces a time-dependent inactivation of rat liver microsomal monooxygenase activity. NADPH and O_2 are required for inactivation and gel filtration only partially reactivates the enzyme. Treatment of liver microsomes from phenobarbital-treated rats with [1-^{3}H]- or 1-*p*-nitrophenyl-1-[1,2-^{14}C]dichloroacetamido-1,3-propanediol (i.e., labeled chloramphenicol) was found by Pohl and Krishna[105] to lead to the incorporation of the same amount of [^{14}C] or [^{3}H] into protein. No radioactive incorporation into microsomes from untreated or 3-methylcholanthrene-pretreated rats was observed. Incorporation is time-dependent and NADPH and O_2-dependent. The mechanism for activation proposed[105] is shown in Scheme 79 (R = *p*-NO$_2$-PhCH(OH)CH(CH$_2$OH)NH).

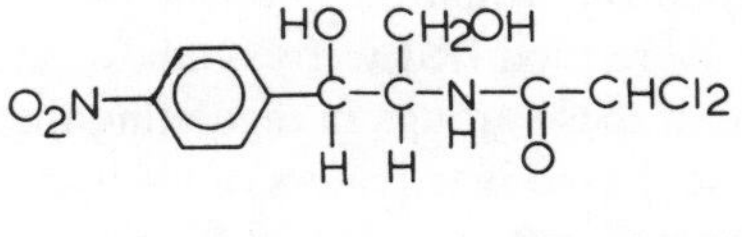

Scheme 78.   Structural Formula 10.46.

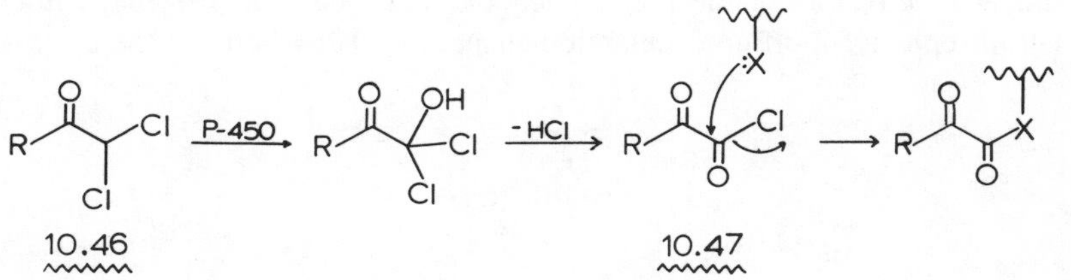

Scheme 79.   Containing Structural Formulas 10.46 and 10.47.

The 7-ethoxycoumarin deethylase and benzphetamine demethylase activities of cytochrome P-450 of a reconstituted monooxygenase system purified from liver microsomes of phenobarbital-treated rats also are irreversibly inactivated by chloramphenicol in the presence of NADPH.[106] Inactivation does not involve heme destruction or loss of cytochrome P-450 which is detected as the carbon monoxide complex. This suggests modification of amino acids rather than heme. [^{14}C]-Chloramphenicol ((1*R*,2*R*)-(+)1-*P*-nitrophenyl-2-[1,2-^{14}C]dichloroacetamido-1,3-propanediol) led to incorporation of 1.5 equivalents of radioactivity into the protein. In the presence of dithiothreitol Halpert and Neal[106] found only 1.2 equivalents were bound. The adduct is stable to treatment with neutral 1 *M* hydroxylamine or performic acid, conditions which would cleave thioesters. Inactivation of cytochrome P-450 by [1,2-^{14}C]chloramphenicol results in a heterogeneous distribution of radioactivity covalently bound.[107] NaDodSO$_4$ denaturation following by neutral hydroxylamine or mild alkaline treatment results in release of about 50% of the radioactivity. Furthermore, dithiothreitol or performic acid treatment also releases about 50% of the radioactivity. The hydroxylamine-stable material was Pronase digested by Halpert[107] and the radioactively labeled amino acid was identified as *N*-ε-chloramphenicol oxamyl lysine by cochromatography with an authentic sample. Saponification gives chloramphenicol acid and lysine, indicating that the active site nucleophile is a lysine residue. About five turnovers of chloramphenicol are required for inactivation. Proteolytic digestion of the labeled enzyme also releases the hydroxylamine-labile material, most of which (40%) was identified as oxalic acid. The minor material was believed to be a cysteine adduct. A more detailed study of the hydroxylamine-labile adducts from inactivation of P-450 by [^{14}C]chloramphenicol was carried out by Halpert.[108] Several adducts were obtained upon hydroxylamine treatment which were partially characterized but their structures were not elucidated. It is not clear why oxalic acid is released upon hydrolysis if the labile material is derived from attack by a nucleophile other than lysine on the chloramphenicol oxamyl chloride (Structural Formula **10.47**, Scheme 79); model compounds under these conditions are converted to chloramphenicol oxamic acid, not oxalic acid. Therefore, it was concluded[108] that the labile products are not derived from attack of nucleophiles other than lysine on **10.47**. The mechanism shown in Scheme 79, however, was suggested as that which leads to the stable adduct. The same metabolic pathway for cytochrome P-450 inactivation was observed by Halpert et al.[109] in vivo as in vitro except proteins other than those associated with cytochrome P-450 also are labeled. The same products, namely *N*-ε-chloramphenicol oxamyl lysine, oxalic acid, and chloramphenicol oxamic acid, were obtained.

Analogues of chloramphenicol in which the propanediol side chain, the *p*-nitro group, and the dichloromethyl moiety are modified were synthesized by Miller and Halpert[109a] in order to assess the importance of these groups in regulating the effectiveness and selectivity of the compounds as mechanism-based inactivators of rat liver cytochrome P-450 isozymes. 1-*p*-Nitrophenyl-2-dichloroacetamidoethane (Structural Formula 10.47a, R = NO_2, Y = $CHCl_2$), 1-*p*-nitrophenyl-2-dibromoacetamidoethane (**10.47a**, R = NO_2, Y = $CHBr_2$), and 1-phenyl-2-dichloroacetamidoethane (**10.47a**, R = H, Y = $CHCl_2$) (see Scheme 80) are time-dependent inactivators of the major phenobarbital-inducible isozyme, cytochrome P-450$_{PB-B}$; 1-*p*-nitrophenyl-2-difluoroacetamidoethane does not inactivate the enzyme.

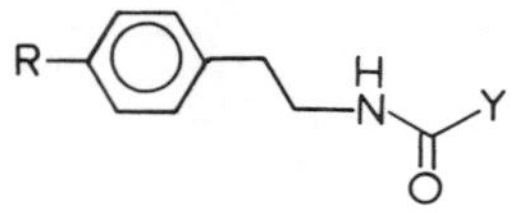

Scheme 80.   Structural Formula 10.47a.

Unlike chloramphenicol, **10.47a** (R = NO_2, Y = $CHCl_2$) is a potent time-dependent inactivator of the ethoxycoumarin deethylase activity from β-naphthoflavone-inducible rat liver cytochrome P-450$_{BNF-B}$. Inactivation of cytochrome P-450$_{PB-B}$ and cytochrome P-450$_{BNF-B}$ with [^{14}C]-labeled **10.47a** (R = NO_2, Y = $CHCl_2$) leads to the incorporation of 1.5 and 1.0 nmol of [^{14}C] per nanomole of enzyme, respectively. Protease digestion of either radioactive isozyme produces compounds **10.47a** (R = NO_2, Y = COOH) and **10.47a** (R = NO_2, Y = CONHLysyl). Alkaline hydrolysis of liver microsomes from phenobarbital- or β-naphthoflavone-treated rats pretreated with **10.47a** (R = NO_2, Y = $CHCl_2$) yields two major products: **10.47a** (R = NO_2, Y = COOH) and oxalic acid. Corresponding results were obtained when [^{14}C]chloramphenicol was used. These results indicate that although the propanediol side chain and the *p*-nitro group are not essential for inactivation of cytochrome P-450, they are important to the selectivity of inactivation of cytochrome P-450 isozymes. A dichloromethyl or dibromomethyl moiety, however, is required for inactivation.

A series of dichloromethyl compounds (mostly RNHCOCHCl$_2$) was studied by Halpert et al.[109b] in order to determine the minimal structure required to inactivate rat liver microsomal cytochrome P-450$_{PB-B}$ and to examine their effect on various other isozymes. The difference in the acetylenic and olefinic inactivators vs. chloramphenicol is that the former generally attach to the heme and the latter to the protein. This may, in part, be a result of the molecular structure. *N*-Monosubstituted dichloroacetamides in which the side chain is phenyl or *n*-octyl inactivate cytochrome P-450$_{PB-B}$ without heme destruction. As the alkyl chain decreases in size, increased heme destruction occurs, which also is evident with the dichloromethyl non-amides. Various compounds show selectivity for the different P-450 isozymes.

2. 19,19-Difluoroandrost-4-ene-3,17-dione

19,19-Difluoroandrost-4-ene-3,17-dione (Structural Formula 10.48) was shown by Marcotte and Robinson[110] to be a time-dependent inactivator of human placental aromatase in the presence of NADPH. Inactivation is irreversible to washing of microsomes. Dithiothreitol in the preincubation buffer does not prevent inactivation. The mechanism proposed is shown in Scheme 81. The acyl fluoride (Structural Formula 10.49) was synthesized and shown to produce 50% inactivation in 30 min. Presumably, the compound is nonenzymatically hydrolyzed, which results in the incomplete inactivation. Consistent with the mechanism proposed is the fact that the monofluoro analogue is oxidized without inactivation of the enzyme to estrone (Scheme 82). Compound **10.48** was synthesized by the route in Scheme 83. The monofluoro analogue was prepared the same way starting from the 19-hydroxy compound.

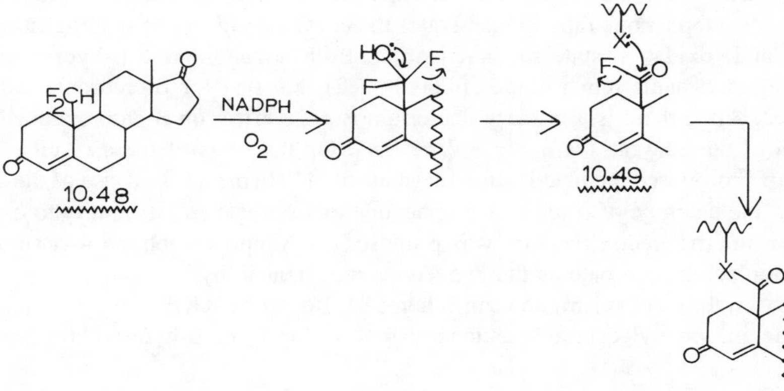

Scheme 81.   Containing Structural Formulas 10.48 and 10.49.

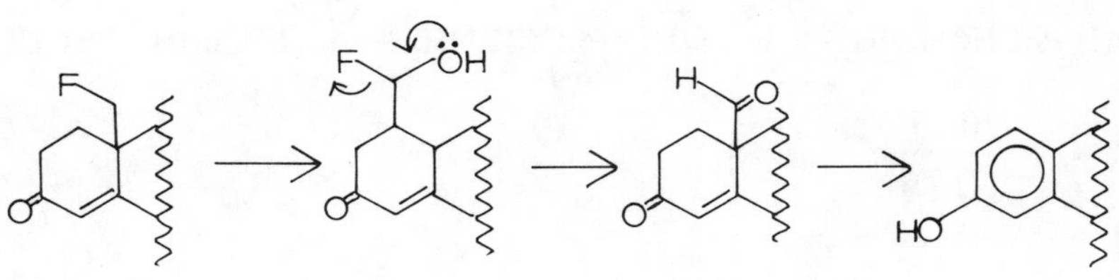

Scheme 82.

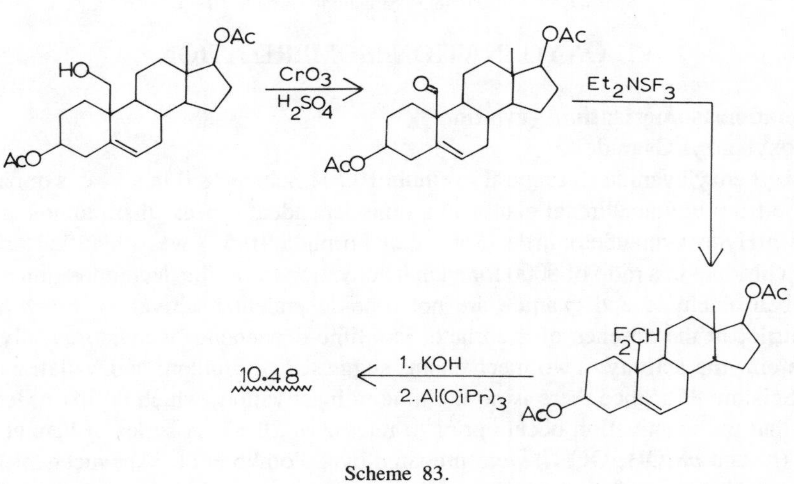

Scheme 83.

C. Oxygenation/Elimination/Isomerization

1. β-Chlorophenethylamine

β-Chlorophenethylamine (Structural Formula 10.50, see Scheme 84) was observed by Klinman and Krueger[111] to be a time-dependent inactivator of dopamine β-hydroxylase from bovine adrenal glands; inactivation occurs only once in every 12,000 turnovers. Compound **10.50** was synthesized by thionyl chloride treatment of β-hydroxyphenethylamine. Both β-hydroxy- and β-chlorophenethylamine are converted into α-aminoacetophenone by dopamine β-hydroxylase, but α-aminoacetophenone does not inactivate the enzyme in the presence of ascorbate.[112] Only the chloro analogue is an inactivator. However, in the absence of ascorbate, α-aminoacetophenone rapidly inactivates the enzyme, suggesting that the enzyme must be in the Cu(II) oxidation state for inactivation. Both dopamine and β-hydroxyphenethylamine have small deuterium isotope effects (~1.2), but that for β-chlorophenethylamine is 6.2 ± 1.2. Also, there is a primary deuterium isotope effect on β-chlorophenethylamine inactivation of the enzyme ($^D k_{inact}$ = 3.7 ± 0.4). On the basis of these results, Mangold and Klinman[112] proposed the inactivation mechanism in Scheme 84. Because of the structural similarity of the enamine of α-aminoacetophenone to ascorbate and its ability to oxidatively dimerize, an internal redox reaction was proposed. α-Aminoacetophone was shown to be capable of replacing ascorbate as the exogenous reductant in hydroxylation.

In order to test the mechanism shown in Scheme 84, Bossard and Klinman[112a] studied phenylacetaldehyde and phenylacetamide as inactivators of dopamine β-hydroxylase (see Section VII.A.4.).

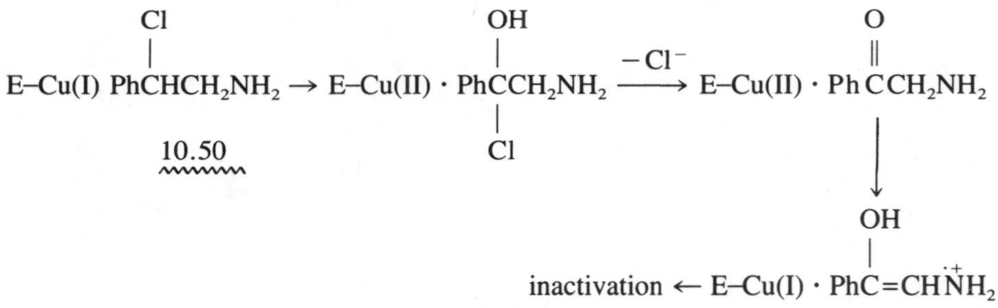

$$\underset{\substack{\text{10.50}}}{\text{E–Cu(I) PhCHCH}_2\text{NH}_2} \rightarrow \underset{\substack{|\\ \text{Cl}}}{\overset{\substack{\text{OH}\\|}}{\text{E–Cu(II)} \cdot \text{PhCCH}_2\text{NH}_2}} \xrightarrow{-\text{Cl}^-} \overset{\text{O}}{\overset{\|}{\text{E–Cu(II)} \cdot \text{Ph C CH}_2\text{NH}_2}}$$

$$\text{inactivation} \leftarrow \text{E–Cu(I)} \cdot \underset{}{\overset{\substack{\text{OH}\\|}}{\text{PhC}=\text{CH}\overset{..+}{\text{N}}\text{H}_2}}$$

Scheme 84.   Containing Structural Formula 10.50.

VI. OXYGENATION/ISOMERIZATION

A. Oxygenation/Isomerization/Acylation

1. p-Hydroxybenzyl Cyanide

p-Hydroxybenzylcyanide (Structural Formula 10.51, Scheme 85) inactivates dopamine β-hydroxylase from bovine adrenal glands in a time-dependent process that requires ascorbate and O_2.[113] p-Hydroxymandelonitrile (Structural Formula 10.52) was identified as the hydroxylation product in a ratio of 8000 for each inactivation event. Its decomposition products, p-hydroxybenzaldehyde and cyanide are not time-dependent inactivators, but p-hydroxymandelonitrile, in the absence of ascorbate, is a time-dependent inactivator. Dialysis does not restore enzyme activity. Two mechanisms suggested by Baldoni and Villafranca[113] are shown in Scheme 85. Since there is no lag time to inactivation, which is first-order, it was suggested that the inactivation occurs prior to release of **10.52**. A series of benzyl cyanide analogues (p- and m- OH, OCH₃) were prepared by Colombo et al.[114] as mechanism-based inactivators of dopamine β-hydroxylase. All of the compounds inactivate the enzyme. The partition ratio decreases with increasing pH for m-hydroxybenzyl cyanide. The products are

R-mandelonitriles. Maximum inactivation occurs when an active site group with a pK_a of 6.0 is ionized. Inactivation by *m*-hydroxybenzyl cyanide was stopped by lowering the pH from 6.4 to 5.0 in the presence of tyramine.[115] When tyramine is omitted, enzyme reactivation results. Reactivation is time dependent, but return of enzyme activity to 100% is not attained. It was suggested by Colombo et al.[115] that the tyramine stabilizes the enzyme-inactivator complex. The results indicate that two different adducts form during inactivation; one is tightly bound or a covalent adduct, the other is reversible, resulting from cyanide binding to the Cu(II) at the active site. The products formed during inactivation of dopamine β-hydroxylase by *p*-hydroxybenzyl cyanide were monitored by HPLC.[116] Initially, *p*-hydroxy-mandelonitrile is formed which decomposes to *p*-hydroxybenzaldehyde and cyanide. [*ring-^3H*]*p*-Hydroxybenzyl cyanide inactivation results in the incorporation of a significant amount of radioactivity only when 100% O_2 is present. Even under those conditions only 0.34 mol of tritium is incorporated per mole of tetramer. This amount does not correlate with the total amount of enzyme inactivated. To test whether cyanide is binding in later stages of inactivation, Colombo et al.[116] treated the enzyme with $K^{14}CN$. In this case, 2 mol of cyanide bind per tetramer. ESR studies indicate a quaternary complex of enzyme-Cu(II)-tyramine-cyanide. The mechanism proposed[116] is in Scheme 86 (Ar = *p*-hydroxyphenyl).

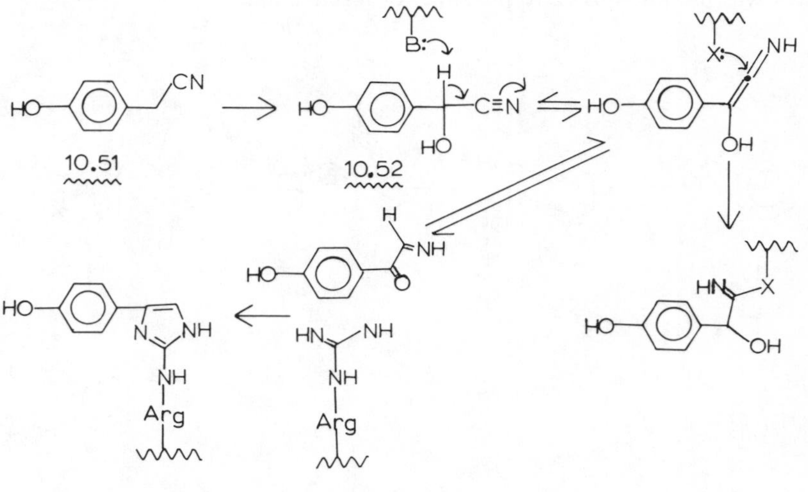

Scheme 85.   Containing Structural Formulas 10.51 and 10.52.

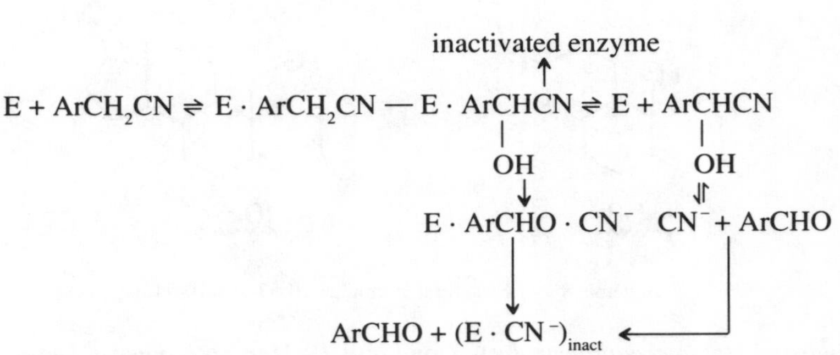

Scheme 86.

VII. OXYGENATION/OXIDATION

A. Oxygenation/Oxidation/Addition
1. Phenol

Phenol irreversibly inactivates mushroom tyrosinase (polyphenol oxidase).[117] Radioactivity from [1-¹⁴C]phenol becomes incorporated into the protein upon enzyme-catalyzed oxidation. A mechanistic hypothesis was offered by Wood and Ingraham[117] that the phenol becomes hydroxylated and oxidized to a quinone that acts as a Michael acceptor for an amino acid in the active site (Scheme 87). Tyrosinase is known to convert tyrosine to dopaquinone.[118] The inactivation of tyrosinase by phenol was reinvestigated with the enzyme from *Neurospora crassa*.[118a] Time-dependent, irreversible inactivation by [U-¹⁴C]phenol was observed. However, in contrast to the work of Wood and Ingraham[117] only insignificant incorporation of phenol was found by Lerch.[118a] Amino acid analysis showed that one histidine residue (His-306) was lost. Inactivation also results in the loss of one atom of copper; consequently, it was proposed that His-306 is liganded to one of the two active-site copper ions. These results suggest that catechol is not the inactivating species in this case; rather, an activated form of O_2 (e.g., hydroxyl radical) may react with His-306. Since hydroxyl radical trapping agents did not prevent inactivation, a copper-bound rather than free radical was proposed to be responsible for inactivation.

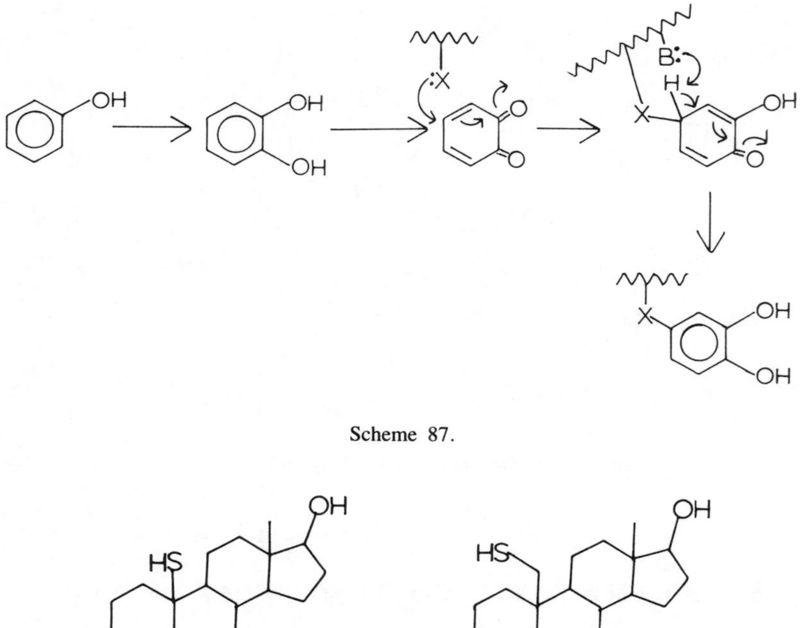

Scheme 87.

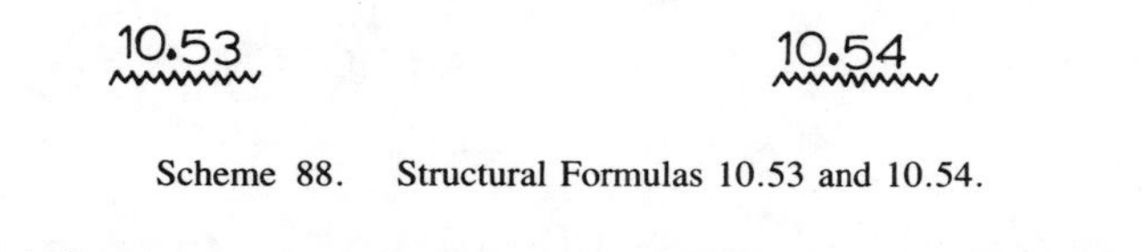

Scheme 88.   Structural Formulas 10.53 and 10.54.

2. 17β-Hydroxy-10β-mercaptoestr-4-en-3-one and 19-Mercaptoandrost-4-ene-3,17-dione

17β-Hydroxy-10β-mercaptoestr-4-en-3-one (Structural Formula 10.53) and 19-mercaptoandrost-4-ene-3,17-dione (Structural Formula 10.54, see Scheme 88) were synthesized as mechanism-based inactivators of aromatase.[119] Both compounds are time-dependent inactivators of human placenta microsomal aromatase which is dependent on NADPH and O_2.

The presence of cysteine has no effect on inactivation rates. Dialysis does not restore enzyme activity. Bednarsky et al.[119] suggest that the thiol is hydroxylated and oxidized to a sulfenic acid which is attacked by an active site nucleophile. Compound **10.53** was synthesized by the route shown in Scheme 89 and **10.54** was prepared by the route in Scheme 90.

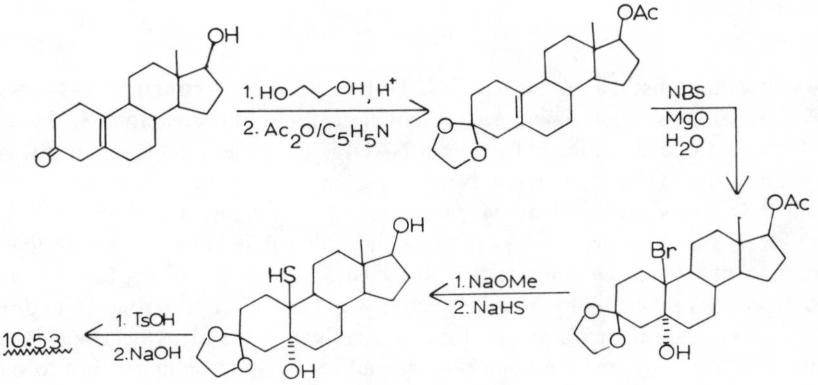

Scheme 89.

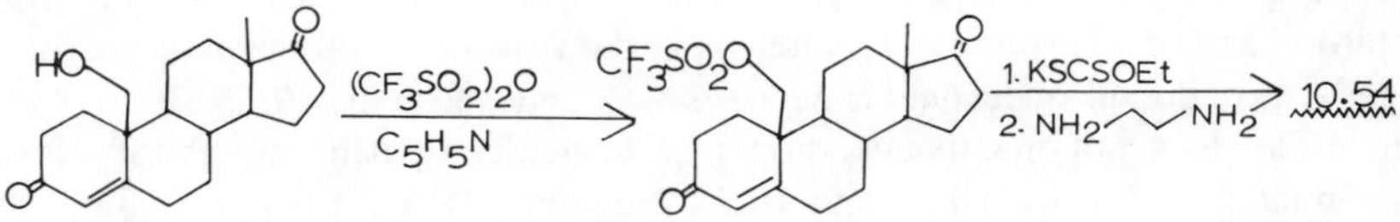

Scheme 90.

3. Spironolactone and Related

Spironolactone, (Structural Formula 10.55, R = Ac; see Scheme 91), a diuretic and antihypertensive steroid, and its deacetylated analog (**10.55**, R = H) destroy cytochrome P-450, in guinea pig adrenal microsomes.[120] The inactivation is time-dependent and requires NADPH. Guinea pig and dog adrenal and testicular microsomal cytochrome P-450 undergo time-dependent inactivation by **10.55** (R = Ac and H) with loss of the heme.[121] NADPH is required for inactivation. Other analogues were tried, and it was found that the presence of the sulfur atom in the 7α-position is required for inactivation. The relative amounts of cytochrome P-450 associated with specific steroid hydroxylases was determined in adrenals, testes, and liver microsomes with the use of **10.55** (R = H) and 7α-thiotestosterone (Structural Formula 10.56, Scheme 92).[122]

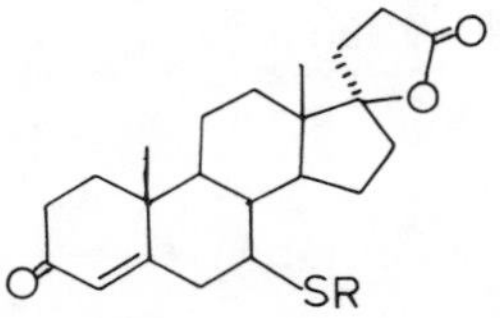

Scheme 91.   Structural Formula 10.55.

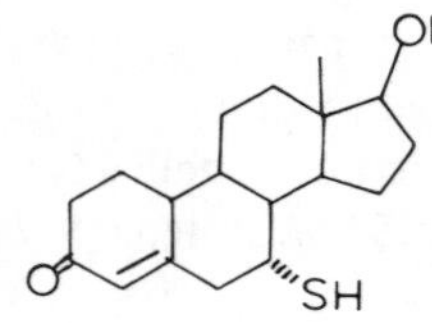

Scheme 92.   Structural Formula 10.56.

At low concentrations, **10.56** causes a NADPH-dependent destruction of hepatic cyto-chrome P-450 that is associated with a preferential decrease in testosterone 7α-hydroxylase activity. At higher concentrations, there is a NADPH-dependent decrease in 2β, 6β, and 16α-testosterone hydroxylase, but not benzo[a]pyrene hydroxylase. No loss of testicular cytochrome P-450 was observed by Menard et al.[122] Compound **10.55** (R = H), on the other hand, causes a much greater loss of testicular rather than hepatic, cytochrome P-450. The destruction of adrenal cytochrome P-450 by **10.55** (R = H) occurs concurrently with a decrease in progesterone 17α-hydroxylase activity, but not of progesterone 21-hydroxylase. When [³⁵S]deacetylspironolactone was used to inactivate adrenal cytochrome P-450, the sulfur atom was found to become covalently bound to the apoprotein and not to the heme portion of the microsomes. A mechanism related to that for parathion (see Section V.A.1.) or CS_2 (see Section V.A.2.) inactivation of cytochrome P-450 was suggested.[122] However, parathion produces high-molecular-weight aggregates in which the sulfur released, cross-links with the apoenzymes of hepatic cytochrome P-450. These aggregates do not form with deacetylspironolactone. Furthermore, whereas carbon disulfide produces a necrotic effect on liver cells, no adrenal or testicular necrosis was detected with **10.55** (R = Ac). Also, carbon disulfide does not inactivate adrenal or testicular cytochrome P-450. Possibly a mechanism related to that for **10.53** and **10.54** (Section VII.A.2.) is applicable.

4. *Phenylacetaldehyde and Phenylacetamide*

In order to test the inactivation mechanism proposed by Mangold and Klinman[112] for β-chlorophenethylamine (see Scheme 84 in Section V.C.1.), phenylacetaldehyde and phen-ylacetamide were studied as inactivators of bovine adrenal medulla dopamine β-hydroxyl-ase.[112a] Phenylacetaldehyde (Structural Formula 10.56a, Scheme 92A) is a mechanism-based inactivator with kinetic parameters similar to those of β-chlorophenethylamine, thus lending support for an intramolecular redox reaction. The actual inactivating species is β-hydrox-yphenylacetaldehyde (**10.56b**). The inactivation mechanism proposed by Bossard and Klinman[112a] is shown in Scheme 92A.

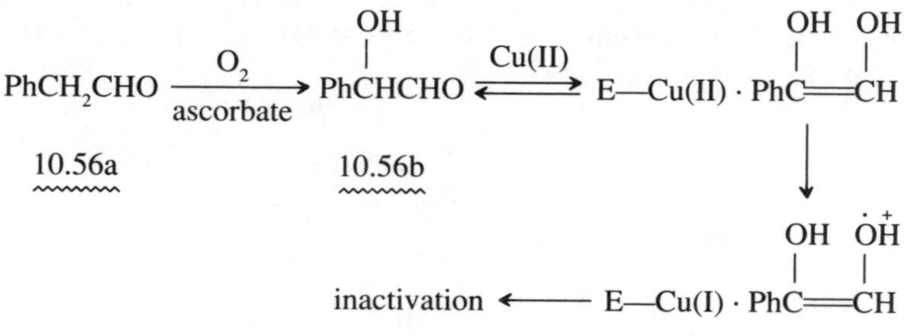

Scheme 92A.   Containing Structural Formulas 10.56a and 10.56b.

Since a key feature of this mechanism is enolization of β-hydroxyphenylacetaldehyde, it suggests that arylacetamides should not inactivate the enzyme because β-hydroxyamides are not expected to enolize. Phenylacetamide and *p*-hydroxyphenylacetamide are substrates and mechanism-based inactivators, but the products, β-hydroxyarylacetamides, are neither inhibitors nor electron donors in tyramine hydroxylation. Therefore, there are at least two types of inactivation mechanisms for dopamine β-hydroxylase, one which involves hydroxylation prior to inactivation (e.g., phenylacetaldehyde) and one which only requires hydrogen atom abstraction (e.g., phenylacetamide). However, both pathways would lead to a similar reactive radical cation intermediate (Scheme 92B).

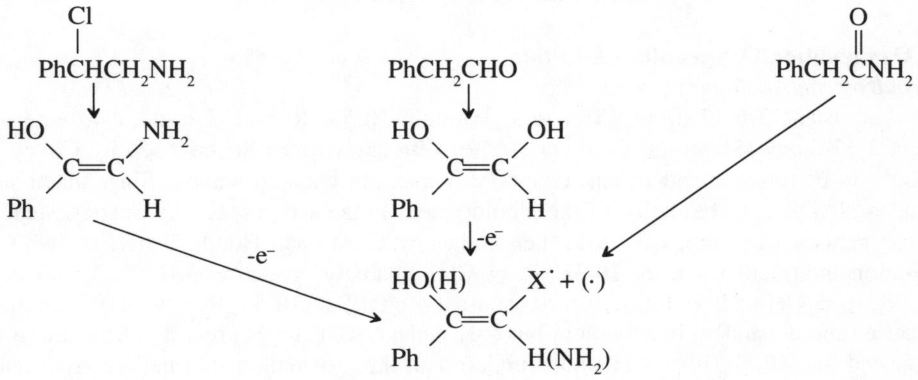

Scheme 92B.

B. Oxygenation/Oxidation/Acylation

1. 4α-(Cyanomethyl)-5α-cholestan-3β-ol

Oxidative demethylation of lanosterol is important in cholesterol biosynthesis. The enzyme responsible for this reaction is 4-methylsterol oxidase, a NADPH-dependent enzyme that does not involve cytochrome P-450. The reaction pathway is believed to involve 3 successive oxygenations to the 4α-carboxylic acid, then decarboxylation by a NAD$^+$ dehydrogenase/decarboxylase.[123] 4α-(Cyanomethyl)-5α-cholestan-3β-ol (Structural Formula 10.57) was synthesized by Bartlett and Robinson[124] as a mechanism-based inactivator of rat liver 4-methylsterol oxidase. Pseudo first-order inactivation occurs only the presence of NADPH. Reduced glutathione has no effect on inactivation. The inactivation mechanism proposed is shown in Scheme 93. Compound **10.57** was synthesized by the route in Scheme 94. The corresponding 4α-difluoromethyl analogue is a competitive inhibitor, but shows no inactivation.

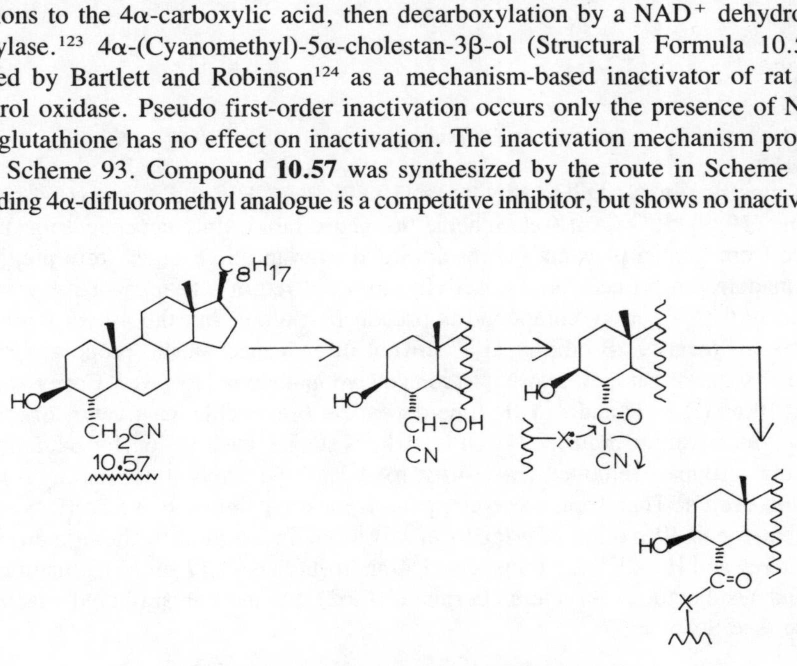

Scheme 93.   Containing Structural Formula 10.57.

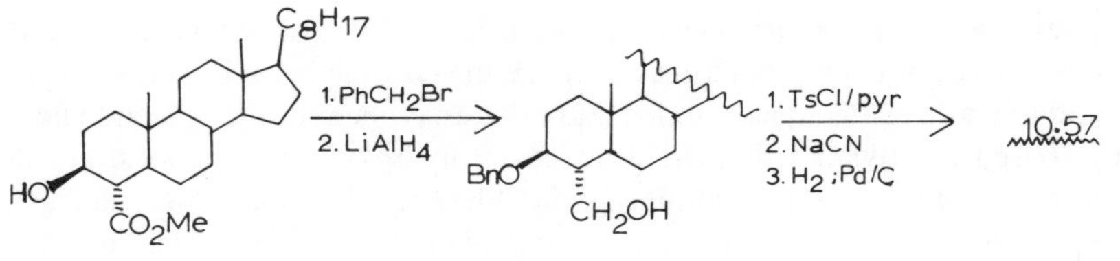

Scheme 94.

VIII. OXYGENATION/OXYGENATION

A. Oxygenation/Oxygenation/Addition

1. Androstene Analogues

4-Androstene-3,6,17-trione (Structural Formula 10.58, R = CH$_3$) and 1,4,6-androsta-triene-3,17-dione (Structural Formula 10.59) were shown (see Scheme 95) by Covey and Hood[125] to be time-dependent inactivators of human placental aromatase. Since inactivation requires NADPH, an oxidation of these compounds to the active species was suggested, but no mechanism was proposed. In further studies by Covey and Hood,[126] the first two C-19 oxidation intermediates from **10.58** (R = CH$_3$), namely, 4-androsten-19-ol-3,6,17-trione (**10.58**, R = CH$_2$OH) and 2,6,17-trioxoandrost-4-en-19-al (**10.58**, R = CHO), were found to cause time-dependent inactivation, but only when NADPH was present. Therefore, it was suggested that **10.58** (R = CH$_3$) must undergo all three oxidations normally carried out by the enzyme before it can produce inactivation.

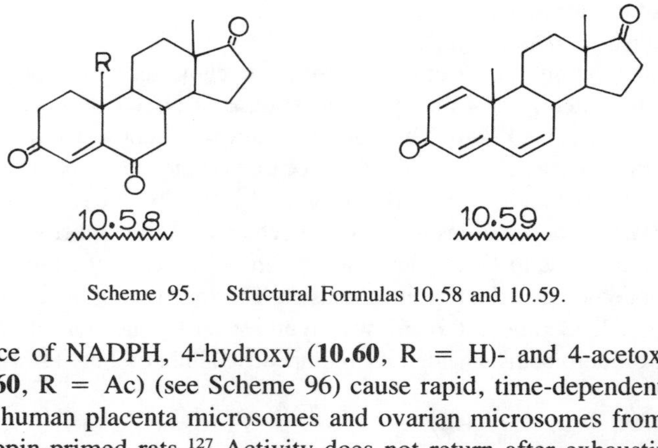

Scheme 95.   Structural Formulas 10.58 and 10.59.

In the presence of NADPH, 4-hydroxy (**10.60**, R = H)- and 4-acetoxy-4-androstene-3,17-dione (**10.60**, R = Ac) (see Scheme 96) cause rapid, time-dependent inactivation of aromatase from human placenta microsomes and ovarian microsomes from pregnant mare serum gonadotropin-primed rats.[127] Activity does not return after exhaustive washing. Inactivation by the 4-acetoxy compound is pseudo first-order, but the 4-hydroxy compound deviates from linearity. Brodie et al.[127] do not offer a mechanistic proposal. Brodie and Longcope[128] suggest that they are simple competitive inhibitors. However, Covey and Hood[126] show that **10.60** (R = H and Ac) are time-dependent irreversible inactivators of microsomal aromatase. Inactivation requires NADPH, which suggests activation by oxidation. If the C-19 methyl group is removed (i.e., 4-hydroxy- and 4-acetoxy-4-estrene-3,17-dione), no inactivation occurs. Therefore, these compounds must be similar to **10.58** (R = CH$_3$) in a requirement for C-19 oxidation prior to inactivation. In addition to these androstene analogues, Covey and Hood[129] also found that 1,4-androstadiene-3,17-dione (Structural Formula 10.61) and testolactone (Structural Formula 10.62) are mechanism-based inactivators of aromatase (see Scheme 97).

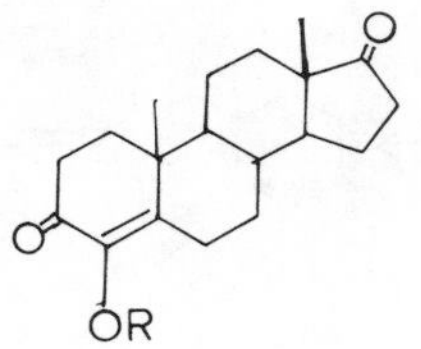

Scheme 96.   Structural Formula 10.60.

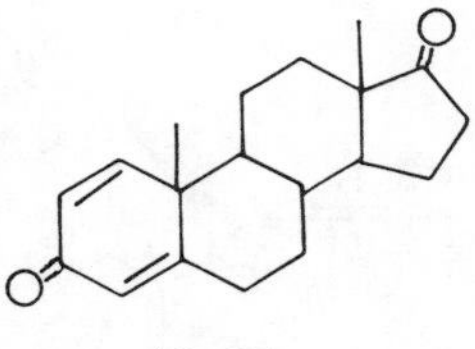

10.61 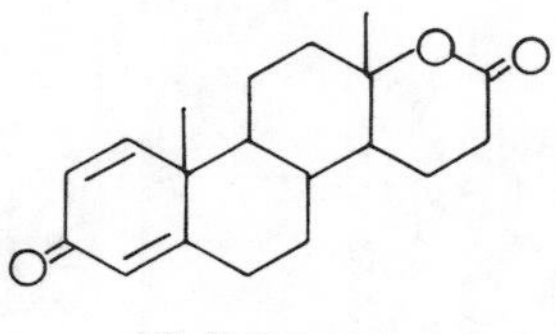   10.62

Scheme 97.   Structural Formulas 10.61 and 10.62.

The results with these five compounds indicate that the 4-hydroxy group and 6-keto group are important to inactivation. The C-1, C-2 double bond, but not the C-4, C-5 double bond, also is required. On the basis of inactivation of aromatase by these five compounds, a new hypothesis was proposed by Covey and Hood[129] for the mechanism of action of aromatase (Scheme 98). With this mechanism as a working hypothesis, the inactivation of aromatase by the five compounds mentioned were rationalized.[129] The first two hydroxylations of **10.58** (R = CH$_3$) would produce the C-19 aldehyde compound (**10.58**, R = CHO). This intermediate could inactivate the enzyme by the route shown in Scheme 99. It is not clear, however, why the X group in Scheme 99 would not be eliminated in a vinylogous fashion, since elimination was proposed in Scheme 98 to occur with the substrate. The vinylogous elimination should be at least as favorable to produce an α,β,γ,δ-unsaturated ketone product (Scheme 100). The rationalization[129] for the inactivation by compounds containing a C-1, C-2 double bond (**10.59, 10.61, 10.62**) arises from a potential aromatization prior to elimination of the enzyme nucleophile (Scheme 101). Inactivation by **10.60** (R = H) would result from competition between elimination of the enzyme nucleophile and H$_2$O (Scheme 102). Although there is no proof for the mechanism of aromatase proposed (see Scheme 98) it cannot yet be excluded. Hahn and Fishman[130] suggested that the third molecule of O$_2$ in the enzyme-catalyzed reaction produces 2β-hydroxy-10β-formylandrost-4-ene-3,17-dione which collapses to estrone. However, Caspi et al.[131] showed that this is not an obligatory intermediate in estrogen biosynthesis.

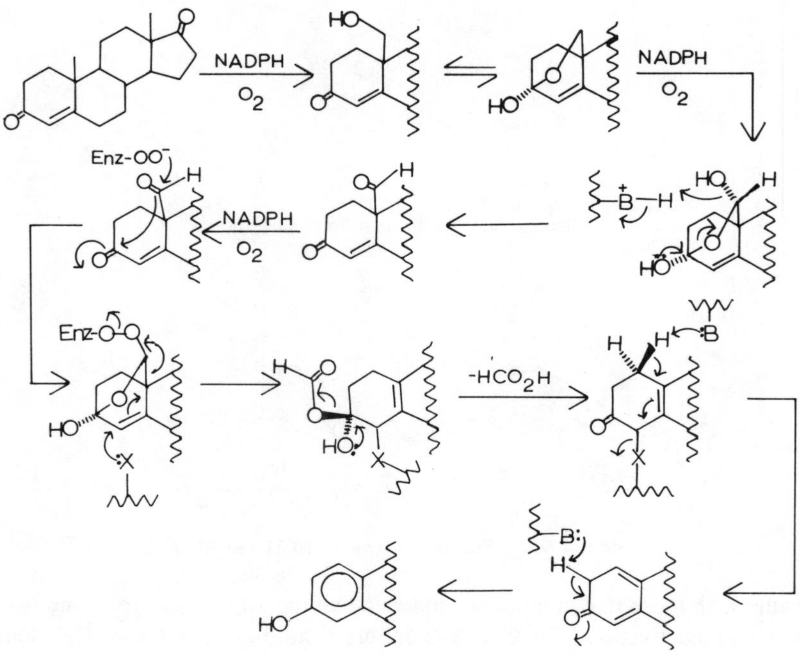

Scheme 98.

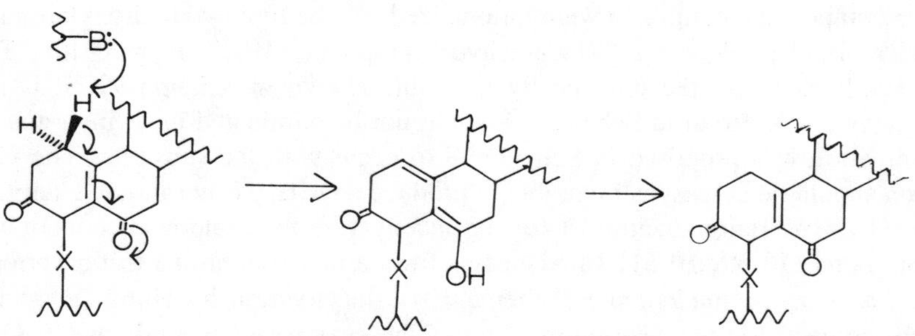

Scheme 99.

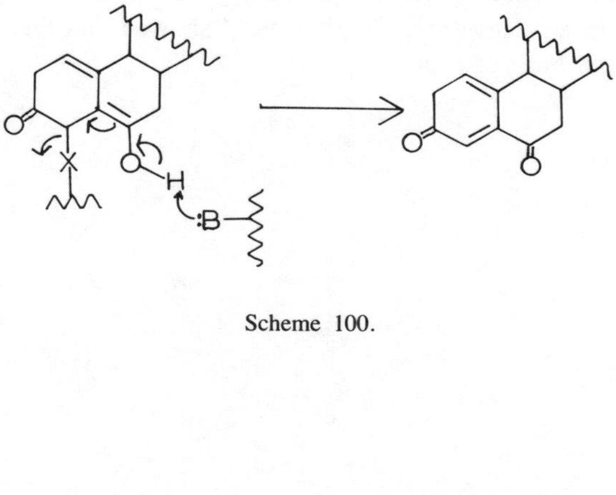

Scheme 100.

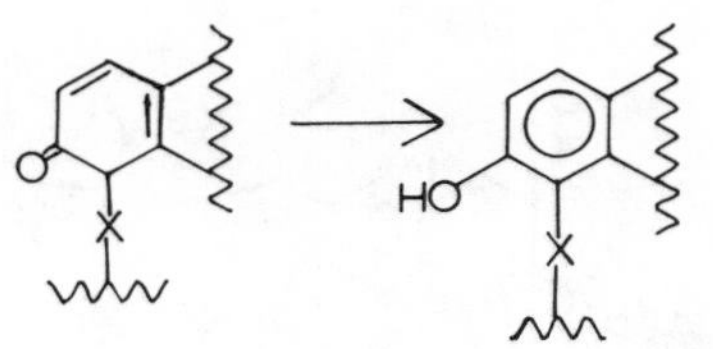

Scheme 101.

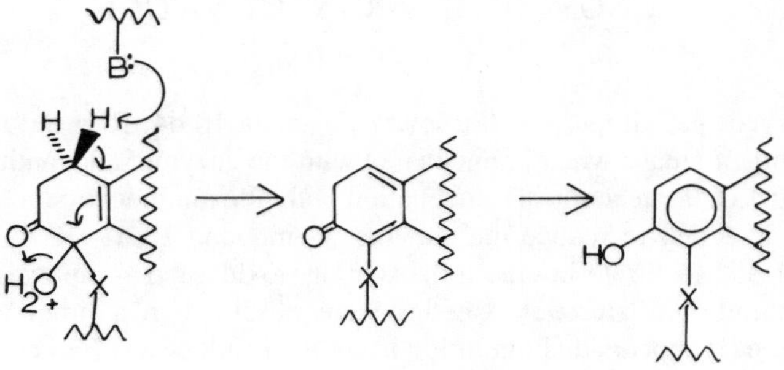

Scheme 102.

Aromatase from human placenta and from ovaries of PMSG-primed rats undergo time-dependent inactivation by a series of substituted 4-androstene-3,17-diones, 1,4-androstadiene-3,17-diones, 4,6-androstadiene-3,17-diones, 1,4,6-androstatriene-3,17-diones, and 2-androstene-3,17,diones in the presence of NADPH.[131a]

IX. NOT MECHANISM-BASED INACTIVATION

A. 1,1,1-Trichloropropene-2,3-oxide

1,1,1-Trichloropropene-2,3-oxide (Structural Formula 10.63, Scheme 103) inactivates hepatic microsomal cytochrome P-450 from phenobarbital-treated rats only in the presence of an NADPH-generating system.[132] An equivalent amount of heme is modified in the process, but the levels of cytochrome b_5 are not altered. However, since there is a lag time for inactivation, it is proposed that the reactive species diffuses from the form of cytochrome P-450 that produces it, and then reacts with a different form of the enzyme.

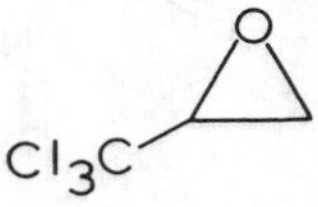

Scheme 103.   Structural Formula 10.63.

B. 5α,6α-Methanocholestan-3β-ol

5α,6α-Methanocholestan-3β-ol (Structural Formula 10.63a Scheme 103A) was designed by Houghton et al.[132a] to be a mechanism-based inactivator of cholesterol 7α-hydroxylase on the basis of a radical mechanism, however, **10.63a** is a substrate, being converted to the 7α-hydroxy derivative without enzyme inactivation.

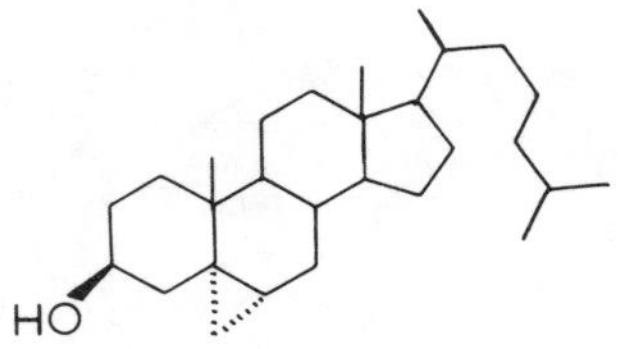

Scheme 103A.   Structural Formula 10.63a.

X. NONCOVALENT INACTIVATION

A. Allopurinol

Elion[133] observed that allopurinol (Structural Formula 10.64, R = H; see Scheme 104) inactivates xanthine oxidase when preincubated with the enzyme; alloxanthine (**10.64**, R = OH) is the product of the reaction. Preincubation with alloxanthine produces no inactivation unless xanthine is added to reduce the enzyme. Compound **10.64** (R = H) was used by Spector and Johns[134] to titrate bovine milk xanthine oxidase; 1.4 equivalents are required for complete inhibition. Saturation kinetics were observed in a time- and temperature-dependent reaction. It appears that inhibition is not the result of just E · I complex formation, but rather a subsequent internal rearrangement to a less readily dissociable form is responsible. Treatment of xanthine oxidase with allopurinol under anaerobic conditions results in reduction of the enzyme.[135] Reoxidation results in a modified form of the enzyme that is incapable of further reduction by either xanthine or additional allopurinol. Alloxanthine is the inactivator, but only when the enzyme is reduced. The inhibition is reversed by incubation with phenazine methosulfate or $K_3Fe(CN)_6$ anaerobically, followed by dialysis. Ferricyanide titration indicated to Massey et al.[135] that the oxidation state of the active site molybdenum was Mo(IV). It was found that, in general, other pyrazolo[3,4-*d*]pyrimidines that are unsubstituted at the 6 position reduce xanthine oxidase; upon reoxidation, inactivated enzyme results. On the basis of the work of Massey and co-workers,[135,136] a mechanism for the inactivation of xanthine oxidase by allopurinol is shown in Scheme 105. Cha et al.[137] observed that inhibition is slowly reversed upon addition of a high concentration of xanthine and that oxygen is required for inhibition. The hypothesis that the enzyme must be in the reduced state before it can be inhibited by alloxanthine, but not by allopurinol, and that allopurinol can reduce the enzyme itself, was supported.

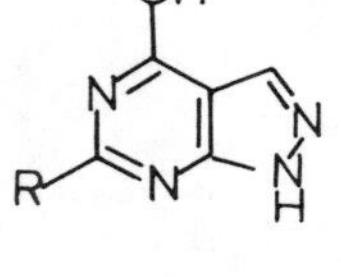

Scheme 104.   Structural Formula 10.64.

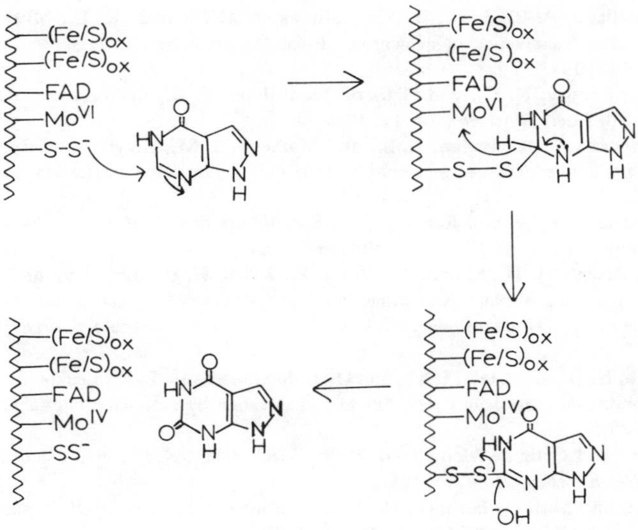

Scheme 105.

The oxidized product of allopurinol, namely oxipurinol, binds to the electronically reduced form of human xanthine oxidase similar to that with the bovine enzyme.[137a]

REFERENCES

1. **Ivanetich, K. M., Lucas, S., Marsh, J. A., Ziman, M. R., Katz, I. D., and Bradshaw, J. J.,** Organic compounds: their interaction with and degradation of hepatic microsomal drug-metabolizing enzymes *in vitro, Drug Metab. Dispos.,* 6, 218, 1978.
2. **Hyman, M. R. and Wood, P. M.,** Suicidal inactivation and labelling of ammonia mono-oxygenase by acetylene, *Biochem. J.,* 227, 719, 1985.
3. **White, I. N. H.,** Metabolic activation of acetylenic substituents to derivatives in the rat causing the loss of hepatic cytochrome P-450 and haem, *Biochem. J.,* 174, 853, 1978.
4. **Ortiz de Montellano, P. R. and Kunze, K. L.,** Self-catalyzed inactivation of hepatic cytochrome P-450 by ethynyl substrates, *J. Biol. Chem.,* 255, 5578, 1980.
5. **Kunze, K. L., Mangold, B. L. K., Wheeler, C., Beiland, H. S., and Ortiz de Montellano, P. R.,** The cytochrome P-450 active site. Regiospecificity of prosthetic heme alkylation by olefins and acetylenes, *J. Biol. Chem.,* 258, 4202, 1983.
6. **Ortiz de Montellano, P. R., Kunze, K. L., Beilan, H. S., and Wheeler, C.,** Destruction of cytochrome P-450 by vinyl fluoride, fluoroxene, and acetylene. Evidence for a radical intermediate in olefin oxidation, *Biochemistry,* 21, 1331, 1982.
7. **Ortiz de Montellano, P. R. and Kunze, K. L.,** Occurrence of a 1,2 shift during enzymatic and chemical oxidation of a terminal acetylene, *J. Am. Chem. Soc.,* 102, 7373, 1980.
8. **Ortiz de Montellano, P. R. and Kunze, K. L.,** Shift of the acetylenic hydrogen during chemical and enzymatic oxidation of the biphenylacetylene triple bond, *Arch. Biochem. Biophys.,* 209, 710, 1981.
9. **Ortiz de Montellano, P. R. and Komives, E. A.,** Branchpoint for heme alkylation and metabolite formation in the oxidation of arylacetylenes by cytochrome P-450, *J. Biol. Chem.,* 260, 3330, 1985.
10. **Gan, L-S. L., Acebo, A. L., and Alworth, W. L.,** 1-Ethynylpyrene, a suicide inhibitor of cytochrome P-450 dependent benzo[*a*]pyrene hydroxylase activity in liver microsomes, *Biochemistry,* 23, 3827, 1984.
11. **Gan, L.-S., Lu, J.-Y. L., Hershkowitz, D. M., and Alworth, W. L.,** Effects of acetylenic and olefinic pyrenes upon cytochrome P-450 dependent benzo[a]pyrene hydroxylase activity in liver microsomes, *Biochem. Biophys. Res. Commun.,* 129, 591, 1985.
12. **Ortiz de Montellano, P. R. and Kunze, K. L.,** Cytochrome P-450 inactivation: structure of the prosthetic heme adduct with propyne, *Biochemistry,* 20, 7266, 1981.

13. **Ortiz de Montellano, P. R., Mico, B. A., Mathews, J. M., Kunze, K. L., Miwa, G. T., and Lu, A. Y. H.,** Selective inactivation of cytochrome P-450 isozymes by suicide substrates, *Arch. Biochem. Biophys.,* 210, 717, 1981.

14. **Feyereisen, R., Langry, K. C., and Ortiz de Montellano, P. R.,** Self-catalyzed destruction of insect cytochrome P-450, *Insect Biochem.,* 14, 19, 1984.

15. **Ortiz de Montellano, P. R., Beilan, H. S., and Mathews, J. M.,** Alkylation of the prosthetic heme in cytochrome P-450 during oxidative metabolism of the sedative-hypnotic ethylchlorvynol, *J. Med. Chem.,* 25, 1174, 1982.

16. **Ortiz de Montellano, P. R. and Reich, N. O.,** Specific inactivation of hepatic fatty acid hydroxylases by acetylenic fatty acids, *J. Biol. Chem.,* 259, 4136, 1984.

17. **Salaun, J. P., Reichhart, D., Simon, A., Durst, F., Reich, N. O., and Ortiz de Montellano, P. R.,** Autocatalytic inactivation of plant cytochrome P-450 enzymes: selective inactivation of the lauric acid in-chain hydroxylase from *Helianthus tuberosus* L. by unsaturated substrate analogs, *Arch. Biochem. Biophys.,* 232, 1, 1984.

17a. **Shak, S., Reich, N. O., Goldstein, I. M., and Ortiz de Montellano, P. R.,** Leukotriene B$_4$ ω-hydroxylase in human polymorphonuclear leukocytes. Suicidal inactivation by acetylenic fatty acids, *J. Biol. Chem.,* 260, 13023, 1985.

17b. **CaJacob, C. A. and Ortiz de Montellano, P. R.,** Mechanism-based in vivo inactivation of lauric acid hydroxylases, *Biochemistry,* 25, 4705, 1986.

18. **White, I. N. H. and Muller-Eberhard, U.,** Decreased liver cytochrome P-450 in rats caused by norethindrone or ethynyloestradiol, *Biochem. J.,* 166, 57, 1977.

19. **Ortiz de Montellano, P. R., Kunze, K. L., Yost, G. S., and Mico, B. A.,** Self-catalyzed destruction of cytochrome P-450: covalent binding of ethynyl sterols to prosthetic heme, *Proc. Natl. Acad. Sci. U.S.A.,* 76, 746, 1979.

20. **Correia, M. A., Farrell, G. C., Olson, S., Wong, J. S., Schmid, R., Ortiz de Montellano, P. R., Beilan, H. S., Kunze, K. L., and Mico, B. A.,** Cytochrome P-450 heme moiety. The specific target in drug-induced heme alkylation, *J. Biol. Chem.,* 256, 5466, 1981.

21. **Osawa, Y., Yarborough, C., and Osawa, Y.,** Norethisterone, a major ingredient of contraceptive pills, is a suicide inhibitor of estrogen biosynthesis, *Science,* 215, 1249, 1982.

21a. **Covey, D. F., Hood, W. F., and McMullan, P. C.,** Studies of the inactivation of human placental aromatase by 17α-ethynyl-substituted 10β-hydroperoxy and related 19-nor steroids, *Biochem. Pharmacol.,* 35, 1671, 1986.

22. **Osawa, Y., Osawa, Y., Yarborough, C., and Borzynski, L.,** Irreversible inhibitors of estrogen biosynthesis (aromatase), *Biochem. Soc. Trans.,* 11, 656, 1983.

23. **Covey, D. F., Hood, W. F., and Parikh, V. D.,** 10β-Propynyl-substituted steroids. Mechanism-based enzyme-activated irreversible inhibitors of estrogen biosynthesis, *J. Biol. Chem.,* 256, 1076, 1981.

24. **Metcalf, B. W., Wright, C. L., Burkhart, J. P., and Johnson, J. O.,** Substrate-induced inactivation of aromatase by allenic and acetylenic steroids, *J. Am. Chem. Soc.,* 103, 3221, 1981.

25. **Johnston, J. O., Wright, C. L., and Metcalf, B. W.,** Biochemical and endocrine properties of a mechanism-based inhibitor of aromatase, *Endocrinology,* 115, 776, 1984.

26. **Marcotte, P. A. and Robinson, C. H.,** Synthesis and evaluation of 10β-substituted 4-estrene-3,17-diones as inhibitors of human placental microsomal aromatase, *Steroids,* 39, 325, 1982.

27. **Johnston, J. O., Wright, C. L., and Metcalf, B. W.,** Time-dependent inhibition of aromatase in trophoblastic tumor cells in tissue culture, *J. Steroid Biochem.,* 20, 1221, 1984.

28. **Covey, D. F., Parikh, V. D., and Chien, W. W.,** 19-Acetylenic, 10-hydroxy-androst-4-en-3,17-diones. Potential suicide substrates of estrogen biosynthesis, *Tetrahedron Lett.,* p. 2105, 1979.

29. **Holbert, G. W., Johnston, J. O., and Metcalf, B. W.,** Synthesis of 11β-hydroxy-18-ethynylprogesterone: an inhibitor of allosterone biosynthesis, *Tetrahedron Lett.,* 26, 1137, 1985.

30. **Nagahisa, A., Spencer, R. W., and Orme-Johnson, W. H.,** Acetylenic mechanism-based inhibitors of cholesterol side chain cleavage by cytochrome P-450$_{scc}$, *J. Biol. Chem.,* 258, 6721, 1983.

31. **Colombo, G. and Villafranca, J. J.,** An acetylenic mechanism-based inhibitor of dopamine β-hydroxylase, *J. Biol. Chem.,* 259, 15017, 1984.

32. **Ahern, D. G. and Downing, D. T.,** Inhibition of prostaglandin biosynthesis by eicosa-5,8,11,14-tetraynoic acid, *Biochim. Biophys. Acta,* 210, 456, 1970.

33. **Downing, D. T., Ahern, D. G., and Bachta, M.,** Enzyme inhibition by acetyleneic compounds, *Biochem. Biophys. Res. Commun.,* 40, 218, 1970.

34. **Kühn, H., Weisner, R., Schewe, T., and Rapoport, S. M.,** Reticulocyte lipoxygenase exhibits both n-6 and n-9 activities, *FEBS Lett.,* 153, 353, 1983.

35. **Kühn, H., Holzhütter, H. G., Schewe, T., Hiebsch, C., and Rapoport, S. M.,** The mechanism of inactivation of lipoxygenases by acetylenic fatty acids, *Eur. J. Biochem.,* 139, 577, 1984.

36. **Kunau, W.-H., Lehmann, H., and Gross, R.,** Chemical synthesis of highly unsaturated fatty acids. III. The total synthesis of polyynoic and polyenoic acids, with four, five, and six triple or double bonds, *Hoppe-Seyler's Z. Physiol. Chem.*, 352, 542, 1971.

37. **Struijk, C. B., Beerthuis, R. K., Pabon, H. J. J., and Van Dorp, D. A.,** Specificity in the enzymic conversion of polyunsaturated fatty acids into prostaglandins, *Recl. Trav. Chim. Pays-Bas*, 85, 1233, 1966.

38. **Sprecher, H.,** The synthesis and metabolism of hexadeca-4,7,10-trienoate, eicosa-8,11,14-trienoate, docosa-10,13,16-trienoate and docosa-6,9,12,15-tetraenoate in the rat, *Biochim. Biophys. Acta*, 152, 519, 1968.

39. **Sok, D.-E., Han, C.-Q., Pai, J.-K., Sih, C. J.,** Inhibition of leutotriene biosynthesis by acetylenic analogs, *Biochem. Biophys. Res. Commun.*, 107, 101, 1982.

40. **Corey, E. J. and Park, H.,** Irreversible inhibition of the enzyme oxidation of arachidonic acid to 15-(hydroperoxy)-5,8,11 (Z), 13(E)-eicosatetraenoic acid (15-HPETE) by 14,15-dehydroarachidonic acid, *J. Am. Chem. Soc.*, 104, 1750, 1982.

41. **Corey, E. J. and Munroe, J. E.,** Irreversible inhibition of prostaglandin and leukotriene biosynthesis from arachidonic acid by 11,12-dehydro- and 5,6-dehydroarachidonic acids, respectively, *J. Am. Chem. Soc.*, 104, 1752, 1982.

42. **Corey, E. J., Lansbury, P. T., Jr., Cashman, J. R., and Kantner, S. S.,** Mechanism of the irreversible deactivation of arachidonate 5-lipoxygenase by 5,6-dehydroarachidonate, *J. Am. Chem. Soc.*, 106, 1501, 1984.

43. **Tai, H. H., Hsu, C. T., Tai, C. L., and Sih, C. J.,** Synthesis of [5,6-³H]-arachidonic acid and its use in development of a sensitive assay for prostacyclin synthetase, *Biochemistry*, 19, 1989, 1980.

44. **Corey, E. J., Park, H., Barton, A., and Nii, Y.,** Synthesis of three potential inhibitors of the biosynthesis of leukotrienes A-E, *Tetrahedron Lett.*, 21, 4243, 1980.

45. **Corey, E. J. and Kang, J.,** Short, stereocontrolled syntheses of irreversible eicosanoid biosynthesis inhibitors. 5,6-, 8,9-, and 11,12-dehydroarachidonic acid, *Tetrahedron Lett.*, 23, 1651, 1982.

46. **Corey, E. J., Kantner, S. S., and Lansbury, P. T., Jr.,** Irreversible inhibition of the leukotriene pathway by 4,5-dehydroarachidonic acid, *Tetrahedron Lett.*, 24, 265, 1983.

47. **Ortiz de Montellano, P. R. and Kunze, K. L.,** Inactivation of hepatic cytochrome P-450 by allenic substrates, *Biochem. Biophys. Res. Commun.*, 94, 443, 1980.

48. **Levin, W., Jacobson, M., and Kuntzman, R.,** Incorporation of radioactive δ-aminolevulinic acid into microsomal cytochrome P_{450}: selective breakdown of the hemoprotein by allylisopropylacetamide and carbon tetrachloride, *Arch. Biochem. Biophys.*, 148, 262, 1972.

49. **Ivanetich, K. M. and Bradshaw, J. J.,** An investigation of the degradation of cytochrome P-450 hemoproteins using SDS gel electrophoresis, *Biochem. Biophys. Res. Commun.*, 78, 317, 1977.

50. **Ortiz de Montellano, P. R., Mico, B. A., and Yost, G. S.,** Suicidal inactivation of cytochrome P-450. Formation of a heme-substrate covalent adduct, *Biochem. Biophys. Res. Commun.*, 83, 132, 1978.

50a. **Davies, H. W., Britt, S. G., and Pohl, L. R.,** Carbon tetrachloride and 2-isopropyl-4-pentenamide-induced inactivation of cytochrome P-450 leads to heme-derived protein adducts, *Arch. Biochem. Biophys.*, 244, 387, 1986.

51. **Liem, H. H., Johnson, E. F., and Muller-Eberhard, U.,** The effect *in vivo* and *in vitro* of allylisopropylacetamide on the content of hepatic microsomal cytochrome P-450 2 of phenobarbital treated rabbits, *Biochem. Biophys. Res. Commun.*, 111, 926, 1983.

52. **Levin, W., Sernatinger, E., Jacobson, M., and Kuntzman, R.,** Destruction of cytochrome P-450 by secobarbital and other barbiturates containing allyl groups, *Science*, 176, 1341, 1972.

53. **Sweeney, G. D. and Rothwell, J. D.,** Spectroscopic evidence of interaction between 2-allyl-2-isopropylacetamide and cytochrome P-450 of rat liver microsomes, *Biochem. Biophys. Res. Commun.*, 55, 798, 1973.

54. **Brinkschulte-Freitas, M. and Uehleke, H.,** The effects of 2,2-diethylallylacetamide on hepatic cytochromes in rats and *in vitro*, *Arch. Toxicol.*, 42, 137, 1979.

55. **Guzelian, P. S. and Swisher, R. W.,** Degradation of cytochrome P-450 haem by carbon tetrachloride and 2-allyl-2-isopropylacetamide in rat liver *in vivo* and *in vitro*. Involvement of non-carbon monozide-forming mechanisms, *Biochem. J.*, 184, 481, 1979.

56. **Ortiz de Montellano, P. R., Yost, G. S., Mico, B. A., Dinizo, S. E., Correia, M. A., and Kumbara, H.,** Destruction of cytochrome P-450 by 2-isopropyl-4-pentenamide and methyl 2-isopropyl-4-pentenoate: mass spectrometric characterization of prosthetic heme adducts and nonparticipation of epoxide metabolites, *Arch. Biochem. Biophys.*, 197, 524, 1979.

57. **Ortiz de Montellano, P. R. and Mico, B. A.,** Destruction of cytochrome P-450 by allylisopropylacetamide is a suicidal process, *Arch. Biochem. Biophys.*, 206, 43, 1981.

57a. **Davies, H. W., Britt, S. G., and Pohl, L. R.,** Carbon tetrachloride and 2-isopropyl-4-pentenamide-induced inactivation of cytochrome P-450 leads to hemederived protein adducts, *Arch. Biochem. Biophys.*, 244, 387, 1986.

58. **Loosemore, M. J., Wogan, G. N., and Walsh, C.,** Determination of partition ratios for allylisopropyllacetamide during suicidal processing by a phenobarbital-induced cytochrome P-450 isozyme from rat liver, *J. Biol. Chem.,* 256, 8705, 1981.

59. **Loosemore, M., Light, D., and Walsh, C.,** Studies on the autoinactivation behavior of pure, reconstituted phenobarbital-induced cytochrome P-450 isozyme from rat liver, *J. Biol. Chem.,* 255, 9017, 1980.

60. **Ortiz de Montellano, P. R., Stearns, R. A., and Langry, K. C.,** The allylisopropylacetamide and novonal prosthetic heme adducts, *Mol. Pharmacol.,* 25, 310, 1984.

61. **Waxman, D. J. and Walsh, C.,** Phenobarbital-induced rat liver cytochrome P-450. Purification and characterization of two closely related isozymic forms, *J. Biol. Chem.,* 257, 10446, 1982.

62. **Ivanetich, K. M., Ziman, M. R., and Bradshaw, J. J.,** Reaction scheme for the degradation of cytochromes P-450 by allylisopropylacetamide and fluroxene, *Biochem. Pharmacol.,* 29, 2805, 1980.

63. **De Matteis, F., Gibbs, A. H., Jackson, A. H., and Weerasinghe, S.,** Conversion of liver haem into *N*-substituted porphyrins or green pigments. Nature of the substituent at the pyrrole nitrogen atom, *FEBS Lett.,* 119, 109, 1980.

64. **Ortiz de Montellano, P. R. and Mico, B. A.,** Destruction of cytochrome P-450 by ethylene and other olefins, *Mol. Pharmacol.,* 18, 128, 1980.

65. **Ortiz de Montellano, P. R., Kunze, K. L., and Mico, B. A.,** Destruction of cytochrome P-450 by olefins: *N*-alkylation of prosthetic heme, *Mol. Pharmacol.,* 18, 602, 1980.

66. **Ortiz de Montellano, P. R., Beilan, H. S., Kunze, K. L., and Mico, B. A.,** Destruction of cytochrome P-450 by ethylene. Structure of the resulting prosthetic heme adduct, *J. Biol. Chem.,* 256, 4395, 1981.

67. **Ortiz de Montellano, P. R., Mangold, B. L. K., Wheeler, C., Kunze, K. L., and Reich, N. O.,** Stereochemistry of cytochrome P-450-catalyzed epoxidation and prosthetic heme alkylation, *J. Biol. Chem.,* 258, 4208, 1983.

67a. **Mansuy, D., Devocelle, L., Artaud, I., and Battioni, J. -P.,** Alkene oxidations by iodosylbenzene catalyzed by iron-porphyrins: fate of the catalyst and formation of *N*-alkyl-porphyrin green pigments from monosubstituted alkenes as in cytochrome P-450 reactions, *Nouv. J. Chim.,* 9, 711, 1985.

67b. **Mashiko, T., Dolphin, D., Nakano, T., and Traylor, T. G.,** *N*-Alkylporphyrin formation during the reactions of cytochrome P-450 model systems, *J. Am. Chem. Soc.,* 107, 3735, 1985.

67c. **Collman, J. P., Hampton, P. D., and Brauman, J. I.,** Stereochemical and mechanistic studies of the "suicide" event in biomimetic P-450 olefin epoxidation, *J. Am. Chem. Soc.,* 108, 7861, 1986.

67d. **Artaud, I., Devocelle, L., Battioni, J. -P., Girault, J. -P., and Mansuy, D.,** Suicidal inactivation of iron porphyrin catalysts during alk-1-ene oxidation: isolation of a new type of *N*-alkylporphyrins, *J. Am. Chem. Soc.,* 109, 3782, 1987.

67e. **Guengerich, F. P.,** Covalent binding to apoprotein is a major fate of heme in a variety of reactions in which cytochrome P-450 is destroyed, *Biochem. Biophys. Res. Commun.,* 138, 193, 1986.

67f. **Traylor, T. G., Nakano, T., Miksztal, A. R., and Dunlap, B. E.,** Transient formation of *N*-alkylhemins during hemin-catalyzed epoxidation of norbornene. Evidence concerning the mechanism of epoxidation, *J. Am. Chem. Soc.,* 109, 3625, 1987.

68. **Testai, E., Citti, L., Gervasi, P. G., and Turchi, G.,** Suicidal inactivation of hepatic cytochrome P-450 in vitro by some aliphatic olefins, *Biochem. Biophys. Res. Commun.,* 107, 633, 1982.

69. **Gervasi, P. G., Citti, L., Testai, E., and Turchi, G.,** Metabolism of vinylcyclooctane and partition ratio between epoxide formation and cytochrome P-450 destruction, *Toxicol. Lett.,* 20, 243, 1984.

70. **Levin, W., Jacobson, M., Sernatinger, E., and Kuntzman, R.,** Breakdown of cytochrome P-450 heme by secobarbital and other allyl-containing barbiturates, *Drug Metab. Dispos.,* 1, 275, 1973.

71. **Elcombe, C. R., Bridges, J., Nimmo-Smith, R. H., and Werringloer, J.,** Cumene hydroperoxide-mediated formation of inhibited complexes of methylenedioxyphenyl compounds with cytochrome P-450, *Biochem. Soc. Trans.,* 3, 967, 1975.

72. **May, S. W., Mueller, P. W., Padgette, S. R., Herman, H. H., and Phillips, R. S.,** Dopamine β-hydroxylase: suicide inhibition by the novel olefinic substrate, 1-phenyl-1-aminomethylethene, *Biochem. Biophys. Res. Commun.,* 110, 161, 1983.

73. **Padgette, S. R., Wimalasena, K., Herman, H. H., Sirimanne, S. R., and May, S. W.,** Olefin oxygenation and *N*-dealkylation by dopamine β-monooxygenase: catalysis and mechanism-based inhibition, *Biochemistry,* 24, 5826, 1985.

74. **Klemm, L. H., McGuire, T. M., and Gopinath, K. W.,** Intramolecular Diels-Alder reactions. X. Syntheses and cyclizations of some *N*-(cinnamyl and phenylpropargyl)cinnamamides and phenylpropiolamides, *J. Org. Chem.,* 41, 2571, 1976.

75. **Barger, T. M., Broersma, R. J., Creemer, L. C., McCarthy, J. R., Hornsperger, J.-M., Palfreyman, M. G., Wagner, J., and Jung, M. J.,** Unsaturated heterocyclic amines as potent time-dependent inhibitors of dopamine β-hydroxylase, *J. Med. Chem.,* 29, 315, 1986.

75a. **Fouin-Fortunet, H., Tinel, M., Descatoire, V., Letteron, P., Larrey, D., Geneve, J., and Pessayre, D.,** Inactivation of cytochrome P-450 by the drug methoxsalen, *J. Pharmacol. Exp. Ther.,* 236, 237, 1986.

75b. **Letteron, P., Descatoire, V., Larrey, D., Tinel, M., Geneve, J., and Pessayre, D.,** Inactivation and induction of cytochrome P-450 by various psoralen derivatives in rats, *J. Pharmacol. Exp. Ther.*, 238, 685, 1986.

75c. **Tinel, M., Belghiti, J., Descatoire, V., Amouyal, G., Letteron, P., Geneve, J., Larrey, D., Pessayre, D.,** Inactivation of human liver cytochrome P-450 by the drug methoxsalen and other psoralen derivatives, *Biochem. Pharmacol.*, 36, 951, 1987.

76. **Ivanetich, K. M., Aronson, I., and Katz, I. D.,** The interaction of vinyl chloride with rat hepatic microsomal cytochrome P-450 *in vitro*, *Biochem. Biophys. Res. Commun.*, 74, 1411, 1977.

77. **Guengerich, F. P. and Strickland, T. W.,** Metabolism of vinyl chloride: destruction of the heme of highly purified liver microsomal cytochrome P-450 by a metabolite, *Mol. Pharmacol.*, 13, 993, 1977.

78. **Ivanetich, K. M., Marsh, J. A., Bradshaw, J. J., and Kaminsky, L. S.,** Fluroxene (2,2,2-trifluoroethyl vinyl ether) mediated destruction of cytochrome P-450 in vitro, *Biochem. Pharmacol.*, 24, 1933, 1975.

79. **Ivanetich, K. M., Bradshaw, J. J., Marsh, J. A., and Kaminsky, L. S.,** The interaction of hepatic microsomal cytochrome P-450 with fluroxene (2,2,2-trifluoroethyl vinyl ether) in vitro, *Biochem. Pharmacol.*, 25, 779, 1976.

80. **Marsh, J. A., Bradshaw, J. J., Sapeika, G. A., Lucas, S. A., Kaminsky, L. S., and Ivanetich, K. M.,** Further invesitgations of the metabolism of fluoroxene and the degradation of cytochromes P-450 *in vitro*, *Biochem. Pharmacol.*, 26, 1601, 1977.

81. **Liebler, D. C. and Guengerich, F. P.,** Olefin oxidation by cytochrome P-450: evidence for group migration in catalytic intermediates formed with vinylidene chloride and *trans*-1-phenyl-1-butene, *Biochemistry*, 22, 5482, 1983.

82. **Miller, R. E. and Guengerich, F. P.,** Oxidation of trichloroethylene by liver microsomal cytochrome P-450: evidence for chlorine migration in a transition state not involving trichloroethylene oxide, *Biochemistry*, 21, 1090, 1982.

83. **White, I. N. H.,** Metabolic activation of saturated aldehydes to cause destruction of cytochrome P-450 in vitro, *Chem. Biol. Interact.*, 39, 231, 1982.

83a. **Rajashekhar, B., Fitzpatrick, P. F., Colombo, G., and Villafranca, J. J.,** Synthesis of several 2-substituted 3-(*p*-hydroxyphenyl)-1-propenes and their characterization as mechanism-based inhibitors of dopamine β-hydroxylase, *J. Biol. Chem.*, 259, 6925, 1984.

83b. **Colombo, G., Rajashekhar, B., Giedroc, D. P., and Villafranca, J. J.,** Mechanism-based inhibitors of dopamine β-hydroxylase: inhibition by 2-bromo-3-(*p*-hydroxyphenyl)-1-propene, *Biochemistry*, 23, 3590, 1984.

83c. **Ash, D. E., Papadopoulos, N. J., Colombo, G., and Villafranca, J. J.,** Kinetic and spectroscopic studies of the interaction of copper with dopamine β-hydroxylase, *J. Biol. Chem.*, 259, 3395, 1984.

83d. **Hosoya, T.,** Effect of various reagents including antithyroid compounds upon the activity of thyroid peroxidase, *J. Biochem.*, 53, 381, 1963.

83e. **Davidson, B., Soodak, M., Strout, H. V., Neary, J. T., Nakamura, C., and Maloof, F.,** Thiourea and cyanamide as inhibitors of thyroid peroxidase: the role of iodide, *Endocrinology*, 104, 919, 1979.

83f. **Davidson, B., Soodak, M., Neary, J. T., Strout, H. V., Kieffer, J. D., Mover, H., and Maloof, F.,** The irreversible inactivation of thyroid peroxidase by methylmercaptoimidazole, thiouracil, and propylthiouracil *in vitro* and its relationship to *in vivo* findings, *Endocrinology*, 103, 871, 1978.

83g. **Taurog, A.,** The mechanism of action of the thioureylene antithyroid drugs, *Endocrinology*, 98, 1031, 1976.

83h. **Engler, H., Taurog, A., and Nakashima, T.,** Mechanism of inactivation of thyroid peroxidase by thioureylene drugs, *Biochem. Pharmacol.*, 31, 3801, 1982.

83i. **Ohtaki, S., Nakagawa, H., Nakamura, M., and Yamazaki, I.,** Reactions of purified hog thyroid peroxidase with H_2O_2, tyrosine, and methylmercaptoimidazole (goitrogen) in comparison with bovine lactoperoxidase, *J. Biol. Chem.*, 257, 761, 1982.

83j. **Nagasaka, A. and Hidaka, H.,** Effect of antithyroid agents 6-propyl-2-thiouracil and 1-methyl-2-mercaptoimidazole on human thyroid iodide peroxidase, *J. Clin. Endocrinol. Metab.*, 43, 152, 1976.

83k. **Morris, D. R. and Hager, L. P.,** Mechanism of inhibition of enzymatic halogenation by antithyroid agents, *J. Biol. Chem.*, 241, 3582, 1966.

83l. **Taurog, A. and Howells, E. M.,** Enzymatic iodination of tyrosine and thyroglobulin with chloroperoxidase, *J. Biol. Chem.*, 241, 1329, 1966.

83m. **Edelhoch, H., Irace, G., Johnson, M. L., Michot, J. L., and Nunez, J.,** The effects of thioureylene compounds (goitrogens) on lactoperoxidase activity, *J. Biol. Chem.*, 254, 11822, 1979.

83n. **Michot, J. L., Nunez, J., Johnson, M. L., Irace, G., and Edelhoch, H.,** Iodide binding and regulation of lactoperoxidase activity toward thyroid goitrogens, *J. Biol. Chem.*, 254, 2205, 1979.

83o. **Nakamura, S., Nakamura, M., Yamazaki, I., and Morrison, M.,** Reactions of ferryl lactoperoxidase (Compound II) with sulfide and sulfhydryl compounds, *J. Biol. Chem.*, 257, 7080, 1984.

83p. **Doerge, D. R.,** Mechanism-based inhibition of lactoperoxidase by thiocarbamide goitrogens, *Biochemistry*, 25, 4724, 1986.

83q. **Doerge, D. R., Pitz, G. L., and Root, D. P.,** Organosulfur oxygenation and suicide inactivation of lactoperoxidase, *Biochem. Pharmacol.,* 36, 972, 1987.

83r. **Pascal, R. A., Jr. and Huang, D. -S.,** Mechanism-based inactivation of catechol 2,3-dioxygenase by 3-[(methylthio)methyl]catechol, *J. Am. Chem. Soc.,* 109, 2854, 1987.

83s. **Goodhart, P. J., DeWolf, Jr., W. E., and Kruse, L. I.,** Mechanism-based inactivation of dopamine β-hydroxylase by *p*-cresol and related alkylphenols, *Biochemistry,* 26, 2576, 1987.

83t. **Snider, C. E. and Brueggemeier, R. W.,** Potent enzyme-activated inhibition of aromatase by a 7α-substituted C_{19} steroid, *J. Biol. Chem.,* 262, 8685, 1987.

84. **Flynn, G. A., Johnston, J. O., Wright, C. L., and Metcalf, B. W.,** The time-dependent inactivation of aromatase by 17β-hydroxy-10-methylthioestra-1,4-dien-3-one, *Biochem. Biophys. Res. Commun.,* 103, 913, 1981.

85. **Parli, C. J., Krieter, P., and Schmidt, B.,** Metabolism of 6-chlorotryptophan to 4-chloro-3-hydroxyanthranilic acid: a potent inhibitor of 3-hydroxyanthranilic acid oxidase, *Arch. Biochem. Biophys.,* 203, 161, 1980.

85a. **Latham, J. A. and Walsh, C.,** Mechanism-based inactivation of the flavoenzyme cyclohexanone oxygenase during oxygenation of cyclic thiol ester substrates, *J. Am. Chem. Soc.,* 109, 3421, 1987.

85b. **Walsh, C. and Latham, J.,** Mechanism-based inactivation of the flavoprotein cyclohexanone monooxygenase by *S* oxygenation, *J. Protein Chem.,* 5, 79, 1986.

86. **Ptashne, K. A., Wolcott, R. M., and Neal, R. A.,** Oxygen-18 studies in the chemical mechanisms of the mixed function oxidase catalyzed desulfuration and dearylation reactions of parathion, *J. Pharmacol. Exp. Ther.,* 179, 380, 1971.

87. **McBain, J. B., Yamamoto, I., and Casida, J. E.,** Mechanism of activation and deactivation of Dyfonate® (*O*-ethyl *S*-phenyl ethylphosphonodithioate) by rat liver microsomes, *Life Sci.,* 10(2), 947, 1971.

88. **McBain, J. B., Yamamoto, I., and Casida, J. E.,** Oxygenated intermediate in peracid and microsomal oxidations of the organophosphonothioate insecticide Dyfonate®, *Life Sci.,* 10(2), 1311, 1971.

89. **Ptashne, K. A. and Neal, R. A.,** Reaction of parathion and malathion with peroxytrifluoroacetic acid, a model system for the mixed function oxidases, *Biochemistry,* 11, 3224, 1972.

90. **Hunter, A. L. and Neal, R. A.,** Inhibition of hepatic mixed-function oxidase activity *in vitro* and *in vivo* by various thiono-sulfur-containing compounds, *Biochem. Pharmacol.,* 24, 2199, 1975.

91. **De Matteis, F.,** Covalent binding of sulfur to microsomes and loss of cytochrome P-450 during the oxidative desulfuration of several chemicals, *Mol. Pharmacol.,* 10, 849, 1974.

92. **Norman, B. J., Poore, R. E., and Neal, R. A.,** Studies of the binding of sulfur released in mixed-function oxidase-catalyzed metabolism of diethyl *p*-nitrophenyl phosphorothionate (parathion) to diethyl *p*-nitrophenyl phosphate (paraxon), *Biochem. Pharmacol.,* 23, 1733, 1974.

93. **Kamataki, T. and Neal, R. A.,** Metabolism of diethyl *p*-nitrophenyl phosphorothionate (parathion) by a reconstituted mixed-function oxidase enzyme system: studies of the covalent binding of the sulfur atom, *Mol. Pharmacol.,* 12, 933, 1976.

94. **Yoshihara, S. and Neal, R. A.,** Comparison of the metabolism of parathion by a rat liver reconstituted mixed-function oxidase enzyme system and by a system containing cumene hydroperoxide and purified rat liver cytochrome P-450, *Drug Metab. Dispos.,* 5, 191, 1977.

95. **Yoshihara, S. and Neal, R. A.,** Comparison of the metabolism of parathion by a rat liver reconstituted mixed-function oxidase enzyme system and by a system containing cumene hydroperoxide and purified rat liver cytochrome P-450, *Drug Metab. Dispos.,* 5, 191, 1977.

96. **Davis, J. E. and Mende, T. J.,** A study of the binding of sulfur to rat liver microsomes which occurs concurrently with the metabolism of *O,O*-diethyl *O-p*-nitrophenyl phosphorothionate (parathion) to *O,O*-diethyl *O-p*-nitrophenyl phosphate (paraxon), *J. Pharmacol. Exp. Ther.,* 201, 490, 1977.

97. **Halpert, J., Hammond, D., and Neal, R. A.,** Inactivation of purified rat liver cytochrome P-450 during the metabolism of parathion (diethyl *p*-nitrophenyl phosphorothionate), *J. Biol. Chem.,* 255, 1080, 1980.

98. **De Matteis, F. and Seawright, A. A.,** Oxidative metabolism of carbon disulfide by the rat. Effect of treatments which modify the liver toxicity of carbon disulfide, *Chem. Biol. Interact.,* 7, 375, 1973.

99. **Dalvi, R. R., Poore, R. E., and Neal, R. A.,** Studies on the metabolism of carbon disulfide by rat liver microsomes, *Life Sci.,* 14, 1785, 1974.

100. **Dalvi, R. R., Hunter, A. L., and Neal, R. A.,** Toxicological implications of the mixed-function oxidase catalyzed metabolism of carbon disulfide, *Chem. Biol. Interact.,* 10, 349, 1975.

101. **Catignani, G. L. and Neal, R. A.,** Evidence for the formation of a protein bound hydrodisulfide resulting from the microsomal mixed function oxidase catalyzed desulfuration of carbon disulfide, *Biochem. Biophys. Res. Commun.,* 65, 629, 1975.

102. **Obrebska, M. J., Kantish, P., and Parke, D. V.,** The effects of carbon disulphide on rat liver microsomal mixed-function oxidases, *in vivo* and *in vitro, Biochem. J.,* 188, 107, 1980.

102a. **Ortiz de Montellano, P. R. and Grab, L. A.,** Inactivation of cytochrome P-450 during catalytic oxidation of 3-[(arylthio)ethyl]sydnone: *N*-vinyl heme formation via insertion into the Fe-N bond, *J. Am. Chem. Soc.,* 108, 5584, 1986.

102b. **White, E. H. and Egger, N.,** Reaction of sydnones with ozone as a method of deamination: on the mechanism of inhibition of monoamine oxidase by sydnones, *J. Am. Chem. Soc.,* 106, 3701, 1984.

103. **Dixon, R. L. and Fouts, J. R.,** Inhibition of microsomal drug metabolic pathways by chloramphenicol, *Biochem. Pharmacol.,* 11, 715, 1962.

104. **Reilly, P. E. B. and Ivey, D. E.,** Inhibition of chloramphenicol of the microsomal monoxygenase complex of rat liver, *FEBS Lett.,* 97, 141, 1979.

105. **Pohl, L. R. and Krishna, G.,** Study of the mechanism of metabolic activation of chloramphenicol by rat liver microsomes, *Biochem. Pharmacol.,* 27, 335, 1978.

106. **Halpert, J. and Neal, R. A.,** Inactivation of purified rat liver cytochrome P-450 by chloramphenicol, *Mol. Pharmacol.,* 17, 427, 1980.

107. **Halpert, J.,** Covalent modification of lysine during the suicide inactivation of rat liver cytochrome P-450 by chloramphenicol, *Biochem. Pharmacol.,* 30, 875, 1981.

108. **Halpert, J.,** Further studies of the suicide inactivation of purified rat liver cytochrome P-450 by chloramphenicol, *Mol. Pharmacol.,* 21, 166, 1982.

109. **Halpert, J., Näslund, B., and Betnér, I.,** Suicide inactivation of rat liver cytochrome P-450 by chloramphenicol *in vivo* and *in vitro, Mol. Pharmacol.,* 23, 445, 1983.

109a. **Miller, N. E. and Halpert, J.,** Analogues of chloramphenicol as mechanism-based inactivators of rat liver cytochrome P-450: modifications of the propanediol side chain, the *p*-nitro group, and the dichloromethyl moiety, *Mol. Pharmacol.,* 29, 391, 1986.

109b. **Halpert, J. R., Balfour, C., Miller, N. E., and Kaminsky, L. S.,** Dichloromethyl compounds as mechanism-based inactivators of rat liver cytochromes P-450 in vitro, *Mol. Pharmacol.,* 30, 19, 1986.

110. **Marcotte, P. A. and Robinson, C. H.,** Inhibition and inactivation of estrogen synthetase (aromatase) by fluorinated substrate analogues, *Biochemistry,* 21, 2773, 1982.

111. **Klinman, J. P. and Krueger, M.,** Dopamine β-hydroxylase: activity and inhibition in the presence of β-substituted phenethylamines, *Biochemistry,* 21, 67, 1982.

112. **Mangold, J. B. and Klinman, J. P.,** Mechanism-based inactivation of dopamine β-monooxygenase by β-chlorophenethylamine, *J. Biol. Chem.,* 259, 7772, 1984.

112a. **Bossard, M. J. and Klinman, J. P.,** Mechanism-based inhibition of dopamine β-monooxygenase by aldehydes and amides, *J. Biol. Chem.,* 261, 16421, 1986.

113. **Baldoni, J. M. and Villafranca, J. J.,** Dopamine β-hydroxylase inactivation by a suicide substrate, *J. Biol. Chem.,* 255, 8987, 1980.

114. **Colombo, G., Rajashekhar, B., Giedroc, D. P., and Villafranca, J. J.,** Alternate substrates of dopamine β-hydroxylase. I. Kinetic investigations of benzyl cyanides as substrates and inhibitors, *J. Biol. Chem.,* 259, 1593, 1984.

115. **Colombo, G., Giedroc, D. P., Rajashekhar, B., and Villafranca, J. J.,** Alternate substrates of dopamine β-hydroxylase. II. Inhibition by benzyl cyanides and reactivation of inhibited enzyme, *J. Biol. Chem.,* 259, 1601, 1984.

116. **Colombo, G., Rajashekhar, B., Ash, D. E., and Villafranca, J. J.,** Alternate substrates of dopamine β-hydroxylase, III. Stoichiometry of the inactivation reaction with benzyl cyanides and spectroscopic investigations, *J. Biol. Chem.,* 259, 1607, 1984.

117. **Wood, B. J. B. and Ingraham, L. L.,** Labelled tyrosinase from labelled substrate, *Nature (London),* 205, 291, 1965.

118. **Rolland, M. and Lissitzky, S.,** Oxydation de la tyrosine et de peptides ou protéines la contenant, par la polyphénoloxydase de champignon. I. Oxydation de la tyrosine et de peptides la contenant, en présence d'acide ascorbique, *Biochim. Biophys. Acta,* 56, 83, 1962.

118a. **Lerch, K.,** *Neurospora* tyrosinase: structural, spectroscopic and catalytic properties, *Mol. Cell. Biochem.,* 52, 125, 1983.

119. **Bednarski, P. J., Porubek, D. J., and Nelson, S. D.,** Thiol-containing androgens as suicide substrates of aromatase, *J. Med. Chem.,* 28, 775, 1985.

120. **Menard, R. H., Bartter, F. C., and Gillette, J. R.,** Spironolactone and cytochrome P-450: impairment of steroid 21-hydroxylation in the adrenal cortex, *Arch. Biochem. Biophys.,* 173, 395, 1976.

121. **Menard, R. H., Guenthner, T. M., Kon, H., and Gillette, J. R.,** Studies on the destruction of adrenal and testicular cytochrome P-450 by spironolactone. Requirement for the 7α-thio group and evidence for the loss of the heme and apoproteins of cytochrome P-450, *J. Biol. Chem.,* 254, 1726, 1979.

122. **Menard, R. H., Guenthner, T. M., Taburet, A. M., Kon, H., Pohl, L. R., Gillette, J. R., Gelboin, H. V., and Trager, W. F.,** Specificity of the *in vitro* destruction of adrenal and hepatic microsomal steroid hydroxylases by thiosteroids, *Mol. Pharmacol.,* 16, 997, 1979.

123. **Rahimtula, A. D. and Gaylor, J. L.,** Investigation of the component reactions of oxidative sterol demethylation. Partial purification of a microsomal sterol 4α-carboxylic acid decarboxylase, *J. Biol. Chem.,* 247, 9, 1972.

124. **Bartlett, D. L. and Robinson, C. H.,** Mechanism-based inactivation of 4-methylsterol oxidase by 4α-(cyanomethyl)-5α-cholestan-3β-ol, *J. Am. Chem. Soc.,* 104, 4729, 1982.

125. **Covey, D. F. and Hood, W. F.**, Enzyme-generated intermediates derived from 4-androstene-3,6,17-trione and 1,4,6-androstatriene-3,17-dione cause a time-dependent decrease in human placental aromatase activity, *Endocrinol.*, 108, 1597, 1981.

126. **Covey, D. F. and Hood, W. F.**, Aromatase enzyme catalysis is involved in the potent inhibition of estrogen biosynthesis caused by 4-acetoxy- and 4-hydroxy-4-androstene-3,17-dione, *Mol. Pharmacol.*, 21, 173, 1982.

127. **Brodie, A. H. M., Garrett, W. M., Hendrickson, J. R., Tsai-Morris, C. H., Marcotte, P. A., and Robinson, C. H.**, Inactivation of aromatase *in vitro* by 4-hydroxy-4-androstene-3,17-dione and 4-acetoxy-4-androstene-3,17-dione and sustained effects *in vivo*, *Steroids*, 38, 693, 1981.

128. **Brodie, A. M. H. and Longcope, C.**, Inhibition of peripheral aromatization by aromatase inhibitors, 4-hydroxy- and 4-acetoxyandrostene-3,17-dione, *Endocrinol.*, 106, 19, 1981.

129. **Covey, D. F. and Hood, W. F.**, A new hypothesis based on suicide substrate inhibitor studies for the mechanism of action of aromatase, *Cancer Res. (Suppl.)*, 42, 3327s, 1982.

130. **Hahn, E. F. and Fishman, J.**, Immunological probe of estrogen biosynthesis. Evidence for the 2β-hydroxylative pathway in aromatization of androgens, *J. Biol. Chem.*, 259, 1689, 1984.

131. **Caspi, E., Wicha, J., Arunachalam, T., Nelson, P., and Spiteller, G.**, Estrogen biosynthesis. Concerning the obligatory intermediary of 2β-hydroxy-10β-formylandrost-4-ene-3,17-dione, *J. Am. Chem. Soc.*, 106, 7282, 1984.

131a. **Marsh, D. A., Brodie, H. J., Garrett, W., Tsai-Morris, C. -H., and Brodie, A. M. H.**, Aromatase inhibitors. Synthesis and biological activity of androstenedione derivatives, *J. Med. Chem.*, 28, 788, 1985.

132. **Ivanetich, K. M., Ziman, M. R., and Bradshaw, J. J.**, 1,1,1-Trichloropropene-2,3-oxide: an alternate mechanism for its inhibition of cytochrome P-450, *Res. Commun. Chem. Pathol. Pharmacol.*, 35, 111, 1982.

132a. **Houghton, J. D., Beddows, S. E., Suckling, K. E., Brown, L., and Suckling, C. J.**, 5α,6α-Methanocholestan-3β-ol as a probe of the mechanism of action of cholesterol 7α-hydroxylase, *Tetrahedron Lett.*, 27, 4655, 1986.

133. **Elion, G. B.**, Enzymic and metabolic studies with allopurinol, *Ann. Rheum. Dis.*, Suppl. 6, 608, 1966.

134. **Spector, T. and Johns, D. G.**, Stoichiometric inhibition of reduced xanthine oxidase by hydroxypyrazol[3,4-*d*]pyrimidines, *J. Biol. Chem.*, 245, 5079, 1970.

135. **Massey, V., Komai, H., Palmer, G., and Elion, G. B.**, On the mechanism of inactivation of xanthine oxidase by allopurinol and other pyrazolo[3,4-*d*]-pyrimidines, *J. Biol. Chem.*, 245, 2837, 1970.

136. **Edmonson, D., Massey, V., Palmer, G., Beacham III, L. M., and Elion, G. B.**, The resolution of active and inactive xanthine oxidase by affinity chromatography, *J. Biol. Chem.*, 247, 1597, 1972.

137. **Cha, S., Agarwal, R. P., and Parks, R. E., Jr.**, Tight binding inhibitors. II. Non-steady state nature of inhibition of milk xanthine oxidase by allopurinol and alloxanthine and of human erythrocytic adenosine deaminase by corformycin, *Biochem. Pharmacol.*, 24, 2187, 1975.

137a. **Spector, T., Hall, W. W., and Krenitsky, T. A.**, Human and bovine xanthine oxidase. Inhibition studies with oxipurinol, *Biochem. Pharmacol.*, 35, 3109, 1986.

Chapter 11

POLYMERIZATION REACTIONS*

I. INTRODUCTION

In this chapter there are two examples of inactivators of DNA polymerase I that first get incorporated into the DNA strand prior to the inactivation of the enzyme.

II. POLYMERIZATION

A. 9-[(2-Hydroxyethoxy)methyl]guanine (Acyclovir) Triphosphate

Type 1 DNA polymerase from herpes simplex virus undergoes time-dependent inactivation by 9-[(2-hydroxyethoxy)methyl]guanine (acyclovir) triphosphate (Structural Formula 11.1, Scheme 1).[1] This compound does not inactivate the resting polymerase, but only while being processed as a substrate. Pseudo first-order rates of inactivation are obtained with no lag time. Sephadex does not restore enzyme activity. Since there do not appear to be latent reactive groups, Furman et al.[1] suggested that inactivation results from the incorporation of this nucleotide (drawn above in a conformation that shows its structural similarity to 2-deoxyguanosine triphosphate) into the DNA primer-template, which induces a nonproductive conformational change in the enzyme. The enzyme may become trapped while trying to excise this bogus nucleotide. Another possibility is that the enzyme may become frozen while searching for a 3'-hydroxyl group which is missing in the inactivator molecule; this group is necessary for elongation. Compound **11.1** is only a weak reversible inhibitor of DNA polymerase α.

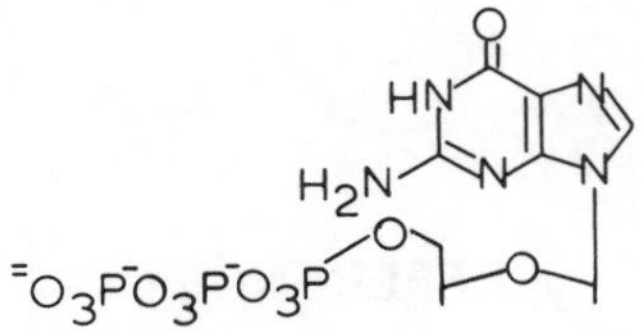

Scheme 1.   Structural Formula 11.1.

III. POLYMERIZATION/ADDITION

A. Adenosine 2',3'-Riboepoxide 5'-Triphosphate

Adenosine 2',3'-riboepoxide 5'-triphosphate (Structural Formula 11.2, see Scheme 2) irreversibly inactivates DNA polymerase I from *Escherichia coli*.[2] Although this compound contains a reactive functional group (an epoxide), evidence is presented by Abboud et al.[2] that suggests it is a mechanism-based inactivator rather than an affinity-labeling agent. Inactivation is first order and shows saturation kinetics. It requires Mg(II) **and** the complementary template. Inactivation by [³H] inactivator ([³H] in the 2 and 8 positions) followed by gel filtration leads to comigration of the enzyme, inactivator, and DNA template with a stoichiometry of 1 mol of [³H] per mole of inactivated enzyme. This suggests that the inactivator serves first as a substrate for the polymerase reaction with elongation of the DNA chain, and then alkylates a base at the primer terminus. Epoxy ATP (**11.2**) also inactivates

* A list of abbreviations and shorthand notations can be found prior to Chapter 7.

human and viral DNA polymerases, but not *E. coli* RNA polymerase or rabbit muscle pyruvate kinase. The synthesis of **11.2** is shown in Scheme 2.

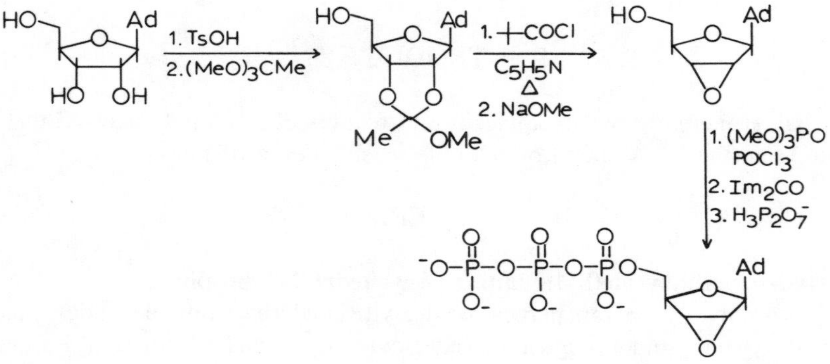

Scheme 2.   Containing Structural Formula 11.2.

IV. EPILOGUE

Although the myriad of inactivation mechanisms described in this book seems quite extensive, only relatively few enzymes have been inactivated by this approach. Furthermore, there are new enzymes with unusual cofactors still being discovered. Assimilation of the information in this book will be most beneficial, if it can be applied to other enzymes or if it sparks ideas for future novel approaches to the design of mechanism-based enzyme inactivators.

REFERENCES

1. **Furman, P. A., St. Clair, M. H., and Spector, T.,** Acyclovir triphosphate is a suicide inactivator of the herpes simplex virus DNA polymerase, *J. Biol. Chem.,* 259, 9575, 1984.
2. **Abboud, M. M., Sim, W. J., Loeb, L. A., and Mildvan, A. S.,** Apparent suicidal inactivation of DNA polymerase by adenosine 2′,3′-riboepoxide 5′-triphosphate, *J. Biol. Chem.,* 253, 3415, 1978.

INDEX

A